In riveting prose, Ott unpacks the consequences of the Exxon Valdez. *Her six years of relentless investigation shines a clear light on the lies, the pseudo-science, the sickness, and the death that have spread as thickly as the oil itself in the wake of this monumental tragedy. Her practical and visionary recommendations challenge us all with the really big question: are we ready to close the curtain on the Age of Oil? Ott's rigorous research suggest that a positive future may depend on our willingness to say "yes."*

—Fran Korten
Executive Director, Positive Futures Network

Through compelling first-hand accounts of the Exxon Valdez oil spill, Riki Ott reminds us that the dire environmental consequences of this disaster live on—in the form of increased rates of asthma, allergies, and cancer. Hers is an eloquent warning that we must not look at Exxon Valdez as an isolated incident, but as a signal of our over-dependence on oil and a premonition of what can and will happen again. 'Sound Truth' movingly cautions us that we can no longer ignore the public health threat of fossil fuels.

—Robert K. Musil, PhD, MPH
Executive Director and CEO, Physicians' for Social Responsibility

Throughout the turbulent waters of this complex story, Riki Ott keeps a steady hand on the tiller and a steady eye on her navigation. Where many authors would be dismasted and shipwrecked by their own fury, she stays clear-eyed and communicative, keeping this story proceeding firmly on course.

—Carl Safina,
author of *Song for the Blue Ocean* and *Eye of the Albatross*

Riki Ott takes you to the womb of Alaska and meticulously documents the insidious devastation of unleashed crude oil on one of the world's most prolific wildlife birthing grounds. She paints a panorama of bedlam, frustration, and corruption on a canvas of blatant corporate disregard for the long term impacts of its actions. Like the crude oil still invading wildlife in Alaska's Prince William Sound, I could not help but reflect on the vast numbers of synthetic chemicals derived from the cracking of crude oil that are invading human and wildlife wombs all over the world, undermining the potential health, behavior, and intelligence of future generations. In closing, Riki Ott poses a series of practical actions that convincingly could turn around society's course of self destruction and in the meantime prevent further catastrophes.

—Theo Colborn, PhD
Professor, Zoology Department, University of Florida
Adjunct Graduate Faculty, Texas A&M
Senior Fellow, World Wildlife Fund

Sound Truth and Corporate Myth$

The LEGACY of the
EXXON VALDEZ Oil Spill

Published by
Dragonfly Sisters Press
P.O. Box 1271
Cordova, AK 99574
www.soundtruth.info

Produced for Dragonfly Sisters Press by
Lorenzo Press
20 Elm Avenue
Norwich, CT 06360
www.lorenzopress.com

ISBN 0-9645-2266-7

Library of Congress Control Number 2004104088

Cover design by Susan Ogle and Lisa Marie Jacobs

FIRST EDITION, First Printing

Printed in the U.S.A. on recycled paper

Ott, Riki.

  Sound truth and corporate myth$: the legacy of the Exxon Valdez
oil spill / Riki Ott. -- 1st ed. -- Cordova, Alaska : Dragonfly Sisters
Press, 2005.

    Includes bibliographical references and index.
    ISBN: 0-9645-2266-7

    1. Exxon Valdez Oil Spill, Alaska, 1989. 2. Oil spills--
Environmental aspects--Alaska--Prince William Sound Region.
3. Restoration ecology--Alaska--Prince William Sound Region.
4. Wildlife rehabilitation--Alaska--Prince William Sound Region.
5. Tankers--Accidents--Environmental aspects--Alaska--Prince
William Sound Region. 6. Exxon Valdez (Ship) 7. Oil spills--Health
aspects. 8. Polycyclic aromatic hydrocarbons--Physiological effect.
9. Petroleum industry and trade--Government policy--United
States. 10. Energy policy--United States. 11. Oil spills--Prevention.
I. Title.

QH545.O5 O88 2005   2004104088
363.738/2/097983--dc22   0502

A portion of the proceeds from the sale of this book goes to the
Oiled Regions of Alaska Foundation for the *Exxon Valdez* Oil Spill
Education Legacy Fund (EVOS ELF).

Sound Truth and Corporate Myth$

The LEGACY of the EXXON VALDEZ Oil Spill

Riki Ott, PhD

DRAGONFLY SISTERS PRESS
Cordova, Alaska

A Note to the Reader

Since the time of Tacitus, historians have written to record and interpret the past in order to know what is a wise course for the present. This book is such an investigative history. As such, it interprets the events surrounding and following the *Exxon Valdez* disaster. In reconstructing events and attempting to explain them, this heavily annotated manuscript reconstructs from court records, official reports, and personal reminiscences, among other things, the events from 24 March 1989 to the present.

In interpreting these events, like all historians, I have had to make analyses, deductions, and speculations. The motivations and actions of Exxon personnel and other players have been subject to the refraction of this historian's prism and, thus, no absolute certainty can be claimed by this author.

However, I believe this to be an accurate record of the whats and whys of the *Exxon Valdez* affair. The reader is left to assess the interpretation of this catastrophe, its consequences, and its lessons.

Permissions:

Figure 3. Reproduced with permission from *Chemical and Engineering News*, 21 September 1998, 76(38):57-67. Copyright 1998, American Chemical Society.

Table 4. Reprinted with permission from Peterson, Rice, Short, esler, Bodkin, Ballachey, and Irons. 2003. Long-term ecosystem response to EVOS. *Science* 302:2082-2086. Copyright 2003 AAAS (American Association for the Advancement of Science).

Quotation from *Generations at Risk: Reproductive Health and the Environment* by Schettler, Solomon, Valenti, and Huddle (1999), pp. 74-75, used with permission from the MIT Press.

Quotations from *Marine Mammals and the Exxon Valdez* edited by Laughlin. Copyright 1994, pp. 52, 277, used with permission from Elsevier.

Dialogue from *Calvin and Hobbes* courtesy of Universal Press Syndicate.

To the healing of all beings harmed by the Exxon Valdez oil spill;

To Malani, Rory, Zak, Zeben, and the Youth,
for they are the ones who will lead us
out of our current oil dependency;

To my father, Fred Ott, Walt Parker, Virginia "Ginny" Wood, and the
Elders for instilling the courage to live our values;

and

In loving memory of Celia Hunter and my mother,
Elizabeth "Jolly" Ott.

"A scientific truth does not triumph by convincing its opponents
and making them see the light, but rather because its opponents
eventually die and a new generation grows up that is familiar
with it."

Max Planck

Contents

A Note to the Reader vii
Illustrations xiii
Tables xiii
Sidebars xiv
Foreword xv
Preface xix
Acknowledgements xxi
Abbreviations xxv
Introduction 1

Part 1: Sick Workers
Exposed: A Cover Up of Mass Chemical Poisoning
1. Exxon's Cleanup from a Worker Health Perspective 21
2. Exxon's Failed Worker Safety Program 39
3. Dangerous Chemicals at Dangerous Levels 55
Exposure and Health Problems
4. Ron Smith and Randy Lowe 71
5. Phyllis "Dolly" La Joie 85
6. A Collection of Stories 97
7. Sara Clarke and Captain Richard Nagel 113
Buried: Workers' Health Claims
8. Vanishing Claims 125
9. Toxic Torts and Justice Denied 139
An Occupational Health Disaster
10. Investigating a Disaster 159

Part 2: Sound Truth
Pre-Spill Studies
11. 1970s Science 181
Early Oil Spill Studies (1989 to 1992)
12. Tracking the Oil 191
13. Coastal Ecology 201
14. Marine Mammals 215
15. Marine Birds 231
16. Fish 249
Ecosystem Studies (1993 to 2003)
17. Sound Ecosystem Assessment (SEA) Program 273
18. Nearshore Vertebrate Predator (NVP) Project 293
19. Apex Predators 317
20. Fish and Oil Toxicity 343
21. Habitat is Where It's At! 359
2004 Status of the Sound
22. Sound Truth 373

Part 3: The Legacy and Beyond
23. The Legacy of the *Exxon Valdez* Oil Spill: Emerging
    Science and Policy 387
24. Beyond the *Exxon Valdez* Oil Spill: Recommendations
    for Strengthening Oil Pollution Prevention 417

Epilogue—Reflections from the Sound 445
Appendix A: Cleanup Workers' Exposure and Illness Data . . .449
Appendix B: Contact Information .455
Appendix C: Recommendations for the Alaska Workers'
Compensation Program 459
Appendix D: The People—Where are They Now? 461
Appendix E: Chronology of the Exxon Valdez *Oil Spill Legacy* 471
Notes 473
References 481
Index 549
About the Author 561

Photographs follow page 276

Illustrations

Figures

Map   Alaska: Trans-Alaska Pipeline Corridor xviii
Map   Spread of Oil from *Exxon Valdez* Oil Spill xx
Map   Prince William Sound, Alaska xxvii
1   Exxon's Partial Release of Liability for Indemnity 33
2   Total Upper Respiratory Infections Reported 57
3   Common Symptoms of Chemical Sensitivity 79

Tables

1   How Much Oil Spilled 7
2   Health Symptoms of Overexposure to Chemicals
        Present during the EVOS Cleanup 163
3   Status of Recovery in Marine Ecosystems as of 2003 375
4   Changing Paradigms in Understanding Oil Effects 394
5   The Legacy: Key Lessons Learned from the EVOS 414
A.1   Exposure Levels of Some Hazardous Compounds
        Present during Cleanup 450
A.2   Exxon's Clinical Data 451
A.3   Reported Injuries and Illnesses by Part of Body Affected .452
A.4   Nature of Reported Injuries and Illnesses 453
A.5   Source of Reported Injuries and Illnesses 454

xiv

Riki Ott

Sidebars

1   How Much Oil Did the *Exxon Valdez* Spill? 4
2   Secrets and Settlements 9
3   Material Safety Data Sheet Excerpts 12
4   Teamwork—ACAT and AFER 38
5   PELs, RELs, and Worker Health .63
6   Misdiagnoses of Chemical Injuries 93
7   Material Safety Data Sheet Excerpts for Citra-Solv 115
8   The Chemical Deluge 144
9   Solvents and Chemical Injuries .149
10 Controversy, Confidentiality, and the Public Interest 206
11 Contrasting Life Histories of Pink Salmon and
    Pacific Herring 250
12 Changing the Course of History: Fishermen's
    Blockade and Ecosystem Studies 275
13 Interpreting Oil Spill Effects on Wild and Hatchery Pink
    Salmon 290
14 Economic Impacts of the Disrupted Salmon Fisheries . . .376
15 Economic Impacts of the Lost Herring Fisheries 380
16 Picking Up the Pieces of a Failed Worker Safety
    Program 438

Foreword

Historically, the most efficient way for society to handle those individuals who point out its faults is to brand them as raving lunatics and crucify them if possible. Humans seem to be so invested in "not being wrong" that they will go to any lengths to reinforce their beliefs even in the face of clear information to the contrary. When those beliefs are challenged and especially when there is clear evidence that their beliefs are wrong the strength of their resistance only seems to increase. Historical examples abound, but one of the most often spoken of in medical circles is Ingatz Phillip Semmelweiss.

Semmelweiss noted that women in labor, who were attended to by physician-students who had just been doing cadaver dissection, often died of fever after childbirth. He reasoned that some infective organism was carried from dissection to the mother. He began the practice of disinfection between procedures and found that the rate of post-delivery mortality dropped dramatically. When he gave this information to his colleagues—other physicians who apparently were doctors because they wanted to help save lives—he found himself being ostracized. No matter how noble-sounding their profession, people just don't like to be told that they are wrong.

Fortunately, individuals with the conviction of Semmelweiss continue to populate our earth. We are truly blessed that in the arena of environmental medicine there are several such persons. Riki Ott is one and neither she nor the individuals whose stories she tells in this book should be labeled as crackpots. But, there are those who would rather call them all nuts than see the picture that she is clearly painting. Riki is telling the truth; a truth that each resident of the United States and certainly those in positions of responsibility need to hear. If only we could lead a horse to water *and* make it drink!

The sad truth is that the major industries associated with petroleum, including agriculture and the chemical manufacturers, have been conducting a huge science experiment on mankind. It is well known that petroleum-based chemicals, among others, are now in all of us (Ashford and Miller 1998; National Academy of Sciences 1991). Each of you reading this book has at least 70 different persistent toxins in your body. This alarming number is actually an underestimate as the most comprehensive study found an average of 91 toxic compounds out of only 210 that were sought. An alarming 43 percent of the toxins tested for were found in persons who *did not* work around chemicals!

How did they get to be so toxic? Simple—they all had bad habits of eating, drinking, and breathing on a daily basis. Currently there are over 80,000 chemicals registered for use in this country, with about 3,800 of them listed as "high use" compounds. Hopefully, in the broader context of all possible chemical exposures, we would *not* find 43 percent of all chemicals used in each person. That is almost too frightening to contemplate.

The proponents of chemical manufacture will grudgingly concede the point that chemical and heavy metal residues are now ubiquitous in humans and wildlife. Yet, they will say that no evidence exists that these toxins are really harmful to human health. On this point they are skating on very thin ice. There is actually ample evidence to show that the same persistent chemicals found in humans are toxic to the nervous system (brain function), immune system (infection fighting), and the hormonal system. The association of these compounds with death is clearly shown in the case of outdoor air pollution. Their association with increased rates of asthma and allergies are also clear. Their association with certain cancers (primarily lymphomas and leukemias) is presently clear and is growing in regards to autoimmune disorders. Being exposed to diesel exhaust increases your chances of both asthma and of cancer. Recent research has also shown that the polychlorinated biphenyl (PCB) levels in women leads to a reduction in the IQ of their children. Exposure to certain dioxins can even change one's chances of having a male child. Unfortunately, consuming farm-raised salmon and non-organic butter increases one's exposure to these very same PCB and dioxin compounds. This list could go on and on.

In addition to clear disease states, there are a host of environmental toxin-related health conditions that don't fit into any specific diagnoses. My practice has been filled with such individuals who have gone to the best physicians in the allopathic world and have been told: "It is all in your head." These are persons who get adverse physical, mental, or emotional symptoms when they breathe in very low levels of everyday chemicals. In the pages that follow, you will be introduced to stories that dramatize this problem. As with my patients, they found that health care providers, employers, friends, and family typically turn away from them and label them as a nut case. The current treatment of choice in the allopathic medical world for this condition is anti-depressants. Having such a condition will certainly be depressing, but these individuals are not living in their own personal hell because they are deficient in Prozac! Their multiple symptoms are due to a buildup of toxins in their body.

While the average person's body already has excessive amounts of persistent pollutants, some have far more than their share. The persons talked about in this book probably had the same amount of compounds *before* working the *Exxon Valdez* cleanup as you and I. Then they worked on an oil spill that added high levels of compounds each day to their existing load. This elevated chemical load likely resulted in a multi-system breakdown that led to chronic problems. Conventional medical wisdom would say that the body should be able to clear all these things out over time. So, if that is right, these people cannot be ill from chemical exposure. The only other explanation, according to conventional medical wisdom, must be that they are mentally imbalanced.

The trouble is that the compounds they were exposed to, as well as the compounds already in all of us, are *not* easily cleared from the body. Our body is designed to rapidly clear compounds that dissolve in water and to absorb all of the fat-soluble items that enter it. The body has no good mechanism for clearing out fats or fat-soluble compounds because they are all expected to be highly useful for the body. Fat-soluble compounds are used for fuel, for tissue integrity, hormones, and a host of beneficial functions. When your body stops absorbing fats or oils, lots of health-related problems begin. So, the mechanisms that were designed for self-preservation become a mechanism of self-destruction when the fat-soluble compounds are not nutrients, but toxins.

Dr. Ott has done a fantastic job of bringing the real story of the *Exxon Valdez* disaster into a very clear, entertaining, and disturbing focus. Read the stories, weep for those whose lives have never been the same since, and be grateful that you do not at present have the same toxin load as the persons in these pages. Then you must realize that it is only a matter of time and exposure until you may begin to have a similar tale. So, begin to change your life now to reduce your own daily exposures and do what you can to reduce the use of chemicals globally.

WALTER J. CRINNION, ND
Director, Environmental Medicine
Center of Excellence
The Southwest College of
Naturopathic Medicine
Tempe, Arizona

Preface—The Truth

"Do you solemnly swear to tell the truth, the whole truth, and nothing but the truth?"

This question is posed to everyone from ordinary people to presidents taking the witness stand. We put our right hand on a bible and answer, "yes," almost without thinking.

But in the province of Quebec, Canada, during the late 1970s at a public hearing on the proposed James Bay Hydro-Electric Project, a Cree Elder gave a different answer. When asked this standard question, he spoke at length to his interpreter who then explained to people gathered for the hearing that this Elder could not speak the whole truth. He could only speak his truth.

This legacy of the *Exxon Valdez* oil spill is the collective story of the people who shared a part of their lives for this book. These are scientists who studied Prince William Sound; people who came to help cleanup the spill; doctors and lawyers who defended people harmed by the oil and cleanup; and myself. Although these people played a role in creating this story, as author, I was entrusted with maintaining the integrity and honor of the individual stories as I wove these strands into our collective truth.

The world would be greatly served if everyone touched by the spill, especially the people who lived through and with this spill, were to share their truths. Only in sharing our truths will we be able to strip away the virtual reality created by the corporate and government spin doctors, identify the real problems that led to it, and find the courage and solutions to prevent another.

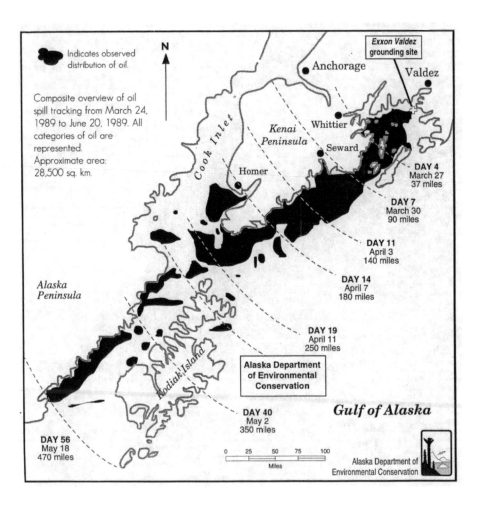

Spread of Oil from the Exxon Valdez Oil Spill
(24 March to 20 June 1989)

Acknowledgements

In some indigenous cultures, people believe that stories are Beings and when they are ready to be told, they will seek out the Storyteller. I am grateful that this story found me and thankful for the inspired six years of adventures to put this story into writing. In this journey alone lies the seed for another story.

My most heartfelt thanks go to those who lent their time and talents during my journey. Laura Honkola was among the first to share the vision of this book. She was a Cordova High School student when she first organized my entire library after the whirlwind of a court-summoned deposition; she is now a college graduate who still transcribes my interviews from tapes. The scientists and cleanup workers gifted me with their stories from the heart; these very human stories balanced my scientific tendencies and set the overall tone for this book. The tape transcribers—Laura Honkola, Robin Holloway, and Linda Kelly—lightened my load. The staff with the Cordova Public Library, the Cordova Public Museum, and the Alaska Resource Library and Information Services (ARLIS), especially Carrie Holba, the "Keeper of the EVOS Files," helped ferret out the details to anchor the personal stories.

A very special thanks goes to two special organizations—the Alaska Forum for Environmental Responsibility (AFER) and the Alaska Community Action on Toxics (ACAT). Interns, staff, and volunteers of these two groups conducted much of the research for this book. I especially wish to acknowledge and thank the hard work and dedicated professionalism of ACAT Executive Director Pam Miller and AFER board member Dan Lawn. Special thanks to ACAT staff Lydia Darby for her healing energy and powerful focus—along with her labor of love in collecting information on cleanup workers.

Others who provided timely support include pre-med student Christina Cinelli, Annie O'Neill (Master's in Public Health, Yale), general researchers Gabriel Scott and Lauren Padawer, Cordova High School student and future environmental medicine doctor Rory Merritt, Gershon Cohen, PhD, and Barbara Williams, executive director of the Alaska Injured Workers Alliance. Hats off in thanks to the many individuals and foundations that supported this research—the Alaska Conservation Foundation (ACF), an anonymous donor through ACF, the Brainerd Foundation, Common Stream, Inc., Environmental Defense, The Fred Gellert Family Foundation, and the Public Welfare Foundation.

Very welcome infusions of legal advice and suggestions were provided freely by Cook Inlet Keeper Executive Director Bob Shavelson and students with the Yale Environmental Protection Clinic under the tutelage of Natural Resource Defense Council's Senior Attorney Dale Bryk.

I wish to acknowledge and thank "the others" for their work and support. "The others" come in many forms: this includes the hundreds of scientists, graduate students, and support crews who were not featured as "characters" in this story, but whose labor contributed to the whole; my family and friends who provided loving support; the many professionals who offered insights and truths; and the Cordova community for encouragement every step of the way.

In the final labor to finish the book, critical reviews from a technical perspective were provided by Stanley "Jeep" Rice, PhD, Bruce Wright, PhD, Charles "Pete" Peterson, PhD, William Rea, MD, Patricia Smith-Willis, MD, and Walter Crinnion, ND. The Home Team Reviewers—Nancy Musgrove, MSc (Fisheries), with Management of Environmental Resources, Inc., Cordovan-at-heart Karen Yoshitomi, and Cordovans Barclay Kopchak and Susan Ogle—reviewed most of the manuscript with an eye towards general readability. My publishing consultant Anthony Maulucci with Lorenzo Press orchestrated all the necessary sections to deliver the book on a very tight schedule. It took a village of family and friends to make this book a reality. And finally, my enthusiastic tactician and agent Don Gastwirth was everywhere at once to ensure the broadest possible market for this book. I am humbled by this showing of strong support.

Despite all the careful reviews and edits, there are bound to be some mistakes or misinterpretations. I assume full responsibility for any such errors and apologize in advance to anyone whom I might offend. Please know that any offense was not intentional.

Finally, I am most grateful for the teachings and support of my two companions: Ilarion (Larry) Merculieff who constantly challenged me to write from the heart; and my friend who worked patiently with me on this book from 4:00 A.M. to noon daily for the past six years—my cat, Tsunami.

List of Abbreviations

ABL	NOAA/NMFS Auke Bay Laboratory
ACAT	Alaska Community Action on Toxics
ADEC	Alaska Department of Environmental Conservation
ADFG	Alaska Department of Fish and Game
ADNR	Alaska Department of Natural Resources
ADOL	Alaska Department of Labor
AFER	Alaska Forum for Environmental Responsibility
AK	Alaska
AkPIRG	Alaska Public Interest Research Group
AMA	American Medical Association
APEX	Alaska Predator Ecosystem Experiment
APSC	Alyeska Pipeline Service Company
ATSDR	Agency for Toxic Substances and Disease Registry
AWCB	Alaska Workers' Compensation Board
BAT	Bioremediation Application Team
BP	British Petroleum
CAT	Computerized axial tomography
CDFU	Cordova District Fishermen United
CERCLA	Comprehensive Environmental Response, Compensation, and Liability Act

CFR	Code for Federal Register
CNS	Central nervous system
DDT	Dichloro-diphenyl-trichloroethane
DECON	Decontamination unit
EMT	Emergency Medical Technician
EPA	U.S. Environmental Protection Agency
EVOS	*Exxon Valdez* Oil Spill
GEM	Gulf Ecosystem Monitoring Program
GGT	Gamma glutamyl transferase
MSDS	Material Safety Data Sheet
NGOS	North Gulf Oceanic Society
NIOSH	National Institute of Occupational Safety and Health
NMFS	National Marine Fisheries Service
NOAA	National Oceanic & Atmospheric Administration
NRDA	Natural Resource Damage Assessment
NVP	Nearshore Vertebrate Predator (Project)
OPA 90	Oil Pollution Act of 1990
OSHA	U.S. Occupational Safety and Health Act
PAH	Polycyclic (or polynuclear) aromatic hydrocarbon
PBT	Persistent bioaccumulative toxic (pollutant)
PCB	Polychlorinated biphenol
PEL	Permissible Exposure Limit
PPE	Personal Protective Equipment
ppb	Part per billion
ppm	Part per million
psi	Pounds per square inch
PVC	Polyvinyl chloride
PWS	Prince William Sound
QA	Quality Assurance
QC	Quality Control
REL	Recommended Exposure Limit
R/V	Research Vessel
SEA	Sound Ecosystem Assessment (Program)
SPECT	Single-photon-emission computed tomography
USDOL	U.S. Department of Labor
USFWS	U.S. Fish & Wildlife Service

TAPS	Trans-Alaska Pipeline System
TILT	Toxicant-induced loss of tolerance
T/V	Tank Vessel
UA	Urinalysis
UAF	University of Alaska Fairbanks
URI	Upper respiratory infection
USCG	U.S. Coast Guard
USDOI	U.S. Department of Interior
USDOL	U.S. Department of Labor
USFS	U.S. Forest Service
USFWS	U.S. Fish & Wildlife Service
USGS	U.S. Geologic Survey
VHS	Viral hemorrhagic septicemia
VOC	Volatile organic carbon
WSF	Water soluble fraction

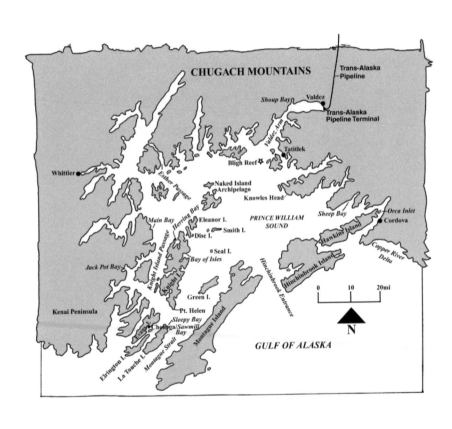

CHUGACH MOUNTAINS

Trans-Alaska
Pipeline

Shoup Bay   Valdez

Trans-Alaska
Pipeline Terminal

Valdez Arm

Bligh Reef   ☆   Tatitlek

Whittier ●

Esther Passage

Naked Island
Archipelago

Knowles Head

Main Bay

Herring Bay

Eleanor I.

PRINCE WILLIAM
SOUND

Sheep Bay

Orca Inlet

Cordova

Disc I.   ⊙ Smith I.

Hawkins Island

Knight Island Passage

⊙ Seal I.

Bay of Isles

Copper River
Delta

Jack Pot Bay

Knight I.

Hinchinbrook Entrance

Hinchinbrook Island

Kenai Peninsula

Green I.

Pt. Helen

Sleepy Bay

Chenega/Sawmill
Bay

Montague Island

Montague Strait

0   10   20mi

N

Elrington I.

La Touche I.

GULF OF ALASKA

Introduction

The Pledge

"BANG! BANG! BANG!" I shoot up, heart beating wildly. It is 7:15 A.M. on 24 March 1989. Someone is pounding on the front door! I race downstairs, barefoot in my nightgown, and throw open the door to find Jack Lamb, acting president of Cordova District Fishermen United (CDFU).

"How long will it take you to get dressed?"

"Five minutes. Why?"

"We've had the Big One. There's a tanker aground on Bligh Reef. It's lost ten million gallons, but there's four times that still on board."

We stare into each other's eyes for a moment, then I gaze past him up Orca Inlet and across Hawkins Island to the northwest. For an instant my mind goes blank, then a tidal wave of emotion floods back in—denial, a hot white flame of anger, a surge of adrenaline, a cascade of ideas.

"I'll get dressed. You start a fire." In ten minutes, we are headed to the CDFU office. During the day, the fire slowly burned out.

It would be a week—and a lifetime—before I returned.

Within the hour, I catch a ride with bush pilot Steve Ranney out

1

to the Sound. I leave Jack Lamb with executive director Marilyn Leland at the CDFU office where they are calling fishermen to get a fleet together to go help. Their voices swirl in my head: "... ten million gallons ... *Exxon Valdez* ... midnight ... alcohol may be involved ...."

Ten million gallons of oil in our beautiful Sound.

Steve Ranney is flying local scientist Chuck Monnett to conduct an aerial sea otter survey before the oil spreads. Just past Knowles Head we spot the stricken tanker at the apex of a black stain on the deep blue waters. We count 130 sea otters and 30 sea lions tucked along the rugged shorelines of islands and bays and several concentrated groups of porpoise and sea birds, but see little else.

We fly to Valdez to refuel and I call the CDFU office. Jack Lamb asks me to stay in Valdez to relay information and represent the fishermen's interests until CDFU forms a response plan. I set up the fishermen's first command post at the Valdez U-Drive rental office at the airport.

Valdez airport, 11 A.M. The atmosphere is charged. People arrive in droves with each plane: black-suited, grim-faced Exxon officials; scientists lugging field gear and computers; reporters bristling with cameras and clipboards. Reporters latch onto me ("Dr. Ott"), drag me outside, and stuff cameras into my face. My camera etiquette is terrible—we repeat sequences until the cameramen are satisfied.

Finally, dazed and overwhelmed, I slip alone outside to think. The din of air traffic recedes as my gaze fixes on the white peaks of the Chugach Mountains, sparkling against the bright blue sky. A question forms in my mind: *I know enough to make a difference, but do I care enough for the Sound to commit my life to this?*

Suddenly the Chugach Mountains vanish and I am back in Wisconsin, watching myself at thirteen as I stand in front of our dining room window with my father and two younger siblings. My father's face is etched with sorrow as state trucks drive up our street spraying great white sweet-smelling clouds of DDT.

My life streams forward like a fast flowing river. There is the blank stare of a robin in my cupped hands—the bird is dying from the neurotoxin DDT. The adolescent girl is reading Rachael Carson's *Silent Spring* to try to understand what is happening in the adult world. My father galvanizes people to action. The Environmental Defense Fund biologist and lawyer are working at our dining room table, organizing their arguments for court. Wisconsin bans DDT that year, 1968, and the rest of the nation follows suit in 1972 (Rogers 1990) as the young woman heads off for college to become a marine biologist—like Rachael Carson.

A path unfolds for me to study marine pollution—oil pollution—from the decks of research vessels in different oceans, from labs in Bermuda, England, Malta, the Carolinas, and Washington State. Thirteen years, five colleges and universities, and three degrees later, including a master's in marine biology and oil pollution from the University of South Carolina and a doctorate in fisheries and marine toxicology from the University of Washington, I hesitate and look north to Alaska.

There is Prince William Sound as I first saw it—from the deck of a fishing boat in early May. The air is alive with the sight and sounds of birds—sea ducks, puffins, gulls, murrelets, cormorants, gulls, bald eagles. The water is alive with sea otters, porpoise, seals, and sea lions. The land is a soul-feast of rugged snow-capped mountains, glaciers stretching to the sea, and a thick ribbon of spruce-hemlock rainforest at the water's edge.

I see my resolve to take just one summer off from my imagined career vanish instantly upon my first glimpse of the coastal fishing town of Cordova. I am in love. I am a commercial fisherman, pulling silvery salmon from the sea on the Copper River Delta, in Prince William Sound, in the wind and waves, in the adrenaline-rushing breakers, in the gentle cradle-rocking calm, in the lashing rain, under a searing sun, under a luminous moon, under the northern lights. I am a fisherman, but I am feeling guilty about not using my education.

It is fall 1987 and I am on the boards of CDFU and United Fishermen of Alaska, picking up where other fishermen before me left off—a seamless handoff. I am working on chronic air and water pollution problems that plague the tanker terminal in Port Valdez, our fishing ground (Ott 1989a). I can see what is wrong and how to fix it to stop the pollution, but the fixes are elusive.

I am learning about politics in Alaska where 85 percent of the state's operating revenue flows from the 800-mile long Trans-Alaska Pipeline System (TAPS) that delivered 25 percent of the nation's domestic oil from the North Slope to the tankers in Port Valdez. I am becoming a player in this drama that intersects lives of ordinary citizens with state and federal regulators, politicians, and scientists; oil industry scientists; the U. S. Coast Guard; the seven oil company owners and their consortium, Alyeska, that operates the TAPS.

It is spring 1989. I am in Juneau, Alaska; Washington, DC; then Dallas, Texas, warning politicians, oilmen, and federal agencies of the high risk of an oil spill in Prince William Sound.

It is the evening of 23 March, sixteen hours ago. I am speaking by teleconference to the Mayor's Oil Action Committee in Valdez.

"Given the high frequency of tankers into Port Valdez, the increasing age and size of that tanker fleet, and the inability to quickly contain and cleanup an oil spill in open water of Alaska, fishermen feel that we are playing a game of Russian roulette. When, not if, 'The Big One' does occur and much or all of the income from a fishing season is lost, compensation for processors, support industries and local communities will be difficult if not impossible to obtain ..." (Ott 1989a).

I wake up to find fishermen's worst fears realized—'The Big One' has happened in Prince William Sound.

And suddenly the Chugach Mountains are straight in front of me again. *This cannot be an accident that I am here now.* I realize that choosing to become involved will change my life forever—it will become my life. In the deafening silence, I find my answer. *Yes, I care enough. I will not turn my back on the Sound.*

With a roar, my womb of silence cracks and the sounds of disaster response flood back. With new resolve, I step back into chaos.

How Much Oil did the *Exxon Valdez* Spill?

I was not the only one who heard 11 million gallons was the low-end estimate of spill volume and 38 million gallons was the high-end estimate during my first 24 hours in Valdez after the spill. A year later, two separate newspaper accounts reported volumes up to 27 and 38 million gallons (Hennelly 1990; Spence 1990). One article (Hennelly 1990) reported that the Alaska Department of Environmental Conservation (ADEC) and Exxon jointly contracted Caleb Brett, a company that gauges tanker loads for the oil industry, to do tank soundings to estimate the volume spilled. However, Caleb Brett "closed ranks with Exxon" and refused to make public documentation of its findings, citing its employer-client relationship (Hennelly 1990, 14).

I found that the Alaska Department of Law conducted a separate investigation of the spill volume in preparation for a civil lawsuit against Exxon over damages to wildlife and habitat. When this lawsuit was settled in 1991 (Sidebar 2), the State of Alaska shelved its investigation. In 1994, in response to repeated public requests,

(continued)

the state released its investigation files to the Alaska Resources Library and Information Services (ARLIS) in Anchorage, Alaska (AK Department of Law 1991).

According to the one-page report filed by Caleb Brett (1989), the *Exxon Valdez* left the tanker terminal in Port Valdez on 23 March 1989 with 53.04 million gallons of oil on board. The tanker grounded on Bligh Reef around midnight. Caleb Brett reported 42.2 million gallons of cargo was transferred to three Exxon tankers and a fleet of barges. Since the *Exxon Valdez* had been carrying a total cargo of 53.04 million gallons, simple math led Caleb Brett to conclude that the total spilled was the difference—10.8 million gallons; however, this volume was never independently verified. In the absence of other information at the time (1989), the media used—and still uses—Exxon's self-reported spill volume estimate.

According to one of the State of Alaska's independent surveyors, "much more oil than the 258,000 barrels [10.8 million gallons] reported [by Caleb Brett] was spilled" (Murchison 1991, ACE 9486070). He continued, "It is my opinion that the major differences [in spill volume estimates] are due to the tremendous 'churning forces' that occurred as the oil gushed from the vessel and the seawater forced itself in, due to the hydraulic pressure[. T]his resulted in the emulsification of oil and water" (ibid.). He found errors in the calculations of oil on board the grounded *Exxon Valdez* and noted, "Most of the water cuts or soundings are questioned because of the holes in the vessel[. A]pparently the method used to differentiate between oil and water was providing inaccurate results, which resulted in grossly under-estimating the water and over-estimating the oil" (Murchison 1991, ACE 9486069).

State investigators tracked each of the three Exxon tankers used to lighter (transfer) oil from the *Exxon Valdez* (Alaska Department of Law 1991). These three tankers went to three different Exxon refineries to offload their cargo. Exxon insisted the cargo was 100 percent oil, however the evidence suggests otherwise. For example, according to the deposition of Claude Wendell Dees, an Exxon Shipping officer in charge of the lightering operation aboard the Baton Rouge (Dees 1991), "We had more water than normal and Captain Solywoda knew that. He left Exxon and took early retirement" (ibid., 193). Exxon never disclosed its shore tank records from each of the three discharge port: these records would show the percentage of water in the offloaded cargo, water which was known to exist. For example, Dees stated, "None of the oil discharged at Hawaii was used because of high water content" (ibid., 164).

However, Exxon did not control all the cargo records of the ill-fated *Exxon Valdez*, because not all of the cargo was offloaded at each of the refineries. For example, Dees stated, "On the trip to Hawaii, we gauged our tanks to determine water content" (ibid., 222). "I gravitated some of it out into one tank and kept it aboard as dirty ballast. We went back to Valdez with this dirty ballast" (ibid., 46). Similarly, cargo that remained on board the other two lightering tankers also ultimately was returned as ballast water to the tanker terminal. According to papers on file with the state's investigation, the oil content of the former cargo (now ballast water) was estimated as a percentage of the total volume by eyewitnesses before the tankers sailed to Valdez (Alaska Department of Law 1991, ACE 10864138–10864143). The amount of dirty ballast, and its tank location, is recorded on official ADEC Ballast Water Discharge Certificate Affidavits at the tanker terminal in Port Valdez. *Exxon never accounted for the known volume of water from the* Exxon Valdez *in the cargo of its three tankers; the State of Alaska investigators did* (Table 1, p. 7).

In light of this evidence, Exxon's self-reported spill volume of 11 million gallons is incorrect. I conclude that the State of Alaska's conservative estimate of 30 million gallons should be used when referring to the *Exxon Valdez* oil spill volume.

The Spill—The Day the Water Died

*"The excitement of the season had just begun, and then we
heard the news: oil in the water, lots of oil killing lots of water.
It is too shocking to understand.
Never in the millennium of our tradition
have we thought it possible for the water to die, but it is true."*

Chief Walter Meganack
Traditional Village Chief
Port Graham Native Village, Kenai Peninsula, Alaska
(National Wildlife Federation et al. 1990, 44)

On 24 March 1989 the *Exxon Valdez* gutted her hull on Bligh Reef,
spilling about 30 million gallons of crude oil—56 percent of her
cargo—into Prince William Sound, according to papers on file with
an investigation conducted by the State of Alaska (Sidebar 1, p. 4;
Table 1, p. 7).[1] After two days of calm weather, a fierce storm swept
the oil through the Sound and out into the Gulf of Alaska. Ultimately
over 3,200 miles of Alaska's shorelines were oiled from the Sound,
past the Kenai Peninsula and Kodiak Island, to parts of the Alaska
Peninsula some 1,200 miles distant from Bligh Reef. A spill of similar
magnitude on the East Coast would have oiled shores from New York
to Cape Canaveral, Florida.

The runaway slick left devastation in its wake. Exxon's oil killed
more wildlife than any other oil spill in the world—ever. Among the
victims were thousands of marine mammals—sea otters, seals, and
even orcas—and hundreds of thousands of marine birds—murres,
marbled murrelets, harlequin ducks, scoters, buffleheads, goldeneyes,
cormorants, and others. Untold millions of salmon and herring were
killed by an invisible cloud of dissolved and dispersed oil that spread
underwater, shadowing the path of the surface slick and hanging offshore from oiled beaches in the Sound.

The killing did not stop with the oil. Exxon's shoreline cleanup
continued to kill plant and animal life that had survived the initial oiling. The pressurized hot water wash laid bare huge swathes of beaches, normally covered with rich communities of sea plants and animals. By the end of 1989, the cleanup had killed *by weight as much
plant and animal life as the initial oiling* (Mearns 1996).

Table 1
How Much Water Was in the Lightered Cargo? (in million gallons)?

Source	Lightered Cargo[a]	Water in Cargo
T/V Exxon Baton Rouge	19.40	5.5
T/V Exxon San Francisco	16.91	3.6
T/V Exxon Baytown	5.01	10.1
5 barges	0.87	?
Estimated total	**42.20**	**19.2**

Spill volume calculations:[b]
Exxon: 53.04 - 42.20 = 10.8 million gallons
(*Exxon alleged the lightered cargo was 100 percent oil.*)
State of Alaska files: 10.8 + 19.2 = 30 million gallons

Sources: Alaska Department of Law 1991; Caleb Brett 1989.

[a]Lightered cargo = oil/water emulsion off-loaded from *T/V Exxon Valdez.*
Estimates of water in the lightered cargo (Alaska Department of Law 1991, ACE
10864138–10864143) are as follows:
•   *Exxon Baton Rouge* retained onboard approximately 9.2 million gal-
lons of an oil-water mixture with an estimated water/oil content of 60/40. That
means about 5.5 million gallons of "oil" supposedly lightered from the *Exxon
Valdez* was *water* (9.2 x 0.6).
•   *Exxon San Francisco* retained onboard approximately 5.6 million gallons of an oil-water mixture with an estimated water/oil content of 65/35. That
means about 3.6 million gallons of "oil" supposedly lightered from the *Exxon
Valdez* was *water* (5.6 x 0.65).
•   *Exxon Baytown* retained most of her cargo onboard. She lightered
approximately 6.6 million gallons of an water-oil mixture with an estimated
water/oil content of 70/30 to the *Exxon Baton Rouge,* which returned to the
tanker terminal in Port Valdez. She lightered approximately 7.9 million gallons
of an oil-water mixture with an estimated water oil content of 70/30 to the
Exxon San Francisco, which returned to the tanker terminal in Port Valdez. That
means about 10.1 million gallons of "oil" supposedly lightered from the *Exxon
Valdez* was *water* [(6.6 x 0.7) + (7.9 x 0.7)].

[b]Volume spilled: State of Alaska papers show that investigators used an error
margin of ± 20 percent to account for unknowns such as the percentage of
water in the cargo offloaded at the refineries and the amount of water in the
cargo lightered by the five barges. Thus, the range of volume spilled, according
to the records on file with the State's investigation, is between 24 to 36 million
gallons. State investigators stress that the low number is "a minimum amount;
apparently much more was lost before the oil was completely offloaded"
(Murchison 1991a, ACE 9486068).

The killing did not stop in 1989. Nearly half of the spilled oil stranded on beaches in the central and southwest regions of Prince William Sound during a fierce three-day storm (Spies et al. 1996).This oil hit the beaches with such force and in such quantity that relatively fresh oil was buried in the intertidal zone under mussel beds and dense forests of seaweed.Thick surface oil lay for months in bays and small fjords of the Sound, repeatedly oiling beaches with each tidal cycle.These repeated strandings garnered media attention and frustrated cleanup crews.

By 1990, the surface oil had hardened into tarry mats and thick encrustations that posed little threat to wildlife. But the buried oil, largely unnoticed, retained much of its toxicity (Short et al. 2004).As this buried oil spread and redistributed over time, it leached poisons that were picked up by sealife that lived, spawned, or foraged in the shallow intertidal zone and nearshore seas. Trapped liquid oil continued killing wildlife for years after the spill.

Unlike most spills worldwide, the effects of the *Exxon Valdez* spill and cleanup on habitat and sealife were well studied—separately—by public-trust (non-Exxon funded) scientists and by Exxon and Exxon-funded scientists (Rice et al. 1996; Wells, Butler, and Hughes 1995). During 1990 and 1991, Exxon was virtually the lone voice in the news, because public-trust scientists were under a secrecy order due to pending litigation over the loss of wildlife and their habitat (Sidebar 2, p. 9; Cummings 1992). Exxon waged a highly publicized media campaign, barraging the public with messages that Prince William Sound had recovered rapidly and everything was back to normal in the wake of the spill (Matthews 1993). Gradually, Exxon's story became the popular understanding of the spill, the cleanup, and the biological effects of this tragedy.

By the time the public-trust story of the spill emerged in 1993, it was too late (Solomon 1993).The popular version was—and still is—Exxon's story. For the next ten years those of us living in the Sound watched in amazement as what was going on in the Sound—what I call "Sound Truth"—departed further and further from Exxon's spill lore. Essentially, Exxon papered over the extent of spill effects in the Sound, a charade that Exxon perpetuates to this day.

Upon publication of the public-trust scientists' summary reports from 2000 forward, I realized it was time to write a book to share their startling discoveries, and the Sound Truth, with the public. In pursuing the story of the effects of Exxon's oil on the Sound's sealife, I uncovered the story of pervasive health problems suffered by cleanup workers. Sick wildlife and sick workers: together, these two stories are perfect mirrors of the harmful effects from oil.

Secrets and Settlements

After the *Exxon Valdez* spilled over half its cargo into Prince William Sound, the United States government and the State of Alaska filed criminal charges against Exxon and civil claims for damages to and recovery of wildlife and public lands harmed by the oil spill.

It took over two years for the federal and state governments and Exxon to reach a settlement that resolved all the issues. During this time (1989–1991) the governments gathered evidence, such as the investigation into spill volume, to support their case. The governments and Exxon conducted their investigations and scientific studies on extent of damages to wildlife and public lands (habitat) in secret as both sides tilted towards litigation. Government scientists were under a U.S. Department of Justice-imposed gag order not to share their work with anyone (Cummings 1992). The public did not know the extent of injury—except for what information Exxon shared with the media or produced for the public in numerous glossy brochures, which proclaimed rapid recovery (Baker, Clark, and Kingston 1990, 1991; Exxon 1991a 1991b; Neff 1990, 1991; Owens 1991a, 1991b).

On 9 October 1991, the U.S. District Court approved the settlement among the federal and state governments and Exxon (*U.S.A. v. Exxon* [1991b]; *State of Alaska v. Exxon* [1991]). Under the Criminal Plea Agreement (*U.S.A. v. Exxon* [1991a]), Exxon was fined $150 million and forgiven $125 million in recognition of Exxon's good corporate behavior in cleaning up the spill. The remaining $25 million was paid to the North American Wetlands Conservation Fund ($12 million) and the national Victims of Crime Fund ($13 million). For criminal restitution, Exxon paid $100 million, which was divided evenly between the state and federal governments. Under the civil settlement, Exxon agreed to pay $900 million in annual payments over a ten-year period. (This was three times lower than the minimum estimate for damages calculated from public surveys and it amounted to about $500 after accounting for inflation, tax-breaks, and the ten-year payment period [Lancaster 1991; Schneider 1991; U.S. Congress, House, 1991a].)

The 1991 civil settlement created a council of state and federal trustees—the *Exxon Valdez* Oil Spill (EVOS) Trustee Council—to oversee restoration of injured wildlife and habitat through the use of the $900 million. The Memorandum of Agreement and Consent Decree (*U.S.A. v. Alaska* [1991]) guides the use of this money. In this book, I refer to scientists funded with public money through the EVOS Trustee Council or the various government agencies as "public-trust scientists." The stories of their work and findings—and of Exxon's spill science—are told in Part 2.

Section 17 of the 1991 civil settlement contains a "Reopener for Unknown Injury" (*U.S.A. v. Exxon* [1991a], 18–19; *State of Alaska v. Exxon* [1991b], 18–19). This requires Exxon to pay the governments up to an additional $100 million to restore wildlife or habitat for harm that "could not reasonably have been known nor . . . anticipated" based on the scientific understanding of oil effects at the time. The settlement may be reopened by any of the three parties to the settlement for additional claims from 1 September 2002 to 1 September 2006. The "$100 million reopener clause," as it is commonly known, is discussed in Part 3.

The Cleanup—"A Tragedy Beyond Belief"

Hazardous by Nature

Oil spill response began the winter before the spill in 1988–1989 when federal scientists, state regulators, and CDFU fishermen, including myself, agonized over the dispersant guidelines proposed by oilmen and supported by Coast Guard. Dispersant use is an emotionally charged issue. Oil companies manufacture and sell dispersants as a first line of defense in response to oil spills. Dispersants act like liquid soap to break up surface oil slicks into tiny droplets that must then be driven by wind and wave action into the water column and diluted with huge volumes of water (Lethcoe and Nurnberger 1989). Dispersants are the oilmen's way of solving pollution by dilution. Essentially, the oil slick "vanishes" from the sea surface. Out of sight, out of mind, and definitely away from being a public relations nightmare.

The trouble is that the tiny droplets of dispersed oil are very toxic to fish and other creatures in the water column under the sea surface. Dispersants are also only 10 to 15 percent effective in cold water with viscous North Slope crude, which leaves the bulk of the oil, potentially, to strand on beaches (Fingas, Bobra, and Velicogna 1987). Why risk fish *and* the beaches, fishermen asked. CDFU preferred mechanical cleanup of oil—physically removing as much as possible from the water—instead of using our fishing water to dilute and disperse toxic oil.

The Initial Guidelines for Dispersant Use that were approved just two weeks prior to the spill is a compromise (ADEC 1993; Lethcoe and Nurnberger 1989). It divides the Sound into three zones, following recommendations by the National Research Council (1989). Deep open water along the tanker traffic lanes and most of the central Sound is designated as Zone 1, where dispersant use is preapproved, except when sensitive marine life like herring is in the water column. Beaches and the intertidal and shallow nearshore areas are Zone 3, where dispersant use is banned to avoid concentrating the toxic dispersed oil in this biologically productive region. In Zone 3, dispersants can be used on a case-by-case basis with approval from the EPA and the state. The bulk of the Sound between deep water and shallow beaches is designated as Zone 2, where dispersant use is conditional in order to protect sensitive wildlife and requires approval by the EPA and the state.

Two weeks later—when Captain Joe Hazelwood radioed the Coast Guard at half past midnight to report his tanker was aground on Bligh Reef and "evidently . . . leaking some oil"—our dispersant guidelines were put to the test.

So began the first oil spill cleanup in the history of the United States to be conducted under the Occupational Safety and Health Act's Hazardous Waste Operations and Emergency Response standard (OSHA 1989). The ink was barely dry in the regulation change. Less than three weeks earlier, OSHA administrators and organized labor had successfully argued against Exxon, among others, that crude oil and petroleum products pose significant health and safety hazards (U.S. Congress, House, 1989a, 1056-1057). The Hazardous Waste Operations and Emergency Response (Hazwoper) training requires forty hours of safety training, special procedures and equipment, medical surveillance, and long-term record keeping.

Under hazardous waste cleanup standards, *all* workers are supposed to get information on *every single* hazardous compound they might encounter on the job. Basic information on a compound of concern is provided in its "Material Safety Data Sheet (MSDS)," which is required to be given to workers as part of normal Hazwoper training.

MSDS excerpts for some of the compounds of concern present during the 1989 cleanup, including oil and three representative solvents—a dispersant, a "bioremediation" product, and an all-purpose degreaser (Sidebar 3, p. 12)—warn of potential short-term (acute) and long-term (chronic) symptoms. Acute symptoms of exposure include dermatitis, headaches, dizziness, nausea, and central nervous system problems. Chronic symptoms of exposure include anemia and other blood disorders (such as leukemia), fetal defects, liver and kidney damage, and "toxic systemic" effects (or total body poison). MSDS excerpts warn to avoid exposure to vapors and aerosols and, in two cases, to keep the product out of sewers and watercourses.

One would think that 'forewarned is forearmed,' but sadly, this was not the case.

Initial Spill Response to Oil on the Water

As succinctly stated by "Ernie" Piper, author of the 1993 Alaska Department of Environmental Conservation (ADEC) final report on the state's role in spill response: "Oil spill response is most effective

Material Safety Data Sheet (MSDS) Excerpts for Some Hazardous Chemicals Present during the 1989 EVOS Cleanup

Crude Oil (Exxon Shipping Company 1988)

Exposure limit for total product: Not established for total product [emphasis added].

Health studies have shown that many petroleum hydrocarbons pose potential human health risks, which may vary from person to person. As a precaution, exposure to liquids, vapors, mists or fumes should be minimized.

High vapor concentrations are irritating to the eyes and the respiratory tract, may cause headaches and dizziness, are anesthetic, may cause unconsciousness, and may have other central nervous system effects including death.

Prolonged or repeated contact with this product at warm or ambient temperatures tends to remove skin oils, possibly leading to irritation and dermatitis.

Contact with this product at warm or ambient temperatures may cause eye irritation . . .

This product may contain benzene . . . Benzene can cause anemia and other blood diseases, including leukemia (cancer of the blood-forming system), after prolonged or repeated exposures at high concentrations . . . It has also caused fetal defects in tests on laboratory animals.

[T]here may be a potential risk of skin cancer in humans from prolonged and repeated skin contact with this product in the absence of good personal hygiene [emphasis added; EVOS cleanup conditions often prevented good personal hygiene.]

Minimize breathing vapors. Minimize skin contact. Ventilate confined spaces.

Corexit 9527™ (Exxon Company, USA, 1992)

Hazardous ingredient: 2-butoxyethanol ["bue-TOX-e-ETH-an-ol"].

OSHA Hazard: eye and skin irritant, vapors irritating to eyes and respiratory tract, toxic systemic via ingestion, inhalation and skin.

May be absorbed through skin to produce hemolytic anemia and kidney damage evidenced by paleness and possible red coloration of the urine.

Vapors and/or aerosols, which may be formed at elevated temperatures, may cause systemic effects.

Chronic effects: Overexposure by inhalation and/or dermal contact may result in damage to the blood and kidneys.

Prevent liquid from entering sewers, watercourses, or low areas. Contain spilled liquid . . . [emphasis added; Prince William Sound is a watercourse.]

Inipol EAP 22™ (Exxon Company, USA, 1989a)

Hazardous ingredient: 2-butoxyethanol.

Health studies have shown that many petroleum hydrocarbons and synthetic lubricants pose potential human health risks, which may vary from person to person. As a precaution, exposure to liquids, vapors, mists or fumes should be minimized.

Inhalation of high vapor concentrations may have results ranging from dizziness, headache, and respiratory irritation to unconsciousness and possibly death.

Components of this product (2-butoxyethanol) may be absorbed through the skin and could produce blood and kidney damage. Symptoms of overexposure include paleness and red discoloration of the urine. *(continued)* when oil is on the water, rather than stranded on shorelines. The faster responders act, the better chance they have. The effectiveness of most on-the-water techniques drops substantially as the oil weathers, emulsifies, and large slicks break up" (ADEC 1993, 49).

Every oil spill is different, every cleanup is different, but the spill response technologies and techniques are roughly the same—and largely ineffective. Coast Guard Vice Admiral Clyde Robbins, the federal on-scene coordinator for the cleanup in 1989, was shocked to find oil spill response technology no further advanced than what he had seen fifteen years earlier (ADEC 1993, 51).

On-the-water spill response included briefly, burning and disper-

---

Petroleum solvents/petroleum hydrocarbons—skin contact may aggravate an existing dermatitis.

Glycol ethers—persons with a history of blood and/or kidney disease should avoid exposure to this product.

Steps to be taken in case material is released or spilled: Keep people away. Recover free product. Minimize skin contact. Ventilate confined spaces. *Keep product out of sewers and watercourses by diking or impounding* [emphasis added; Prince William Sound is a watercourse.]

Simple Green® (Sunshine Makers, Inc., 2002)
Use of product: an all purpose cleaner and degreaser . . .
Ingredient: 2-butoxyethanol (less than 6 percent)
Note, however, that Butyl Cellosolve [2-butoxyethanol] *is only one of the raw material ingredients that undergo processing and dilution during the manufacture Simple Green®. Upon completion of the manufacturing process, Simple Green® does not possess the occupational health risks associated with exposure to undiluted Butyl Cellosolve* [emphasis in original].

Adverse effects on human health are not expected from Simple Green®, based upon twenty years of use without reported adverse health incidence in diverse population groups, including extensive use by inmates of U.S. federal prisons in cleaning operations.

Repeated daily application to the skin without rinsing or continuous contact of Simple Green® on the skin may lead to temporary, but reversible, irritation.

Simple Green® is a mild eye irritant; mucous membranes may become irritated by concentrate-mist.

The Simple Green® formulation presents no health hazards to the user when used according to label directions for its intended purposes.

(Author's note: The U.S. Environmental Protection Agency (2003) lists 2-butoxyethanol at the top of its list of ingredients to avoid in its Janitorial Products Pollution Prevention Program. The EPA web page states products with the listed ingredients "pose very high risks to the janitor using the product, to building occupants, or to the environment." Comments under chronic effects for 2-butoxyethanol list reproductive and fetal damage, liver and kidney damage, and blood damage. www.westp2net.org/janitorial/tools/haz2.htm)

sant use. The initial dispersant of choice was Corexit 9527, an Exxon product. However, first it was too calm and then it was too stormy for effective dispersant use (Lethcoe and Nurnberger 1989, 44-49). Barrier booms to contain oil and skimmers to pick up the contained oil were used in summer1989, initially in open water and then, with marginal results, nearshore to contain and pick up oil and oil/solvent mixtures draining from treated beaches.

After the fierce storm on the third day, whatever hope had existed to contain the oil was lost. It was clear, even as Exxon mobilized more and more people, vessels, and equipment to respond to the larger slicks and all the shredded pieces moving through the southwest Sound, that shoreline cleanup would become a bigger priority than on-the-water response. The on-the-water spill response extended roughly through mid May when the shoreline treatment began in earnest (ADEC 1993; Harrison 1991).

Before the transition from on-the-water response to shoreline cleanup, visiting Congressman Peter DeFazio (D-4th OR) summed up the situation. "Alyeska has had to rely on a mosquito fleet (i.e., commercial fishermen), sport fishing boats, and the local communities to help them preserve the remnants of a great ecosystem. I think that is a tragedy beyond belief" (U.S. Congress, House, 1989a, 146). Unfortunately, the real tragedy in terms of human health was about to begin.

Shoreline Cleanup

Exxon's shoreline cleanup operation was an evolving experiment the entire summer (ADEC 1993, 61-82). The first week in April, Exxon started—and largely ended—a futile attempt at wiping rocks one by one with absorbent material. The highly publicized effort backfired as it accentuated the hopelessness of the situation.

Next Exxon tried several techniques to wash the beaches. First, crews flooded the upper intertidal reaches with seawater from a perforated hose, while workers tilled the lower reaches with rakes and other tools to agitate and release the oil. This worked but it was slow. Moderate pressure hoses worked better, especially if the seawater was warmed up. High-pressure hoses coupled with seawater heated to scalding temperatures worked even better to dislodge oil.

When Coast Guard Commandant Admiral Paul Yost arrived on scene in mid April, he "made it clear that hot water washing was his preferred method of treatment" (ADEC 1993, 96; Wohlforth 1989). Exxon submitted a plan to "wash about 300 miles of shoreline with cold water, employ about 4,000 people, and be done by September 15" when the winter storms would start in earnest (ADEC 1993, 96). The state and federal governments pushed Exxon "to get crews in the field, doing meaningful cleanup" (ibid. 98) before May 15, the date that biologists expected wildlife to show up in large numbers to rear their young. A consensus was reached to use hot- and cold-water flush systems. "Exxon and the military began bringing in barges, boilers, hoses, and portable pumps to support beach-cleaning crews, and by the first week of May, shoreline cleanup was the focus of the entire operation" (ibid. 120). According to Exxon, cleanup crews surged from some 3,000 people in mid May to more than triple that during peak operations (Harrison 1991; see also Nauman 1991).

When even the pressurized hot water wash proved too slow to remove oil, Exxon received authorization from the state and the EPA to test chemical cleaners—dispersants—first Corexit 9580M2 and, when the public objected, then Inipol EAP 22, a dispersant cloaked in the beguiling name of "bioremediation." Both are Exxon products; both contain industrial solvents.

Thus unfolded the cleanup—or "the money spill" as it was known in Cordova and the twenty-one other oiled communities—a tragedy that would have longer lasting effects on the habitat and wildlife of the Sound, cleanup workers' physical health, and residents' emotional health than the oil spill itself.

Fulfilling a Pledge

After the collapse of the pink salmon and herring fisheries in Prince William Sound in 1993, I decided it was time to apply my energy and talents to address the economic and environmental problems stemming from the oil spill. I reasoned if I couldn't help solve problems in my own backyard, how could I justify working on statewide issues? I sold my half of the boat and permit to my fishing partner and friend and resigned from the boards of CDFU and United Fishermen of Alaska.

Then I helped to found two nonprofit organizations, one to involve area residents in rebuilding our struggling economy through sustainable use of our natural wealth and the second to continue the work I had started with CDFU. Gradually the latter organization, the

Alaska Forum for Environmental Responsibility (AFER), narrowed its focus to holding the oil industry, specifically the TAPS owners, accountable to the laws protecting public health, worker safety, and the environment. It was through AFER that I rallied support to conduct what became a five-year investigation of the social, economic, and environmental aftermath of the *Exxon Valdez* oil spill.

The results of this investigation are told in two books. This first book focuses on what I call "the legacy" of the *Exxon Valdez* oil spill and it chronicles how this legacy came to be. Exxon's spill provides a portal to understanding a startling truth: oil is much more toxic to people and the environment than previously thought. This book shows that one particular fraction of crude oil, polycyclic aromatic hydrocarbons or PAHs, may well be the DDT of the 21st century. Readers are invited to consider some suggestions for change at the end of this book, if we, as a nation of concerned citizens, want to fully recognize and reduce the public health and environmental risks of our dependency on oil.

The second book, *Enough! Paying for the Legacy of the Exxon Valdez Oil Spill* (2007), focuses on the social, economic, and political ramifications of this spill on the fishing community Cordova and the perpetrators of the spill—Exxon, the TAPS oil industry, and the state and federal governments. Again, Exxon's spill clarifies a larger truth: we need to put the needs of people, not corporations, first, if we wish to manage our political, legal, economic, and social systems as if life matters.

One final note to fulfill a promise to people who shared their personal stories for this book: the opinions expressed by individuals in this book are their own and do not necessarily reflect those of the department, agency, or entity for which the individual worked or works. Additionally, I occasionally altered the language—but not the factual content—for literary style.

Part 1
Sick Workers

"Mists and aerosols containing polynuclear aromatic hydrocarbons (PAHs) are, in principle, a cancer risk . . ."

John Park and Michael Holliday (1999)

Exposed:
A Coverup of Mass Chemical Poisoning

Chapter 1.

Exxon's Cleanup from a Worker Health Perspective

Worker Safety and Health Problems during On-the-Water Response

Fishermen were among the first spill responders. When I flew over the stricken tanker nine hours after the accident, I saw a swirling smoky bluish haze over the blackened sea surface—aromatic hydrocarbons vaporizing into the air. The stench of hydrocarbons was so strong it made our heads and stomachs reel. We flew up to a higher altitude to clear our heads. I ended up in Valdez for five days and witnessed and was part of the initial chaos optimistically called "spill response."

Headaches, Nausea, Dizziness, and Desperate Measures

Even before I set foot in Valdez that first morning, Alyeska, the party responsible for initial spill response, and Exxon had made a formal request to the Coast Guard to drop 50,000 gallons of dispersant on the leading edge of the slick, starting immediately (ADEC 1993, 60).

The pre-approved dispersant guidelines called for demonstrating dispersant effectiveness prior to approval, even in Zone 1. The Coast Guard authorized a test, not whole-scale application, because the weather conditions were too calm to mix the chemical with the oil and disperse the slick—and because Alyeska did not have proper spray equipment. Alyeska planned to use a helicopter and drop dispersant from buckets.

In desperation I called Harvard Professor Dr. Jim Butler for advice. Dr. Butler had chaired the National Research Council's dispersant review study. He said, "It's important to note that all the talk about the use and effects of dispersants on big spills is hypothetical. It's based on lab data and a few small field tests. No one has successfully treated a spill of this size."

It turned out that the oilmen really didn't know much about the chemical nature of the dispersants either. Transcripts of internal telephone conversations at the Alyeska Emergency Center, recorded for several days after the spill, were quite revealing.[1] When the owner of the tank truck called to ask how to clean out his truck after delivering the dispersant Corexit 9527 to the Anchorage airport, Alyeska scientist Dick Mikkelsen didn't know.

He said, "We're taking I-don't-know-how-many hundreds of thousands of gallons of this stuff (and) we're going to be spreading it all over Prince William Sound. It can't be too damn toxic." He instructed the driver, "[W]hen that thing gets empty . . . you fill her about three-quarters full of water and drive it around for half [an] hour or forty-five minutes, and then dump the damn stuff down a sewer" (Alyeska Pipeline Service Company [APSC] 1989, KWY001042402–403).

The sewer plan was approved by Exxon's dispersant expert Gorden Lindblom, also on the telephone, who commented, "It's a surfactant. You may not want to dump it right in one big whomp [in the municipal sewer system] . . . You know, you can [put] it in over a period of time and it might be all right" (APSC 1989, KWY001042534–535).

Mikkelsen reviewed the material safety data sheet briefly and mused, "It says, 'don't put the raw stuff in a waterway' . . . ." (APSC 1989, KWY001042403). The irony of the warning on the data sheet and the fact that Alyeska wanted to put "the raw stuff" in the biggest waterway in the area—Prince William Sound—seemed lost on Mikkelsen.

Coast Guard officials authorized three tests in three days before they deemed the dispersant was effective (Lethcoe and Nurnberger

1989). By then, Alyeska had a C-130 aircraft with proper spraying equipment—but it was too late. A fierce spring storm whipped the oil into a mousse, with up to 80 percent water. Strong winds shoved it through the Sound and into sensitive nearshore and intertidal habitat.

Despite very limited dispersant effectiveness because of the oil condition, Exxon applied more dispersants in Zone 1 and in Zone 2 near the tanker, dousing the deck of the *Exxon Valdez*—and Coast Guard personnel and tanker crew—in the process.

Desperately, Exxon requested approval to apply dispersants on the heavy oiling in Knight Island Passage—in Zone 3. The state denied Exxon's request: it wrote, "Because Exxon has failed to demonstrate its ability to accurately and effectively target the dispersant Corexit 9527 in Zone 1, the state cannot approve dispersant application in Zone 3" (Lamoreaux and Baker 1989). This and the storm ended the use of dispersants for on-the-water response, but it did not end the debate over whether the chemicals *should* be used, which resurfaced later during the shoreline cleanup.

Meanwhile, while the dispersant debate raged, around 9 P.M. on 25 March Exxon conducted a test burn, torching a small portion of the slick (12,000 to 15,000 gallons) near Goose Island, well away from the grounded tanker—but directly upwind of Tatitlek village, ten miles distant (ADEC 1993, 56). The next morning two Alaska Native men from Tatitlek told me, as we walked up the dock in the Valdez harbor, that they had come in to see what was going on. "People in the village are sick, throwing up in the street," one said.

I was incredulous. "Exxon didn't tell you about the test burn? They didn't evacuate the pregnant women and children? This stuff is toxic!"

One of the men spun around, dashed down the dock, leapt into a skiff, and took off. The other shook his head sadly: "His wife is pregnant."

Later the Natives asked, "Who is responsible for sickness there? Who is responsible for evacuating us?" The Natives' questions went unanswered and they were never evacuated—or treated for their symptoms, but there were no more burns. The storm on the third evening foreclosed that option.[2]

While Alyeska, then Exxon, experimented with burn and dispersant tests, fishermen sped to Bligh Reef and vainly tried to boom the tanker to contain the spilled oil. Some became nauseous and dizzy—classical symptoms of over-exposure to oil fumes that hung thick in the air around the tanker. The first workers on site were given no protective equipment by Alyeska or any warning that the oil and its fumes could be extremely hazardous to their health (APSC 1989, KWY001043337-338; Hill 1989).

The storm changed everything. Exxon's attention shifted to mechanical pickup of oil on the water. My fishing partner and I chose not to work on Exxon's cleanup. Instead, much later, I collected stories from people who did.

Skin Rashes and Headaches

Cordova fisherman Lynn Thorne (Weidman 2001) was raised by a fishing family and married to a fisherman at sixteen. She had gillnetted, seined, and tendered (bought fish for canneries) in the Sound most of her life. Her reaction to the spill was "to get out there to fix this thing that happened. We knew we wouldn't be herring fishing, and we knew we needed to clean this up before the salmon came. I believed that we could actually clean it up."

Her husband, Skip Thorne, immediately unloaded their herring gear and joined the mosquito fleet—the rag-tag armada of commercial and sport fishing boats working to keep oil out of the salmon hatchery in Sawmill Bay. Lynn Thorne pulled their young daughter out of school in Wasilla, where the family had relocated, and caught the last state ferry to Cordova, which is off Alaska's road system, on 3 April before the ferry was reassigned to spill cleanup. From the ferry deck she looked for the "usual birds and animals, but there was nothing. It was just totally quiet." In Cordova she arranged for her daughter to stay with relatives, signed a now-mandatory Exxon cleanup contract at the CDFU office, and flew out on 5 April to join her husband on their tender, the *New Adventure,* in Sawmill Bay.

For the next month the Thornes and their two crew worked in Elrington, Latouche, and Knight Island Passages in the southwest Sound, following the state ferry, which was being used as living quarters for most of the skiff people who collected dead and injured wildlife. The Thornes took orders from ADEC's Les Leatherbury and others who assumed that the Thornes had a contract with the state since they had worked with the state to defend the salmon hatchery in Sawmill Bay. Their job was to fuel the little skiffs and gillnet boats and to provide a hospice for dying sea otters. Dead otters and birds were taken to the ferry but according to Lynn

Thorne, "The ferry was a madhouse. There were droves of reporters. We gave the otters a safe place where it was quiet and calm, away from all the flashing bulbs and the chatter and noise."

There were others who sought sanctuary on the *New Adventure.* Lynn Thorne recalled, "You'd think having reporters out there would be a good thing, but it was so invasive to have the cameras stuck in your face when you were mentally in so much pain from being here and seeing the devastation . . . The only chatterboxes out there were people who were in Prince William Sound for the first time. People who had been here forever couldn't talk. They were devastated."

As the *New Adventure* worked its way north towards Eleanor Island, Lynn Thorne began to have health problems. She remembered the sickening smell of spilled crude when she had first landed in Crab Bay near Sawmill Bay, but that smell had been mild in comparison to what the *New Adventure* and its crew encountered further north. She recalled, "The stench of oil and the thickness of oil in the water were just horrid. All the rocks and beaches were totally coated and there was no escaping the smell. About two days after we got to Eleanor, I had a rash on my chest and a headache. The whole time out there up on that end of the island, I was not feeling well . . . It felt like the oil went right through your body. You could taste it, smell it. You couldn't get away from it. My body didn't like it, that's for sure."

Lynn Thorne noticed that she was not the only one having problems: "All the skiff people were distressed—headaches, red eyes, coughs." She heard from the skiff people who stayed on the state ferry that some of the first spill workers on the boats and beaches only lasted a few days before they were flown away for medical treatment. She observed none of the skiff people wore gloves or respirators—just raingear.

Gradually, as Exxon and VECO, the primary cleanup contractor to Exxon, gained control of cleanup operations from the early chaotic days, someone realized the *New Adventure* was actually an Exxon charter, and it was called back to Cordova, recommissioned, and sent to Valdez to work for VECO. Lynn Thorne's headache and chest rash cleared when she arrived back in Cordova on 1 May after a month of spill work. She recalled, "I was going to have (the rash) checked out by a doctor, but by the time I hit town, there was nothing to be seen. I thought it was great because it was gone."

The *New Adventure*'s new job was hauling freight and supplies to staging boats in the Sound. Lynn Thorne was the captain because

Skip stayed in Cordova to get ready for the Copper River fishery. She described the new assignment as "absolute bullshit. We had a slow boat compared to the gillnet boats. We were asked to do really stupid stuff like run four rolls of fax paper or a box of groceries out to Green Island—on our 47-foot boat!" What really angered her was when VECO ordered them out into the Sound without a destination. She felt VECO "just wanted us to look busy when the Coast Guard or Exxon were flying important people around. VECO didn't want us sitting in the harbor doing nothing."

When Lynn Thorne stalled or refused to keep up the charade one too many times, VECO reassigned the *New Adventure* to oily waste garbage detail around Eleanor Island and Knight Island. This, she described, "was hell. We were running the *New Adventure* close to beaches that were heavily oiled. There was mist from the sprayers and sludge in the water. The oil was just being spread around. The stench was just as bad as the first time out. And then we had the oily garbage bags to deal with. The waste—mostly rotting *Fucus* (a brown seaweed) and oiled absorbent pads—was delivered in garbage bags, covered with oil inside and out. I wore rain gear, the same rain gear day after day, and it was totally dripping with oil smudges. After the first week, I ran out of clothes that didn't have oil on them. The bags sat on the deck in the sun for a few days before a bigger boat would offload them to take to incinerators in Valdez or Kodiak. The bags stank. It was pretty messy work."

Lynn Thorne's rash, headache, and red eyes all reappeared during the stint of garbage detail. "I looked like hell," she said. "So did everyone else. It wasn't just me."

In early June she ended her cleanup contract so she could get ready for seining. Once again her headache, rash, and other symptoms cleared when she returned to Cordova. Because the symptoms cleared, Lynn Thorne never consulted a doctor or reported her health problems to Exxon, VECO, or the state. In town, after sixty-one days on the spill cleanup, she discovered cleanup workers were supposed to have four hours of hazardous waste training before working on the spill.

Lynn Thorne was left with a haunting feeling that she had been over-exposed to dangerous chemicals that would eventually compromise her health.

Warnings from Health Experts

Health problems became so widespread, so fast, that medical doctors, labor advocacy groups, and others took notice and sounded warnings as early as April 1989. Dr. Robert Rigg, former Alaska medical director for Standard Alaska (BP), warned, "It is a known fact that neurologic changes (brain damage), skin disorders (including cancer), liver and kidney damage, cancer of other organ systems, and medical complications—secondary to exposure to—working unprotected (or inadequately protected) can and will occur to workers exposed to crude oil and other petrochemical by-products. While short-term complaints, i.e., skin irritation, nausea, dizziness, pulmonary symptoms, etc., may be the initial signs of exposure and toxicity, the more serious long-term effects must be prevented." He recommended pulling cleanup crews "off the beaches—and out of the Sound—[to] avoid further tragedy in the form of human suffering, illness and disease . . ." (Rigg 1989).

John Havelock, a former attorney general for Alaska, warned that, "We cannot continue to have Exxon and its contractors substituted for the state in taking responsibility for the recovery. A person run over by a drunken driver does not hand over to that driver the primary responsibility for determining the type of medical treatment, the hiring of the doctor, and the determination of therapy" (CFS 1989 1[26]).

Despite the warnings from experts, the State of Alaska seemed to downplay the health risks to cleanup workers in two public health advisories issued by the Alaska Department of Health and Social Services (1989a, 1989b). The first advisory, issued *after* the Alaska Department of Labor declared the cleanup was a hazardous waste operation, stated: [T]here is no risk of adverse health effects from breathing the air. Risks are greatest to workers heavily exposed to oil during some cleanup activities, but the risks to these workers is considered to be low and with appropriate training and personal protective equipment as required by the hazardous waste regulations, cleanup activities can continue and workers can be confident that their health will not be compromised" (Stuart 1989, 2).

A visiting team from the Laborers' International Union of North America chastised the state officials for their actions. Calling crude oil "toxic and hazardous," the team listed health risks such as skin cancer from contact, leukemia from inhalation of benzene, and respiratory damage from oil-water aerosols from the planned use of hot water sprays (Phillips 1999). The team also pointed out the risk of long-term or chronic effects such as central nervous system, liver, kidney, and blood disorders. Team leader Eula Bingham, who was Assistant Secretary of Labor for Occupational Safety and Health in the Carter Administration, urged the state to declare the cleanup a hazardous waste operation, and she proposed a plan to independently monitor workers for chemical exposure.

After much debate with Exxon officials, the Alaska Department of Labor (ADOL, "a-DOL") shifted its position and declared the cleanup was a hazardous waste operation, which meant Exxon was required to properly train and protect workers. Bingham's team argued that OSHA standards required that workers were supposed to receive forty hours of training on how to handle crude and other chemicals in order to minimize risk of long-term health problems. Exxon proposed four hours of training to accommodate the emergency nature of the cleanup, even though it was evident that the cleanup would go on all summer. The state approved Exxon's request. The National Institute of Occupational Safety and Health (NIOSH, "NY-osh") team deemed the four-hour training classes adequate, although only a small portion of the course was devoted to handling toxic chemicals (NIOSH 1991; Stuart 1989; VECO 1989).

Others who saw the training video disagreed. Matt Gillen of the Occupational Health and Legal Rights Foundation declared the training violated the workers' right-to-know provisions of federal regulations. He said, "They're saying the tarry stuff is not toxic, but that's not true. It can still cause skin cancer" (Lamming 1989b). Physician Scott Barnhart of the University of Washington's Occupational Health Clinic also was concerned. He noted the training program did not mention that inhaling petroleum could cause nausea and dizziness, and there was no information on long-term effects such as leukemia and blood disorders. Workers told the media that Exxon had assured them that they were not breathing toxic amounts of fumes and that the fumes became less toxic as the oil aged. Bingham's team remained concerned that workers just didn't comprehend the long- and short-term effects of working with oil.

The workers' fate was entrusted to Exxon and VECO, Exxon's primary cleanup contractor. VECO had served the oil industry for years. It had a reputation for fighting—and ignoring—environmental regulation. In 1988 the EPA had cited VECO for "failing to maintain proper records related to hazardous waste generation, storage and disposal," and for "failing to conduct personnel training on the handling of and working conditions associated with hazardous waste" at an oilfield support facility on the North Slope (Kelder 1988). Earlier, VECO had been found guilty of illegally encouraging its employees to make pro-oil political contributions (Keeble 1999, 103). VECO's cavalier attitude towards hazardous waste cleanups jeopardized the health of thousands of cleanup workers.

Shoreline Cleanup—Killing Alaska While Trying to Save It

Within the first week of the spill, some 800 people were hired for the cleanup response. Within one month, there were some 3,000 people hired. As Exxon geared up for shoreline treatment, the number surged to 9,000, then over 11,000 (Harrison 1991). However, no amount of people can clean up spilled oil without effective equipment, designed to work in bad weather conditions. This type of equipment doesn't exist. Hordes of people *can*, however, further damage sensitive habitat and wildlife (Davidson 1990, 195-196), as discussed in Part 2 of this book.

Priorities—Worker Safety or 'Miles of Beaches'?

By the end of May, the spill cleanup had become a massive public relations campaign that played out daily on televisions across the nation. A VECO supervisor summed up the situation. "Exxon is trying to save its image. VECO is reaching for a piece of the pie. What we really need is a professional oil spill cleanup company" (Davidson 1990, 190), which, at the time, VECO was not, but all VECO's activity on the beaches made Exxon look good. Exxon could report daily "in exquisite detail the number of vessels deployed, number of workers involved, number and types of various skimmers and boilers—and, finally, miles of beaches treated per day ..." (ADEC 1993, 120).

The 'miles of beaches' became the standard measure of cleanup progress, but the numbers came from VECO and VECO was playing rough. VECO foremen bullied, assaulted, and fired workers for talking frankly with the media, and then the foremen lied about the progress (Davidson 1990, 189-190). ADEC reported, "A foreman at one work site informed an ADEC team member that although they had recovered approximately one barrel of oil and twenty bags of oiled *Fucus* from the site, they reported recovery of eight barrels of oil and forty bags of oiled *Fucus*" (CFS 1989 1[58]).

State ADEC monitors found they were "watching a cleanup effort that did not match the 'miles of progress' reports back at headquarters. Those 'miles of beaches' showing up on the progress reports were not necessarily cleaned, and the oil coming off the beaches wasn't necessarily being picked up" (ADEC 1992; ADEC 1993, 122). As the summer progressed, ADEC insisted that Exxon change terminology from "beaches cleaned" to "beaches treated," so people would understand that the shoreline "cleanup" was not as effective as the word "clean" implied.

Ed Meggert (2001) was one of the ADEC cleanup monitors. He was 44 years old, responsible, and experienced at working in remote sites. Fresh out of graduate school in mid-April, he was stationed on the *Denali*, a boat that worked beaches accessible from Knight Island Passage. Luckily, two other ADEC monitors were onboard to give him a sketchy orientation. His job was to monitor beach cleanup crews from the Knight Island group, east to Smith Island, and south to Latouche and Elrington Islands, to make sure various cleanup crews were following their work plans and picking up oil as it was flushed off beaches. His assignment seemed straightforward. He quickly discovered the job was not.

Most days, Meggert logged hundreds of miles on his skiff, worked twelve- to fourteen-hour shifts, seven days a week, all summer. By April VECO had organized workers into six task forces, each with varying numbers of beach crews and a hierarchy of supervisors. He found some of the beach crews consistently did a good job; some did not. Adjusting booms to catch oil flowing off beaches was tricky with the tides. As the tide dropped, booms had to be let out or they would be stranded high and dry on the beach. Sometimes booms couldn't be pulled right at the end of a shift because they were still catching oil. When Meggert repeatedly reported the same crews for sloppy work to Exxon and VECO supervisors, the crews were reassigned to night shifts. He split his shift so he could continue to monitor these crews at night. He believed his presence and that of the other ADEC monitors made a difference in how VECO and Exxon conducted the cleanup.

Meggert found most lower level VECO and Exxon supervisors really believed in what they were doing. He befriended some of the supervisors who were from outside Alaska, and he took them, surreptitiously, to see unoiled beaches. These people were visibly moved when they realized the magnitude of harm and the hopelessness of cleanup efforts. But the upper level supervisors and the top level beach bosses "were climbers," according to Meggert. "They would tell you black was white." One day as he watched a small older Native woman pick her way across the slippery beach with full trash bags in each hand, an Exxon supervisor told him there was a barrel of oil in each bag. Meggert retorted, "That's one strong woman! Five hundred pounds in each hand!"

VECO's apparent number fudging was not the least of the state and federal governments' concerns. VECO found hot water at high pressure flushed oil off beaches much faster than cold water at low pressure. Exxon successfully persuaded the Coast Guard and state to approve the pressurized hot water wash for shoreline treatment, despite warnings from its own scientists that this method was highly destructive to sealife (Chapter 13).

NOAA scientists also warned that the pressurized hot water wash was cooking beach life and killing the few hardy plants and animals that had survived the initial oiling (Associated Press 1991; Wohlforth 1990a). With no survivors to recolonize and restore life, NOAA scientists warned that the "treated" beaches might actually take longer to recover than beaches that were left alone (ADEC 1993, 63-64). Further, they found that the pressurized hot water wash was mixing oil with fine sediment (glacial flour and *"floc,"* a loose fluffy matrix of fine sediment and bacteria), causing the oil to flush off beaches and sink into more sensitive areas that might have escaped contamination. Exxon scientists maintained this was a "natural cleansing" process for oiled beaches (Bragg and Yang 1995). However, public-trust scientists determined that such particulate matter was an important pathway to transfer absorbed or bound *Exxon Valdez* oil to mussels and other organisms, which filter the water for particulate food (Houghton and Elbert 1991; Juday and Foster 1990).

In July at an oil spill conference in Seattle, Coast Guard Vice Admiral Clyde Robbins, the person ultimately responsible for approving the steam cleaning and overseeing the cleanup, asked, "How much damage do you want to do to a beach to save it? I don't want people to say about me that I killed Alaska trying to save it" (Robbins 1989, 2-3).

NOAA scientists suggested, "Sometimes the best thing to do in an oil spill is nothing" (ADEC 1993, 64). 'Nothing' was not an option for

Exxon. The cleanup operation was politically unstoppable. It had taken on a life of its own.

In its desperate strait to get oil off beaches whatever the consequences, Exxon sought approval to field test chemical products designed to speed removal of oil (Chapter 6). One of Exxon's products, Corexit 9580M2, raised serious concerns among scientists who warned that the product, basically a kerosene-based industrial solvent, would dissolve the oil and carry it into the sediments, increasing ecological damage (ADEC 1993, 69). Exxon, the Coast Guard, NOAA, and the fishermen all knew this, which was why dispersants had been banned in nearshore and intertidal areas in the dispersant guidelines.

Lead Lungs and Respiratory Problems

Ed Meggert (2001) was one of the ADEC monitors at Disk Island, just north of Knight Island, on 24 July for the initial tests with Corexit 9580M2 (CFS 1989 2[14]). He remembered the day as cool and windless with misty rain. First, the chemical was sprayed on the beach and, after a period of time, the beach was rinsed with the pressurized hot water wash. With the low cloud cover, mist from the spraying cloaked everyone on the beach. There was just no avoiding it.

While none of the ADEC monitors had any safety training on hazardous waste cleanups, up to that day Meggert had been careful to stay upwind and out of the funny-smelling mist generated by the pressurized hot water wash. He was suspicious of the mist because cleanup workers had repeatedly asked him—ever since he started in April—to help them get face shields and respirators. Although he constantly relayed their pleas to VECO and Exxon supervisors, he never saw the equipment provided.

No one had respirators during the dispersant test on Disk Island. That evening Meggert's lungs felt heavy and they burned. The next day, out on a different beach, he noticed the beach crew from the Disk Island Corexit test was considerably smaller. He asked the VECO supervisor, Perry Williams, what happened. Williams explained that he had to send fourteen or fifteen people out on Medi-Vac," Meggert recalled. The following spring, he discovered that, after the Disk Island Corexit test, Exxon had circulated a partial release form to workers, disavowing any future health claims against Exxon—and Exxon had paid workers

$600.50 to sign the form (Figure 1, below) (Exxon 1989c).

Shortly after the Disk Island Corexit test, Meggert came down with the "Valdez Crud," a term used by the VECO doctors to describe a variety of spill-related symptoms, including headache, sore throat, sinus infection, and cough. He never completely regained his health. That fall, in a Valdez hardware store, Meggert saw Perry Williams—and barely recognized the gaunt, stooped, sallow person as the strapping former boxer he knew from the spill cleanup. Williams admitted he'd been sick since the spill, but he didn't know what was wrong. It took Meggert ten years to figure out what happened to himself—and everyone else—on the beach at Disk Island.

Figure 1
Exxon's Partial Release Form for Indemnity from Adverse Health Effects from Cleanup Work

)01

PARTIAL RELEASE

FOR AND IN CONSIDERATION of the sum of **•••••••••••••600 DOLLARS AND 50 CENTS** Dollars ($ ••••••600.50) paid to the undersigned, receipt of which is hereby acknowledged, and intending to be legally bound hereby, the undersigned ABSOLUTELY AND IRREVOCABLY RELEASES AND DISCHARGES, Exxon Corporation, Exxon Shipping Company, their directors, officers, employees and agents, and the M/V EXXON VALDEZ, its officers and crew, from any and all claims, demands and causes of action of every kind and character, whether known or unknown, for any and all damages and claims that have been or may be sustained by the undersigned prior to and including the date of this release, and arising out of or in association with the incident involving the M/V EXXON VALDEZ on March 24, 1989, including any oil-containment or clean-up procedures that followed. The undersigned expressly excepts and reserves all claims, demands and causes of action that arise after the date of this release.

The sum stated above is accepted by the undersigned in full settlement of the claim described above. The undersigned understands that this sum was agreed upon as compromise settlement and is not an admission of liability by any party. In further consideration of the payment stated above, the undersigned hereby assigns, sells, transfers and subrogates to Exxon Corporation, without limitation, any and all rights, claims, interests and causes of action known or unknown for Exxon Corporation's own use and benefit, that the undersigned has or may have in respect to the claims described above against any person, corporation or governmental agency, including any liability fund that may be available for the payment of damage claims; by so doing, the undersigned gives Exxon Corporation full power and authority, for Exxon Corporation's own use and benefit, and on such terms and conditions as Exxon Corporation may deem reasonable in the exercise of its sole discretion, to litigate, compromise, settle or otherwise dispose of, in the name of the undersigned or otherwise, any claims described above, including the power and authority to release and discharge, in the name of the undersigned or otherwise, any persons, corporations or governmental agencies.

Executed this _____ day of _____, 1990.

Witness: _____

S. S. No. :

Check No. :   Signature _____

Check Date: Printed Name _____

Claim No. :
                                    _____
                                    Address

Source: ADOL, AWCB 1992a; Exxon 1989c

Two weeks later, larger-scale field tests of Corexit 9580M2 resumed on Smith Island. Workers at Smith Island initially refused to spray the product on the beach, because the barrels were labeled "toxic to fish." The entire beach crew was fired and replaced with one that followed orders (Wells and McCoy 1989).

During the field tests, ADEC monitors caught VECO contractors racing around the booms, using prop wash from their skiffs to hide the elusive reddish-brown solvent plume that escaped under the booms (ADEC 1993, 70). Exxon argued that the oil-solvent mixture was mixing with *"glacial flour"* (extremely fine sediment particles) and losing its buoyancy. However, in subsequent lab tests without glacial flour, the chemical dispersant produced the same subsurface brown plume. Ultimately, due to public opposition and uncertainty about risks to marine life (CFS 1989 2[20], 2[24]a), Corexit 9580M2 was not approved for wide-scale use (CFS 1989 2[24]b), but Exxon never backed off from its position that it should have been approved (ADEC 1993, 71; U.S. Coast Guard 1993, 135, 350).

Control Issues

Exxon took extraordinary steps to control the information about the injured workers. Exxon and its contractors also obstructed federal and state oversight agencies that tried to independently monitor worker health and exposure levels. Exxon refused to allow officials from NIOSH, one of the federal agencies charged with overseeing hazardous waste cleanups, or the Laborers' International Union, access to workers at remote sites (NIOSH 1991, 30; Phillips 1999). Unable to conduct their systematic, record-based field evaluation of worker conditions and health as planned, NIOSH officials were forced to rely on Exxon's exposure monitoring data and clinical data, but at the end of the summer, Exxon and VECO refused to voluntarily release their records to the federal government. NIOSH officials never subpoenaed Exxon or VECO for the records, citing lack of staff to pursue to the case.

Eula Bingham with the team from the Laborers' International Union was critical of the federal government. "Quite frankly, they should have been more aggressive," she said, "but the government just folded" (Phillips 1999).

Sick workers were left to fend for themselves.

An Ending, Many Beginnings
Lingering Oil, Lingering Harm

Exxon's target date to complete the cleanup before the onset of winter storms was 15 September. During the last week of the cleanup, Exxon staged a public relations blitz. Exxon's chairman Lawrence Rawl toured the Sound's shorelines and announced that, "Hundreds of miles of the Prince William Sound and Gulf of Alaska shorelines are certainly what most people would consider clean. A very small percent of the shorelines, which were most heavily impacted, have been cleaned several times, but still have some oil residue. These areas are environmentally stable, however, and pose no risk to fish or wildlife" (CFS 1989 2[39]).

Residents of the Sound and people involved in the cleanup knew better. Just before one of the beach crews was pulled off Point Helen on lower Knight Island, they decided to dig an experimental pit to see how far down the beach was "clean." Digging was relatively easy in the loose cobbles and soon several people were standing in the bottom of a hole 4-foot deep by 8-foot wide. ADEC monitor Meggert watched as subsurface oil from the porous beach started to fill up the hole. The men in the pit scrambled for safety just before the liquid oil flowed over the tops of their 18-inch high rubber boots.

Only time and comprehensive long-term research would tell whether the remaining oil posed a long-term risk to fish and wildlife or not. The Sound itself would reveal the long-term effects of the oil spill to the marine ecosystem over the next decade. This story, discovered by fishermen, Natives, and scientists who monitored the Sound's slow recovery, is told in Part 2 of this book.

About 1,800 people filed claims with the Alaska Workers' Compensation Board ("Workers' Comp Board") for spill-related injuries and illnesses (ADOL 1990a). Meggert was among them. The nature and source of these claims, and what happened to the sick workers, is discussed in Chapter 6.

A handful of cleanup workers filed *"toxic torts,"* personal injury lawsuits for chemical-induced injury. Some didn't live to see the outcome of their case; others faced a slow deterioration of health. Most sick workers did not file any claims or lawsuits because they never suspected their illnesses were related to their cleanup work. After all, Exxon and VECO had repeatedly stated the oil was "safe" — 'just wear the right gear' (Moeller 1989). The saga of some of these sick workers unfolds in the next nine chapters.

Some Statistics

By the end of 1989 cleanup, according to a subsequent paper published by Exxon, the company had employed over 11,000 workers who logged nearly 21 million hours of labor (ADOL 1990a). Exxon reported 45 percent of cleanup workers worked as "Oil Spill Response Technicians," directly on the shoreline treating beaches or immediately offshore operating skimmers and tending boom (Carpenter, Dragnich, and Smith 1991). Another 19 percent were counted as "support vessel crews," which included resupply and trash boats such as the *New Adventure*, operated by the Thornes. Another nearly 9 percent were "shoreline vessel crews," who worked from boats to clean inaccessible beaches. (The remaining 27 percent include security, hotel services, contractor management and support, and other.) Some of these workers share their cleanup stories and subsequent health problems in the next section.

NOAA scientists estimated that 1989 cleanup operations recovered approximately 14 percent of the spilled oil by on-the-water response (8–9 percent) and shoreline cleanup (5–6 percent) (Spies et al. 1996). However, I consider even these low percentages to be extremely optimistic as they are based on Exxon's numbers for recovered oil *and* spill volume. For example, a spill of 35 million gallons would reduce Exxon's self-reported recovered oil to just over 4 percent.

Exxon conducted its cleanup operations for another two years with limited work in 1992 before the company closed down its operations for good. The highly damaging pressurized hot water wash was largely discontinued after 1989. "Bioremediation" was the primary chemical treatment method in 1990 and 1991 (Mearns 1996). Chapters 6 and 7 focus on stories and health effects of workers who used or were exposed to Inipol, one of Exxon's products used on the cleanup.

Profiting from the Cleanup

VECO, the company that helped Exxon repair its battered image after the spill, was paid handsomely for its work (ADN 1990; Tyson 1990): the contractor grossed $800 million from the cleanup alone, including $32 million in profits from the 4 percent cost-plus cleanup contract (Keeble 1990, 103). VECO continued to make Exxon look good over the next few years: in December 1989 with cleanup profits, VECO acquired *The Anchorage Times,* the state's second largest newspaper, and reported on the oil spill cleanup for the next three years before it folded (Frost 1989)."Expecting a powerful newspaper owned by the oil industry to give fair and solid reporting (on oil issues) is like expecting the pope to run fair reporting about abortion in the Vatican newspaper," commented Ben Bagdikian, a former *Washington Post* editor and popular media critic (Postman 1989).

It turned out that Exxon also re-couped much of its cash outlay for its cleanup. Exxon claims to have spent $2.2 billion on its cleanup. But Exxon officials never talk about how much of this expense the company *recovered* through tax write-offs, lawsuits, and reimbursements. An investigative reporter with *The Dallas Morning News* reported that Exxon wrote off more than $2.8 billion for spill-related expenses before 1994 (Curriden 1999). (At Exxon's estimated 24 percent rate on income, this would result in a direct tax savings of about $670 million.) By 1997, Exxon had forced insurers to pick up approximately $1.2 billion of its cleanup expenses (ibid.).[3] While Exxon has made no clear public accounting of the relationship between these two figures, it appears that the company has been allowed to recover at least half of its cleanup expenses, and perhaps more.

Teamwork—ACAT and AFER

Alaska Community Action on Toxics (ACAT) is a statewide organization established in 1997 and dedicated to achieving environmental health and justice. ACAT's mission is: to assure justice by advocating for environmental and community health. We believe that everyone has the right to clean air, clean water, and toxic-free food. We work to eliminate the production and release of harmful chemicals by industry and military sources; ensure community right-to-know; achieve policies based on the precautionary principle; and support the rights and sovereignty of Indigenous peoples. ACAT provides environmental health information, legal and medical referrals at the request of chemically injured workers and veterans. ACAT has four program areas: Military Toxics and Health; Northern Contaminants and Health; Pesticide Right-to-Know; and Water Quality Protection. For more information, visit ACAT's web site: www.akaction.org.

Alaska Forum for Environmental Responsibility (AFER) was established in 1994 with private donations and funds from an out-of-court settlement with the oil companies that own and operate the Trans-Alaska Pipeline System (TAPS) and the security firm Wackenhut, among others, over a covert surveillance operation on private citizens and public employees (U.S. Congress House, 1991b, 1992). AFER's mission is: to hold industry and government accountable to the laws designed to safe-guard the environment, provide a safe and retaliation-free workplace, and foster public education based on current science. AFER focuses primarily on providing citizen oversight of the TAPS oil industry and government regulators. AFER's independent reviews and reports, along with TAPS historical information, are available from: www.alaskaforum.org.

Chapter 2.

Exxon's Failed Worker Safety Program

In 2001, I asked Pam Miller with the Anchorage-based nonprofit, Alaska Community Action on Toxics (ACAT), to help the Alaska Forum for Environmental Responsibility (AFER) investigate Exxon's cleanup and the persistent health problems that plagued many of the former cleanup workers (Sidebar 4, p. 38).

During our investigation we discovered one successful toxic tort, *Stubblefield v. Exxon* (1994). We believe this case largely succeeded for three main reasons: the lawyer was good and dedicated to his client; the lawyer was able to dodge the land mine of chemical sensitivity (Chapter 3); and the lawyer uncovered key failings of Exxon's worker safety program.

In this chapter, we examine the five key missing elements of Exxon's worker safety program as identified and explained by Daniel Teitelbaum, MD, in his expert witness deposition during the toxic tort *Stubblefield v. Exxon* (1994) (Teitelbaum 1994). The same elements are discussed, to a lesser extent, by the federal investigators who wrote the NIOSH Health Hazard Evaluation Report. We realized Exxon's safety worker program was like a ladder missing many of its

rungs—it just wasn't adequate for the task and it actually put workers in harm's way.

*(Author's note: in this chapter, page numbers in parentheses
that follow quotes or other information refer to Teitelbaum 1994—
his deposition; all other quotations are fully cited.)*

Missing Rung No. 1: "A Fundamental Command-in-Control Failure"

As an occupational medicine physician, Dr. Teitelbaum is a problem-
solver. He has organized worker safety programs to respond to chemical spills and airplane crashes. He has been in literally hundreds of
situations where he had started with sick workers and, usually in a
few days, had figured out the chemical cause of the illnesses and cor-
rected the situation.

The doctor defines occupational medicine as "the effect of
health on work and the effect of work on health" (107). To address
"the effect of work on health," proper worker safety programs: are
tightly organized and employ doctors who specialize in work-related
injuries and illnesses; train workers to properly handle the chemicals
of concern; and respond rapidly to minimize health problems. To
cover "the effect of health on work," proper worker safety programs
contain elements such as pre-screening physicals and job descriptions to ensure people's physical abilities and health are matched
with specific tasks.

No Lead Occupational Medicine Physician

Dr. Teitelbaum faulted Exxon's worker safety program for "a fundamental command-in-control failure." He stated there was a central "failure to develop an occupational medicine structure." Dr. Teitelbaum
had written extensively on this and organized entire military units to
respond to chemical spills. In the doctor's opinion, "When there is a
mass disaster, one begins with a person who is responsible . . . for the
occupational health and safety program . . . ." (102).

The lead occupational medicine physician is responsible for
immediate patient care, record keeping, and triage of patients or the
task of assigning a priority order to projects on the basis of where

funds and resources were most needed to minimize disease outbreaks, health problems, and unsafe work places (107). Successful
triage means few workers become sick or injured and disease outbreaks are quickly controlled, because workers are properly diagnosed and treated and work sites are made safe.

Instead of hiring an occupational medicine physician to lead the
worker safety program, Exxon (through its primary contractor
VECO) hired an emergency room physician. Dr. Teitelbaum pointed
out, there was "no longer an emergency" by June, July, August, and
September (141). Emergency room physicians are not trained to
triage work-related health problems. Without the essential triage
function, the ability of Exxon's worker safety program to identify and
respond to any illness outbreaks was nonexistent.

Without a qualified occupational medicine physician in charge,
Exxon's worker safety program foundered. It was as if there were
two trains; one had a lead occupation physician "conductor" and the
other did not. Exxon's medical team boarded the wrong train, so
every station was wrong and the train never reached the destination
of a viable worker safety program. Exxon's train—its worker safety
program—appeared to be all smoke and mirrors.

Faulty Structural Organization

In Dr. Teitelbaum's opinion, there was nothing very hard about the
structural organization of a proper occupational medicine program.
For example, for a cleanup operation involving 10,000-plus workers,
he stated, "I would have had a medical director and four occupational medicine physicians reporting to him so that you had one physician for every 2,500 workers. I would have had two nurses and an
industrial hygienist assigned to each team, and . . . safety personnel,
so that at every beach on every shift you had either an EMT or a certified safety person who could provide first aid, and then I think you
would have had a functioning operation" (109).

Further, Dr. Teitelbaum stated that the lead occupational medicine physician and the lead core team of health care providers
should have remained in residence for the duration of the cleanup.
This way, they would have gained a comprehensive understanding of
the health problems that developed and they would have been able
to provide consistent and timely quality care for the workers
(117–120).

The doctor stated that Exxon's worker safety program was "not a functioning operation . . . ." It was managed by industrial hygienists, "a series of people in rotation and contracted out to contractors, who really never understood what was going on and were not able to bring to focus on the problem all of the resources, which were necessary; all of the laboratory facilities, which were necessary; and all of the equipment, which was necessary. And [the hygienists] had no line of reporting to a central authority or to a medical authority, which might have brought together the issues of human health impacts and the industrial hygiene setting" (99–100). Exxon's safety program was ineffective, in part, because health care providers tried to understand a complex problem from a chronologically disjointed series of pictures rather than a continuous video.

Dr. Teitelbaum placed the blame for the inadequate worker safety program directly on Exxon. He said, "Exxon is a very sophisticated company with an extensive medical department, and I cannot conceive of the fact that they would put 11,000 workers into a work setting and it wouldn't occur to them that they needed a medical support group to deal with that. I find that very difficult to believe" (137–138).

Effect on Workers

Dr. Teitelbaum was certain Exxon's flawed worker safety program made his client sick. He explained, if an occupational medicine physician familiar with the work setting had treated injured worker Garry Stubblefield, that doctor would have recognized that the excessive diesel exhaust and oil mist were causing Stubblefield's health problems and he might have been able to help the patient.

Instead Stubblefield wound up in an Anchorage hospital where he didn't get treated properly, because the Anchorage doctor was not trained for an industrial setting and didn't understand the effects of work on health. Stubblefield was allowed to go back to work in the same environment that made him sick in the first place without any changes such as a respirator or air-conditioning to prevent him from becoming even sicker (110–113, 117–118).

Dr. Teitelbaum was also certain it was not just Stubblefield who became sick because of Exxon's erroneous worker safety program. The doctor believed the high incidence of upper respiratory infections (URIs) severe enough to get recorded—6,722 reported cases or more than half of the workers (Exxon 1989b)—pointed to "some kind of epidemic respiratory disease" (113). According to Dr. Teitelbaum, an occupational medicine physician would have recognized there was a problem, particularly with the epidemic outbreak of URIs. The physician would have consulted with the lead hygienists so they could work with their field personnel to find and correct the exposure problem that caused the outbreak. Without a central occupational medicine physician, the worker safety program fell to pieces and thousands of workers became sick (107-109, 113).

Whatever the cause of the epidemic of respiratory problems, Dr. Teitelbaum faulted Exxon and its flawed worker safety program for *never* identifying the disease or correcting the problems that led to it. Instead Exxon allowed hundreds of people to get sick every week of the cleanup operations. In the doctor's experience, a good industrial hygiene program should have been able to identify and correct the problems within a matter of days or a couple weeks, but as the doctor stated Exxon's program "was not designed to be responsive to the problem" (111-112).

Missing Rung No. 2: No Comprehensive Industrial Hygiene Plan

A good industrial hygiene plan is designed to monitor work setting exposures to chemical contaminants; prevent or intervene in illness outbreaks; and track injury and illness reports to minimize chronic health risk. The essential triage function of disaster response programs depends on a good industrial hygiene plan. Triage also depends on efficient implementation and constant monitoring of the industrial hygiene plan; that is, there is a plan *and* the plan is followed.

Dr. Teitelbaum stated another "essential failure" of Exxon's worker safety program was "a failure to develop a comprehensive industrial hygiene program" (98-99). The doctor stated there was "no consistent and effective execution of an industrial hygiene . . . program" in Exxon's worker safety program (102). The doctor faulted Exxon's program for faulty assumptions and design; inadequate monitoring; no quality assurance/quality control (QA/QC); and no preventative care plan to curtail illness outbreaks.

Faulty Assumptions and Design

Exxon's industrial hygiene plan was based on the commonly-held assumption that the two primary exposure risks to workers from oil spills are skin contact with weathered crude oil and inhalation of volatile organic carbons (VOCs) or "oil vapors." But everyone— Exxon, the federal and state health monitors, and the labor union representatives—expected the volatile compounds from the oil to evaporate quickly into the air and disperse. This was the main reason that the federal officials had allowed Exxon to reduce the hazardous waste training time from forty to four hours in 1989. Exxon sampled for oil vapors, nonetheless, supposedly to evaluate worker exposure to these compounds and make sure the oil vapors dissipated quickly as expected.

The pressurized hot water wash altered expectations and invalidated these assumptions.

NIOSH investigators realized that the high-pressure wash generated crude oil mists and particulates from splash-back off the rocks and that this created a potential health hazard, which had not been previously considered. To address the federal investigators' concern, Exxon collected 114 samples of oil mist and 30 samples of polycyclic aromatic hydrocarbon (PAH) aerosols for analysis (Med Tox 1989b, 1989c). This was not nearly adequate for statistical tests to define exposure risk from what Dr. Teitelbaum referred to as "a pervasive experience" (76) for thousands of workers—the cloud of oily sea spray generated by the pressurized hot water wash.

In comparison, Exxon collected 1,611 samples for oil vapors (Med Tox 1989a). This was fifty-four VOC samples for every one airborne PAH sample. This sampling regime did not accurately reflect the risk of exposure to workers. Exxon expended a lot of effort sampling compounds that were not expected to be present, the oil vapors, and a little effort sampling oil mist and PAH particulates, which were pervasive experiences for virtually everyone on or near the beaches. Dr. Teitelbaum found Exxon's worker safety program "to be defective" because it was based on incorrect assumptions regarding worker exposure to hazardous chemicals (101).

To make matters worse, the few PAH samples that were collected were not handled properly, according to Dr. Teitelbaum. In order to measure PAHs, which are not found in oil mist samples, one had to collect particulates and measure specifically for PAHs. Exxon did not follow the NIOSH manual for establishing a statistically valid sampling process. NIOSH officials collected particulates, but they did not measure for PAHs. OSHA personnel used outdated equipment when they sampled for PAHs, which the doctor found "absolutely appalling" (125). Dr. Teitelbaum did not believe any organization had done a rigorous survey for PAHs that would allow exposure assessments to be broadly interpreted as "safe," or not, for cleanup workers (125–130).

Inadequate Monitoring

Dr. Teitelbaum observed that few samples of oil vapors were collected in areas where the cleaning agents or solvents were used. Exxon's air quality data set shows fewer than 100 of 1,611 VOC samples were collected for these agents (Med Tox 1989a). Calling this "a striking oversight," Teitelbaum stated, "You go out and you look for benzene in a place where there is unlikely to be any benzene, and in a place where there is likely to be benzene, no assessment was made. It is just a huge error in assessment of the work setting." He stressed, "Something . . . significant . . . was unassessed" (75).

Exposure of the DECON laundry crews to airborne PAHs was not routinely assessed by anyone, yet these crews steam-cleaned workers' rain gear with the same pressurized hot water wash equipment that was used to steam-clean the rocks. Fumes in the small laundry rooms also made the DECON crews sick as described by Phyllis "Dolly" La Joie in Chapter 5. NIOSH investigators observed DECON crews using "[n]atural ventilation (open doors) . . . to dilute the air concentrations of these vapors in the DECON cleaning areas" (NIOSH 1991, 29). However, DECON crews kept doors shut on cold or rainy days to retain heat to dry the clothes and PPE *("personal protective equipment"*—trade lingo for rain gear, boots, gloves, hard hats, eye goggles, and respirator masks). La Joie said she never saw anyone from the federal or state government, or Exxon contractors, monitoring her crew's workstation during her night shift (La Joie 1996, 125).

The NIOSH team only evaluated DECON operations for two of the six Task Forces (cleanup crew teams) and their findings were widely disparate: in one operation, "A number of PPE items were not being decontaminated at all. Workers' street clothes were visibly contaminated" (NIOSH 1991, 28). The other DECON team did a better job.

Dr.Teitelbaum and NIOSH investigators also found that monitoring for cleaning solvents was deficient. For example, according to the doctor, Exxon did not test for excessive levels of limonene, the active ingredient in De-Solv-It, which was used by the DECON crews to clean workers' rain gear (Phillips 1999). Dr. Teitelbaum found no measurements of exposure to this material and he stated, "That's a major failure [of the monitoring plan]. This is not a benign material" (51). Other cleaning solvents such as CitroKleen and Simple Green® were used with equal abandon by workers, as described by Ron Smith and Dolly La Joie (Chapters 4 and 5, respectively), and with no monitoring by Exxon contractors or government oversight agencies. The NIOSH Health Hazard Evaluation does not mention these products or solvents.

Quality Assurance /Quality Control (QA/QC) Program

In his deposition Dr. Teitelbaum stated there was a quality control problem with the entire data set, because Exxon's air quality samples had been analyzed by several different labs without a QA/QC program (56). The doctor stated "that one of the most dangerous things you can do is to use a whole series of different laboratories" without an inter-laboratory quality assurance program (124). In reference to Exxon's 1,611 samples, he stressed, "The number of samples which were taken means really very little" (124). The inter-laboratory quality assurance program was a necessary step, designed to calibrate procedures among the different labs: without it, there was no way to compare or relate the results from different labs. As the doctor stated, the monitoring program was "missing . . . the cable that tied all of these [sample results] together, and that should be a quality assurance cable" (125). Without this cable, Exxon's claims that exposure levels were "safe" for workers are meaningless.

Nonexistent Preventive Care Plan

In the case of the *Exxon Valdez* oil spill (EVOS), court records reveal there was, in Dr. Teitelbaum's words, "some kind of epidemic respiratory disease" that ran unchecked for the five-month duration of the cleanup (113). A good functional industrial hygiene plan contains a preventative care plan, which is essentially a feedback loop to minimize disease outbreaks. Had Exxon's plan had this feedback loop, according to Dr. Teitelbaum, the lead occupational medicine physician would have recognized the epidemic outbreak of URIs was a health concern. The lead occupational medicine physician would have consulted with the lead industrial hygienists so they could work with their field personnel to find and correct the exposure problem that caused the outbreak, thus intervening and arresting the disease outbreak. Dr. Teitelbaum explained with such a feedback loop, intervention of illness outbreaks usually occurs within a matter of days to a few weeks: diseases are certainly not allowed to run unchecked for months (113–118).

But in Exxon's worker safety program, according to Teitelbaum, the industrial hygienists "were not adequately informed about the nature of the materials being used." They didn't understand "the bounce-back from the rocks created . . . aerosols . . ." They didn't understand "the implications of using solvent-containing materials or [2-] butoxyethanol on skin" (132–133). "[T]here was no one . . . that had the faintest idea that limonene could be an allergen and could cause asthma . . . There seems to have been no understanding of what [a] crowd[ed] . . . living . . . condition does to respiratory function, spread of disease" (134). They knew little "about the exposures to diesel fuel or other engine exhaust . . . or the implications of exposures to aerosols, including seawater mist . . ." (154). The industrial hygienists did not even take Exxon's four-hour safety training course (133)!

Since the hygienists on the beaches weren't trained to recognize work-related health problems specific to cleanup operations, Dr. Teitelbaum surmised "There was no feedback from the field [hygienists] to the industrial hygiene authority, which did exist, so that [the authorities] had no idea what was actually going on. It was as if there were a vertical arrangement with a break in the ladder so that somewhere below the superintendent level, there was no or little contact between industrial hygiene and the folks who were doing the work" (101).

Oversight agencies took little action to identify, correct, or arrest the "epidemic respiratory disease" outbreak, because Exxon did not report any of the 6,722 claims of respiratory problems to state or federal health officials, the state or federal compensation boards, the U.S. Coast Guard, or the Alaska epidemiologist (Chapter 3; ADOL 1990a; Wilson 1991). In their Health Hazard Evaluation, NIOSH investigators recommend that in the future, monitoring systems should track *illnesses* as well as injuries (NIOSH 1991, 33). This seems to indicate NIOSH investigators likely were aware that Exxon was not reporting illnesses. Yet, somehow, Exxon failed to adequately protect worker safety. As a result, thousands of workers became sick.

Missing Rung No. 3: Lack of Proper Training

A proper worker safety program for a disaster response should train both the medical team and the workers to recognize chemicals of concern, properly handle those chemicals, and recognize symptoms of overexposure to prevent disease outbreaks. As discussed above, Dr. Teitelbaum opined that Exxon did not adequately train its medical team, especially the industrial hygienists who were actually on the beaches with the workers, to recognize the health hazards specific to the cleanup. The doctor held a similar opinion of the lack of training for the workers.

Dr. Teitelbaum had looked at the training manuals and watched a video of the four-hour training program, which federal health officials had approved. He judged the training program "inadequate" (101), "just really, completely unsatisfactory" (165). It failed to meet the basic requirements of the federal law's Hazard Communication Standard, which the doctor had helped write (OSHA 1994).

According to the doctor, the whole purpose of this standard was "to make it known to the worker that the information is available" on hazardous chemicals in the work place, and that "he has a right to know" about the chemicals he will encounter. The doctor explained the federal standard *requires* that workers be provided with what is known about the chemical "to take away the classic question about the one who is not educated enough to know what to ask" (166–167).

Yet Dr. Teitelbaum observed in Exxon's safety training video that the workers did not have any materials—not even notepaper—in their hands. He questioned, "How could they possibly have learned anything?" (170). He pointed out there were no pre-tests or post-tests to determine if the workers had actually learned anything about the chemicals on the cleanup or safety precautions for avoiding potential health problems from chemical exposure (167).

Dr. Teitelbaum felt that many workers probably did not report potential work-related health problems, because they didn't have "the faintest idea of what the risks were" (101–102). The doctor said

Stubblefield, for example, never really understood the complexity of the exposure that he faced. "He had been told that the materials they were using on the beach were benign . . . [and] that the materials they were using to degrease his clothing and so on couldn't hurt him, and it's not true. [H]e wasn't given any material safety data sheets [MSDS] or access to anything where he could have gone and checked it if he had a question" (166). He and the other workers weren't trained to report their health problems or recognize their illness could stem from chemical exposure (117-120). The doctor said, "They didn't know what they should do to protect themselves. They didn't know about medical data. They didn't know about anything" (164-165).

The cleanup workers themselves perhaps best summed up their understanding of the health risks and effects. Dale Herrick of Soldotna, Alaska, told a reporter that "he wasn't too worried about his close contact with the oil. He said Exxon had done tests on the workers and found they weren't breathing toxic amounts of the fumes, which become more benign as the oil aged. Masks were available if the workers wanted them . . ." (Lamming 1989b). Smith of Soldotna said his trainers "told us we could eat that stuff [the oil] on our pancakes" (Phillips 1999). Tim Robertson from Seldovia, Alaska, who had worked on the spill for five months, stated he had never seen medical data on health hazards of crude oil or chemicals; he knew of no baseline urinalyses done on workers before sending them out on beaches and he knew the workers considered respirators "optional" (McDowell 1989; Spence 1989).

Missing Rung No. 4: Failure to Execute Program
Failure to Implement Program

It's one thing to plan a worker safety program for a disaster response; it's another to actually carry out the program. Often things don't go as planned, sometimes because early assumptions are found not to be true or not to work, and adjustments need to be made to protect workers and minimize health problems.

In the case of the *Exxon Valdez* cleanup, little went as planned, but little effort was made to adjust the program for cleanup realities. One of the biggest reality checks was the unanticipated need for protection from airborne oil PAHs and mists, as demonstrated by Exxon's air sampling program. Most of the air quality monitoring for

PAHs was done in August, four months after the pressurized hot water wash was initiated and really too late to help workers, who were never informed of the test results—and risk—anyway.

Common sense and Exxon's four-hour training program indicate the wisdom using of respirators, but there were never enough to meet demand. Shortages of gloves and respirators, among other supposedly protective gear, were common throughout the entire 1989 cleanup, according to both workers and observers. Six weeks after the spill, former cleanup worker Lisa Jones' statement was included in a congressional oversight committee hearing: "We were promised [respirator] masks. The hoses splash [oil] on your face and [it] burns after awhile. I have a constant headache still . . ." (Jones 1989, 1142-1143; see also Olsen 1989; Ward 1989).

La Joie's court deposition describes her experiences with supply of respirators, goggles, and gloves. As DECON crew, she said, "We tried respirators for awhile, but we couldn't—couldn't get enough" (La Joie 1996, 98). "We ran out . . . in a couple of weeks totally. The suppliers couldn't keep up on gloves and respirators" (ibid., 100). She had the same difficulty getting respirators when she worked on the beach crews. She said, "We asked why we couldn't get them, and [the VECO foremen] said, 'Well, you go home if you don't want to [work without respirators]' " (ibid., 139). Every morning on her own time, she said she spent "at least an hour trying to scrounge up respirators" (ibid., 151), and she provided what she could to her fellow beach workers.

As for wearing their protective gear, La Joie reported that her DECON crew had "a lot of problems [wearing safety goggles] because they'd get steamed up and we couldn't see . . ." (ibid., 98). Her beach crew also had problems with their gear. She described that on sunny days, her rain gear was "like being in a sauna" (ibid., 146), and the beach crews stripped by noon, operating the sprayers in just underwear, coveralls, and life jackets. "You just got sprayed" (ibid., 146), said La Joie. The oil residue and mist "always" splashed on her face and sometimes in her eyes. She said the beach crews were supposed to wear Tyvek® suits under their raingear "every day" to prevent the oil from contacting skin. But she said, "We saw . . . maybe one of those [Tyvek suits] each during the whole oil spill [cleanup] . . . we didn't get any of them in our crews" (ibid., 147).

During their three site visits, NIOSH investigators found, "Wearing of PPE was not consistently enforced from work site to work site. Although many workers were in the proper gear, many exceptions were noted. These usually involved not wearing eye protection, gloves, or PVC garments [rain gear]. The hands and forearms of many workers were contaminated with "weathered" crude oil. During warm weather, ORTs [Oil Spill Response Technicians; i.e., beach cleanup workers] were frequently observed taking off the tops of the PVC rain gear" (NIOSH 1991, 31). The NIOSH team spoke with nurses who reported treating rashes—which usually occurred on the hands, forearms, face, or neck—from not wearing protective gear (NIOSH 1991, 29).

Dr. Teitelbaum noted, Exxon's medical staff "made a series of assumptions at the beginning [of the cleanup] that there wouldn't be certain kinds of exposures; that medical surveillance was not necessary; that people would wear their protective clothing and their respirators and their barrier creams and all . . . [but] there was no concerted effort to evaluate whether that, in fact, was happening, and what the consequences was of its not happening in many places . . ." (Teitelbaum 1994, 100-101). It was as if no one in charge cared about protecting the workers.

Failure to Monitor Program Effectiveness

Once a worker safety program is up and running, someone needs to check on a regular basis that the program is working as planned to effectively protect worker health and safety. In the trade, this is called "program effectiveness monitoring."

There were many examples that Exxon's worker safety program *was not* working. NIOSH investigators found that "Decontamination of PPE was not consistently effective in the prevention of skin contact with the weathered crude oil . . ." (NIOSH 1991, 31). In her statement to the congressional oversight committee, cleanup worker Jones explained that wet and oily "[c]lothes were hung on top of one another on nails [in] the barge . . . Our clothes never did get washed. The rain gear was not cleaned at all. Our boots were wet. After the third day we were going to wash our own clothes, [but] were told we would be fired immediately" (Jones 1989, 1142).

NIOSH investigators did not evaluate the effectiveness of the PPE to protect the dispersant and Inipol crews from skin contact with the chemical solvents. One reason was because Inipol use came into vogue in August after the federal investigators left.

Workers discovered that Inipol dissolved the rubber gaskets in the spray packs and leaked out; it dissolved their plastic raingear; and when it came into direct contact with skin, it caused skin rashes, blood in the urine, and other health problems (Chapters 6 and 7). Workers with symptoms were pulled from the Inipol crews—after they had been overexposed. Despite the obvious health problems, Exxon's medical team and field supervisors told workers repeatedly that Inipol was not hazardous as long as they "wear the right gear" (Moeller 1989). Workers were not told, however, that plastic raingear was *not* the "right gear."

As a result of their experience on the cleanup, NIOSH investigators recommended, "First-line supervisors should ensure that workers are: (1) provided with and wear appropriate PPE (including respiratory protection and protective clothing); (2) trained in the proper procedures for wearing PPE; and (3) required to inspect their PPE before beginning work each day. Also, one individual should be assigned the daily responsibility for the proper cleaning, storage, and inspection of PPE" (NIOSH 1991, 50).

The fact was these NIOSH recommendations were supposedly already part of Exxon's safety training program. Exxon did not adequately implement or monitor its own program to see if it was working as designed to protect workers.

Missing Rung No. 5: Lack of Proper Medical Attention

The proof of the pudding is in the eating: given all the safety training, monitoring, medical specialists, and personal protective gear, the real test of a worker safety program is how many workers get sick. To pass this test, worker safety programs for hazardous waste cleanup require proper medical attention before and during the exposure. If a large number of workers get sick during the response operation, follow-up medical attention is also required after the exposure. Unfortunately for the workers, Exxon flunked this test on all three counts.

NIOSH investigators never evaluated this critical element of a worker safety program in their Health Hazard Evaluation. However, Dr. Teitelbaum did.

Regarding pre-screening medical attention, the doctor explained that the lead occupational medicine physician should have determined "the effect of health on work" through pre-employment physicals and job descriptions with work assessments for each task to match people's physical ability and health with specific tasks (139–140).

The doctor used Stubblefield as an example: maybe Stubblefield had a pre-existing condition or was not physically fit enough to handle the job assigned to him. The MSDS for crude oil and products used during the cleanup warn that persons with pre-existing conditions of dermatitis, allergies, asthma, and diseases of the blood, liver, and kidney may find these conditions aggravated by exposure. How could Exxon know whom to hire and for which job without basic medical information and a screening process? (110–113).

Dr. Teitelbaum maintained that an occupational medicine physician familiar with the work setting would have been much more likely to recognize symptoms of occupational asthma and other work-related diseases than a physician who had no background on the cleanup and who was operating within a traditional non-industrial setting (117–118).

As for follow-up medical care, it was nonexistent—at least the type of comprehensive long-term monitoring at the spiller's expense that is lawfully required when hazardous waste cleanups go awry. Elsewhere, I explore the possibility that public health officials knew that Exxon's cleanup was an occupational health disaster, but no one admitted it (Chapter 8).

Pam Miller and I quickly realized that, in our opinion, Exxon's worker safety ladder was broken from the top down, because the lead health care professionals were not able to identify and treat chemical illnesses from cleanup work. And, it was broken from the bottom up, because the workers and their supervisors were not trained to identify chemicals of concern specific to the cleanup (such as PAH aerosols) or recognize symptoms from overexposure.

It was obvious to us that, under such circumstances, workers were exposed to dangerous chemicals. But the critical question, which we set out to examine next, was "were workers *overexposed* to these chemicals?" In other words, were workers exposed to *dangerous levels* of dangerous chemicals (Chapter 3).

Chapter 3.

Dangerous Chemicals at Dangerous Levels

In this chapter, Pam Miller and I examine the medical and exposure evidence from *Stubblefield v. Exxon* (1994) to support our contention that most workers, not just Stubblefield, were exposed to dangerous levels of dangerous chemicals during the cleanup.

(Author's note: in this chapter, page numbers in parentheses that follow quotes or other information refer to Teitelbaum 1994— his deposition. Six-digit numbers refer to motions, orders, or other material from Stubblefield v. Exxon [1994].*)*

'Occupational Asthma,' Damning Evidence, and Serendipity

In 1991 Garry Stubblefield of Granbury, Texas, sued Exxon, VECO, and VECO's contractor, Norcon, for severe injury to his breathing system and brain damage sustained during the oil spill cleanup. Stubblefield claimed Exxon, VECO, and Norcon were negligent for not warning workers of the health risks of steamy oil-seawater mists;

for not adequately protecting workers from the mists; and, in his case, excessive diesel fumes.

Stubblefield worked as a crane operator on various barges in 1989, directing a hose to spray hot water on oily beaches (Phillips 1999). He worked for only two months before he was incapacitated by coughing, wheezing, and shortness of breath, caused, he felt, by the diesel fumes from a nearby generator and from the oily spray that daily misted his operator's cab. When his health problems didn't clear up after he quit the cleanup, he went to doctors who told him his lungs had been permanently damaged by chemical exposure and he was at risk of developing cancer. Stubblefield hired Dennis Mestas, an Anchorage-based attorney, who filed his toxic tort case in Alaska Superior State Court.

Exxon and VECO immediately requested—and the court granted—a sweeping court order for confidentiality. The order limited use of confidential documents to Stubblefield's litigation; restricted access of documents to parties in the litigation; limited copying of documents, and required the return of all confidential documents at the end of the case (No. 910919).

Mestas was president of the Alaska chapter of Trial Lawyers for Public Justice. He realized this order would affect other injured workers, each of whom would have to individually request production of documents from Exxon and VECO. He argued repeatedly against confidentiality on the grounds of public access and free speech. He wrote that federal courts allowed plaintiffs to share documents because "such cooperation promotes a speedy and inexpensive determination of actions ..." and it allowed access of records to individuals who couldn't afford to take on large corporations (No. 930604, 6). The Anchorage district court called his arguments "meritorious," but decided to uphold the original protective order (No. 931018, 3).

Mestas requested a list of documents from Exxon and VECO including the standard safety procedures and manuals used by Exxon on board its company tankers and during the cleanup; the air quality monitoring data (collected by Med-Tox in 1989) that Exxon had refused to give to government officials; and the medical records from the cleanup (No. 911030).

Despite the promised confidentially, Exxon and VECO were extremely reluctant to give Mestas any information. A year passed in a flurry of motions and counter motions before Exxon released its safety procedures and manuals. It took another year, three requests for production, three motions to compel, two court hearings, and a court order before Mestas was finally given the monitoring and medical records he sought No. 930604, 3).

In 1993, when he reviewed the safety procedures and manuals, he realized why Exxon had been so resistant. Court evidence suggest that Exxon knew that Prudhoe Bay crude oil is a health hazard—the corporation has an extensive library on toxicity of crude oil and proper worker protection. For example, Exxon's "Ocean Fleet Safety Manual" demonstrates the company's knowledge of and ability to detect diesel, oil, and other noxious vapors and the procedures and precautions to protect workers from exposure to these chemicals (No. 920906, 9-16). An entire section of the manual is devoted to various breathing apparatus such as air-purifying respirators "to protect against exposure to low concentrations of toxic and/or noxic gases in airborne particulates" such as the oil mist in the cleanup (No. 920906, 8).

Figure 2
Total Upper Respiratory Infections (URIs) Reported by Workers during 1989 Cleanup

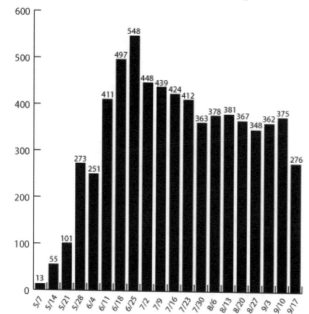

Source: Exxon Company, USA, 1989b. In *Stubblefield v. Exxon* (1994).
(Author's note: see Appendix Table A2 p. 451 for data on these figures.)

Yet during the cleanup, most of the crews were not given respirators—or even told they needed this gear. The atmospheric testing devices used to assure safety of Exxon personnel aboard company vessels were not provided for any of the cleanup vessels. Mestas interpreted this as "clear evidence of Exxon Shipping's negligence" and the reason why the company had tried to conceal this type of safety information from his client (No. 920906,10).

In 1993 when Mestas finally received the medical data and monitoring reports, he had Carl Reller, an Anchorage environmental researcher, organize and review all the records that were jumbled and stuffed into over a hundred boxes (No. 921231, 2). Reller was thorough, meticulous, and persistent (pers. comm., 7 December 2000). He found a graphic summary of the clinical data, which showed 300 to 500 workers were treated every week of the cleanup from late May through mid-September for "URIs"—upper respiratory infections with cough and flu-like symptoms (Figure 2, p. 57). He tallied the columns and found *a total of 6,722 reported cases of URIs*, yet during the cleanup, Exxon had claimed a near zero work-related illness rate (ADOL 1990a; Wilson 1991).

To his amazement, the URI summary charts and the Med-Tox master list of air sampling data were some of the few documents of thousands that were not marked confidential. Seizing the opportunity, Reller asked his boss if he could use these documents to analyze exposure levels of cleanup workers to oil mists, vapors, and aerosols (PAHs), as well as other compounds of concern, for the Alaska Health Project, a public nonprofit organization. Reller suspected the air quality data might reveal why so many workers had respiratory problems.

Mestas filed a motion with the court, arguing the information was not subject to discovery constraints because privilege had not been requested. Exxon attorneys misplaced the request for omission and missed the deadline to respond. When VECO attorneys discovered these errors, they persuaded the court that the critical medical records were in VECO's custody and subject to confidentiality (No. 930324), but it was too late. Reller had already copied the documents for personal use in the brief window of opportunity when the records were public. His analysis of Exxon's air quality data, made public in 1993, is discussed below under "Exposure Levels."

Dangerous Chemicals on the *Exxon Valdez* Cleanup

Mestas had to prove that his client's health problems stemmed from overexposure during the cleanup. In May 1994, he hired Dr. Daniel Teitelbaum, the Denver toxicologist and occupational medicine physician, as his expert witness to review Stubblefield's medical records and Exxon's documents. As a toxicologist, Dr. Teitelbaum (1994) founded and directed Poisonlab, a large industrial hygiene lab. He has done thousands of samples personally and he has treated hundreds of people with occupational asthma from chemical exposures. He is still the primary caregiver for thousands of asbestos workers.

Occupational Asthma and Airborne Chemicals

It was clear to Dr. Teitelbaum from the medical records that Stubblefield had asthma, which was a type of reactive airways disease. The medical records showed that Stubblefield's symptoms included severe shortness of breath, wheezing, constricted lung airways, and outpouring of fluids into his lungs. In Stubblefield's case, it was a permanent condition, and it had progressed over the years into a "twitchy" or hyper-reactive airways disease, where any irritant—cigarette smoke, perfume, diesel exhaust—could unexpectedly set off a horrible case of wheezing (30, 54–55). Although the physician did not believe in chemical sensitivity, he acknowledged that people with twitchy or spasmodic airways are hyper-vigilant, because so many chemicals can set off an asthma-like attack (26, 69–71).

Dr. Teitelbaum believed that Stubblefield was primed for the disease by exposure to excessive diesel fumes during the cleanup (33). Unlike other barges on the cleanup, Stubblefield's barge had a generator with a horizontal exhaust stack, which blew diesel fumes directly into his crane cab (160–161). According to the doctor, diesel exhaust is a complex mixture of oxides of sulfur and nitrogen, carbon monoxide, large numbers of particulates, and soot with polycyclic aromatic hydrocarbons (PAHs), metals, and insoluble organic compounds such as aldehydes and ketones (24–26).

Dr. Teitelbaum thought of diesel exhaust as a short-term irritant and a long-term carcinogen delivery system (36–41). It is known to cause occupational asthma, and it contributes to an increased risk of developing cancer. Inhaling diesel exhaust damages the cilia of the respiratory epithelium and suppresses the lung's ability to clear itself of irritants such as soot, oil mist, and other particulates. NIOSH investigators report that diesel soot is composed of solid carbon cores, which act as tiny air filters capable of absorbing "as many as 18,000 different substances from the combustion process" (1991, 13-14). Diesel exhaust also damages the immune system of the lung, which makes a person more susceptible to the effects of other toxic chemicals and more likely to get viruses and respiratory infections.

Dr. Teitelbaum thought that the so-called oil mist "was a far more complicated and complex exposure than has been appreciated, and that ... this mist at one or another time consisted of weathered oil, seawater, De-Solv-It, Inipol, and other materials, which were used for cleaning . . . ." (46-47). He called oil mist an "an extremely irritating particulate material," and he described the potential health problems from excessive exposure to each of the individual components (53-56).

Weathered oil he described as "a sticky, higher molecular weight, oily substance, mostly free of volatile [organic carbons], containing a fair number of metals and other contaminants, probably capable of supporting bacterial growth . . . ." (47). Weathered oil contains varying concentrations of polycyclic aromatic hydrocarbons (PAHs) such as chrysenes, benzanthracene, and benzopyrene (56), which the EPA considers compounds of concern. For example, benzanthracene is known to cause skin cancer, lung cancer, and bladder cancer (62-64).

Dr. Teitelbaum thought the pressurized hot water wash released weathered oil (PAHs) into the air as an aerosol, which was carried deep into the lungs or coated exposed skin. Dr. Teitelbaum had seen videos of the clouds of oily-seawater steam that enveloped the beach workers, blew off the beaches, and misted Stubblefield's crane cab— and Stubblefield himself who often stood outside of his cab to avoid the diesel fumes.

Exposure to oil mist, the doctor thought, contributed to Stubblefield's occupational asthma and put him at increased risk for developing cancer. The doctor believed that weathered oil, when inhaled, caused inflammation of the bronchial tubes in the lungs, which provokes a twitchy airways response, so that when the inflammation cleared, the lungs remain prone to spasms, as in Stubblefield's case.

The doctor was also concerned that Stubblefield's exposure to oil mist was a *"carcinogenic initiation event,"* which he described as "a genetic interaction between a molecule of (a cancer-causing chemical) and a molecule of DNA" (84). This event is a necessary first step in developing cancer, and it is "permanent," "irreversible," and "hereditable" (84–85). This initiation event increased Stubblefield's risk for developing cancer. As Dr. Teitelbaum noted, the link between pulmonary, skin or bladder cancer and oil mist is well-established (83–87).

Exposure to seawater mist raised major health concerns for Dr. Teitelbaum, who pointed out that seawater is "a very irritating material to the lungs" (48). When British navy firefighters who used seawater as their water source reported trouble breathing, similar to what Stubblefield experienced, studies found that seawater decreased the lung's ability to transfer oxygen. In cases of drowning or near drowning, seawater causes a severe inflammatory reaction in the lungs. Rescued victims often come down with bronchitis and pneumonia; ones who recover often are left with a lifetime of asthma and twitchy airways—from just a single exposure to seawater inhalation. Dr. Teitelbaum was also concerned that bacteria and microorganisms, which live in the seawater, might cause health effects when carried deep into people's lungs, but no studies have been done in this area (75–78).

Plenty of studies have been done, however, on the "detergent solutions," as Dr. Teitelbaum called the solvents used during the cleanup. The doctor was alarmed by the widespread use of De-Solv-It and Simple Green to clean clothing, skin, skiffs, and other things as if these solvents were completely benign substances (72). Inipol, Corexit, and De-Solv-It all contain volatile napthas, which are potential sources of the carcinogen benzene and its chemical derivatives. Inipol, Corexit, and Simple Green contain 2- butoxyethanol, which the doctor described as quite an irritant when inhaled at any level (174). De-Solv-It also contains limonene, "a very potent allergen" in the doctor's opinion (49), and one capable of causing acute poisoning, asthma, *dermatitis* (inflammation of the skin) (49), *lipoid pneumonitis* (inflammation of the lungs), *albuminuria* (presence of blood proteins in the urine often symptomatic of kidney disease), and *hematuria* (presence of blood or blood cells in the urine) (127).

Choice of Standards

Dr. Teitelbaum was adamant that the permissible exposure limits

(PELs) Exxon chose to use during its cleanup were not adequately protective of workers (Sidebar 5, p. 63). For example, he pointed out there is actually no standard for whole crude oil itself (60). OSHA has only established PELs for individual components in crude oil, but these do not account for the potential synergistic effect of multiple simultaneous exposures of various aromatic hydrocarbons.

Researcher Reller (1993) found that Exxon followed a practice common in the oil industry—Exxon used the OSHA PEL standard for *mineral oil,* a highly purified and nontoxic product, as a surrogate to indicate safe exposure limits to crude oil mist, which has known toxic and carcinogenic properties. Reller found no evidence that Exxon ever compared actual samples of Prudhoe Bay crude oil mist to the samples collected on the beaches to determine what difference this inappropriate choice of standards made for the workers. Further, NIOSH currently recommends an exposure limit *ten times lower* than the OSHA safety standard to protect worker health (NIOSH 2004; OSHA 2004b; Table A.1).

To make matters even worse, Exxon also used a surrogate standard exposure limit for PAH aerosols—the OSHA PEL for "nuisance dust." Dr. Teitelbaum pointed out that by using this surrogate, Exxon totally ignored the irritants and metals in PAH aerosols, the immunotoxic and carcinogenic properties of PAH aerosols, and the interactions among chemicals in PAH aerosols, all of which occur, the doctor noted (188-189).

Effect of Extended Work Hours

In addition to his concerns about Exxon's choice of standard PELs, Dr. Teitelbaum explained that extended work shifts disrupt normal circadian rhythms and dramatically alter a person's ability to deal with infection or other kinds of stresses or to take medications—independent of any exposures to hazardous substances (148-150). Extended work hours result in shift work syndrome or *"dyssynchronicity"*(152).

The doctor called the cleanup work shifts of eighteen hours a day, seven days a week, "intolerable" and "a highly stressful environment" (149, 153). He stated, "Shift work is among the most dangerous and disruptive things you can possibly do, and making people work eighteen hours a day is calculated to make people sick, [even] if nothing else happened" (149). The doctor compared it to taking a person "who is physically capable of lifting 60 pounds twenty times a day and putting [him] in an environment where he is lifting 100 pounds thirty times a day." He insisted, "You're going to get a problem. ... there is no question about it" (144).

Dr. Teitelbaum advised that *just based on shift work alone*, acceptable exposure limits should have been 20 percent of the OSHA standards (191)—five times lower than the standard Exxon chose to use. He pointed out other situations where companies voluntarily reduce the OSHA standard to protect all their workers, including ones who work overtime or who weigh less than the "standard" male (Sidebar 5). For example, IBM uses 10 percent of the OSHA PEL in its plants; Amoco has a company-wide policy of 25 percent of the OSHA PEL (192).

In addition, Reller (1993) found that Exxon had full knowledge of the dangers to workers posed by extended work shifts. Three years before the spill, Exxon (1986) had published a study showing standard PELs need to be reduced to adequately protect workers with "unusual work schedules." Applying Exxon's own model, Reller calculated Exxon's supposedly safe PEL for oil mist exposure should have been reduced by at least half to compensate for an 84-hour workweek.

PELs, RELs, and Worker Health

According to the Occupational Safety and Health Administration (OSHA) web site, OSHA "sets permissible exposure limits (PELs) to protect workers against the health effects of exposure to hazardous substances. PELs are regulatory limits on the amount or concentration of a substance in the air. They may also contain a skin designation. OSHA PELs are based on an 8-hour time weighted average (TWA) exposure. OSHA PELs are enforceable" through the U.S. Department of Labor (www.osha.gov).

"The National Institute for Occupational Safety and Health (NIOSH) is the federal agency responsible for conducting research and making recommendations for the prevention of work-related injury and illness. NIOSH is part of the Centers for Disease Control and Prevention (CDC) in the Department of Health and Human Services" (www.niosh.gov). Based on its research, NIOSH advises recommended exposure limits (RELs) to protect workers. NIOSH RELs are always more conservative than OSHA PELs, however, NIOSH RELs are not enforceable.

The reason for the difference in worker protection standards originates in the OSHA standard-setting procedure, which involves public comment and risk assessment to weigh the benefit and risk of using the intended product or substance. Thus, OSHA PELs are a compromise among user groups and potentially affected parties, while the NIOSH RELs are based more on science and less on economic concerns. Companies often voluntarily chose to use PELs that are more stringent than the OSHA PELs to accommodate unusual work hours or physically smaller employees than the standard 70 kilogram (154 pound) person used in risk assessment formulas.

Dr. Teitelbaum succinctly stated that everyone—Exxon, its contractors, and the federal investigators—was working in "a genuine vacuum" (60) of information about appropriate exposure limits for crude oil and other substances present during the cleanup.

Dangerous Levels of Dangerous Chemicals

When the Exxon examiner asked Dr. Teitelbaum whether he had an opinion on Stubblefield's level of exposure to oil mist, PAHs, diesel exhaust, De-Solv-It, Inipol, or other harmful substances, the doctor stated, "There is no point to having opinions about things that should have been measured and weren't" (32). He explained that the monitoring programs by Exxon and the federal government were completely inadequate. He said, "There was no attempt to collect the kind of samples that would give you the answer and allow you to estimate individual people's exposure" (57).

However, a great deal could be said about the average and maximum (high-end) exposure levels to the general population of cleanup workers. On any given day beach workers were exposed to a wide range of concentrations for oil vapors, oil mists, oil aerosols, and other dangerous chemicals. Exxon's contractor, Med-Tox, monitored exposure levels of different jobs by sampling devices attached to individual worker's outer gear. For each chemical of concern, the data show a range of exposures from zero (no exposure) to a maximum level.

The data were averaged, but Dr. Teitelbaum explained that in a work setting such as the EVOS cleanup, where the primary exposure routes are through inhalation of an aerosol or coating of skin, it is important to look at the high-end exposures. He explained, "[T]he concentration [of oil and/or chemicals] within the droplet is extremely high, although the concentration in any cubic meter may be low, because the droplets may be widely dispersed. Those droplets, when they hit, are usually very highly contaminated. It's part of the problem of trying to look at a particulate or an aerosol. Aerosols tend to be very concentrated" (174). In other words, the risk of inhaling droplets is usually low, but the risk of poisoning is high when one inhales contaminated particulates. During the Exxon's cleanup, even the risk of inhaling droplets was high; oil mist was so thick that maximum exposure levels exceeded—by two times—OSHA's permissible limit for dust! (Table A.1, p. 450)

In fact, the Med-Tox (1989c) statistical summary shows *the maximum exposure levels* for oil vapors (benzene), oil mist, PAH aerosols, 2-butoxyethanol, carbon monoxide (diesel exhaust), and hydrogen sulfide (decomposing oiled debris and carcasses) *all exceeded OSHA's permissible limits* (Table A.1, p. 450). The maximum exposures also *exceeded the NIOSH recommended standards* for everything except PAH aerosols—and this only because NIOSH has not set an REL for this substance. This means, for example, if PAH aerosols were an intoxicant, the cleanup workers most exposed to the clouds of oily mist would have been driving drunk with blood levels twice the acceptable limit, because the maximum exposure levels exceeded by two times OSHA's PEL for PAH aerosols.

It is also worth mentioning that OSHA attempted to set more stringent PELs for all these substances and the new standards, in fact, went into effect on 1 March 1989; however, these more stringent standards were subsequently vacated.[1] More stringent NIOSH RELs also went into effect on 1 March 1989 to reflect the OSHA proposal; *these NIOSH RELs were not subsequently vacated.* Had the more stringent OSHA standards remained in effect, the maximum overexposures would have been even higher than what is listed in Table A.1, p. 450—closer to the listed overexposures for the NIOSH RELs.

Average exposures tell a different story—they were all well within the legal permissible limits, but as Dr. Teitelbaum explained, in a situation with "all respirable material, we would have a problem" (1994, 59). Average exposures should be viewed as mere paper statistics rather than as realistic exposure levels for cleanup workers breathing the oily mists. The maximum overexposure for oil mist was on a beach where workers used the pressurized hot water wash. All this to say that these data show there was a significant risk of overexposure to dangerous chemicals for workers actively engaged in beach cleanup.

NIOSH investigators point out in their Health Hazard Evaluation Report (1991) that Exxon's data for PAH aerosols are conservative, because the analytical procedures used by Exxon's contractor were 10 to 100 times *less likely* to detect PAHs than the methods used by the federal team (11, 13, 26). Even worse, the PAHs analyzed by both Exxon and the federal team were seventeen priority pollutants, which *under*-represented exposure to workers as many airborne PAHs were excluded from the analysis (Med-Tox 1989c, Table 11.2.1).

In light of the evidence, Dr. Teitelbaum was quite certain that Exxon's cleanup workers, in general, particularly those on the

beaches and nearshore areas, were overexposed to dangerous levels
of dangerous chemicals.

An Expensive and Very Secret Settlement

In reading the thousands of documents himself, attorney Mestas concluded that Dr. Teitelbaum had clearly demonstrated Exxon's and
VECO's neglect in exposing workers to dangerous levels of dangerous chemicals. As discussed in Chapters 4 and 5, other cleanup workers also spoke of unreported illnesses because VECO required sick
workers to take three days of leave without pay, but then workers
received pink slips on their third sick-leave day, so their injury or illness was never recorded. From Alaska's workers' compensation
records (ADOL 1990a), Mestas knew that Exxon and VECO never
reported any of the 6,722 respiratory claims to NIOSH or the Alaska
epidemiologist. He also knew that failure to report work-related illnesses and injuries within ten days is a direct violation of federal law
(33 USCS 930) and Alaska state law (AS 23.30.070). Under federal
law, failure to report injuries within ten days of notification carried
penalties of up to $10,000 per case.

Mestas wondered what the legal difference was between a
"recordable" illness and other illnesses. He hunted through the thick
OSHA regulations for an answer. He found in the OSHA regulations
(29 CFR Part 1904) a description of exemptions to reporting work-
related injuries and illnesses. A specific subsection (1904.5(b)(2)),"
states: "Exclude from record keeping those injuries/illnesses that do
not provide information useful to the identification of occupational
injuries and illnesses and thus would skew natural injury/illness
data." The list of exemptions, number (viii), states: "Colds and flu will
not be considered work-related" (U.S. Dept. of Labor, OSHA, 2004a).

Finally Mestas understood the "URI" classification. By classifying
all 6,722 respiratory claims as *infections*—Upper Respiratory
Infections—rather than work-related *illnesses*, Mestas theorized that
Exxon evaded the reporting requirements *and* the long-term medical monitoring requirements under federal law. Apparently, Exxon
used a one-line exemption to vacate its health care provider responsibilities to thousands of cleanup workers.

Mestas was convinced there was no "Valdez Crud"—the phan-
tom virus rumored to have caused the epidemic of cold and flu-like
symptoms reported weekly by hundreds of workers (Stranahan

2003). Mestas had not found any science or medicine to support such assertions or rumors. The State epidemiologist also confirmed lack of an identified virus (Phillips 1999). Exxon's chief of medicine told a news reporter that illnesses didn't respond to antibiotics (ibid.).

Meanwhile, an average of 385 weekly cases of upper respiratory illnesses was *only* reviewed by VECO Safety—there was no medical review or industrial hygiene review. The whole preventative care portion of the medical plan, the vital feedback loop between the medical staff and the field hygienists, never took place. No medical surveillance program was ever put into effect. VECO Safety, under contract to Exxon, simply recorded thousands of work-related respiratory illnesses as "colds" or "flu" and gave workers Tylenol and antibiotics, which, as cleanup workers noted, generally had little effect (La Joie 1996, 121).

Under the weight of evidence thoughtfully and thoroughly assembled by Mestas, Exxon and VECO eventually settled. The terms of the settlement were confidential, but the same reporter who was following the case found that the insurance companies representing Exxon and VECO ended up in a separate lawsuit over which company was responsible for payment in Stubblefield's case. Exxon spent one million dollars fighting VECO before paying Stubblefield two million dollars (Phillips 1999).

Miller and I concluded from our investigation that if NIOSH investigators had subpoenaed Exxon's monitoring data and clinical records, they could have realized that respiratory illnesses far exceeded any physical injuries during the cleanup. And the federal investigators might have concluded, as we did, based on available evidence, that long-term health monitoring and medical care was—*and still is*—desparately needed for *Exxon Valdez* oil spill cleanup workers.

The incriminating documents obtained by Mestas were returned to Exxon and VECO. The documents were not discovered by any other lawyers in the small flurry of toxic torts filed immediately after the *Exxon Valdez* oil spill. These documents in *Stubblefield v. Exxon* (1994) are still protected by confidentiality.

Exposure and Health Problems

Chapter 4.

Ron Smith and Randy Lowe

(Author's note: Ron Smith could not be interviewed by the author because of a gag order, stemming from settlement of a 1994 personal injury lawsuit against Exxon and VECO (Roberts v. Exxon [1999]). Instead, I reconstructed Smith's story from court records. Page numbers in parentheses refer to Smith 1996—his deposition; Other materials fully cited.)

Acute Health Problems (1989)

Immediately after the oil spill in March 1989, Ron Smith from Soldotna traded his log skidder and truck for a neighbor's twenty-three-foot aluminum set net skiff and outboard motor. His wife Shirley approved of the deal. In their mid-thirties, the couple had been through a lot together since they had married in their late teens—different jobs, moves, three children now in their teens. Smith was a welder by trade, but an entrepreneur at heart. He enjoyed physical labor and being outdoors. Over the years he worked as a logger, heavy equipment operator, mechanic, and at other odd jobs to provide for his family. Lured by the prospect of good paying spill work, he headed to Valdez, just like a lot of other people from the Kenai Peninsula.

In Valdez, Smith landed two contracts with Exxon, one for his skiff, the "*JB*," and one for a friend's skiff, the "*Tank*," which he had brought with him. He kept an eye on his friend's skiff. He made sure the person assigned to run the *Tank*, Randy Lowe, took orders directly from himself. Smith discovered that he and Lowe were practically neighbors in Soldotna.

Smith and Lowe were "skiff people," and they worked together all summer in the Sound. They set absorbent boom in coves to contain oil flushed off beaches by cleanup crews. They tended boom to make sure tie lines to shore and anchors set offshore held through tidal cycles. They moved oily boom, coiled in the bow of their skiffs, from cove to cove. They hauled oily trash off the beaches. They shuttled beach crews with all their gear, porta-potties, and lunches between beach and barge. They cleaned their skiffs daily with strong solvents CitroKleen and Simple Green—two, three, five gallons a day—to wipe off the slippery oil and keep the boats safe to carry people. They spent nights together on the housing barges, first the *Coastal Star*, then the smaller *Crystal Star*, then the big *Greens Creek* barge. The two became fast friends as the summer progressed.

Because Lowe and Smith had signed on the cleanup early, neither received any training from Exxon or VECO about hazardous chemicals or protective clothing. Working in the open small skiffs was dirty work—most of the time they wore plain cotton gloves, a float coat, rain pants, and rubber boots. When Smith started getting headaches—blazing, consuming headaches—he attributed them "to the [oil] vapors and working long, hard hours . . . in the sun and on the water" (93). On windless sunny days, he had the worst headaches. He could look across the coves and see the oil vapors rising like heat waves off "acres and acres and acres" of water (178).

When Smith returned to Soldotna that fall, he was planning on taking a few months off because, as he said, "I worked enough that summer to do me for two years" (48). He figured he "should have been real happy to be back at home and settled in," but a lot of times, for no reason he could see, he'd "get real unhappy . . . just all of a sudden get real depressed and down and out" (125-126). Mood swings and depression were something entirely new to him. He had outbursts of temper, which was very unlike him (Didriksen 1993, 2-3). He had trouble remembering things and concentrating. Some days he felt like he was in a complete fog. Headaches continued to plague him. The blazing headaches from the cleanup stayed with him after the work ended and became a "regular" part of his life (94).

Lingering Symptoms (1990 to 2003)

By January 1990, Smith felt he would go crazy unless he did something. He called Lowe and suggested they go to the Virgin Islands to find some carpentry work repairing hurricane-damaged buildings. Lowe was game. When they returned a couple months later, Smith started commercial fishing for halibut, then herring. Lowe fished herring with him in the spring, then the two worked separately during the summer.

Sometime that fall Smith got an emergency phone call from Lowe, who was sick, really sick. Smith rushed his friend to the hospital. It was the first time he'd ever seen Lowe like that. To the alarm of both men, Lowe's health deteriorated rapidly. Smith watched his friend seemingly "on his deathbed a couple of times" that fall and winter (88). It worried Smith that his buddy, whom he had worked with every day on the spill, was "falling to pieces and going into the hospital on a regular basis" (107). He remembered his own headaches and mood swings, and it dawned on Ron that "something was affecting us." The two concluded "that it was probably chemical exposure" from the spill (107). Lowe moved to California to be with his parents and to seek medical treatment. The two stayed in touch by phone.

In February 1991 Smith sustained a severe neck injury during an emergency airplane landing in a whiteout. The next year and a half were a nightmare of pain, sleepless nights, drugs, doctor visits, and therapy. Frustrated with the slow healing, he grew depressed. He found this more than he could handle on top of his crazy mood swings since the spill cleanup. He took antidepressants for the first time in his life, drugs that he had shied away from taking after the cleanup.

In July 1991 he attended a six-week "work hardening program" to rebuild his strength and endurance, but his progress was limited (Hopkins 1991). A CAT scan revealed his neck had a herniated disc, which had not been evident in the initial X-rays. His doctor encouraged him to think about other work because of permanent impairment—a 14 percent loss in motion range from his neck injury (Dowler 1992). For Smith, working as a "car salesman," as he put it, was "completely out of even my realm of imagination to do" (225).

Smith stopped going to see his therapist, but he continued stretching and strengthening exercises, and he adjusted his life to fit his physical limitations. He realized he would have a tough time

finding employment because of his physical condition, but he
thought he could start his own business. He could still weld and
maybe do light labor or supervise work crews.

In June 1992 Smith started his own scrap metal recycling business with the help of his wife, Shirley. On his first job since the accident, he became nauseated—"queasy" as he put it—cutting up scrap
metal at the Swanson River oilfield service company yard (69). Smith
had worked as a certified welder off and on for all of his adult life. He
had never before had a reaction to metal smoke or acetylene torch
fumes. There seemed to be nothing unusual about the scrap metal: it
was extremely thick—and coated with oil residue.

At first, Smith just thought the incident was odd—so odd he
mentioned it to Shirley, but then he started reacting to other things
as well. When he fueled his vehicles or equipment, the gas or diesel
smell made him dizzy and it was worse, he noticed, on sunny hot
days when there were more fumes. Gas and diesel exhausts were
obnoxious—they nauseated him and shortened his breath. Welding
smoke became intolerable unless he positioned himself just right and
the wind blew the smoke away. Wood smoke left him gasping for air.
Waiting in line at the grocery store one day, he started feeling queasy
and he realized he was reacting to the perfume of the woman in
front of him. He moved to the back of the line. Then he became
aware that spray paint, colognes, and perfumes all made him ill.

Troubled by these odd reactions to things he had been around
most of his life, in mid June, he shared his experiences with Lowe,
who was still down in California. Lowe wasn't surprised. He was
being treated by William Rea, MD, at the Environmental Health
Center-Dallas. Dr. Rea is a practicing cardio-thoracic surgeon, who
had pursued a matter of personal interest to become one of the few
doctors in the country, at the time, specializing in treating chemically contaminated people. Lowe explained that Dr. Rea treated people
for *"chemical sensitivity,"* the medical term for people who shared
strange reactions or sensitivities to different chemicals normally
found in their home, work place, or other places like Smith
described—the grocery story, the gas station, or heavy traffic.

Encouraged by his friend, Smith called Dr. Rea and agreed to a
blood test with Accu-Chem Laboratories, a company that measures
toxic organic hydrocarbons in blood and fat. When his lab work
came back in August, Dr. Rea told him that he had "very high levels of
some pretty dangerous chemicals" (91). What concerned Dr. Rea in
particular were the high levels of petroleum hydrocarbons such as

trimethylbenzene and methylpentanes. The doctor had seen similar levels recently in people working around crude oil—the *Exxon Valdez* oil spill and the oilfields in the Gulf War (Rea 1998, 46). Dr. Rea encouraged Smith to come to the clinic for tests and treatment.

Smith was frightened and torn. He felt he was up against something that he couldn't control, and he didn't want to get as sick as his friend Randy, but this was an enormous financial commitment. His insurance company wouldn't cover treatment for chemical sensitivity. He talked with Lowe, who told him that Belli's law firm in San Francisco was accepting injured cleanup workers as clients and paying for their initial evaluations at Dr. Rea's clinic. Lowe and several other Alaskans had just sued Exxon and VECO for personal injury in September 1992, and they were hoping to receive compensation for medical expenses and pain and suffering. Smith talked with Shirley, who supported his need to seek treatment. The couple decided if there was a chance to recover medical expenses, as well as damages for suffering and injury, they should take it. Smith became a client of Belli's law firm.

Chemical Detoxification Treatment

Dr. William Rea recognized the oil spill cleanup workers as one of the four primary groups of people at high risk for developing heightened sensitivity to chemicals—industrial workers (Ashford and Miller 1998). The other three groups include residents of communities with contaminated air or water, individuals with personal and unique exposures to chemicals, and occupants of "tight buildings" where poor ventilation affected mostly office workers and schoolchildren.

Dr. Rea had founded the Environmental Health Center-Dallas in 1975 to treat what he saw as a growing problem—exposure to a rapidly increasing number of chemicals in our food, water, air, soil, homes, and workplaces. By 1992, Dr. Rea had worked with over 20,000 patients who had heightened sensitivity—similar to allergic reactions—to extremely small amounts of previously tolerated chemicals, drugs, foods, and food/drug combinations. He knew a great deal about chemical sensitivity, the varied symptoms associated with it, how to rid the body of chemical poisons, and how to reduce risk of chemical exposure and associated health problems.

Among his colleagues Dr. Rea is widely recognized as one of the leaders in the emerging field of environmental medicine, which deals with chemical exposures and sensitivity. The doctor is utterly devoted to practicing environmental medicine and helping people heal. When he is not treating people, he spends hundreds of hours writing up his findings to help guide other doctors in this field. In 1992 he had just published the first volume in a four-volume, 3,000-page definitive treatise on chemical sensitivity. When other environmental doctors can't help their patients, they send them to Dr. Rea. Part of Dr. Rea's gift is a genuine compassion for people discarded by the medical world, because their illnesses didn't fit traditional diagnoses—and because of the implications of their illnesses for society.

Smith knew that Dr. Rea's treatment took at least two months, and he was anxious to get started. His initial visit with Dr. Rea was 30 November 1992. Before evaluating Smith, Dr. Rea explained the basic concept of chemical sensitivity (Rea 1992): each person has a unique and finite capacity to tolerate pollution. The doctor compared a person's capacity to tolerate chemicals and environmental stress such as pollen, mold, dust or viruses with the capacity of a rain barrel to hold water. Chemical sensitivity happens in two stages: initiation and triggering. During initiation, the rain barrel becomes full following acute (sudden massive) or chronic (slow, ongoing) exposure to various pollutants. Once the rain barrel is full, a person loses the tolerance (adaptation) for environmental pollutants. Subsequent exposure to even extremely low levels of previously tolerated chemicals, drugs, and food, or combinations of these, can trigger symptoms such as Smith was experiencing or even other problems.

The problem, Dr. Rea explained, is that most physicians base their diagnoses on the triggering agents, the substances causing the most recent complaints, and they completely miss the substances that started the problems by filling the rain barrel. Dr. Rea noted that Smith's chief complaints were "stomach and bowel problems, back and joint pain, sore muscles and tiredness" (Rea 1993). He explained that environmental doctors try to determine what causes a person's physical illnesses or loss of tolerance to chemicals. If there were chemical triggering agents, as Dr. Rea usually found in people who had become odor sensitive, then he could prescribe treatment to reduce the chemical load in the body and restore some degree of health, depending on the individual.

In order to find clues about the initiating substances, Dr. Rea asked Smith to review his life—health problems, jobs, family history, everything—in chronological order, using the detailed twenty-seven-page questionnaire that Smith had filled out in preparation for his evaluation (Rea 1997a, 2061-2088). He found the doctor genuinely compassionate and a good listener. (He didn't realize this was in part because the doctor himself had become chemically sensitive from a pesticide exposure and had been treated by Theron Randolph, MD, who had first described chemical sensitivity in medical journals in 1951 [Rea 2001].) He liked how Dr. Rea stopped him every now and then to explain some aspect of chemical sensitivity. This demystified his illness and gave him hope that he could get better.

After Smith talked about his childhood kidney ailment, which was cured, and his asthma, which wasn't, Dr. Rea told him chronic asthma usually indicates allergies to dust and pollen. Dr. Rea explained that health problems with dust, pollen, and mold indicate a human body is sensitive to biological pollutants. When Smith talked about welding and living in a trailer home, Dr. Rea stopped him again. The doctor considers welding a high hazard occupation for chemical exposure and trailer homes notorious for off-gassing formaldehyde. Smith realized he was exposed to chemical pollutants at both his home and workplace.

Smith's stories of the oil spill cleanup astonished Dr. Rea. The doctor had heard similar stories of massive exposure to crude oil fumes, diesel fumes, and solvents with minimal or no protective gear from Tom Pickworth, Lowe, Tim Burt and others who had worked in Prince William Sound and had come to his clinic for treatment. When Smith mentioned his headaches, mood swings, and depression after the cleanup, Dr. Rea explained this response was similar to a caffeine, tobacco, or drug addict going through withdrawal from the high doses of chemicals (Rea 1992). This made sense to Smith.

He told Dr. Rea his stomach problems and heartburn had also started sometime after the cleanup and had progressively worsened. These new symptoms concerned Dr. Rea, because he suspected it meant Smith's body was so overloaded with pollutants—the rain barrel was full—that symptoms had switched to different body systems in response to this traumatic chemical exposure (Rea 1992).

Dr. Rea listened carefully to Smith's story about the plane accident. The doctor suspected this severe physical trauma further stressed Smith's body, slowing his natural ability to heal from his first trauma—the massive chemical exposure during the oil spill cleanup. Dr. Rea told him, "medicated wellness is not wellness." The doctor explained the multiple drugs taken since the plane accident only masked his symptoms—his depression, mood swings, pain, stomach upset, and headaches—and allowed him to adapt or become used to his loss of tolerance (Rea 1992), initiated, Dr. Rea suspected, by the cleanup. Dr. Rea told him that people who took three or more medications usually were chemically sensitive (Rea 1997a, 2053).

When Smith said he had first noticed a reaction to welding smoke upon returning to work after the plane accident and then he noticed he now reacted to perfumes, gasoline, diesel, and car exhaust, this confirmed Dr. Rea's suspicions. Odor sensitivity is one of the cardinal signs of chemical sensitivity (Rea 1997a, 2042-43). It indicates a person is overloaded with chemicals and unable to process previously tolerated substances. Even small amounts of substances—chemicals, drugs, or certain foods—can trigger an exaggerated reaction as in Smith's case. The doctor explained that when Smith had initially reacted to welding smoke, he was in a "de-adapted" state since he had not been around these substances for a year and a half. During this time he had lost his tolerance to cope with these chemicals (Rea 1992).

When Smith finally talked about his joint and muscle aches, fatigue, and stomach problems, which were his most recent complaints, Dr. Rea was not surprised. The doctor explained that Smith's symptoms were spreading within his body and switching to other organs and systems (Rea 1992). This is typical in the second stage of chemical sensitivity when the body becomes so overtaxed that even minute exposures of any substance can trigger a response. Responses can erupt anywhere within the body.

Smith thought the plane accident was causing some of his problems like his back pain. Dr. Rea agreed and explained that severe physical trauma could also trigger chemical sensitivity, but in Smith's case, secondary symptoms indicative of chemical sensitivity—the mood swings, stomach problems, for example—had occurred before the plane accident. Also, Smith's blood work indicated high levels of compounds associated with crude oil.

Smith asked why Lowe and the other cleanup workers whom Dr. Rea was treating all had different problems, and why they were all having problems now, several years after the cleanup. Dr. Rea explained another aspect of chemical sensitivity was that different symptoms commonly developed in different individuals who are exposed to the same toxic event, because every individual is unique (Rea 1992). Dr. Rea said variations in response depend on nutritional state, genetic susceptibility, and how many pollutants the individual already had in their body at the time of new exposure. This uniqueness also explains why a significant length of time might occur between an initial toxic event and the development of symptoms. For example, a relatively fit, healthy individual might be able to

Figure 3
Common Symptoms of Chemical Sensitivity

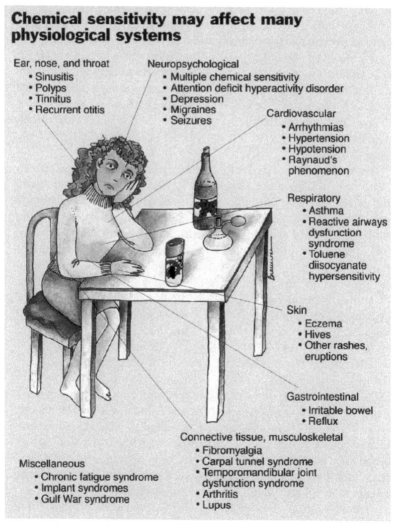

Chemical sensitivity may affect many physiological systems

Ear, nose, and throat
• Sinusitis
• Polyps
• Tinnitus
• Recurrent otitis

Neuropsychological
• Multiple chemical sensitivity
• Attention deficit hyperactivity disorder
• Depression
• Migraines
• Seizures

Cardiovascular
• Arrhythmias
• Hypertension
• Hypotension
• Raynaud's phenomenon

Respiratory
• Asthma
• Reactive airways dysfunction syndrome
• Toluene diisocyanate hypersensitivity

Skin
• Eczema
• Hives
• Other rashes, eruptions

Gastrointestinal
• Irritable bowel
• Reflux

Connective tissue, musculoskeletal
• Fibromyalgia
• Carpal tunnel syndrome
• Temporomandibular joint dysfunction syndrome
• Arthritis
• Lupus

Miscellaneous
• Chronic fatigue syndrome
• Implant syndromes
• Gulf War syndrome

Reprinted with permission from *Chemical and Engineering News* 76(38):57. Copyright 1998 American Chemical Society.

repair tissue damage and break down toxic chemicals without any noticeable symptoms for a while—maybe years, while a person who was more susceptible might get sick relatively quickly. Over the years of treating people, Dr. Rea had become familiar with the symptoms commonly characteristic of chemical sensitivity. He told Smith that his symptoms and those of the other cleanup workers fit this list (Figure 3, p. 79).

After going through his history and questionnaire with the doctor, Smith felt comfortable with his decision to come to the clinic. Dr. Rea's explanations for his symptoms made sense to him, and he was ready to proceed with whatever came next—or so he thought.

He survived a four-day fast, drinking only spring water, which Dr. Rea explained was necessary to clear his body of enough pollutants to allow him to run accurate and reliable laboratory tests in order to confirm his initial evaluation (Rea 1997a). Smith found the series of lab tests, including as he put it, "about 6,000 shots" in his arm, a brain scan, eye tests, and some other neurological exams, a test of his patience (87).

From the test results, Dr. Rea determined Smith was sensitive to a variety of substances including diesel, natural gas, formaldehyde, and men's cologne. The doctor diagnosed Smith with "chemical exposure, chemical sensitivity, edema, myalgia, myofibrositis, fatigue, neutrotoxicity, and gastrointestinal malfunction" (Rea 1993). He concluded Smith was "totally disabled" and recommended a treatment plan of "avoidance of inciting agents and heat chamber depuration/physical therapy . . . for six to eight weeks."

During the next eight weeks, Smith followed a rigorous regime of dry heat saunas, massage, nutrients, and exercise (Rea 1997a). He had chiropractic manipulation of his neck and back to relieve the tingling sensation in his arms, hands, and fingers (Johnson 1993). He fasted periodically, ate a controlled diet, and drank pure water. Clinic staff carefully monitored his release of pollutants. He learned to self-monitor his moods and symptoms to become more aware of his body's needs. He learned ways to make his own home and workplace more pollutant-free. His tests and therapy occurred in specially constructed, less polluted rooms at the Environmental Health Center–Dallas, and he lived in similarly constructed housing in Seagoville, a community about twenty miles southeast of Dallas where Dr. Rea sent patients to avoid the Dallas ambient air pollution.

If it weren't for Ecology Housing in Seagoville, Smith might not have stayed for the duration of his eight-week treatment at Dr. Rea's clinic. Smith was used to medication to make him feel better immediately, and he found the daily workout at the clinic boring and bordering on ridiculous. But his friend Lowe and Tom Pickworth were also staying at Ecology Housing while being treated at the clinic. The three spent their evenings together, sharing experiences at the clinic and swapping oil spill stories.

The cleanup workers noticed they were different from the other patients, many of whom were consumed by fears about their poor health being delusional, because they couldn't pinpoint a single definite exposure that made them sick (Ashford and Miller 1998). Smith and his friends knew the cleanup work had made them sick, so they were able to retain their sense of normalcy in the destabilizing world of chemical sensitivity. The other patients had come to Dr. Rea as a last hope after being rejected by medical specialists, friends, family, and healthy co-workers. At Seagoville these patients also shared time and stories amongst themselves, and they began to regain their self-confidence and believe they could get better. Smith found these patients more needy than he could handle, so he stuck with his two friends and sometimes the Goodwins, a couple who managed Ecology Housing and who enjoyed the company of the Alaskans (Goodwin 2001).

As Smith finished up his treatment at the clinic, he completed a second Accu-Chem blood test and a psychological evaluation. The second blood test indicated that the levels of hydrocarbons in Smith's body had dramatically decreased, but the trimethylbenzenes and some methylpentanes were still present, "parked" as Dr. Rea said, in Smith's system. Dr. Rea had expected to see remnants of these compounds because they were some of the hardest pollutants to clear out of the body.

Dr. Rea used Dr. Nancy Didrikson's psychological exit evaluation to give him some idea of his patient's overall wellbeing because he believed that the body, mind, and brain all interacted to produce a state of health (Didrikson 1997). The evaluation showed that Smith was conscientious and dominated by a sense of duty, but that he was still anxious about his compromised physical abilities and that he was easily overwhelmed and frustrated by everyday challenges (Didrikson 1993). The examiner recommended stress management, relaxation training, and other therapy to improve his ability to cope with his chemical sensitivities. Dr. Rea told Smith some follow-up work would be helpful to him, but when Smith left the clinic at the end of January 1993, it was for good.

He returned to Alaska, feeling like "a million bucks" (228), but once back in the real world, with its constant chemical bombardment, he soon became ill. He decided to apply, within reason, what he had learned at the Dallas clinic to decontaminate his own home, minimize chemical exposure at work, and change his diet—and he decided to sue Exxon and VECO for his health problems (Chapters 8 and 9).

Living with Chemical Sensitivities

Smith's experience at the Environmental Health Center-Dallas changed his life. Dr. Rea was the only doctor who had satisfied his questions (Smith 1996, 106): why—at forty—he was losing his health and his ability to work. Now, Smith framed his body's reactions to things either in the context of his chemical sensitivities or his plane accident, which in his mind were two distinct events. This new understanding helped him deal with his problems in a way that, he hoped, would eventually restore his health. But it was not easy living inside a body that seemed bent on falling to pieces. His wife Shirley was incredibly supportive. Together, the couple dealt with his health problems as best they could, and they adjusted their lives to fit his needs.

When he had first returned to Soldotna a year earlier, Smith had felt worse than before he went to the clinic. But he knew he was not used to being in a "dirty environment," compared to the environmental housing in Seagoville, and his chemical sensitivities flared until his body adapted—or until the couple figured out what was causing the problem and fixed it. For a brief time, his skin felt prickly and irritated. Shirley switched to a petrochemical-free laundry soap like Ron had used in Dallas, and the problem vanished. Then his feet developed itchy rashes. He remembered the cotton clothing at the Dallas clinic, and he switched to cotton socks. The rashes stopped. He became increasingly sensitive to gas and diesel fumes and exhaust—as more time passed, he noticed it took less to make him feel worse. He didn't do any mechanical work on his cars, and when he filled his car, he stepped away from the pumps.

When Smith started doing light work again, his back problems flared. Some days he woke up feeling like he could "go and conquer the world" (228), but after working a couple hours, he had a lot of pain and difficulty. His doctor prescribed drugs—muscle relaxants and painkillers, but Smith remembered what he had learned from Dr. Rea about drugs—"medicated wellness is not wellness." He took the medicine only sparingly, and he worked instead to regain his health.

Chapter 5.

Phyllis "Dolly" La Joie

(Author's note: the following story is based on a deposition obtained through federal court records in Roberts v. Exxon *[1999]. Page numbers in parentheses refer to La Joie 1996—her deposition; other material is fully cited.)*

In 1996 Dolly La Joie arrived at her deposition for her personal injury toxic tort lawsuit against Exxon and VECO with a large scrapbook, which she had painstakingly put together for her children and grandchildren. The scrapbook had color pictures from her six-week stint with the decontamination or "DECON" crew on the *Greens Creek* barge. Although both Exxon and VECO had forbidden cameras on the beaches, a Native man named Denty, who worked as a carpenter on the barge, had brought one with him and had been allowed to take pictures. He posted his pictures on a note board on the barge and filled orders for duplicates.

La Joie had carefully labeled all her pictures and mixed them with other spill memorabilia—news clips, magazine articles, and official Exxon and VECO safety and orientation pamphlets and cleanup injury charts. She often used her scrapbook as an aid for telling stories when she gave talks about her cleanup experiences. As she answered questions during her deposition, she paged through her scrapbook, lending dramatic images and a very real presence to her stories as did her labored breathing.

Acute Health Problems (1989)

La Joie had quite a history with the oil industry. In the early 1970s she had worked five years for Sohio Construction in the Prudhoe Bay oil-fields at the construction facilities as an engineering aide. She described her job as "like being a librarian" (51), keeping track of all the building procedures, blueprints, documents, and papers and find-ing them for the engineers who were constructing the pipeline, the pump stations, the gathering centers, and other facilities. Sometimes she was sent out with her papers to accompany the engineers "to make sure they were building according to the specifications" (53). She pointed to the pictures showing where she worked.

She was proud of her work "up there, in a city in another world" (55), as she said, and proud of the industry. That was why, at fifty-eight, when she learned of the oil spill in Prince William Sound from the news, she packed her bags in her apartment in Hawaii and head-ed back to Alaska where she usually spent her summers with her grown children. She figured she was part of the industry that had taken the oil out of the ground and she might as well help clean it up (67), especially in the Sound—her old stomping ground.

In a whirlwind two days, La Joie was hired by VECO in Anchorage on 26 May 1989, bused to Valdez for an orientation class, loaded onto a boat with other new hires, shuttled to the *Greens Creek* barge anchored near Knight Island, and immediately put to work on the night shift in the DECON unit after two days with no sleep. She recalled being "absolutely, totally exhausted" (74).

La Joie "begged" to be a skiff driver. During the late 1970s she and her husband had spent years in the Sound, hunting seal under the state's bounty program and commercial fishing. The VECO foreman took one look at the petite wiry older woman and assigned her to washing duty on the barge. She recalled that "the women mostly got put in decontamination where . . . [VECO] felt they'd be safer" (38) than out on the slippery oil-coated beaches, doing some of the "most dangerous, dirtiest work in the world" (41).

La Joie was one to make the best of any situation. The women on the evening shift of the DECON crew worked well together. Each evening the DECON crew gathered up the oily rain gear, rubber boots and gloves, hard hats, life jackets, and underclothing of the beach crews after they stripped for their showers. Using steam guns like the beach crews used to spray oily rocks, the DECON crews sprayed all the oily gear that couldn't go in a washing machine. As the night air cooled, clouds of hot oily mist and salt water steam engulfed the DECON crews as they worked outside. Clean gear was hung in the drying rooms, which were *"conexes,"* an industry term for box-car-shaped units designed to carry freight. When the DECON crews found the gear wasn't drying, she and the others asked VECO to order huge industrial heat blowers. The blowers kept the drying rooms hot and steamy, but at least the gear dried.

The women on DECON crews realized right away they were getting chemicals steamed into their lungs. When the Exxon lawyer asked her who told her this, she replied, "[N]o one had to tell us. We were already told [during orientation] all the chemicals that were in crude oil, and we were steaming it, and we knew we were breathing it. We thought we better take some precautions" (99). The safety goggles steamed up "really fast," but the women tried to wear them to keep from getting splashed with oil and hot water. "You always got splashed anyway," La Joie recalled (98). They tried to wear respirators, but they "could never get enough" and the VECO supply people "ran out totally in a couple of weeks" (98, 100). Whatever they got from supply, they used, including little paper masks like those sold in drug stores, which she knew did not really block any chemicals. She had worked to decontaminate and monitor radiation of nuclear submarines crews at the Pearl Harbor Navy Shipyard before she started working on the oil spill cleanup.

When the Exxon lawyer asked her if she ever used any respirators that covered her entire face including her eyes, she responded, "I saw those, but I never wore them. Those were used by the crews that sprayed that fertilizer [Inipol], or whatever it was, on those test shores to eat up oil. . . . They had gas masks because that was really lethal stuff" (102). The Exxon lawyer responded, "I will move to strike[delete] that . . ." (102).

The DECON crew took turns doing laundry. No one liked this job. The laundry room was small, enclosed, and the strong oil vapor from the clothes in the dryer made the DECON crew sick. Whenever La Joie felt dizzy from the fumes and like she was going to pass out, she would step outside for fresh air.

The women quickly found the laundry detergent Tide was not up to the job of removing oil, especially with the saltwater wash. La Joie explained to the Exxon lawyer that Tide was not a degreaser. The VECO supply people sent over stronger solvents, "many different kinds and one that almost looked like gasoline," as La Joie recalled (94). Finally VECO supply sent Simple Green®, which the women found worked well when added with Tide to each wash load. The DECON crew was unaware of the health hazards of using Simple Green, which contained 2-butoxyethanol, the same active ingredient in Inipol.

The job that absolutely everyone refused to do—except La Joie—was turning gloves. The DECON crews started machine-washing gloves instead of throwing them out as the crew had been told to do because the crew didn't have any way to replace the gloves—the suppliers couldn't keep up with the demand. The thick heavy rubber gloves had to be turned inside out so the liners would dry in the dryer—and they had to be turned back right side out once they were dry.

La Joie explained, "It was an excruciating job that had to be done because the guys had to have gloves . . . They just had to. It was so cold" (109–110). She said, "[T]he young girls . . . wouldn't do it anymore because they were afraid it would ruin their fingernails like mine from all the solvents and all the oil" (110). She cried some nights because of the terrible pain in her hands. Her fingernails started to rot and disintegrate, and her wrists and forearms became swollen, so badly one day that she couldn't open doors or dress herself. Yet, as long as she was physically able, she never refused to turn gloves, because "No one else would do it and it had to be done" (112).

La Joie told her VECO supervisor that the glove job hurt her arms, and she went to see the clinic doctors—Exxon or VECO, she didn't differentiate—who finally bound her hands to reduce the swelling. She had pictures of her bandaged hands. The doctors told her to rest her hands and shoulders, but VECO wouldn't let people take more than two sick days leave without pay before firing them, according to La Joie (118). This was substantiated in other personal injury lawsuits (Chapter 2).

La Joie also went to the medical clinics for broken toes, skin rashes, and an endless string of sore throats with, as she said, "all the cold symptoms; sinus, headaches . . . coughing" (121). She tried gargling with salt water and taking Tylenol. Nothing worked. The Exxon lawyer asked, "Shall we call that what, visiting a doctor for a cold?" Not aware of the symptoms of overexposure to oil vapors, mists and aerosols, she responded, "I guess that's the closest thing I can think of" (121–122).

Finally, after working twelve-hour shifts, seven days a week, for six weeks or so, she was given a week off, which she spent in Anchorage with her grown children. En route back to the Sound, she and the other workers stopped at the VECO office in Valdez for another orientation. She recalled being given a lot of "stuff" (261). Among the papers were reports and charts showing the different types of cleanup jobs, the exposure levels for each job, and the number and types of injuries on the cleanup. She kept these charts and put them in her scrapbook.

La Joie noticed right away that her job in the DECON unit, specifically working in the laundry room with the dryers and all the fumes, was not listed, and there was no exposure information. This didn't surprise her as she had never seen any federal or state OSHA inspectors, Coast Guard people, state ADEC people, or any Med-Tox monitoring people during her night shift (124–125).

When La Joie returned to the *Greens Creek* barge, she eagerly volunteered for beach cleanup duty so she could get out into what she "thought was going to be fresh air," as she said (108). On the beaches, she found the stench of crude oil "sickening" and "horrible" (140). There were also dead, decaying animals, mostly seals, seal pups, and birds, which the beach crews put in trash bags to be collected by the skiff people.

She survived her beach crew initiation. The supervisors handed her a fireman's high-pressure hose and turned on the cold water full blast. The hose flipped the 100-pound woman through the air, much to everyone's amusement. Then they gave her a steam spray gun and that also knocked her flat the first time, but she quickly learned to wrestle the hose to do her bidding. She found that everyone preferred to use the steam spray guns because the high-pressure hoses were so hard to hold down. The beach crews took turns working with the high-pressure hoses. As the crews worked with the spray guns, they were engulfed in clouds of oil mist and saltwater steam.

La Joie tried to wear a respirator all the time as she had been taught in orientation to protect her from the oil fumes, but respirators were in short supply. She recalled, "We asked why we couldn't get them, and . . . [the VECO foremen] said, 'Well, you go home if you don't want to . . . [work without respirators]' " (139). Every morning on her own time, she spent "at least an hour trying to scrounge up respirators" (151) and she provided what she could to her fellow beach workers.

The beach job was dangerous and dirty as the VECO supervisors had forewarned. Each day the crews had to carry the 100-pound water pumps, hoses, and other equipment to the beaches and set them up. The oil was three feet deep in places, and the rocks were slippery and sharp. La Joie described it as "a world of constant falls, slips. This [oil] stuff was unbelievable. [Y]ou were just constantly on your knees" (148). She explained, "[E]very time you fall, you had to crawl in it. You had it all over you" (147).

La Joie tried taping her rain jacket cuffs to her gloves and her pants' legs to her boots to reduce her oil exposure, but she found, "[T]he oil just ate away at the tape and the edges got open . . . ." (146). She was constantly wet from rain, spraying, or splashing. On sunny days, the beach crews stripped by noon and "just got sprayed" (146). Her beach crew never was supplied with Tyvek suits to wear under their rain gear to protect them from the oil as she had been told during VECO orientations that beach crews needed to do (147).

After several weeks with the beach crews, La Joie went to work on the barges used for treating shoreline inaccessible to beach crews. The beach crews considered these jobs as premium, and she described her new job as "pretty cushy" (151). The high-pressure hoses and spray guns were already on the Omni and Maxi barges, so the crews did not have to haul the heavy equipment over oily rocks to set it up every day. The barges usually were maneuvered up to sheer rock walls, which the crews sprayed directly from platforms that could be raised or lowered.

While on the barges, La Joie did a variety of odd jobs: housekeeper, nurse, spray crew. As housekeeper, she cleaned toilets, which "was a constant thing" (144), as she recalled because of the oily crews. She used strong solvents such as Simple Green, disinfectants, and grease solvents, "too numerous to even begin to name" (145), as she told the Exxon lawyer. She never mentioned using a respirator while doing housecleaning. She took care of people who fell in the water, too, making sure they changed into dry clothes before they got hypothermia.

Her final job on the cleanup was DECON again, this time helping clean all the equipment and the USS *Duluth*, an old navy transport ship that served as a berthing vessel during the cleanup. La Joie and the DECON crew used the usual detergents, degreasers, solvents—Tide, Simple Green and others—that they had used all summer. They used hot water on the floors and in the showers to make sure they cleaned everything "really good" (155) as she said, for the Navy.

When La Joie left the cleanup work on 15 September 1989, she rested, or as she put it, "collapsed" (157), in Anchorage for a few weeks. She described it as "kind of a cultural shock . . . It was worse than jet lag. It was January before I even started to look for a job. I was that exhausted" (158).

As La Joie rested to muster her energy, she filled the idle hours at her home in Hawaii by putting together her scrapbook of cleanup memorabilia. She also took a daily water aerobics class to prevent her acute tendonitis from stiffening. She found her arms ached from her fingers up into her shoulders. Tendonitis was just the beginning of her health problems.

Lingering Symptoms (1990 to 2003)

When she tried to work again in 1990, she found she was still too tired to work a full shift, so she worked part-time; first at duty free shops at the airport, then as a census taker, then substitute teaching, then a banquet food server at the Royal Hawaiian Hotel. In the spring of 1990, she started getting frequent colds and sinus problems. She wasn't sure what brought on the colds, but she thought her whole immune system was down. She began to experience brain fog in the mornings, and she couldn't seem to get functioning until later in the day. Her health seemed to gradually get worse and worse.

In December 1991 she became "really, really sick," as she said, with "some kind of sinus infection, or something" (172, 173). She couldn't get her temperature down, and she had a terrible sore throat. She thought it was a horrible flu, but it went on for months. Whenever she tried to work or exert herself, she realized she just became sicker. In the mornings, she woke up with severe nausea and excruciating headaches. She felt her head was going to explode. Several times the pressure was so bad she had to crawl to the bathroom. With the nausea, she experienced a badly bloated stomach, no matter what she ate. She described it as feeling "pregnant all the time" or like somebody blew up her stomach with a tire pump (176, 194). She thought she was going to die.

Her health insurance provider, Kaiser, assigned Steven Hong, MD, a new and relatively young internal medicine specialist, to her case. Dr. Hong gave her antibiotics, which didn't help, and he sent her to a food allergist, who found nothing.

By chance on 1 May 1992, La Joie received a call from Jacqueline Payne, one of her former supervisors in the DECON unit on *Greens Creek* barge. When she told Payne that she'd been in bed sick for nearly half a year, Payne told her that others were getting sick and that she should get her blood tested by Accu-Chem Laboratories.

La Joie recalled, "that was a big fight" with Dr. Hong (183). The doctor thought he knew what chemicals to test for, he had already performed a series of blood tests, and he had found nothing save high cholesterol. La Joie insisted.

Her Accu-Chem blood tests came back with elevated levels of same compounds that were elevated in Smith's blood work (Chapter 4), only La Joie had extremely high levels of methylpentanes (solvents). Dr. Hong took one look at the Accu-Chem blood work and, according to La Joie, said he "didn't know anything about chemical poisoning . . . He didn't know what it meant; what it could cause—nothing, absolutely nothing, about it. He said he wasn't trained in it" (186). He referred La Joie to occupational medicine doctors.

When La Joie saw her Accu-Chem blood work in June 1992, she realized for the first time that her health problems were probably caused by her work on the cleanup. She remembered during the cleanup orientations, the workers were told there could be long-term health consequences, but the exposure levels would not be high enough to make anyone sick. No one ever described what the symptoms of chemical poisoning might be.

Once she felt she understood the cause of her strange health problems, she started to regain some control over her life. After a diligent search, she found an allergist, George Ewing, MD, who knew about chemical poisoning. She first visited him in October 1992 with problems of recurrent headaches and respiratory problems, chronic sinus problems, inability to concentrate, fluid retention, and pain in her hands, arms, and shoulders.

Dr. Ewing found she was intolerant of alcoholic beverages, fats, sugars, and spices. The doctor showed her pamphlets describing her symptoms and listing them as among the known effects of chemicals. Dr. Ewing diagnosed La Joie with chemical sensitivity, secondary to massive crude oil exposure, and recommended she avoid all chemical exposures. She continued to see both Dr. Hong and Dr. Ewing until 1994 when she could no longer afford her Kaiser insurance.

In January 1993 Dr. Hong diagnosed La Joie with adult onset diabetes. Since the cleanup and mostly during 1992, she had gained weight rapidly, up to 170 pounds from her cleanup weight of 100 pounds. She did not know that crude oil aerosols and some solvents are endocrine disrupters that can disrupt thyroid function and cause rapid weight gain. The prescription medicine for diabetes gave her some relief. She lost 20 pounds. Diuretics helped reduce the swelling in her legs and ankles. Dr. Hong diagnosed her stomach problems as acute gastroenteritis and put her on medicine to reduce the stomach acids and pain. She had pictures in her scrapbook of her strangely bloated stomach. Dr. Hong also found her liver greatly enlarged. La Joie knew that the liver works to break down harmful chemicals and is susceptible to their effects.

Dr. Hong sent La Joie to numerous other specialists at Kaiser Permanente, none of whom could explain his client's symptoms. The chief of industrial medicine did not believe the Accu-Chem blood test explained her illnesses (Hong 1997, 46). A radionuclide study and a CT (computerized tomography) scan came back normal (ibid.,21, 48–49). The dermatologist attributed her skin rashes to aging (ibid., 52). The allergist thought it "highly unlikely" that her

---

Misdiagnoses of Chemical Injuries

Aching joints. Headaches. Tiredness. Dizziness. Shortness of breath. Persistent coughing. These are all symptoms that could be attributed to colds, flu, stress, or other medical conditions. These symptoms can also be some of the first signs of a chemical exposure. Doctors often lack the medical training to properly diagnose or treat occupational diseases that might be caused by exposure to toxic chemicals.

The U.S. Occupational Safety and Health Administration (OSHA) recognizes that "lack of knowledge about health effects associated with chemical exposures contributes to chronic under-reporting of occupational illnesses. Misdiagnosis is a problem and often symptoms are treated without realizing that the cause is an occupational chemical exposure" (OSHA 1994).

While some chemicals can cause acute poisoning that result in the rapid appearance of symptoms, many exposures cause chronic effects where the harm is not apparent for a long period of time—months, even years. These symptoms can include certain cancers and effects on the neurodevelopment of an unborn child of a parent exposed in the workplace. Symptoms can be subtle and difficult to diagnose.

Although knowledge about occupational illnesses is not new, industry has often successfully withheld information about the health effects of certain chemicals or lobbied to prevent official recognition of occupational diseases (Geiser 2001). "Many of the first studies on black lung, asbestos related illnesses, and illnesses caused by exposure to Agent Orange were done by industry, and the results were covered up for years" (Clipp 1993, 73). Asbestos companies "concealed their own studies and resisted compensation claims for some 30 years. Likewise, the paint industry hid their studies of lead paint hazards and the polyvinyl chloride industry fought lawsuits by workers who alleged that exposure to vinyl chloride had led to their angiosarcoma of the liver" (Geiser 2001, 127–128).

complaints were caused by allergies (ibid., 20). The neurologist couldn't find any abnormalities and concluded her symptoms were "likely of a psychological source" (ibid., 36). The psychiatrist concluded she was either a severe hypochondriac or suffering from a "delusional disorder" (ibid., 73).

Dr. Hong finally decided he didn't believe that La Joie was as sick as she described. He thought her high cholesterol and her Type II diabetes were attributable to either her diet or genetic makeup (ibid., 30–31, 55), and her abnormal liver function was caused by her diabetes, high cholesterol, and obesity (ibid., 57). He told her there was no reason for her not to work full time, and he refused to sign her application for Social Security Disability (ibid., 83–85) (Sidebar 6, p. 93). La Joie thought it ironic that Dr. Hong didn't think she was sick from chemicals yet prescribed numerous chemical drugs to make her feel better. La Joie decided he could not admit that he could not treat her because of his "huge . . . ego" (La Joie 1996, 240).

Meanwhile, La Joie grew progressively worse. Her sense of smell seemed oddly magnified. She found traffic exhaust and cigar or cigarette smoke gave her headaches and made her nauseous. Modified air—either air conditioning or heat—caused headaches and sore throats, and it dried her sinuses, making them crack and bleed. Strong cleaning solvents made her cough, choke, and nauseous, and gave her headaches. The odor from a rubber slipper shop caused her to throw up. She had to choose her cosmetics carefully as even some of the non-hypoallergenic ones caused rashes (221–223).

La Joie struggled to understand chemical poisoning and how to treat it. She reduced her headaches and sinus problems by letting the strong Hawaii trade winds blow unchecked through her apartment. She went to meetings of the Human Ecology Action League, a group of people with health problems caused by chemicals. She learned about lymph node cleansing from a woman at a garage sale. She read everything she could, including information she received from Dr. Rea whom Jacqueline Payne had recommended.

La Joie could not afford treatment in Dr. Rea's clinic (Chapter 4), so for a couple years, she devised and actively followed her own detoxification or "detox" program. This consisted of two hours of saunas per day, nutritional supplementation with natural minerals and vitamin antioxidants, aerobic exercise in a swimming pool, lots of herbal detox teas, and the lymph node cleansing treatment for her digestive system. La Joie told the skeptical Exxon lawyer that she believed her self-help efforts saved her life.

Dr. Ewing was very supportive of La Joie (1997a), and he signed her state disability forms (Ewing 1994), which the board took two and a half years to approve. She barely survived those nightmarish years of ill health, no money for treatments, and extremely high costs of Hawaiian housing.

In 1996 she turned sixty-five and her life improved markedly. She moved into government run senior housing and began to receive Medicare and Medicaid. She began to visit Kaiser Permanente again, where she was assigned a new doctor. And she just dealt with her growing litany of illnesses on a day-to-day basis, as she told the Exxon lawyer.

The medical and legal challenges La Joie faced in pursuing a personal injury lawsuit for chemical-induced illnesses stemming from exposure during the cleanup are discussed in Chapters 8 and 9.

Chapter 6.

A Collection of Stories

INIPOL ("IN-a-pole"). The sordid saga of this product's approval and use epitomize corporate greed and the multiple failings of the federal government's regulatory and oversight processes to protect public and worker health. The health consequences of this debacle are tragic for victims and hauntingly disturbing for those who learn what happened on the beaches of Prince William Sound. Readers are forewarned.

Inipol Stories and History

Inipol EAP 22 was one of several chemical products listed by the U.S. Environmental Protection Agency (EPA) for use on oiled beaches (U.S. EPA 2004a). Inipol was marketed as a liquid fertilizer. It was manufactured by Elf Aquitaine, a French company, in partnership with Exxon.

In May 1989 the EPA proposed the addition of fertilizers to oiled beaches to encourage growth of naturally occurring bacteria to speed breakdown of oil (CFS 1989 2[15]). "Bioremediation" was the latest federal craze for treating toxic waste sites, despite test work that had been done mostly in confined and controlled lab conditions (Begley and Waldrop 1989). The ADEC report called the EPA's "Alaska

Bioremediation Project, an unprecedented exercise in applied biotechnological research" (ADEC 1993, 74). The EPA put up about $5 million; Exxon committed additional funding under a special technology development agreement; and the Exxon-EPA liaison plunged headlong into a project to test several kinds of slow-release solid pellets and the liquid Inipol.

The Exxon-EPA team fast-tracked tests. Lab tests showed that Inipol was harmful to small marine invertebrates, but it could be applied in weaker solutions to minimize harm to wildlife and accelerate degradation. A ninety-day field test was initiated at Snug Harbor on Knight Island in early June. According to the ADEC report, "when both the state and federal government gave tentative approval to use of fertilizers, the program stood on a few lab tests, thin field test data, and literature searches that gave only limited evidence about whether the fertilizers were toxic. There was virtually no broader ecological analysis about what the addition of all those nutrients might do" (1993., 75). More lab and field test were initiated in July 1989.

The ADEC report noted there were concerns by other scientists that the tests were moving too fast and that there was no clear evidence the fertilizers actually worked—and some evidence that the fertilizers were more toxic than the oil.[1] These concerns were ignored. There were no public hearings and no independent scientific review. The Exxon-EPA liaison plowed ahead except in Seward where a strong public-private sector group banned Inipol within the City's jurisdiction until its effects on wildlife were better understood (CFS 1989 2[24]a). According to the ADEC report, "approval for widespread use of fertilizers came barely halfway through the tests to determine whether fertilizers worked" (ADEC 1993, 75).

Part of Exxon's enthusiasm for Inipol may have been related to the fact that the chemical base was 2-butoxyethanol—an industrial solvent for grease and oil. Another component was the surfactant laurel phosphate, a detergent that produced dispersant-like effects when sprayed too heavily. Overdosing stripped the oil off the rocks and created the same problems as the banned Corexit 9580M2. Some of the early "successes" of Inipol were most likely the result of such chemical overdoses (ibid., 76–77; CFS 1989 2[34]). Another problem, observed by ADEC monitors including Ed Meggert, was that Inipol separates into chemical phases at the cold ambient temperatures, and this enhances the stripping effect of the solvent phase.

As the ADEC report observed, "how badly one wanted or needed results drove one's judgment about how much risk was acceptable" (ibid., 76). The MSDS for Inipol, supplied by Exxon and dated 28 July 1989, showed that it caused a variety of health effects in humans, ranging from dizziness, headaches, central nervous system depression, liver and kidney damage, and red discoloration of the urine (Exxon 1989a; see also Alaska Health Project 1989) (Sidebar 3, p. 12). Exxon seems to have downplayed the health risks to people and wildlife, noting that the 2-butoxyethanol evaporated quickly within about twenty-four hours. Exxon alleged that crews who applied Inipol to beaches and worked within this toxic window were protected from overexposure by their personal protective gear.

To allegedly protect other people and wildlife from overexposure during this toxic window, Inipol-treated beaches were flagged with scary-faced balloons. However, no monitoring was done to determine if this worked, because Exxon and VECO kept their crews off beaches for at least two tidal cycles (twenty-four hours) after an Inipol application. But others watched the remote beaches. Natives from Chenega Bay reported thick windrows of dead juvenile salmon washed up on one Inipol-treated beach with the first flood tide after treatment (Pete Kompkoff, Native Village of Chenega, pers. comm., August 1989).

The select crews who applied Inipol to beaches, such as Exxon's Bioremediation Application Teams (BAT), trusted that VECO and Exxon would provide adequate protective gear and proper instruction in use of the chemical, but neither was the case. The workers were instructed not to breathe or absorb the compound through their skin, and they were told repeatedly that Inipol was not hazardous as long as they wore the "right gear." Unlike for the majority of beach crews, Exxon and VECO supplied the BAT crews with Tyvek suits, respirators, and goggles as well as raingear, gloves, and boots. Workers were led to believe that this personal protective gear was the "right gear." A Coast Guard safety officer familiar with hazardous material handling and training and who had been at the spill later told the press, "[A] rainsuit is [worthless] as protective equipment except for one chemical: water" (Murphy 2001).

The BAT crews were not told that 2-butoxyethanol was an industrial solvent for resins and plastics. Inipol dissolved the rubber gaskets in the spray backpacks and leaked out; it dissolved the plastic raingear—some of the "right gear"—and it came into direct contact with the workers' skin (Moeller 1989).

When beach crews who followed in the wake of BAT crews became sick from exposure to Inipol-treated beaches, Exxon downplayed the health risks. The most publicized incident occurred on 14

August, when four workers of a twenty-one-member crew in Seldovia reported headaches, skin rashes, blisters, and nausea after applying Inipol to a beach (Spence 1989a). A local doctor, Larry Reynolds, examined the entire crew. Blood work for one of the workers revealed 2-butoxyethanol in his bloodstream.

Shocked and outraged community members held a town meeting (McDowell 1989). Exxon's doctor stated that 2-butoxyethanol "is in a lot of compounds on the shelves" and sold as cleaning agents. However, Dr. Reynolds warned, "We're within a generation of Agent Orange ... There are no good studies to see what things do to you on a long-term basis" (Spence 1989b). Community member Tim Robertson reported that, since he started working on the spill in April, he had never seen medical data on health hazards of crude oil or chemicals; he knew of no baseline urinalyses done on workers before sending them out; and he knew the workers considered respirators "optional" (Spence 1989b). State inspectors with the Department of Labor stated the Seldovia incident was a violation of right-to-know provisions in the Hazard Communication Standard (Spence 1989c).

The Seldovia incident was not isolated. The state was understaffed and unable to follow through with fines and penalties in many instances, according to federal officials (Spence 1989c). One could only wonder how many cases of mass chemical poisoning occurred without any media coverage or knowledge by health oversight officials such as the Disk Island Corexit event (Chapter 1). Even in the more highly publicized events such as Seldovia, there was no follow-up and the cases were largely forgotten, except by the affected workers and communities (Ortega 1989).

Two years later in 1991 during a panel presentation on bioremediation at the 1991 International Oil Spill conference, I queried the Exxon panel about the Seldovia incident. Exxon's scientists responded, "Now is not the appropriate time to discuss this" and they turned off my floor microphone. Seven years later in 1998 at a Coast Guard-sponsored, "information-gathering" meeting on dispersants in Washington, DC, I brought up the Seldovia incident again, with fellow Cordovan David Grimes, as an example of human health risks associated with dispersant use in the field as opposed to controlled conditions in the labs. We found the event had largely been "forgotten" by Exxon representatives and Coast Guard officials present at the meeting. Our concerns were not addressed.

The intense debate over how well bioremediation worked—if at all—continued from 1989 through 1991, along with the application of Inipol to oiled beaches in the Sound and the Kenai Peninsula. Exxon and the EPA claimed fertilizers sped up natural degradation by three to five times, but many independent scientists put the rate lower—one to two times faster—or could find no difference between fertilized and unfertilized beaches. The most definitive comments came from three independent reviewers of the 1990 Bioremediation Monitoring Program *in December 1990*. They "found the report to be misleading and to draw conclusions that were not supported in substance by the data and experimental protocols. Overall, there simply was no significant difference between treated and untreated areas" (Capuzzo, Farrington, and Kellogg 1990a, 1). They concluded, "Given the degradation rates of untreated versus treated being essentially the same, the recommended best course of action is *not* to add fertilizer" (Capuzzo, Farrington, and Kellogg 1990b, 1, emphasis in original).

Perhaps most revealing in this debate was the fact that Jim Clark, the EPA scientist in charge of the Inipol experiments and a staunch supporter of the chemical's use, left the EPA to work for Exxon Research and Engineering Company in Gulf Breeze, Florida, in 1991.

The experiences of several cleanup workers who used Inipol or were exposed to Inipol follow below and in Chapter 7, respectively.

Don Moeller and his Crew
Acute Exposure (1989): Task Force 1

When Don Moeller (2001) heard the news about the oil spill, he went straight to his boss' office at Horizons, a state-operated group home for mentally handicapped people, and asked for time off. He wanted to work on the cleanup to earn money to pay off some land he had bought just outside Valdez. His boss said his job would be waiting for him whenever he returned—he didn't want to lose this big-hearted capable young man who devoted so much time in service to others. Moeller had worked with the handicapped in Valdez, Alaska, ever since he had graduated from high school in 1980. At the time, he said, his parents had threatened to take him back to Minnesota with them unless he got "a respectable job, something other than commercial fishing." The Harbor View job had passed muster and Moeller had settled in Valdez.

Moeller wound up on Task Force 1; his new home was camp housing on the *Glacier Bay Explorer,* one of eleven berthing vessels. At first, he worked on many different beaches, he said, "Mopping up oil with pom-poms, oil pads, anything they had at the time. It was chaotic; it was a mess. They didn't have any idea how to clean up oil."

He could see that the beaches needed to be washed and that the giant fire hoses weren't working all that well. He set to work helping to develop a head system that would generate enough pressure to pump seawater through large perforated hoses. The header devices and hoses were set up at the top of a beach. As water flushed down the beach, crews tilled and stirred the loose cobble and gravel. Oil bleeding off the beaches was corralled by booms and picked up by skimmers—to the extent those crews were able to recover the elusive oil sheen.

Flushing and tilling didn't work on all beach types, so Moeller helped set up generators on the landing craft to heat and pressurize huge volumes of seawater. He said, "The heat was a mess. We test-treated Eschamy Bay at low tide and we ended up 'cleaning' hundreds and hundreds of sand dollars. Everything on the beach was cooked and it all bubbled up."

Hastily drafted operational guidelines for shoreline cleanup, circulated less than a month after the spill, advised crews to avoid spraying the "*green zone,*" the richly productive lower intertidal area covered by algae. However, many crews ignored these restrictions, according to the ADEC final report.

The chaos melded into a routine as the weeks passed. Moeller busied himself with fueling and tending to the generators on the landing crafts; moving the header devices and hoses up and down beaches with the tide; and operating the steam cleaning equipment. The long shifts, seven days a week, were exhausting. Moeller described people as "grungy tired." He said, "We didn't even get hand wipes out there until June" and people didn't always wear their gloves. He watched workers, "sitting there, puffing on cigarettes with their grimy hands, leaving oil marks on the cigarettes, and smoking right past the oil marks. Sheer stupidity."

Another one of the several hundred people on Task Force 1, Evan Lange (2001), a bright observant Alaska high school graduate, who commented on the daily routine. "I don't think the average person realizes what it was like out there. If your hands got oiled in the morning, you had to live with that until you returned to your decontamination barge that evening. So everything you did on the beaches—

going to the bathroom, eating lunch, taking a break—you had to do with oiled hands. It's almost unreasonable to think that somebody working in those conditions could keep from being contaminated— could keep the oil off their skin, off their faces, out of their eyes. I'm sure people did everything they could to keep the oil away from their food, but we sat right there on the oily rocks to eat our lunch. The oil was definitely a source of contamination even for people who did everything right in terms of taking precautions to prevent becoming oiled."

The respirators showed up about "midway through the summer," according to Moeller, but because of the moisture content from the oily seawater mist, they were "no longer good after a couple of hours." He said, VECO "never gave us other filters for the respirators." Absent proper safety equipment, he signed up as a volunteer for the Med-Tox air quality-monitoring program, sponsored by Exxon, hoping that at least workers would be apprised of any dangerous work conditions.

His skills with people and equipment caught the attention of VECO supervisors and after about six weeks, he was rewarded with a new job: general foreman of a beach crew. With the new job came a room of his own. Moeller said, "I was so happy! I could sleep; no one was going in or out; no one was coughing; no one was dragging dirty oily gear into the room." There were rules against bringing oily gear into the rooms, of course, but he explained that not everyone followed the rules.

Then, towards the end of July, VECO rewarded Moeller with another job offer: the chance to supervise "BAT 1," the first Bioremediation Application Team on the cleanup. He was told that people who worked on BAT 1 would be "guaranteed" more time on the cleanup and that he could handpick his crew—just not any women, because menses invalidated the urine tests needed to monitor BAT crews' exposure levels to Inipol.

Acute Exposure (1989): Bioremediation Application Team (BAT) 1

Moeller's BAT 1 crew all attended a Hazwoper class on a big Exxon berthing vessel. He said, "We had full lectures from VECO trainers, explaining everything to us. We were given Material Safety Data Sheets (MSDS). They kept saying it was 'bioremediation'—there's a

bug in there, but according to the MSDS it's a solvent. We had guarantees from VECO that our urine would be tested daily for exposure.
They made it sound like it was nothing to be worried about." He
added, "They made me keep a ledger. The only time during the whole
summer they made me keep a ledger as general foreman was when
we did 'bioremediation'."

The first Inipol application was on Green Island on 31 July 1989.
Moeller noted the following in his ledger (Moeller 1989):

*"1 August 1989—[spray] packs are leaking and breaking
down a lot. Seems that the pump or pump gasket is breaking
down only." A little later that day, "My rain gear fell apart, so
was removed. I worked about two hours around chemical in
only a Tyvek suit and legs were exposed."*

"2 August 1989—more packs have broken down."

*"3 August 1989—sprayers that have motors are doing great.
Guys are doing well, but two have said their throats are sore."*

*"4 August 1989—failed UA [urinalysis] again. Medic is pulling
me from beach." Later that day at a meeting, "Exxon's Richard
Becker said again the same [thing] that [Exxon's] John
Messinger told us about the chemical—no hazard to us. Just
wear the right gear."*

He was sent to "the Miller barge," as the medic barge was called
after its head physician, Dr. Miller with Exxon. They drew blood and
tested him, then sent him to town for more tests with Exxon doctors
who sent him to his personal doctor in Valdez. According to Moeller,
his personal doctor told him, "You just got high blood pressure, monitor it and see how you're doing." His doctor gave him an "OK" to
back out and Moeller said, "that was good enough for Exxon. I went
back out."

Moeller said that no one had any idea how to apply Inipol to the
beaches. He explained, "It was just an experiment as we went." At
first, the heated chemical was pumped under pressure into backpacks outfitted with spray hoses. But, Moeller said, "The backpacks
were made for pesticide. The Inipol dissolved the gaskets between

the tank and the hose. The pump couldn't draw the Inipol through the hose and this caused pressure to build up inside the tank where, a couple of times, the pressure blew the lid off the tank top. It happened to me." The liquid Inipol drenched him, running down his back. It was low tide and there was no way to pull him off the beach to take a shower. By the time he got back to the berthing barge, he said, "The chemical spill left a red mark down my back."

Moeller noticed the beaches seemed to be reacting to the chemical as well. He said, because the backpacks didn't work, "We laid the Inipol on thicker than we should have. The beaches were just froth, when the tide washed back up, from that chemical."

He suggested that VECO get some airless paint sprayers to get around the pressure problem. The airless paint sprayers had heavy generators so he rigged up wheels for the smaller ones, but he said "that was a joke, trying to get those rigs over rocks. And then the Inipol made the rocks even more slippery." He rigged a larger airless generator, powered from a landing craft, with extensions, long hoses, and nozzles. This was time consuming as the hoses had limited range and the landing craft had to be constantly moved and secured.

The respirators also were not working well. Moeller said, "With the moisture in our breath, and stuff, the respirators became so concentrated that they didn't work after a while—within a few hours. You could tell when they weren't working. You'd draw your breath and you'd smell and taste Inipol." He described the taste "just had a 'slickness' to it in your mouth."

The Inipol was carried in a big heated 5,000-gallon chemical tank on board a specially designated chemical boat. The tank had a retaining trough around it to prevent any spillage from reaching the water. Moeller thought this was odd, because they were spraying Inipol on the beaches flushed twice daily with tides. But he never asked about this or whether there were additional health risks associated with heated Inipol. He described the cold liquid as "like honey." The MSDS for Corexit 9527, which contains the same hazardous compound, 2-butoxyethanol, as Inipol, states, "Vapors and/or aerosols, which may be formed at elevated temperatures, may cause systemic effects." The Inipol MSDS, supplied by Exxon, contains no such health warning.

One night there was a big messy spill on the chemical boat. Moeller said, "I don't know whose decision it was to use the chemical boat as garbage scow, but everyone threw their garbage on the boat and then it piled up until the garbage barge picked it all up. One of the garbage bags snagged the valve on the chemical tank and Inipol started flowing out of the tank."

He was getting ready to take a shower on his berthing vessel when his VECO supervisor called to tell him to grab a couple people and go clean up the chemical spill. Moeller said, "When I got there, no one had turned off the valve. The boat people were real apprehensive about this chemical and they didn't want to get anywhere near it. The retaining trough around the chemical tank had already filled, and Inipol was overflowing onto the boat. The garbage bags dissolved—when we picked them up they fell apart. We scooped everything up with shovels and put it in big 55-gallon Rubbermaid® garbage cans. But the chemical cracked the side of one can and everything just ran out all over the boat again. The handle on the other one snapped and broke when we tried to move it. I radioed my boss to ask him what to do. He came back fifteen or twenty minutes later and said, 'just wash it all into the water.' So that's what we did."

The chemical boat was anchored in a small bay on Seal Island. Moeller watched as the "whole little bay just turned milky white that night, all around the landing craft." After they washed down the boat, they decided to steam clean the deck, which was slippery with the cold honey-like Inipol. Hot steamy chemical fumes enveloped the workers.

Moeller and his crew were paid over-time that night. However, his body paid a higher price. He explained, "The Inipol ruined the stitching on my rain gear and popped out the stitches. The Tyvek suit underneath stuck to my skin where the chemical leaked onto it. My legs turned bright pink, like someone had slapped them a few times. And I was real nauseous all night." He was sent back to the Miller barge and this time there was blood in his urine. He was pulled off BAT 1.

Moeller said, "Exxon and VECO said 'we're going to monitor you, keep checking on you. I had all these promises. It was supposed to be one time a year for the next few years—that's what VECO said. They took a couple more blood tests that winter, then I never heard another thing from them."

VECO and Exxon did make good, however, on their promise to keep Moeller on through the winter as a general foreman of the winter cleanup. He picked up oiled cleanup debris—poms-poms, whole sections of boom, and other floating garbage. He transported scientific monitoring crews for the bird studies and some of the *"benthic"* (seafloor) sediment studies.

When spring rolled around, he realized he wasn't ready for another summer shoreline cleanup. He quit and took another year off to build the cabin on his land before he resumed his job at Horizons.

Lingering Symptoms (1990 to 2003)

Moeller did a few odd jobs to supplement his income from Harbor View. He bartended and one night in fall 1991, he said, after he and another employee closed up, "We were going to strip the dance floor. I never even touched the chemical. The next thing I knew I had a bad rash on my arms—they turned bright red. I was having trouble breathing and shortness of breath."

He started having problems with deodorant—it "burned" him until he switched to totally organic natural brands. He had similar reactions to colognes, shaving cream, and shampoos. Shampoos burned his scalp—he switched to natural brands and kept his hair short. He had to give up on Colgate, because, he explained, he'd be brushing and suck in a breath and all of a sudden, his "lungs would just shut down." He went to organic toothpaste. For awhile, he ran his laundry twice just to get rid of residual Tide. Besides prickly skin, he kept getting a rash in his groin area and under his arms. He finally switched to organic laundry soap.

A couple of years passed and Moeller noticed other things. Cigarette smoke and diesel exhaust caused his lungs to close up. One breath of either was all it took. His fingers started getting numb and tingly. His doctor attributed this to frost bite. He would wake up in the middle of the some nights with his pillow drenched with sweat from fever. He gave up on taking flu shots because they knocked him flat for four to six weeks and it felt like he was having a heart attack—his chest felt terribly constricted. His sinuses were "shot." His energy level was down and he put on weight. He thought he was just getting old and fat.

In 1998 Moeller received a phone call from a former cleanup worker, J. T. Beardin, who had sued Exxon and VECO for his health problems. Moeller remembered him from the winter cleanup of 1989 to 1990. Beardin and another person had become sick that winter after spending time in the small bay at Seal Rocks where the Inipol had spilled off the chemical boat several months earlier. There was no record of the Inipol spill, so Beardin asked Moeller to join

their lawsuit to provide evidence of the spill. Moeller agreed to join
the lawsuit, which had been filed in Louisiana.

Their case was eventually settled on 8 October 2001, because
Beardin had passed away and the families, according to Moeller,
"wanted to get this behind them." Moeller's share of the settlement
was $10,000.

Of his cleanup experience, Moeller said, "I would never do it
again." He has adjusted his lifestyle to live with his chemical sensitivities and other chronic health problems.

Evan Lange

*(Author's note: The person I interviewed for this story wished to
have his identity protected. His name has been changed. This story
is used with his permission and review.)*

Acute Exposure (1989 to 1990)

In August 1989 Evan Lange volunteered to join a 'bioremediation'
team (Lange 2003). He had been tending boom all summer from a
skiff and he was ready to try something new. He remembered his job
switch was "kind of on the fly. We just jumped ship from Task Force
1 to this real specialized group of people and support vessels. They
didn't really set us aside to give us much training." He thought it a little strange that he had taken a four-hour training at the beginning of
his summer work—"at least they were trying to do something standardized"—but then Inipol came along and all he received was sec-
ondhand information. No MSDS, no safety training, no nothing.

Lange said he never had the feeling that he "was stepping into a
tried and true program. It was more like 'let's make this work' kind of
thing." He described going out to the beaches with twenty people, or
so, but he "never witnessed twenty people out on the beaches dis-
persing the chemical." Instead, he and maybe one or two others
would spray Inipol on the beaches, while everyone else watched—
from a distance, either above the beach in the grasses or down on
boats in the water. "There was a lot of interest in seeing how this
worked. It was like they were testing these operations and trying to
get workable and useful methodologies."

The first "methodology" didn't work very well. Lange explained,

"We were having problems with Inipol getting cold, because the stuff gets pretty viscous when it's cold and it resists flow. We put it into these plastic backpacks, like a home garden fertilizer sprayer. The person carrying the backpack would have to keep it pumped up. The backpacks had a rubber tube with a metal wand and you pressed a trigger to spray. It was a very small group of Inipol workers and we'd just walk down the beaches, pumping and spraying the Inipol. When the backpack was empty, you just walked back down the beach and get more Inipol."

Lange figured there was "some kind of optimal spray they were looking for. When we pumped the backpacks, the Inipol kind of dribbled out. It was not a nice spray pattern. When it dribbled out, it meant you were just spilling it on the beach. The Inipol would kind of make the oil run off the rocks, but that's not what we were trying to do, really, we were told we trying to get the microbes activated and get their populations up. And the best way to get this to happen was to apply the Inipol in a uniform coating over the entire area of application." Lange noticed that even when he managed to achieve the desired uniform coating, the freshly sprayed Inipol immediately melted oil from the rocks. He wondered if this cosmetic cleansing was an unstated desired effect of Inipol use.

Lange and the other sprayers gave urine samples every time they returned to their home boat after a round of spraying. He explained that VECO was checking for blood in the urine and he understood that Inipol could cause kidney damage. He always wore his protective gear—a Tyvek suit and rain gear—and a cartridge type respirator. He tried to be careful.

Lange only went out with the Inipol crews about six times before he left the cleanup to attend college in Anchorage. The next summer he applied to work on the cleanup again, because he "didn't have anything better lined up." When VECO found out he had experience with Inipol, he said, "They became very interested in me." This time, he took a Hazwoper training course at the VECO headquarters in Anchorage.

Lange said, "VECO definitely had their act together a lot better than they had the first year. It felt very organized and thorough." He worked with an Inipol crew in rotating shifts of two weeks on, one week off. He said, "It was a very, very intensive and directed effort. They knew what they were doing; we knew what we were doing; and we were all doing it. There was none of this experimental trial and error stuff. It was almost like everything was choreographed. We had a routine. It was not like the first year at all."

Lange said, "We were running a real light operation. There was one home boat. We were using a pontoon boat with a large insulated tank for the Inipol, which was heated. We would nose the platoon boat onto the beach and we would jump out with a wand and long high-pressure hose from an airless sprayer, which can pump up to three thousand psi (pounds per square inch). People would guide the hose over the rocks while I used the wand at the end of the hose to spray a very large area very quickly—even an area the size of a football field."

He said, "There was definitely a fine mist associated with our spray operations. It was so fine that it just floated in the air. We wore a down-turned face shield, a respirator, a helmet, Tyvek suits, rubber gloves and rubber boots with Duct tape around the cuffs. The face shields didn't really protect you from getting the mist around your eyes and upper cheekbones and between your nose and mouth. There was no seal."

In between shifts, Lange was sent to get his blood drawn at VECO clinics. He recalled, "The doctors used a very large shafted needle because they didn't want to damage the blood cells. They were using smaller needles at first, but those must have been damaging the cells. I've never seen a needle that big in me before or since." Once when his blood was drawn, his vein exploded and blood splattered everywhere. The memory of those big needles haunts him and he still can't give blood without almost passing out.

Lange worked through August 1990 when he left to continue college. It was his last summer on Exxon's cleanup.

Lingering Symptoms (1990 to 2003)

In fall 1989 Lange started having trouble thinking. He described it as "feeling kind of clouded." He also noticed a lack of energy. He could still think and analyze and function, so he decided these feelings were part of his new college surroundings and new life away from home. He never thought it could be related to his cleanup work. He described himself as "an optimist" and explained that he "likes to make up for any shortcomings by succeeding in other areas." He was determined not to let his physical problems get him down.

In summer 1991 he had a thorough pre-screening physical for a job palletizing food products for a grocery store chain in Anchorage. He was told he was anemic and that he should do a follow-up with a physician to determine how that would affect his energy levels. He thought his anemia was related to diet, so he started taking iron supplements. This raised his iron level slightly but he was still anemic due to low red blood cell and hemoglobin counts. He was also married and he and his wife had their first child. With all the responsibilities of family life and college pressure, he explained, "It never really dawned on me that the oil spill was the reason for me feeling tired and spacey."

His health problems persisted. He described it as feeling "kind of numb in my head. It was difficult to think, kind of like being underneath fluorescent lights—kind of a spacey feeling like I was in some kind of zone or something."

In 1994 he completed college and started a job as a mechanical engineer. He was still searching for the reasons he felt tired all the time. He discovered his grandmother had diabetes, so in 1995 he went in for a physical to check out that possibility. Everything turned out fine except his hemoglobin levels, which were "a bit low." They were still low in 1998 when he went in for a special blood test to determine if his anemia was hereditary. It wasn't. He did more blood work in 2001 and his hemoglobin and hematocrit levels were still low. Further testing indicated there was no internal bleeding. No one could explain any reason for his anemic condition.

It finally began to dawn on him that his cleanup work might be related to his health problems. He realized that he probably had been exposed to Inipol during the cleanup and he knew from the MSDS—which he finally obtained years after the oil spill—that Inipol "could produce blood and kidney damage." He reasoned, "Even if you took all the precautions, which you were told to do, to protect yourself, you could still get exposed to the chemicals. You were out there for half a day—twelve to fourteen hours—in that environment, and your protective shell is not invincible."

At the same time, he realized that his really bad episodes where he "was really zonked out" were becoming fewer and less intense. By 2001 he had two more children. With a full family life and a busy job, he decided to just live with his health problems and make the best of his situation. He never sued Exxon or VECO for his health problems.

Chapter 7.

Sara Clarke and
Captain Richard Nagel

During our investigations, Pam Miller and I discovered two workers with unusual exposures to Inipol (and other solvents and oil) and chronic systemic effects—their bodies crashed. One person was able to stabilize her condition with environmental medicine treatment as in Ron Smith's case. The other person was never treated by environmental medicine doctors and his condition continues to decline. His doctors have told him there is nothing more they can do for him.

Cleaning Up after the Cleanup—Sara Clarke

(Author's note: The person I interviewed for this story wished to have her identity protected. Her name has been changed as has the name of the company she worked for. This story is used with her permission and review.)

Acute Exposure (1989)

On 24 March 1989 Sara Clarke (2003) was doing yoga and watching television in her Anchorage home when the first images of the *Exxon*

113

Valdez spill flashed across the screen. Shocked, she collapsed in front of the television. Her mind raced: "No! Not Prince William Sound!" She had grown up in Alaska and the Sound was her favorite spot in the state to relax and recreate—"the heart of the watermelon," she called it. Now thirty-three, she had worked most of her life for the oil industry, in some form or another, and now she worked for Waste Away, Inc., an oil support contractor that cleans up hazardous wastes.

Waste Away, Inc. did not become involved with the cleanup until midsummer 1989 when Exxon offered the company a contract to monitor the waste disposal sites during the cleanup demobilization in September and to conduct the on-land impact assessments at the lay-down yards in Anchorage.

Clarke said, "One needs to realize that cleaning up after the cleanup also created a huge environmental impact. During demobilization Exxon planned to have cleaning stations to steam-clean and wash boats, booms, and other gear and equipment. Once cleaned, everything was to be hauled to Anchorage for over-winter storage. The storage sites needed to be checked for contamination before use and again after use to make sure there weren't any impacts to the property.

Clarke spent August, September, and October at different cleaning stations, monitoring runoff from the cleaning operations. The work entailed digging pits on the beach in the loose shale between the cleaning area and the sea to monitor the level of oil contaminants from the washing operations. She explained, "If there was significant leachate, we would build a catchment at the base of the slope so the contaminated water wouldn't go into the ocean."

Most of the boom and other gear were steam cleaned with the pressurized hot water wash. Initially the degreaser Citra-Solv was used to speed the process. Clarke had reviewed the MSDS for the solvent Citra-Solv and didn't have any health concerns (Sidebar 7, p. 115). Her company had designed a health and safety protocol for her work based on the use of this solvent.

But as the cleaning stations swelled with miles of boom and tons of other gear, a decision was made by Exxon and VECO to integrate Inipol into the cleaning process. The reasoning was simple: Inipol would both remove the oil and bioremediate (naturally degrade) any oil that flushed out of the cleaning station. By the time cleanup crews started using Inipol, the cleaning stations were in full swing and the monitoring crew had their routine. The health and safety protocol was never changed to reflect the increased risk of exposure to Inipol.

Clarke said, "VECO and Exxon said there weren't any exposure problems with Inipol. We were told it was like Citra-Solv—an organic cleanser. We were told Inipol was an organic cleanser." She believed them. She said, "It smelled like an organic cleanser when it atomized." She knew that Inipol had been applied to beaches in the Sound and elsewhere, so she assumed the chemical was "safe." With so much else to do, she didn't question the matter any further and instead plunged into her work. Literally.

One night she worked on a crew to dig and sample pits to monitor an all-night boom washing operation. From 9 P.M. to 5 A.M., while her co-workers drove backhoe and moved floodlights, Clarke worked in the pits with the Inipol-solvent-oil leachate dribbling down all around her. In the floodlights she could see the mist raining down from the boom steam-cleaning operations. She didn't have a respirator. Initially, she described the stench as "overwhelming," but she grew used to it. She collected a cooler full of samples.

Job completed, she went back to her hotel room and flopped on the bed to rest a bit before she jumped on the early morning plane to Anchorage. She noticed people looked at her strangely during the short plane ride to Anchorage. A co-worker met her at the airport—

Material Safety Data Sheet (MSDS) Excerpts for Citra-Solv

Citra-Solv (Citra-Solv, LLC, 2001)
Hazardous ingredients: D-Limonene
Inhalation: High concentrations are irritating to the respiratory tract; may cause headache, dizziness, nausea, vomiting, and malaise.
Skin: Brief contact may cause slight irritation; prolonged contact may cause moderate irritation or dermatitis.
Eyes: High vapor concentration or contact may cause irritation and discomfort.
Ingestion: May result in vomiting; aspiration of vomitus into the lugs (sic) must be avoided.
Health hazards (acute and chronic): acute effects are possible irritation and discomfort; no know (sic) chronic effects have been established. Harmful or fateful if swallowed. Vapor harmful.
Accidental release measures: . . . confine and absorb into approved absorbent; place material into approved containers for disposal; *do not was(h) (sic) to sewer or waterway* [emphasis added; Prince William Sound *is* a waterway.]

(Author's note: Miller and I are concerned about potential adverse human effects from this product and do not understand why it was used in situations where solvent aerosols were generated, increasing risk of inhalation, or where it could have drained into waterways.)

and burst into tears. "What's wrong?," Clarke exclaimed.

"You smell like you've been dipped in gasoline!" the co-worker said. "You can't smell yourself, can you? Your sense of smell is fried. Look at your hands!"

Clarke's hands were covered in blisters and her arms were starting to blister as well. She recalled thinking, "Whoa! What did I get into?" The co-worker rushed her home to shower and change her clothes.

After what she later described as her "really, really, really acute exposure," she continued monitoring the cleaning stations. She said she "knew atomized emissions were not good for us," but just like most of the other cleanup workers, she was caught up in the moment with all the tasks at hand. She said Waste Away, Inc. had "safety and health plans for everything," but she said those plans were based on "what we were told by our client." She explained, "You have this sort of insane trust that the people you are working for are going to tell you the truth about what you are being exposed to. We were really at the mercy of the information we received." She worked in the pits through October 1989.

Chronic Symptoms (1990 to 2003)

After Clarke's stint on the EVOS cleanup, she noticed she was increasingly sensitive to chemicals. She had increasingly frequent night sweats and trouble breathing. She was diagnosed with adult onset asthma. Her skin also started to peel off from her elbows down to her wrists every spring and fall. She joked about her "Exxon molt" and didn't see a doctor because it didn't hurt or itch and her skin always grew back.

During extended travel in Europe in 2001, she became very sick. At first, she thought it was some European flu bug that kept her in bed for three weeks. But one day she suddenly collapsed. She explained, "You know those little toys where you push the button and the figurine collapses? That was me." When she pulled herself up, she was disoriented and she no longer had control of her left leg. The whole left side of her face was numb and she had bouts of drooling. She thought she had had a stroke, but a MRI and neurological examination found nothing. The doctors told her she was stressed out and banned her from traveling.

At the end of that summer, she was still suffering from odd collapses, lack of energy—as if her "battery pack ran out," and brain fog that totally muted her ability to think. En route home to a new job, Clarke stopped in Washington, DC, to visit an acquaintance, Russ Jaffe, who had developed a health screen protocol for chemical poisoning. He ran a battery of tests on her and told her, "This is a chemical exposure."

She racked her brain. She knew she had been exposed to benzene, but that would have shown up in her blood samples. Russ Jaffe did not have the ability to test for polycyclic aromatic hydrocarbons (PAHs) from crude oil in the blood stream, but he told her that whatever she had been exposed to was "draining her system." Her thyroid was gone, her liver and kidneys weren't functioning properly, her adrenal system had completely shut down, and every essential mineral in her body was nearly non-detectable—at levels too low to measure with sensitive equipment. She described this as "like Rickets." She had lost nearly 60 percent of the bone mass—the calcium—in her jaw and three teeth had to be pulled.

When she returned home, her mother was there to greet her. Her mother took one look at her and burst into tears. "Oh my God!" her mother screamed. "What have they done to my daughter?" Word traveled quickly among Clarke's close-knit friends that she was dying. Her friends rallied their support.

She found she was too sick to work full-time at her new job and she went on medical leave in June 2002. Then, she said, "The race began to save my life." Her friends took her to see Brad Weeks, a medical doctor who treats holistically. At his clinic on Whidbey Island, Washington, Dr. Weeks told her, "I can't tell what you got into, but this is definitely an endocrine disrupter and your system is crashing. Fast. We have to get this stuff out of your system—now; and we have to aggressively rebuild your mineral system until your body can start generating these minerals itself."

Clarke started an intensive program of chelation therapy, which she describes as "homeopathic chemotherapy," and intravenous vitamin and mineral supplementation. Four rounds of chelation put her in bed for two weeks, then she'd be back for more. And she took, as she said, "lots and lots and lots" of different pills, mostly homeopathic, but some western medicine as well. Her body slowly responded.

Unable to work, she lost her job. In March 2003 while visiting friends on Bainbridge Island, she idly flipped through the new issue of *Mother Jones* and chanced upon the article on former *Exxon Valdez* cleanup workers who were struggling with chronic illnesses.

She felt as if she was reading her own medical charts. She realized that she "hadn't looked back far enough" in her exposure history, because in her mind in 1989 she was just exposed to "harmless" cleaning products—Citra-Solv and Inipol. But reading the *Mother Jones* article, the truth struck her "like a bolt of lightening." She finally connected her strange systemic illnesses to Inipol exposure.

Reflecting on her experiences Clarke said, "Corporations prey on public trust." She calls this "the most egregious and insidious emotional abuse and rape of the psyche and emotions of the human population" that she has ever seen. She added, "It's so subtle." She had believed as she had been told that there would be no exposure problems at the cleaning stations. Her naivete robbed her of a rich and wonderful life.

Today she is better, but it is a tenuous better and she will always have to work to maintain her health. For the short-term she hopes to work with Alaska Community Action on Toxics (ACAT) to develop a health recovery protocol for the unconventional health problems caused by chemical-induced illnesses from the *Exxon Valdez* cleanup.

Falling to Pieces—Captain Richard Nagel
Acute Exposure (1989 to 1991)

When the spill hit in March 1989, Richard Nagel, a licensed Master Captain, was discharged from Cook Inlet Marine, which operated mostly tugs and barges in Cook Inlet, and he went to work for General Marine Services (Nagel 2003). Within three days, he was in Prince William Sound, operating one of the company's landing crafts to transport boom where needed. He worked on the cleanup for three years. He said he would not have done that had he not thought he "could make a difference."

In 1989 Nagel worked as an advisor to Exxon and a supervisor for booming operations, because he had his supervisor's training for hazardous waste handling. Initially, he brought boom into Crab Bay, Evans Island, to support the fishermen's defense of the salmon hatchery in Sawmill Bay; then he moved boom to other salmon hatcheries. He secured areas for staging *"supersuckers,"* the large vacuum trucks brought down from the North Slope oilfields and mounted on barges to rapidly off-load oil skimmers. He moved fishing boats, boom, and skimmers into heavily oiled areas in response to VHF radio reports from airplanes or helicopters. All the while he was constantly moving, transporting boom, skimmers, garbage, people, whatever needed to be moved from one place to another.

During that first summer, he ran a dozen vessels or more with crews of six to twenty people. At first, the oil fumes were bad—in some of the bays where oil lay six inches thick on the water surface, he said, "You couldn't breathe right and your eyes would tear constantly." But over the summer, conditions improved. He was one of the last people out on the Sound in September, transporting garbage and old rain suits. He even carried a helicopter on his deck for Exxon officials to run final reconnaissance missions. He worked until his boat was washed and everything was put away for the winter. He kept a meticulous record of where he traveled, and for what purpose, but at the end of his cleanup stint in 1989, Exxon confiscated his logs, claiming his records belonged to the company.

In March 1990 Nagel was back out on the Sound as an advisor to Exxon and a member of the ad hoc Prince William Sound Advisory Committee, a citizen-industry coalition to oversee tanker traffic and terminal operations. He joined several survey teams to assess the status of oiled beaches after the winter storms and before the summer cleanup. He was impressed with Mother Nature's handiwork. He concluded, "The winter storms did more to remove oil from the beaches than the hot water spraying in 1989."

After completing the spring surveys, he worked on the vessel *Columbia*, shuttling people to and from beaches in the Bay of Isles, Knight Island, where they were conducting more field tests of Inipol. Nagel knew that the *Columbia* made its own drinking water by desalinizing water from the bay. He asked Exxon personnel whether the salt filter would also take out chemicals. He was handed the MSDS for Inipol and is sure he was told, "This stuff is harmless. You can eat it and it won't hurt you. Besides none of the chemicals are washing off the rocks—Inipol sticks 'like honey'." He retorted that the *Columbia* was "a stone's throw" away from the spray operations and he was "not as gullible as a lot of people on this boat." He didn't believe that the Inipol stayed on the beach or that the filters removed chemicals besides salt. Shortly after this incident, he broke his right kneecap in a fall on the beach and he was sidelined from cleanup work for several weeks.

When he returned in June, he worked for Pegasus Barge Company and operated the company's 80-foot landing craft, the

Pegasus, for the rest of 1990 cleanup. The boat carried a very large, heated and insulated tank of Inipol, which was used to supply the pontoon boats and the crews with the spray backpacks (Chapter 6). Nagel said, "When we would load up the pontoon boats, there would be spills constantly. But it didn't seem to make any difference, because the crews were going to spray it on the beaches and in the water. Everyone figured I was the supervisor, because I had the captain's license and the hazardous waste training. No Exxon supervisor ever told me what to do in the event of an Inipol spill and Exxon never provided safety training for people involved with transporting Inipol."

There were other Inipol spills as well. Several times when the tank on the *Pegasus* was being refilled by the larger offshore supply vessel, operated by Crowley Marine, the rigid transfer pipes broke and Inipol "went everywhere," according to Nagel. His crew just washed the chemical into the water. Neither he nor his crew wore respirators when they transferred the Inipol or cleaned up the spills. He said, "We knew that some of the people who sprayed Inipol on the beaches were urinating blood, but that never happened to us. We never had our urine checked, but we assumed everything was fine. I was naive then ..."

The *Pegasus* traveled far and wide during 1990, from the Sound to the Kenai Peninsula and beyond to the Barren Islands. It was the only boat that supplied Inipol to the pontoon boats as far as Nagel knew. Again, he kept a meticulous record of his travels and the boats he supplied with Inipol. This time, he kept two sets of records, one for himself and one for Exxon.

In mid-September the *Pegasus* was decommissioned in Seward. Nagel watched as the Inipol tank, half-full, was pulled off the boat with a large crane. Someone had forgotten to loosen one of the pipes and once again, Inipol showered everywhere—this time in the small boat harbor and on the nearby boats. The captain observed an Exxon supervisor shrug and say, "Well, it's harmless."

During the winter of 1990 to 1991, Nagel was sick on and off all winter with the flu and he was sick to his stomach "all of the time." That didn't stop him from returning to the Sound in May 1991 for one final summer of cleanup work. He worked as a second captain on a large vessel that shuttled people around to survey beaches. They did "spot-spraying" with Inipol, but for the most part, the oil had weathered past the point where 'bioremediation' was effective. By August he had had enough of the cleanup work and he quit. He resigned from the PWS Advisory Committee and left Alaska.

Chronic Symptoms (1990 to 2003)

Nagel traveled in Central America off and on for the next three years. His health started to deteriorate. He said, "It seemed like every time there was a flu going around, I always got it and it stayed with me longer than anyone else." He was in Costa Rica in 1994 when he became very, very ill. He checked into a hospital and discovered he had cancerous tumors in his stomach and intestines. During emergency surgery, part of his stomach and intestines were removed.

When he was in Portland, Oregon, teaching a Hazwoper class on early response and hazardous waste handling, a friend jokingly asked him if he would like some Inipol? Nagel responded, "Exxon or French?" His friend handed him the original MSDS on Inipol from the French company, Elf Aquataine. Nagel was stunned—the French MSDS showed that Inipol caused cancer in laboratory mice; the MSDS supplied by Exxon in 1989 did not. Exxon had supposedly altered the chemical composition of the product, however, the captain realized the time frame between product testing and approval was probably too short to determine if the revised product caused cancer in mice.

Instead of returning to Alaska, he moved to the southeast United States to better care for his burgeoning health problems. At the beginning of the cleanup in 1989, he described himself as weighing a "robust 260 pounds;" by 2003 he was down to 172 pounds and still losing weight. His right knee never healed properly despite eight surgeries. He was diagnosed with calcium breakdown and blood disorders including hypocalcemia (too little calcium in the blood) and polycythemia (high white blood count and low red blood count). He was also diagnosed with central nervous system symptoms including seizures, severe depression, acute anxiety, loss of balance, blurred vision, memory loss, severe migraine headaches, night sweats, and hot flashes. His driver's license and captain's license were suspended because of the seizures. Since August 2002 he has received 100 percent disability payments from Medicaid to defray some of his medical expenses.

His doctors have told him that he is dying. He has never been seen by occupational medicine doctors or doctors who specialize in chemical poisoning.

Buried:
Workers' Health Claims

Chapter 8.

Vanishing Claims

In this chapter, Pam Miller and I investigate the nature and source of health claims, reported by thousands of cleanup workers. We examine Exxon's records, which were discovered through one personal injury lawsuit (Chapter 3), and the records on file with the Alaska Workers' Compensation Board. We also find evidence that government health officials dismissed these claims instead of treating them as indicative of major health issues worthy of long-term monitoring.

Exxon's Records

The skimpy six pages of clinical data, obtained from Exxon through an extremely contentious discovery process in *Stubblefield v. Exxon* (1994) (Chapter 3), contain a world of information (Exxon 1989b). The clinical data show that a total of 6,722 cases of "Upper Respiratory Infection" (URI) were recorded by Exxon's medical team from the first week in May through closure of cleanup operations in mid September 1989. In medical circles, URI is a catchall phrase for common colds and other infections of the upper respiratory tract including the nose, throat, and pharynx. Even at first glance at the data, it was obvious something went horribly wrong during the 1989 cleanup, if roughly one of every two workers became sick enough to report respiratory problems.

Exxon refused to allow anyone to review or even glance at its clinical data—not the NIOSH investigators, not the state or federal OSHA representatives, not the Alaska epidemiologist, not the Alaska Workers' Compensation Board, and certainly not workers and the public. No one except Exxon and VECO knew the data even existed and the companies held these cards tightly to their chest. The six pages of data were peeled from Exxon's clenched fists by a court disclosure order in late 1992, well after the state and federal reports on the health effects of the cleanup had been completed. Before Exxon and VECO regained control of the documents and hid them from the public under a confidentiality order, the clinical and exposure data were obtained by a handful of people.

The first public airing of Exxon's horrible secret and the extensive health effects of the cleanup was an article by maritime reporter William Coughlin with the *Boston Globe* in April 1992(a). The issue held the media's attention for about a month, then it vanished from the news for the next seven years. In March 1999, Natalie Phillips with the *Anchorage Daily News* wrote a lengthy investigative article on chronic health problems of cleanup workers and concerns of government officials and occupational professionals about Exxon's worker safety program. In November 2001 the *Los Angles Times* carried another investigative piece by Kim Murphy and in March 2003 *Mother Jones* published a piece by Susan Stranahan. Each of these articles brought to light more wrenching stories on the human effects of the health disaster, but none further analyzed the data available in the hard-won six pages.

I tried and quickly became stonewalled by the word games. The main problem was that the "onshore" and "offshore" designations were legal artifacts based on where workers were housed, not where they worked (Table A.2, p. 451). Health claims from workers who were housed "onshore" were covered through the federal Longshore and Harbor Workers Act. Health claims from workers who were housed "offshore" in Exxon-contracted berthing vessels and other ships were mostly covered through the federal Occupational Safety and Health Act.

These designations were also only very general categorizations of relative risk of exposure. Many offshore crews worked on or near the beaches and were directly exposed to hazardous conditions; however, some offshore jobs, including divers, cooks, fuel barge attendants, and medical teams, were less exposed. Conversely, many of the onshore workers did administrative or other low risk work, while others handled oily waste, worked at boat-cleaning stations, or otherwise exposed themselves to similar risks as offshore workers. Unfortunately, the broad "onshore" and "offshore" designations in the clinical data set made it impossible to decipher, which, if any, of the various cleanup jobs posed more of a health risk to workers.

It seemed, however, that Exxon was also interested in this issue. The six pages of data included Exxon's analysis of respiratory illnesses relative to onshore and offshore (PWS) workers (Table A.2, p. 451). These comparisons show the average weekly incidence of reported respiratory problems was up to four times higher for PWS workers than onshore workers, 10.5 percent and 2.7 percent, respectively. Since about one-quarter of the offshore workers were located outside the Sound, I adjusted for this and the average weekly incidence of URIs in PWS workers dropped to 7.8 percent. The adjusted rate is still over three times higher than for onshore workers. This suggests, *in very general terms*, that cleaning beaches may have posed more of a health risk than "onshore" work in town.

In mulling over the data set, I realized that many more than 11,000 workers most likely were involved with the cleanup. I knew a lot of fishermen such as the Thornes (Chapter 1) simply quit the cleanup when fishing started. Many workers could have become injured or ill and been replaced, as evidenced by VECO's pink slip method of avoiding illness claims (Chapter 3). I concluded that Exxon's oft-cited number for the total cleanup workers, 11,000, is a low-end estimate, much like Exxon's oft-cited number for the spill volume, 11 million gallons. The truth about the numbers of cleanup workers—and potentially sick former workers, assuming some fraction of the total is chronically ill from cleanup exposures—lies in payroll records.

Alaska Workers' Compensation Board

I was determined to learn how 6,722 cases of potential work-related illnesses fell through the safety net of federal and state worker protection and compensation laws. For starters, only about one in twenty-five respiratory illness claims were reported to the Alaska Workers' Compensation Board ("Workers' Comp Board"). About 1,800 people filed claims for spill-related injuries and illnesses, starting as early as

March (Wilson 1991).[1] Roughly two-thirds of these were for injuries, mostly sprains, strains, bruises, or crushing. The remaining one-third or some 600 claims were for illnesses; of these, 264 claims were coded as respiratory problems.

Confusion arose among state and federal OSHA health officials and the U.S. Coast Guard over where and how to file claims. *Where* the claim was filed depended on whether the incident occurred on land, on the water, on the water but docked, or above or below the high water mark. *How* the claim was coded depended on where it was filed because the three systems were all slightly different.

To its credit, the Workers' Comp Board developed a special database for all the oil spill-related claims that allowed data entry of a broader scope of information than any of the "standard" systems (ADOL 1990a). The Alaska Department of Labor (ADOL or "A-doll") analyzed the claims as a data set to determine trends of work-related claims. Pam and I looked carefully at the ADOL report and data set to determine if we could learn anything more about the nature and source of these claims and to see if we agreed with ADOL's conclusions.

Chemical Poisoning Not Recognized

Under federal law (OSHA), work-related injury and illness claims are coded by *"nature"* (type; i.e., sprain, burn, fracture, etc.), *"source"* (cause; i.e., animal, ladder, cold, chemical, etc.), *"type of accident"* (fall, overexertion, etc.), and *"part of body affected"* (i.e., head, upper extremity, trunk, etc.), with breakouts for specific body parts. Occupational data were all new to me, but I noticed right away that the coding system seemed more set up to deal with physical injuries than illnesses. For example, the *"types of accident"* list does not have specific codes for exposures to oil, chemicals, and solvents.

This made sense from an historic perspective. When OSHA was passed in 1970, the concern at the time was to protect workers from physical injuries. Little was known then about symptoms associated with chemical-induced illnesses, which became a concern after the explosive growth of the chemical industry, starting in the mid-twentieth century.

Reported injuries and illnesses are usually categorized as time loss claims or non-time loss claims, with more of the latter in any given year. For example, 518 of 1,797 claims filed for the cleanup were time loss claims. The majority of claims, the non-time loss ones, are usually not coded and analyzed further. However, ADOL recognized the oil spill cleanup as "an unusual event of large scale" and coded all cases received to "provide a better understanding of the oil spill cases" (ADOL 1990a, 28).

ADOL compared the time loss claims from the cleanup to those from 1987 (the last year completed at the time of the spill) to look for unusual patterns of injury or illnesses.[2] With claims grouped "by nature," ADOL's analysis shows injuries occurred at about the same rate or at a lower rate during the cleanup compared to 1987. However, ADOL remarked, "The prevalence of work-related illnesses on the cleanup is notable" (ibid., 29).

I looked carefully at ADOL's analysis of the time loss claims, grouped by *"affected body part."* I saw there are nearly ten times more illness claims in general for cleanup workers (21.8 percent) than for the 1987 workforce (2.3 percent) ("Body System" in Table A.2, p. 451). Respiratory system illnesses are nearly twenty-one times higher among cleanup workers compared to the 1987 workforce; digestive system problems are fourteen times higher; and nervous system illnesses are double the norm.

I also noticed that ADOL seemed to have trouble deciding how to code some of the illnesses. Rates for "unspecified" illnesses of the body system are nearly six times higher for cleanup worker and rates for "non-classifiable" and "non-coded" illnesses combined are eleven times higher (Table A.3, p. 452). These observations suggested that the cleanup was unusual both in terms of nature and source of many health problems.

Using ADOL's grouping of all the illness claims (including non-time loss) "by nature" or type, I resorted the data into categories I would expect from oil inhalation exposures—respiratory damage, chemical symptoms, and central nervous system problems (Table A.4, p. 453). I anticipated that these categories might account for almost all the illnesses, but I was surprised: respiratory symptoms accounted for 43 percent; chemical symptoms for 14 percent; but central nervous system (CNS) problems for only 3 percent of the total. This didn't make sense, as CNS symptoms are one of the key indicators of crude oil inhalation. However, a whopping 35 percent of the symptoms were coded as "undefined."

Following a hunch, I called the author of the ADOL report, Jim Wilson (ADOL, Research and Analysis, Juneau, AK, pers. comm., 15 June 2001), who explained that work-related claims are coded without the benefit of medical personnel. "Undefined" illnesses include headaches, dizziness, burning eyes, ear and nose bleeds, and other classic CNS symptoms of exposure to crude oil and solvent inhalation.

I checked out ADOL's analysis of all the claims "by source," again focusing on illnesses and using my three prime suspect categories of chemical symptoms, respiratory symptoms, and "other" (as a surrogate for CNS symptoms). This time the data showed that the source for nearly 40 percent of the illness was unknown (Table A.5, p. 454). From its source analysis, ADOL concludes, "The major source of all respiratory cases was due to cold weather" (1990, 29). But one can't catch a cold from working in a cold environment. ADOL also concludes that petroleum was *not* a significant source of injury and illness claims, but this conclusion does not consider inhalation of oily vapors, mists, and aerosols as potential sources. According to Wilson, no one questioned whether the large number of predominately respiratory illnesses and "not classified" illnesses could have been caused by chemical poisoning rather than simple colds or flu.

The ADOL report concludes that the accident rate for the cleanup was not "unusual" relative to the 1987 claims. Miller and I disagree. We believe that ADOL overlooked chemical-induced illnesses as a possible cause of the respiratory problems and other illnesses, and we offer the following reasons to support our contention.

I could account for at least three potentially huge sources of error. First, the OSHA system was never designed to recognize or interpret occupational illness data from chemical exposures and it lacks more specific codes for illnesses, as evidenced by the high numbers of "undefined" or "non classified" claims. Second, the OSHA system also fails workers who do not understand the nature of their illness and so reported their claims inaccurately. And third, under the OSHA system illness claims can be inaccurately coded by staff without expertise in occupational medicine or industrial hygiene.

I couldn't help but wonder what ADOL might have concluded had the state known about the 6,722 cases of respiratory illnesses. If these respiratory illnesses had been reported, ADOL might have realized that the accident rate during the cleanup was very unusual and the state might have dealt differently with the sick workers claims.

Potential Chemical Poisoning Cases Dismissed

Miller and I enlisted the help of Barbara Williams with the Alaska Injured Workers' Alliance to determine the fate of the cleanup claims. We suspected that inasmuch as the OSHA system had failed to accurately recognize chemical-induced illnesses from the cleanup, the state system would also fail to adequately compensate workers for these illnesses. We figured if chemical-induced illnesses were *not* adequately compensated through "the system," there would be little incentive to improve workers' protection laws to prevent these illnesses from occurring in the first place. Little did I realize that we were stepping into a hornet's nest in our attempt to investigate workers' claims from the cleanup.

Each state determines the fate of its own injured workers; there is no federal oversight. The Alaska workers' compensation program was set up as a bargain that traded the legal rights of workers for an administered no-fault insurance system. According to a 1995 report, "Working without a Net," by the Alaska Public Interest Research Group (AkPIRG or "ACK-pirg"), the program does not work as planned. Instead of providing immediate compensation to workers without lengthy litigation, the program perpetuates "a grotesque fraud on working people and on an Alaskan public, which assumes that the program adequately addresses work place injuries . . ." (AkPIRG 1995, 3).

Suffice it to say that the Alaska workers' compensation program has been the target of repeated investigations, audits, and reports for the past quarter century (AK State Legislature 1999). The Alaska Workers' Compensation Act underwent sweeping revisions in 1988, one year before the spill, out of concern by businesses that the law favored injured workers at the expense of economic development. Official audits, sponsored by the State of Alaska, and reports sponsored by public watchdog organizations conclude that the current version of this act severely disadvantages workers.

The oil spill cleanup workers who filed claims with the state in 1989 and thereafter fell under the revised act. A quick look at the inner workings of the workers' compensation process, through the eyes of AkPIRG reveals what happened to most cleanup workers' claims. Under this system, injured workers "are placed in the care of insurance adjusters" who are essentially caseworkers, serving as the point of contact for all health care providers" (AkPIRG 1995, 19). Keep in mind that adjusters are responsible for *the insurer's* financial interests, not the worker's. Adjusters are bound by statue to a thirty-day deadline to process claims.

AkPIRG found that "process" means in most cases "controvert" or to determine that the worker no longer needs compensation benefits. According to my tenth edition of the collegiate Webster's dictionary, controvert literally means to engage in controversy, or to dispute a point by reasoning, something which adjusters are paid to do and injured workers are not. The small pool of workers' attorneys, whose fees average less than one-third of those of insurance company attorneys, report being quite familiar with the "wholesale controversion" tactics of insurance adjusters (AkPIRG 1995). Webster's dictionary contains no definition for "controversion." However, in the trade this means that an injured worker has either reached *"medical stability"* and is expected to be able to return to work, or the worker is rated for a permanent impairment.

"Medical stability," as defined under Alaska law (AK 2000), as "the date after which further objectively measurable improvement from the effects of the compensable injury is not reasonably expected to result from additional medical care or treatment . . ." The law recognizes that additional medical care may be *needed*. However, the law states, "Medical stability is presumed in the absence of objectively measurable improvement for a period of forty-five days," absent "clear and convincing evidence" to the contrary [AS 23.30.295 (21)].

According to AkPIRG, if an injured worker doesn't reach medical stability in the forty-five days allowed in state law, then an insurance adjuster can "controvert" (deny) all benefits, while claiming that the worker is able to return to work or permanently disabled. The forty-five-day window is barely enough time for a broken bone to mend and certainly not enough time for persons with chemical-induced illnesses to recuperate.

Chemical-induced illnesses often result in some degree of permanent impairment, sometimes to the extreme of complete disability, as in the case of La Joie and some of the Inipol workers (Chapters 5 to 7). Alaska law calls for the workers' compensation system to use the *American Medical Association (AMA) Guides to the Evaluation of Permanent Impairment* to determine the existence and degree of the impairment [AS 23.30.190 (b)].

While AkPIRG points out extensive failings of reliance on this guide (1995, 32-35), one of the largest problems, at least relative to cleanup workers, is that the AMA, as a whole, is enormously reluctant to acknowledge chemical sensitivity as a disability (Chapter 9). Many of the symptoms of chemical sensitivities and other chemical-induced ill-nesses are not rated in the AMA Guides. In Alaska, if an injury is not addressed in the AMA Guides, it is given a *"zero rating"* by the Workers' Comp Board. In the convoluted and circular reasoning characteristic of the controversion process, the Workers' Comp Board does not treat a condition with a "zero rating" as an impairment! (AkPIRG 1995, 34)

To further stack the deck against chemical-induced illnesses, in 1988 the Alaska legislature eliminated provisions that addressed impairments, which could be medically proven, but which were not found in the rating system. Injuries that are *not addressed* by the current ratings system include: chemical and environmental sensitivity; brain damage (such as brain fog, loss of cognitive abilities, loss of ability to concentrate, depression, and irritability); loss of strength, certain types of skin conditions, headaches, and chronic pain, among others (ibid., 34). Each of these symptoms is known to be associated with chemical sensitivity—at least by doctors who recognize this disease. None of the common chemical sensitivity symptoms, described by cleanup workers' and similar to those diagnosed by medical doctors in Gulf War vets (Ashford and Miller 1998), are recognized by the workers' compensation system in Alaska.

Once workers are controverted, many drop their claims, because they cannot afford to wait the mandatory sixty days without financial and legal assistance for a pre-hearing before the Workers' Comp Board. The state keeps no data on system dropouts.

However, AkPIRG reports, "Of the 30,000 workplace injuries in Alaska each year, 10,000 are serious enough to require recuperation time away from work" (1995, 31). AkPIRG assumed that a large number of these would ultimately seek remedy through the hearing process, because of widespread worker dissatisfaction with the lack of compensation. Between 1993 and 1995, AkPIRG found the average number of pre-hearings was 1,804, but an average of only 347 cases, or *3.5 percent of the total 10,000*, made it through the hearing process to some resolution! And this 3.5 percent includes cases that were ultimately denied by the Workers' Comp Board.

Given this history, I held little hope that *any* of the cleanup workers had been adequately compensated, much less the workers with chemical-induced illnesses. I followed the trail of a handful of them—Scott Roberts (*Roberts v. Veco* [1996]), Ron Smith (ADOL, AWCB 1992b), and Ed Meggert (ADOL, AWCB 1992a), among others—

through the Workers' Comp Board process. Meggert had an illustrative experience under a system that outwardly masquerades to compensate injured workers.

Plagued by poor health since his exposure to Corexit mist in July 1989 (Chapter 1), Meggert filed a claim for respiratory injury with the Workers' Comp Board, the next spring. Meggert was a state employee and he had been on a state job when he was exposed to the solvent aerosol. He harbored no illusions about the workers' compensation system.

Three weeks before he received official notice from the state that it received his claim (ADOL 1990c), he received an application from the state's insurance adjuster, Surety of Alaska, *denying his claim* (ADOL 1990b). When Meggert protested, Surety of Alaska's claim manager wrote that he wanted to make "doubly sure that we have treated you fairly," but the manager insisted that he had "received no medical information that would indicate that your respiratory distress was connected with your claimed exposure to Corexit. Yours is the only claim we are aware of concerning this product used during the oil spill cleanup" (Surety of Alaska 1990).

Meggert sent the insurance adjuster a copy of the disclaimer form that Exxon had paid people to sign (Figure 1, p. 33). He explained to the adjuster, there were "good reasons why you are not aware of any other claims" (Meggert 1991). He charged that Surety of Alaska acted as expected, "though a bit premature," he noted, since he was still undergoing medical examinations for his illness. His first doctor found no bacterial or viral component to his illness and sent him to a specialist, who, according to Meggert (2001), "poked around inside me for awhile and basically told me it was all in my head. There wasn't any damage there." Meggert decided his condition was more "aggravating than life-threatening," so he didn't further challenge the controversion notice. He let the matter drop.

Several years later, Meggert (2001) noticed that exposure to cigarette smoke in enclosed areas would make him "feel like hell" the next day. He became prone to sinus infections, which he never had before the spill cleanup, to the point where, during the winter, he "just pretty much had one steady sinus infection." He developed "very mild asthmatic symptoms" when he walked outside in cold weather. At first, he thought all these problems were "just a normal side effect of getting older." But then he ran into some other people who had been out on the beach during the ill-fated Corexit test and they were having similar—or worse—health problems.

Luckily for Meggert, his health problems, though annoying, were not disabling. Others weren't so "lucky."

Of the original 1,797 *Exxon Valdez* oil spill worker injury claims filed with the state, 440 claims were covered by the federal Longshore and Harbor Workers' Act. Applying the dismal 3.5 percent to the remaining 1,397 claims handled by the state suggests that less than fifty cases may have been "resolved" by the Workers' Comp Board—and not all of these with positive results for workers as in the case of Scott Roberts (Chapter 9). Most injury and illness claims that stemmed from the cleanup simply vanished, like Meggert's case, leaving no clues that there could be persistent health problems.

Federal Investigation Fizzles

I turned next to the NIOSH (1991) Health Hazard Evaluation Report to determine what the federal investigators learned of cleanup workers' exposures and health problems, and what they recommended to address these issues.

In April 1989 at the request of the Laborers' International Union of North America (1989; Laborer's National Health and Safety Fund 1989a, 1989b), the Alaska State Health Department, and the U.S. Coast Guard, the National Institute for Occupational Safety and Health (NIOSH) to decided to conduct a health hazard evaluation during the *Exxon Valdez* oil spill cleanup. NIOSH investigators visited Alaska three times in 1989, during April, June, and July. The team's findings regarding Exxon's worker safety program are discussed in Chapter 2.

According to the NIOSH report, "During July and August 1989, a medical epidemiologist assigned to NIOSH's Division of Safety Research attempted to conduct a systematic record-based review of illness and injury information . . . [H]owever, attempts to collect this type of information were largely unsuccessful" (1991, 6). This was not entirely the investigators' fault. Exxon obstructed the NIOSH investigators' planned study by refusing to provide access to workers on remote beaches (Phillips 1999).

The NIOSH investigators tried other approaches to collect work-related injury and illness data. "A sample of medical records at the hospital in Valdez [the major community provider of health care] revealed a variety of injuries and illnesses among oil spill workers, but the relatively low proportion of VECO [the major contractor involved in the cleanup of the oil spill] employees among these workers suggested that the latter were not representative of the workforce. Records at the hospitals in Anchorage were not filed in a way in which those involving visits related to the oil spill could be readily retrieved. A questionnaire survey of a portion of the oil spill workers was planned, but logistic difficulties prevented its timely implementation" (NIOSH 1991, 29–30).

NIOSH investigators were aware "of the dramatic increase in upper respiratory tract illnesses among workers and residents of Valdez" and they noted in their 1991 report, "Upper respiratory infections among workers were reportedly common." They reported that the Alaska Department of Health seemed to think that this outbreak of illnesses was caused by a virus, although the state epidemiologist later told a reporter that the state's lab had found no unique influenza viruses in Valdez (Phillips 1999). The NIOSH team was aware of "at least one reported incident of acute, self-limited, irritant and neurologic[al] symptoms affecting several workers who may have been exposed to crude oil" (NIOSH 1991, 29). It was certainly not hard to learn of other group chemical exposures (Chapters 1 and 6), but NIOSH chose to believe that the rapid spread of the respiratory illnesses was "presumably facilitated by the crowded living conditions on some of the vessels used for housing" (ibid., 29). Based on available data, the NIOSH team decided, "Routine periodic medical testing of the workers was not conducted and did not appear to have been warranted" (ibid., 30).

The NIOSH team (1991) also toyed with the idea of evaluating worker exposure to weathered crude oil, particularly the hazardous polycyclic aromatic hydrocarbons (PAHs), using *"biomarkers,"* chemical or biological evidence associated with specific chemical exposures and detected by sophisticated blood work analysis. They identified "no plausible technique" for a biomarker analysis and found only "trace concentrations (10–31 parts per million)" of PAHs in the weathered crude oil, so they dropped the matter (30).

It is true that the use of biomarkers to identify minute traces of chemicals parked in body organs and to track the biological effects through analysis of blood chemistry was in its infancy in 1989. However, techniques certainly existed to pinpoint crude oil and other chemical exposure in human blood in 1989. Commercial labs like Accu-Chem Laboratories in Dallas, Texas, had an extensive database on biomarkers indicative of various chemical exposures in humans. In fact, several sick workers used this lab three years after the cleanup to identify evidence of crude oil and solvents in their blood in preparation for personal injury lawsuits (Chapters 4 and 9). Also scientists studying the persistent effects of oil in wildlife also used biomarkers extensively (Chapter 18). Their biomarker work occurred at minute levels of oil—low parts per billion—or *10,000 times lower than the supposedly "trace" concentrations* identified by the NIOSH investigators.

A properly conducted exposure assessment on cleanup workers in 1989 with biomarkers would have been a seminal and precedent-setting study with potentially enormous contributions to understanding human health effects of PAH exposure. But it did not happen.

Instead, unable to conduct their systematic, record-based field evaluation of worker conditions and health as planned, NIOSH officials were forced to rely on Exxon's air quality monitoring data for exposure assessment and Exxon's clinical data for evaluation of health effects. At the end of the summer, Exxon and VECO refused to voluntarily release their records to the federal government. NIOSH officials never subpoenaed Exxon or VECO for the records, citing lack of staff to pursue the case (Phillips 1999). Two years later, VECO officials claimed they had "no recollection now of anyone denying access to medical records" (Murphy 2001).

The NIOSH investigators concluded, "*Based on available data*, there is no basis for recommending long-term medical surveillance of the health of the workers involved in the cleanup of the oil spill" (1991, 30, emphasis added).

This statement reveals the lack of concern rampant among those whose job it was to care for cleanup workers and the circular logic that created no apparent need for follow-up medical care. NIOSH investigators based their conclusions on information that was "available." However, the investigators made little effort to get all of the available information. They failed to subpoena Exxon to obtain its exposure data (discussed in Chapter 3) and its clinical records—the 6,722 respiratory claims. They superficially glanced at the cleanup claims filed with the Alaska Workers' Compensation Board, stating only the obvious in one brief paragraph. And then they parroted the state's conclusion that the respiratory claims "consisted primarily of bronchitis-type, rather than chemical-induced, illnesses" (ibid, 31).

Chapter 9.

Toxic Torts and Justice Denied

In this chapter, Pam Miller and I investigate the history of one of the many ill-fated personal injury lawsuits filed in the aftermath of the 1989 cleanup. Chemical-induced personal injury cases are known as *"toxic torts."* In this case, *Roberts v. Exxon* (1999), we examine the medical and legal hurdles that tripped sick workers, one by one. We find evidence that justice was ultimately denied and learn of the consequences for some of the injured people.

(Author's note: I reconstructed this story from court records on file at the federal courthouse in Anchorage, AK. Two-digit numbers refer to motions, orders, or other material from Roberts v. Exxon *[1999].)*

Against the Odds

In January 1994, Ron Smith and Dolly La Joie joined two other injured workers—Scott Roberts and Richard Merrill—and filed a lawsuit in California through Melvin Belli's office. They sued Exxon corporation, its shipping company SeaRiver Maritime, and VECO for "severe personal and bodily injuries . . . resulting in permanent disability," stemming from the oil spill cleanup (No. 1, 3). Their lawsuit claimed Exxon and VECO were negligent in failing to provide proper protection or supervision, or even warning, that the chemicals, petroleum, and other substances were dangerous and damaging to health. Their lawsuit also claimed Exxon and VECO were negligent under general maritime law for subjecting the plaintiffs to dangerous conditions. Each plaintiff asked for two million dollars.

They were very optimistic about their chance of receiving a favorable award. After all, Belli was the legendary 'king of torts' with a fabled career making millions of dollars defending injured victims against doctors and insurance companies (*California Bar Journal* 1996). Smith and La Joie knew chemical sensitivity was real and they understood that the oil spill cleanup work had made them sick. Smith had one of the best environmental doctors in the country, if not the world. But there was a lot that they didn't know, things that would have a direct bearing on their case. For example, when they filed their lawsuit in 1994, they didn't realize that the justice system simply did not yet support the victims' demand for justice in most toxic tort cases (King 1999). Credible, independent scientific studies on chemical sensitivity were scarce. So were studies definitively linking specific chemicals as the cause of tested proven diseases (Kanner 1999). It often takes years for diseases like cancer to show and factors like smoking, diet, pre-existing conditions, and genetics confound the linkage to a specific substance or event (Rea 1995). The prevailing medical opinion holds that chemicals are harmless unless proven otherwise. Further, there is a deep-seated societal bias against the recognition of chemical illnesses in general, primarily because of liability concerns by the petrochemical industry, health insurers, and the federal government.

The medical profession with its attendant drug industry and insurance providers has grown huge, and it has much to lose by people and doctors who claim that debilitating illnesses are caused by chemicals at trace levels far below regulated levels. The drug industry is part of the revolution in synthetic organic chemical production, which had soared from less than 10 million tons per year in 1945 to over 100 million tons by 1980 (Ashford and Miller 1998). Over 1,000 new chemicals a year were entering U.S. markets by the 1990s. Production—and profit—overwhelm common sense and societal safeguards. The health effects and toxicity of the vast majority of the chemicals in commerce—drugs, pesticides, cosmetics, food additives—are unknown, and what is known was from research mostly sponsored by the drug and chemical industries. People with chemical sensitivities are the undesired and silent side effects of this chemical-based prosperity.

Chemical sensitivity is an emerging area of both medicine and law, and it is extremely controversial in both arenas (Liberman, DiMuro, and Boyd 1999). The often vague and multi-system symptoms associated with chemical sensitivity pose a problem for health care professionals and create a field day for corporate lawyers. Environmental doctors such as Dr. Rea find themselves constantly pitted against legions of other physicians, primarily allergists and psychiatrists who believe that chemical sensitivity is psychological, a view that serves corporate interests and infuriates patients. Professional jealousies and rivalry from the traditional practices, inflamed by corporate lawyers, hamper understanding of the new disease process and hold patients and public policy hostage (Hileman 1991).

Few succeed in overcoming these huge odds. Those who do find that it takes dedicated lawyers, putting in lots of time and effort and working cooperatively with plaintiffs, to win these cases. Further, successful plaintiffs find they, themselves—not their lawyers—are their own best advocates. Constant vigilance and active participation are critical to winning. As other chemically poisoned people discover, "Polite clients end up paying!" (King 1999, 11).

Smith and La Joie learned this the hard way. They made several critical mistakes. They trusted their lawyers to handle their case, while they watched expectantly from the sidelines. They didn't know that Belli's law firm was imploding from a series of suits and counter suits between the aging Belli and his younger partners over the direction of the firm and financial management.

The California court where the injured workers filed *Roberts v. Exxon* (1999) entered a stay in July 1994, ruling that Alaska was the only appropriate venue for the case. When their attorneys took no further action during the next eighteen months, Exxon and VECO tried to have the case dismissed for lack of activity. The case was on the verge of being dismissed when their lawyers finally moved it to Alaska federal court in February 1996.

The lawyers' tardiness was foreboding and it shook Smith's confidence, because he knew the same lawyers had also been similarly inactive in his friend Lowe's case—with disastrous consequences. After the lawyers had failed to comply with a court order to compel responses to Exxon's discovery requests, the judge had dismissed Lowe's case *Payne v. Exxon* (1997) in its entirety in December 1995.

An appeal to the Ninth Circuit Court was unsuccessful.

Then, during the same week that Lowe lost his lawsuit, Belli filed for bankruptcy protection to reorganize his failing law firm (Carlsen 1995). When the plaintiffs' lawyers finally moved *Roberts v. Exxon* (1999) to Anchorage in early 1996, the crumbling Belli empire was consuming much of their California lawyer's attention.

However, the injured workers picked up Alaska co-counsel with the change in venue and, for the next several months, the new lawyers with Houston & Henderson fought off many attempts by Exxon and VECO to have the case dismissed on various legal technicalities. Exxon and VECO were so tenacious in their effort that Judge Russel Holland, who had been assigned the case, finally wrote, "[I]t is preferable that cases be decided upon their merits rather than technicalities" (No. 18, 4). Judge Holland was familiar with Exxon's legal tactics, because he was also handling the massive consolidated case brought by fishermen, Natives, communities, and others against Exxon for oil spill damages.[1]

In July, Belli passed away quietly from pancreatic cancer. Shortly after his death, a federal court transferred seventy-six of Belli's civil cases, including *Roberts v. Exxon* (1999), to Daniel Stenson with the law firm of John E. Hill, Belli's former partner, which had been handling the case all along (Brazil 1996).

Playing the Odds—Corporate Defense Strategies
Statute of Limitation

When Exxon and VECO couldn't get *Roberts v. Exxon* (1999) dismissed in its entirety, company lawyers switched tactics and focused on dismissing the individual plaintiffs one at a time. Their first target was Scott Roberts, who claimed health problems from being hit in the ear with Customblen during the cleanup. Customblen is a bioremediation product with slow-release pellets that contain nitrogen and phosphorus fertilizer. Roberts had filed a claim with the Alaska Workers Compensation Board ("Workers' Comp Board"), but the board had denied his claim after a hearing (ADOL, AWCB 1995).

In September 1996, Exxon and VECO moved to dismiss Roberts' claims for the same reasons that the Workers' Comp Board had dismissed his case earlier (No. 26). Exxon and VECO argued—again—that Roberts' claim was time-barred; that is, it exceeded the two-year statute of limitations for filing a claim once a person was aware of a job-related injury. Exxon and VECO also argued that since Roberts' claim had already been litigated and decided by the Workers' Comp Board, it couldn't be litigated again.

It took a year, but ultimately Exxon and VECO were successful. In 1997, Roberts' claim was dismissed with prejudice on these legal technicalities—and only oblique mention of the injuries that triggered the lawsuit in the first place (No. 37). A year later in 1998, the Ninth Circuit Court upheld the lower court's decision (No. 83). Meanwhile, Richard Merrill, another plaintiff in the case, saw the handwriting on the wall and quietly settled out of court for $10,000 (No. 57).

Pinning the Blame on Something Else

La Joie and Smith, the only remaining plaintiffs in *Roberts v. Exxon* (1999), proved to be more of a challenge for the Exxon and VECO lawyers. The central challenge involved the heavy use of solvents during the oil spill cleanup. Neither Exxon nor VECO had produced any records for monitoring worker exposure to the cleaning solvents De-Solv-It, Citra-Solv, CitroKleen, Simple Green, and others. This created a potentially huge liability problem that was difficult to dodge.

Occupational medicine practitioners widely recognize health problems from solvents (Ashford and Miller 1998; Wilkinsen 1998). Solvents are "lipophilic," literally "fat-loving" compounds that readily dissolve into fatty tissue and nerves, which are insulated with a fat compound called myelin. Solvents are known to affect the autonomic nervous system, especially the respiratory center, and cause breathing to become labored. Respiratory distress was the chief complaint of cleanup workers in 1989 (Chapter 8). Solvents are also known to act directly on the brain and central nervous system; affect thought processes, particularly memory and concentration; and cause changes in mood and personality—precisely what Smith and La Joie, among others, had experienced after the cleanup.

To prevail, toxic tort litigation does not require proof of medical causes, but rather proof that the person is disabled and that *some* action of the defendants caused the disability (Prosser and Keeton 1984, section 52, Chap. 8 passim). Corporate defense lawyers have devised several strategies to maneuver around the disability problem—sick people are often convincing to juries (Rachel's

Environment and Health Weekly 1999). They troll for information during *"discovery,"* (a legal process during which evidence is collected) to buttress their arguments. In *Roberts v. Exxon* (1999), the defense lawyers were searching for any other possible cause of the plaintiffs' health problems, on which the lawyers could pin the plaintiffs' disabilities and for which Exxon and VECO were not responsible. Discovery is a bit like a fishing expedition: one knows what one is looking for, but one never knows what one will catch.

The corporate lawyers' strategy for dealing with Smith was to confuse his cleanup-related health problems with other chemical exposures prior to the cleanup and with physical injuries since the cleanup. Smith's occupational exposure history was a virtual smorgasbord for the corporate lawyers (Smith 1996). The list included pesticides and insecticides from a farm job; chemical, metal, paint, and oil smoke and dust from welding; sandblasting dust; gas, diesel, and machine oils from working with heavy equipment; PCBs from digging contaminated dirt, among other things. Exxon's lawyer also found Smith had been exposed, at home or elsewhere, to other substances known to trigger reactions in chemically sensitive people. This list included spray paints; formaldehyde off-gassing from his trailer home; cleaning agents such as Mr. Clean, Formula 409, and Gunk; battery lead, when he poured lead fishing sinkers as a kid; paints of all sorts; snuff; insecticides sprayed around the home; pets; house plants . . . . The list went on and on (Sidebar 8, below).

The Chemical Deluge

The Occupational Safety and Health Administration (OSHA) estimates that there are 650,000 chemical products, with hundreds of new products introduced annually. About 32 million workers handle and are potentially exposed to one or more chemical hazards (OSHA 1998, 1). Current scientific research demonstrates that exposures to extraordinarily low doses of certain chemicals can result in serious health problems including nervous system damage, reproductive disorders, endocrine disorders, immune system effects, and birth defects. Virtually nothing is known about health hazards from chemical mixtures.

Congress enacted the Occupational Safety and Health Act in 1970 to ensure safe and healthful living conditions for every American worker over the period of his or her lifetime. Under provisions of the law, employers must inform employees about the identities and hazards of workplace chemicals to which they are exposed. Yet, OSHA has set permissible exposure limits (PELs) for only 600 industrial chemicals (Gordon 1998, 65). OSHA has developed fewer than 30 health standards since 1971 (Schettler et al. 1999, 252). A 1997 study by the Environmental Defense Fund reported that basic information was available on environmental and health effects for no more than 29 percent of the highest production-volume chemicals that are produced or imported into the United States (Geiser 2001, 179).

Then, there was his plane accident. In response to a series of transparent questions designed to shift the responsibility for Smith's health problems from Exxon and VECO to the plane accident, Smith finally exploded in exasperation. "I'm claiming that not long after the spill my whole body started going downhill at a high rate of speed, and it ain't stopped yet" (Smith 1996, 98). When the lawyer persisted and asked him point blank if he was claiming that his occasional chest pains had anything to do with his oil spill cleanup work, Smith exploded again. "I feel it all adds up to be tied together. I think I get under stress. I feel it causes me to have chest pains. I feel that's . . . because it's deteriorated my health, and I can't do what I want to do. In other words, the mind is willing, but the body is not, and sometimes I get real frustrated with myself, frustrated with other people, and just . . . highly stressed out." "I feel like [the oil spill cleanup work] has broken down my whole system. Just run me down. The chemicals are slowly eating my whole system away" (ibid., 104–106).

The Exxon and VECO lawyers knew exactly what they were doing. Despite Smith's very convincing story, the lawyers, acting shrewdly within the law, had found plenty of loopholes that would work in court to relieve their clients of responsibility for Smith's injuries. They next turned their attention to Smith's only expert witness.

Legal Hacking of Chemical Sensitivity Claims

In January 1998, the Exxon and VECO lawyers took Dr. Rea's deposition in Dallas. Dr. Rea was a reluctant expert witness. He regarded trials and depositions as counterproductive. They stole time from his precious work, and the court process was biased to advantage wealthy corporations and deny justice to those whom he devoted his life. During the previous decade, the doctor had given only some two hundred depositions for his patients, and he had a good sense for which ones were complete wastes of his time based solely on the lawyers. Dr. Rea didn't hold much hope for Smith, because his lawyer did not even bother to show up at the doctor's deposition (Rea 1998, 4–5).

Dr. Rea's deposition proceeded in a manner very familiar to the doctor. Corporate defense lawyers were always very good, well prepared, and well-versed on chemical sensitivity. They knew how to play the odds to win.

The Exxon and VECO lawyers focused on the two medical tests, which showed that Smith's injuries resulted from chemical exposure during the oil spill cleanup. The Accu-Chem blood test (1992b) showed high levels of chemicals associated with exposure to crude oil and solvents (trimethylbenzenes, 1,1,1-trichloroethane, 2-methylpentane, 3-methylpentane, and n-hexane). The single-photon-emission computed tomography (SPECT) brain scan showed darkened areas of Smith's brain where these chemicals had cut off blood flow, interfered with the brain's function, and produced some of Smith's symptoms such as short-term memory loss (Hickey 1992). Dr. Rea had not lost a court case since he and a colleague had pioneered using the SPECT brain scans for chemical sensitivity testing. The Exxon and VECO lawyers knew this, so they tried to get the doctor to admit that a multiplicity of other things could have led to the same test results and symptoms, and that Smith's tests weren't all that unusual compared to control populations.

It was hard for the lawyers to get around the high levels of solvents and trimethylbenzene in Smith's blood work with Dr. Rea insisting that "those levels in a person who's not in the middle of an oil field or oil refinery don't accumulate like that" (Rea 1998, 56). Trimethylbenzene and n-hexanes are associated with crude oil exposure. Dr. Rea had seen these compounds in Gulf War vets. So had other physicians (Chapter 10).

The Exxon and VECO lawyers were well aware of the sick Gulf War vets. They switched tactics. They knew people could get solvents, the methylpentanes, from a wide variety of common sources—glues, gasoline, degreasers such as Gunk—all of which Dr. Rea confirmed. The lawyers specifically avoided mentioning degreasers used on the cleanup—De-Solv-It, CitroKleen, and Simple Green. Trimethylbenzenes are found in paint thinners, perfumes, dyes, and fuel additives, and n-hexane, which is in glues, air fresheners, degreasers, paints, penetrators for pesticides, varnish, solvents, and gasoline. Exxon's lawyer asked specifically, "How does 1,1,1-trichloroethane get into people *who aren't oilfield workers*?" (Rea 1998, 48, emphasis added). Dr. Rea answered, "Degreasers, dry cleaning fluids, paint," and even decaf coffee up until at least 1993 (ibid., 48–49).

The lawyers fired a volley of diversionary questions. The medications Smith took after his plane accident—the steroids, painkillers, mood elevators—could they add to his body load? Welding smoke—could that cause neurological and respiratory symptoms, and even an abnormal SPECT scan, in someone who was already chemically sensitive? Tobacco, specifically snuff—could that contribute to chemical sensitivity? Could a combination of painkillers and welding fumes cause chemical sensitivities in a susceptible individual? Could not wearing any kind of respiratory protection while digging around PCB-contaminated dirt cause chemical sensitivities? Wasn't the average home the most polluted place commonly encountered in the environment? Dr. Rea confirmed that all these things could contribute to chemical sensitivities (Rea 1998, 23–25, 57).

Dr. Rea knew control populations were a problem. People are exposed to so many chemicals in their daily lives that it is almost impossible to find unexposed individuals. Dr. Rea had tried to find what he thought of as "normal" people to develop a baseline of control brain scans for his SPECT test. But finding such people—no flu, no other sicknesses, no medication, no dope, no chemicals—in today's society was much harder than he thought. Out of three hundred people from the University of North Texas, only twenty-five proved to be "normal."

Accu-Chem Laboratory's control population makes it easy for corporate lawyers to make it look like the "average" person should have symptoms similar to Smith's symptoms, which Dr. Rea interpreted as indicative of chemical sensitivity. Dr. Rea had treated over 26,000 people and he had seen a lot of diverse symptoms at very low levels of chemicals, including at the Accu-Chem control levels.

As for the Accu-Chem control population, Dr. Rea insisted, "[I]t's meaningless," "a totally invalid set of statistics" (Rea 1998, 15, 44), and he had tried to get Accu-Chem to take it off of its reports. It is the patient population, not "normal" individuals. It was the several thousand sick people who had blood tests from all over the country, people from industrial settings, housewives, students, hospitalized people, white collar workers. Smith's blood work, for example, became part of Accu-Chem Laboratory's "control" population average.

For example, at the Accu-Chem control levels for the solvents in Smith's blood (toluene, 2-methylpentane, and 3-methylpentane), Dr. Rea had observed people with neurological symptoms such as short-term memory loss, headaches, and imbalance; heart arrhythmias or irregularities; gastrointestinal upset, gas, and bloating—and those with no physical symptoms at all. The symptoms depended, of course, on the person's general state of nutrition, total body load of chemicals, and genetic make-up, among other things, as Dr. Rea—and the Exxon and VECO lawyers—knew.

Finally, the corporate lawyers wanted to know if a person worked with hypothetical "sulfur" compounds (their surrogate term for oil and solvents) in 1989 and started having headaches in 1993, would that period of time be so long that the doctor would question whether the exposure caused the headaches? And could intervening exposures, activities, and life events possible have caused the headaches?

Dr. Rea affirmed most of these questions (Rea 1998, 46–50), but he went on to state firmly that it is possible—"almost always" (54)—to determine what is causing a patient's sensitivity problems by their environmental exposure history and diagnostic tests.

Psychiatric Diagnoses and Other Remedies

Exxon and VECO lawyers devised different strategies for dealing with La Joie, whose health problems seemed to center around her exposure to solvents (Sidebar 9, p. 149; Figure 3, p. 79). VECO lawyers managed to dodge the solvent issue entirely by arguing La Joie had "other remedies" to pursue her claims. She had filed a Workers' Comp claim against VECO with the federal Department of Labor in January 1996. When the VECO lawyer asked why she had waited so long, she responded that she had been told that it was already too late to file a claim when she first realized in 1992 that her health problems were connected to the oil spill cleanup (La Joie 1996, 66). She decided to do it anyway, but not then, when she was too sick or cash-strapped to deal with the claims forms.

Despite the fact that her Workers' Comp claims were filed too late to be legally valid, the VECO lawyer used the very act of filing the claims as an excuse to dismiss her toxic tort claim against VECO. He successfully argued that La Joie had "other remedies," even though she clearly did not: the Workers' Comp Board formally denied her claim *four months before her deposition* (ibid., 64–65).

Meanwhile, the Exxon lawyer planned to use La Joie's own doctors to undermine her powerful story. The Exxon attorney was particularly interested in Dr. Hong, because the doctor had been the primary caregiver for La Joie for nearly two years and he did not believe that her health problems had anything to do with the *Exxon Valdez* oil spill. Dr. Hong also believed that his patient was absolutely convinced that her problems originated from the cleanup and so, in a sense, she was not lying to him about her symptoms (Hong 1997, 71). Rather, Dr.

Hong had concluded that his patient's health problems were delusional (ibid., 73–75). This diagnosis played right into Exxon's hand.

During Dr. Hong's deposition, the Exxon lawyer made a point to show that the doctor had not come about this diagnosis lightly. He had followed the standard general protocols he had been taught in medical school and practiced by Kaiser physicians when he treated La Joie (ibid., 18). First, he listened as his new patient described her history of health problems, then he tried to objectively diagnose her symptoms through a battery of laboratory tests. Next he assessed the nature of the problem, based on the subjective information and the lab results. Then he worked up a plan for La Joie to treat her symptoms.

Despite a series of lab tests, referrals to other specialists, and prescribed treatments, Dr. Hong could find no medical explanation for the symptoms La Joie described. The neurologist suggested that she

Solvents and Chemical Injuries

In the book *Generations at Risk*, the authors review the physical properties of solvents that enable them to readily enter the human body: "They evaporate in air at room temperature and are therefore easily inhaled; they penetrate the skin easily; and they cross the placenta, sometimes accumulating at higher doses in the fetus. In addition, many solvents enter breast fat and are found in breast milk, sometimes at higher concentrations than in maternal blood. Solvents contaminating drinking water enter the body through skin absorption and inhalation in the shower, as well as through drinking. In fact, the total exposure from taking a ten-minute shower in contaminated water is greater than the exposure from drinking two quarts of the same water. Solvents are generally short-lived in the body, lingering for no more than several days. On the other hand, exposures may occur daily" (Schettler et al. 1999, 74–75).

Synthetic hydrocarbon-based (organic) solvents accumulate in fatty tissues because they are soluble in fat. They are used as degreasers and are found in dry cleaning chemicals, paints, pesticides, and pharmaceuticals. Organic solvents include alcohols, acetone, aromatic hydrocarbons such as benzene, chlorinated hydrocarbons such as trichloroethylene, and glycol ethers such as 2-butoxyethanol.

Exposure to solvents can cause a range of ill effects including damage to the skin, liver, central nervous system, lungs, and kidneys (Harte et al. 1991, 110). Certain solvents can inhibit blood cell production. Many solvents are carcinogenic. Recent animal studies show that glycol ethers can cause birth defects, testicular damage, infertility, and failed pregnancies. In people, exposed workers experience lowered sperm counts. Women experienced reproductive problems including increased infertility and risk of miscarriage (Schettler et al. 1999, 91). Exposure to solvents, especially through inhalation, can lead to systemic symptoms (experienced throughout the body), as the MSDS for Corexit 9527, for example, warns (Sidebar 3, p. 12). Individual tolerances to chemical exposures vary as do the symptoms (Figure 3, p. 79; Rea 1992, 35–40).

had "hypersomatization syndrome" (ibid., 36) which Dr. Hong interpreted to mean that the vague symptoms "were likely of a psychological source" (ibid., 36) or in her mind rather than something
physical. Dr. Hong diagnosed La Joie with adult onset diabetes, likely caused by a combination of her being overweight and her genetic make-up. However, he was not able to control her diabetes with
medication (ibid., 56).

At his patient's insistence, Dr. Hong had sent blood samples to
the Accu-Chem lab in Texas. When the blood tests showed abnor-
mally high levels of hydrocarbons and other chemicals, he referred La
Joie to a Kaiser Permanente industrial medicine doctor. This specialist reported that he didn't think the elevated levels of chemicals
could account for her symptoms and he recommended more psychological testing. The psychiatrist concluded, "Despite certain blood
findings I think of her as either severe hypochondriasis or a delusional disorder, somatic type" (ibid., 73). Dr. Hong explained this basically meant that this specialist thought many of her illnesses were,
essentially, all in her head (ibid., 24, 73).

There were several things La Joie's attorney, Daniel Stenson,
could have pointed out in her defense—if he had been present. Dr.
Hong had done the best he could based on what he had been trained
to see and the information he was given. He had never seen La Joie
when she was really sick with bouts of flu-like symptoms, fever, nausea, and pounding headaches. She was too sick to be seen then.
Instead, she always waited until she was feeling better before she
went to the clinic, where she appeared to Dr. Hong to be "asympto-
matic," or not experiencing the horrendous problems she described.

Dr. Hong was not trained in chemical poisoning or occupational
medicine and he didn't know to look for or recognize chemical-
induced illnesses. This is not a reflection of the inadequacy of his
individual training, but rather it is illustrative of the fact that medical
training was far from adequately preparing doctors to meet the
needs of a population exposed to thousands of chemicals. Lacking
the necessary training to properly diagnose and treat a chemically
sensitive patient, he had tried to fit her symptoms and descriptions
into his repertoire of choices and the best one that fit was "delusional." Ironically, to reach this diagnose, the doctor had to delude
himself into believing La Joie's symptoms weren't bona fide physical
entities.

The Exxon attorney used Dr. Hong to make it appear that La Joie
was trying to get disability compensation when she was not sick. The

Exxon attorney framed his questions to make it appear that she started seeing Dr. Ewing, because the new doctor agreed to sign her disability papers, not because he treated people for chemical sensitivity and understood how the disease prevented a person from functioning in a normal work environment.

Once the Exxon attorney had the information he wanted from Dr. Hong, the attorney turned his attention to Dr. Ewing. The elderly Dr. Ewing trusted his patients to tell him how they felt. He had seen La Joie's scrapbook, heard her stories, and understood that she had had a "massive crude oil exposure" (Ewing 1997a, 79-80). He also had her records from Kaiser. He did not do the thorough diagnostic exam and evaluation for chemical sensitivity that Dr. Rea performed on his new patients, because Dr. Ewing did not have access to an environmental treatment clinic. Dr. Ewing thought that La Joie would have benefited from treatment with Dr. Rea, but knew that she couldn't afford it. So he had tried to help her to the best of his abilities.

For the Exxon attorney, Dr. Ewing was relatively easy to discredit. Dr. Ewing was not as thorough or disciplined as Dr. Rea in his approach, diagnosis, and treatment of chemical sensitivity. Dr. Ewing operated mostly on trust, intuition, and experience, backed up by diagnostic tests performed by others. This was simply not enough to prove in a court of law that chemical exposure during the cleanup caused La Joie's health problems.

The Exxon lawyer knew that the "psychiatric" diagnosis economically devastated patients suffering from chemical sensitivity. It makes it difficult for patients to secure health and disability insurance benefits, Social Security disability, and other such benefits, because of limited coverage for psychiatric disorders in most insurance plans. It also eliminates the chance for compensation through personal injury lawsuits and it makes workers' compensation unavailable in many states.

There was only one last matter for the Exxon attorney to attend to—La Joie's scrapbook. Its silent pages were powerful testimony to corporate negligence and reckless behavior. The Exxon lawyer requested that the entire scrapbook be marked as an exhibit—and then kept it with the excuse of wanting to reproduce the photographs in color (La Joie 1996, 262-263).

(Author's note: The document was retrieved almost five years later by La Joie with the author's assistance.)

On Why the Plaintiffs Lost

Smith and La Joie were not happy with Daniel Stenson, the lawyer from the John E. Hill firm. Like Smith, La Joie had signed on with Melvin Belli and was upset to suddenly find herself assigned to another firm. She said later (pers. comm., January 2001), "Dan was so discouraging—he didn't believe we had a case." It appeared to the plaintiffs that their attorney's lack of confidence in winning their case may have impaired his advocacy. Court records show Smith and La Joie had plenty of good reasons to be unhappy with their lawyer.

Lack of Diligence?

Barely a month after deposing the plaintiffs' doctors, Exxon and VECO sought summary judgement against Smith and La Joie to dismiss their case on the grounds that their expert witnesses were incompetent. At the same time, Exxon and VECO moved to exclude Dr. Rea's testimony, and Exxon moved to exclude Dr. Ewing's testimony, in complex briefs that included numerous supporting exhibits, key legal cases, and medical affidavits of allergists hostile to the concept of chemical sensitivity (Nos. 61–65).

In toxic tort claims, medical causation must be established through expert scientific testimony. In *Roberts v. Exxon* (1999), if the court agreed with the Exxon and VECO lawyers that the plaintiffs' expert witnesses were incompetent and excluded the doctors' testimony, Smith and La Joie would have no expert witnesses and no case. Their claims would have to be dismissed.

This was a critical juncture in the case. To survive the motions and preserve their plaintiffs' claims, Stenson and his co-counsel could have countered each of the defendants' five main arguments with supporting legal cases and medical affidavits.

Instead, Stenson and his co-counsel filed a meager six-page response to the motions by Exxon and VECO (No. 74). With more effort, it seemed Stenson and his co-counsel could have countered the arguments raised by Exxon and VECO attorneys and at least have given the energetic representation their case merited.

One main argument was that Dr. Rea failed to establish medical causation. VECO argued Dr. Rea never testified that Smith's "alleged" exposure to chemicals on the oil spill cleanup caused his "alleged" health complaints. Significantly, Exxon's lawyers had never asked Dr.

Rea about the SPECT brain scan tests during his deposition—and they carefully avoided all mention of this powerful test in all their motions and briefs.

Stenson and his co-counsel could have argued that Dr. Rea's initial evaluation of Smith and subsequent screening procedures ruled out other possible non-spill related causes of Smith's symptoms in the years immediately following the cleanup. The lawyers could have argued that the combination of the Accu-Chem blood tests and the SPECT brain scan proved that chemicals associated with crude oil and solvents caused Smith's health problems.

The second main argument used by Exxon and VECO was that Dr. Rea's expert scientific testimony failed to meet the minimum federal standards for reliability. Under a 1993 U.S. Supreme Court decision known as *"Daubert"* (*Daubert v. Merrell Dow Pharmaceuticals* 1993), trial judges were assigned the role of screening expert scientific testimony on the basis of general acceptance in the scientific community (King 1999; Tellus report 2003). *Daubert* is the undoing of many toxic tort cases simply because chemical sensitivity is not well accepted within the traditional medicine community. *Daubert* forces the chemical sensitivity research, evaluation procedures, and treatments to be judged by the standards of traditional medicine. It tries to fit the round peg of chemical sensitivity into the square hole of traditional medicine. It is inappropriate and, worse, *Daubert* actually discourages innovation.

Despite the problems poised by *Daubert* (Chapter 10), Stenson and his co-counsel might have pointed out there are double-blind challenge tests, a method accepted by traditional medicine, and other new reliable reproducible tests developed by environmental doctors. There are peer-reviewed papers; they just are not published in traditional medical journals, which refused to accept articles on chemical sensitivity. But articles are published by the American Academy of Environmental Medicine, the professional society for environmental medicine physicians, and by other organizations as well, which indicates a growing scientific acceptance of chemical sensitivity. Many of these papers were published after 1992, which shows a growing interest by the broader medical community. The tables are slowly turning in favor of chemical sensitivity as a new disease paradigm. The plaintiffs' lawyers could have pointed out that some of the loudest critics of chemical sensitivity were those with financial ties to the chemical and drug industries—and the traditional allergists.

The third main argument put forward by Exxon and VECO was

that Dr. Rea was a doctor retained by lawyers, a so-called "hired gun," rather than a treating doctor, one who actually saw patients. Stenson and his co-counsel could have easily refuted this claim. Dr. Rea had treated over 26,000 patients and performed hundreds of studies over the span of nearly a quarter century. It was true that Belli's firm had paid for Smith's initial evaluation at Dr. Rea's clinic, but Smith himself had paid for his treatment.

The fourth main argument brought by Exxon and VECO was that the plaintiffs' lawyers had failed to prove negligence or that actions by Exxon and VECO had caused Smith and La Joie to become sick. Negligence raises complex issues of industrial hygiene and environmental toxicology that have to be addressed—and were successfully addressed in another toxic tort litigation from the *Exxon Valdez* cleanup (Chapter 2). In the critical response to motions by Exxon and VECO for summary judgment in the case of *Roberts v. Exxon* (1999), Stenson and his co-counsel did not adequately show that Exxon and VECO were negligent.

The fifth and final argument raised by Exxon, and one that was irrefutable, was that the so-called "expert witness" reports had been written by Stenson.

Imprudence

During discovery, the Exxon attorney had found that Stenson had drafted Dr. Ewing's expert witness report (Ewing 1997b) and that Dr. Ewing had accepted it essentially without change after several telephone conversations with Stenson to discuss the report's contents (Ewing 1997a, 61–71). When questioned further, Dr. Ewing admitted that it was the first time in his experience that the lawyer wrote the expert witness report in a court case. Exxon's attorney didn't tell Dr. Ewing that this was unwise. He would save this bit of information—like an ace up his sleeve—to play in his final hand in court.

Much to the Exxon attorney's delight, he found Dr. Rea's expert witness report (1997b) looked remarkably similar to the one written by Stenson for Dr. Ewing. When the Exxon lawyer presented Dr. Rea with "his" three-page expert witness report that he had supposedly written about Smith's injuries related to chemical sensitivity, the doctor was stunned (Rea 1998, 7–12). The report was dated 24 October 1997. Dr. Rea had not seen Smith since Smith left his clinic in early 1993. The report was not printed on his letterhead and it was not in his case files or his secretary's computer, but his name was affixed to the last page. Dr. Rea refused to authenticate the report. The doctor realized Stenson had written the report without his permission or knowledge.

Polite Clients Pay

Exxon and VECO lawyers attacked the six-page response to their motions as totally inadequate. The VECO lawyers observed that the plaintiffs' lawyers "failed to provide one shred of support" for chemical sensitivity or the admissibility of Dr. Rea's testimony—"not one affidavit, not one reference to a medical journal or publication of any kind, nothing . . . ." (No. 87, 4-5). The plaintiffs' lawyers simply stated in their response, "The defendants don't like the opinions of Dr. Rea and Dr. Ewing . . ." (No. 74, 4). The plaintiffs' lawyers even offered the excuse of a "filing error" (No. 87, 8) to explain the inauthentic expert witness reports (Stenson 1998), insisting that this "short term confusion" was not "an out and out effort to compromise the court process" (No. 74, 2). By not countering the points raised by the corporate medical experts and lawyers, the plaintiffs' lawyers left Judge Holland with little choice.

Although court proceedings dragged on for another year and a half, the case was essentially over in 1998. In February 1999 Judge Holland ruled that the "purported expert reports of Doctors Rea and Ewing were prepared by counsel for the plaintiffs" (No. 90, 3) and, therefore, inadmissible. "The judge found this "quite extraordinary," "imprudent," and "in violation of the spirit of" a federal rule of civil procedure, Rule 26(a)(2)(b) (No. 90, 3, 4, 5).[2] With this case history, the judge had to agree with the defendants that the plaintiffs' expert witnesses were retained, not treating, experts. Lacking any evidence to the contrary, Judge Holland had to rule that the expert witness reports were inadequate and failed to meet the reliability requirements imposed by *Daubert*.

Once the judge ruled the expert witness testimony was inadmissable, he had little choice but to dismiss the entire case. He wrote, "Without admissible expert testimony, which tends to establish that exposure to crude oil while involved in the *Exxon Valdez* oil spill cleanup produced plaintiffs' injuries, defendants are entitled to summary judgment as a matter of law" (No. 91, 5). The plaintiffs' lawyers failed to request an oral argument.

Roberts v. Exxon (1999) was over for La Joie, but for Smith, the worst was yet to come. Smith found he owed Exxon and its shipping company SeaRiver Maritime a total of $1,585.15 for costs taxed against him by the clerk of courts (No. 101). He initially refused to pay, but he realized this was a mistake when the federal marshal was ordered to seize his personal property to satisfy the judgment. He paid.

Then in March 1999, VECO requested an award of attorneys' fees against Smith. VECO recognized that attorneys' fees are not allowed in maritime law, but argued there is an exception for cases in which "a party has acted in bad faith, vexatiously, wantonly, or for oppressive reasons" (No. 96, 9). VECO argued that the act of drafting Dr. Rea's report by Smith's lawyer, among other things, was "designed to extract a settlement from VECO and Exxon despite the fact that he had no valid claim" (No. 96, 9).

The VECO lawyers figured Smith's share of their total fees for the case was $30,258.25, which they wanted as an award to sanction Smith for his "frivolous" claim against VECO. "An award of attorneys' fees against Smith will also serve as an example to others that they should not engage in similar conduct," the VECO lawyers wrote (No. 96, 11–12).

VECO's request was unopposed by Stenson and his co-counsel and approved by Judge Holland who awarded VECO's total attorneys' fees—$57,418—against Smith "to bring this troublesome case to a conclusion" (No. 105, 2).

Smith found himself left in the lurch by his attorneys. He simply didn't have that kind of money. It had taken him over three years just to pay the $4,176.50 he owed to Dr. Rea. Faced with a staggering bill, he was forced to accept VECO's offer: VECO would waive the outstanding debt if Smith signed a gag order regarding his case. Reluctantly, he signed.

In an effort to put the whole awful mess behind him and get on with his life, he never sued his own lawyers for malpractice. Neither did La Joie.

An Occupational Health Disaster

Chapter 10.

Investigating a Disaster

When Pam Miller and I discovered that many of the EVOS cleanup workers had been exposed to dangerous levels of dangerous chemicals and that many had become sick as a result of this overexposure, we realized that hundreds, if not thousands, of workers could have potential chronic health problems.

In this chapter we conducted a literature review of the compounds of concern to learn the nature of these potential chronic health problems. We then initiated an independent worker health survey through Yale University Medical School's Department of Epidemiology and Public Health to learn if these potential health problems are being realized. We also examined the latest medical findings and legal treatment of chemical-induced injury—chemical sensitivity or, more precisely, "toxicant-induced loss of tolerance" or TILT.

Paper Trails
Symptoms of Overexposure—A Literature Review

Miller and I reviewed the technical literature for crude oil; crude oil mist; polycyclic aromatic hydrocarbons (PAHs); diesel fumes; seawater mist; the dispersants Inipol EAP 22, Corexit 9527, and Corexit

9580M; and the cleaning solutions Simple Green and De-Solv-It. We searched diligently for human health effects of crude oil mist, aerosolized weathered oil, and seawater mist, because we were concerned that the pressurized hot water wash had aerosolized the beached oil. We found that all of the commercial solvents have proprietary and thus confidential "surface active agents" and other chemical constituents for which we could not evaluate human health effects.

Our findings verify Dr. Teitelbaum's concerns, as expected, and uncover other supporting evidence that the chemicals present during the cleanup are potent human health hazards and that overexposure to these chemicals can lead to both the short- (acute) and long-term (chronic) health problems.

We found that studies of human health effects of crude oil mists and aerosols from other tanker spills were very limited and fairly recent. Researchers examined the health effects in the 1993 *Braer* accident in Shetland, Great Britain and the 1999 *Sea Empress* spill in Milford Haven, Wales, and the *Nakhodka* wreck in Japan's Oki islands. People living near the *Braer* and *Sea Empress* spills were exposed to oil vapors, mists, and aerosols, generated by fierce storm winds and waves. A significant number of exposed Shetland residents reported headaches, throat irritation, skin irritation, itchy eyes, mood change and, to a lesser extent, tiredness, diarrhea, nausea, wheezing, cough and chest ache (Campbell 1993; Campbell et al. 1993). In a follow-up study five months later, a significant number of exposed residents reported their health had altered and deteriorated since the spill (Campbell et al. 1994).

Similarly, exposed Wales residents reported feeling generally ill, headache, nausea, sore eyes, runny nose, sore throat, cough, itching skin, rash, shortness of breath, and weakness immediately after the spill. Researchers conclude that these symptoms are consistent with those expected from the known toxicological effects of oil and suggested there was a direct health effect of the spill on the population (Lyons et al. 1999).

In the *Nakhodka* spill, residents who volunteered to cleanup the spill were exposed to storm-generated oil vapors, mists, and aerosols (Morita et al. 1999). Although the average length of time spent on the cleanup was only four to five days, workers reported back and leg pain, headache, and eye and throat irritations. Urine samples indicated possible exposure to oil vapors or food additives, a confounding variable. Urine samples returned rapidly to normal after the cleanup.

Researchers who investigated health effects from these three spills conclude that further studies should be undertaken to assess long-term effects of these exposures, but no such follow-up studies were discovered in the literature.

We found the recent literature review by researchers Park and Holliday to be most helpful. These authors found few studies on inhalation effects from exposure to crude oil as a single entity, but they point out, "There are literally thousands of papers dealing with the chronic effects of inhalation exposure to specific crude-oil components" (Park and Holliday 1999, 120). They review the effects by organ system, beginning with the toxicity of specific hydrocarbons to the blood and blood-forming organs *("hematotoxicity")*, the kidneys *("nephrotoxicity")*, and the nervous system *("neurotoxicity")*.

From their review, Park and Holliday conclude that mists and aerosols "provide a route by which non-volatile components of crude oil (both fresh and weathered) may enter the body and give rise to toxic responses" (ibid., 123). They determine that mists and aerosols probably provide a penetrating vehicle for PAHs to enter deep into the lungs and make contact with the sensitive lung tissues and membranes involved with organ function. Regarding oil mists, the authors note, "[I]t is well known that even very small quantities of aliphatic hydrocarbons in contact with lining of the lung . . . give rise to a severe chemical pneumonitis [inflammation of the lungs], characterized by pulmonary edema, hemorrhage and tissue necrosis, which can be fatal" (ibid., 122). Regarding PAH aerosols, they conclude from studies of systemic effects in mice and rats that "components of crude oil inhaled as an aerosol can pass through the alveolar membrane and therefore have the potential to cause toxic responses in other organs" (ibid., 122). Further, this "suggests that *systemic effects from exposure to crude-oil aerosols cannot be ruled out*" (ibid., 123, emphasis added).

Miller and I realized from the Park and Holliday paper that crude oil is much more hazardous to human health than the *Exxon Valdez* cleanup workers were led to believe. Inhalation of oil mists creates a health hazard whereby the aliphatic hydrocarbons become potent chemicals capable of damaging lung tissue. Inhalation of oil aerosols transfers deadly PAHs to and through the lungs, then into the body, creating the potential to harm other organs and other organ systems.

Our investigation of health effects of the 'detergent solutions'—the solvents—present during the *Exxon Valdez* cleanup revealed additional concerns beyond the hazards and symptoms listed in the Material

Safety Data Sheets (MSDS) (Introduction). In general, we found no studies of health effects of these products on oil spill cleanup workers, but
the laboratory studies with specific chemicals in the products, usually
tested on animals, provided ample grounds for concern.

For example, Park and Holliday concluded from animal studies
on skin exposure to crude oil that health effects are also associated
with the solvents used to remove the oil from the skin. Researchers
found from studies with 2-butoxyethanol that persons who expose
large portions of their skin are at risk of absorbing toxic doses of this
systemic poison (Johanson, Boman, and Dynesius 1988). Further, they
found that *water facilitates absorption of this chemical* through the
skin of guinea pigs (Johanson and Fernstrom 1988). This is relevant
to the EVOS cleanup because beach workers were enveloped in an
oily seawater mist—and they also removed their raingear on sunny
days. Another clinical study suggests that prolonged exposure in
mammals, to oil, the dispersant Corexit 9527, or in combination
could alter the ability of intestinal microfauna to break down toxic
chemicals (George et al. 2001). This could explain the high incidence
of gastrointestinal distress claims reported by cleanup workers to the
ADOL (ADOL 1990).

Miller and I discovered from our investigation that symptoms for
crude oil exposure show a considerable, and not surprising, overlap
with symptoms from exposure to degreasers and dispersants, including the liquid "fertilizer" Inipol (Table 2, p. 163).

Inhalation of most compounds to which workers were exposed
during the 1989 cleanup can cause immediate and long-term respiratory problems and central nervous system (CNS) disorders.
Overexposure to oil mist, PAHs, and 2-butoxyethanol, (an ingredient
in Inipol, Corexit, and Simple Green), can cause chronic effects such
as liver, blood (anemia), and kidney disorders and endocrine disruption (with reproductive effects). Overexposure to oil mist and PAHs
can cause blood disorders (leukemia).

Overexposure to oil through skin contact can cause cancer
(epitheliomas), endrocrine disruption (hormone imbalances), and
immune suppression, which weakens one's ability to fight disease,
infection, and chemical poisons. Overexposure to Inipol through skin
contact can cause dermatitis. Studies on skin exposure to PAHs of
animals and humans suggest that the combination of sunlight (UV
exposure) and oil is more irritating than oil alone (Gomer and Smith
1980). This is relevant because it was common for EVOS cleanup
workers to remove their rain gear on hot days.

Table 2
Health Symptoms of Overexposure to Some of the Hazardous Chemicals Present during the 1989 EVOS Cleanup

(Acute=a; Chronic=c; Both=•; †=effects reported, duration unknown)

Symptom	Oil on Skin	Oil Mist	PAH Aerosol	Seawater Mist	Diesel Exhaust	2-butoxy ethanol	Corexit Products*	Inipol*	Simple Green*
Respiratory damage		•	•		•	a	†	†	
CNS disorders		•				a		†	
Liver disorders	c	c	c			c			
Blood disorders (leukemia)		c	c						
Blood disorders (anemia)	c	c	c			c			
Kidney disorders	c	c	c			c	†	†	
Skin disorders	c			a	c	a	†	†	
Endocrine disruption	c	•	•			c	†	†	†
Immune suppression	•	•	•		c				
Toxic systemic						c	†		
Fetal effects						a		†	

Sources: Boffetta, Jourenkova, and Gustavsson 1997; Burnham and Bey 1991; Burnham and Rahman 1992; Campbell 1993; Campbell et al. 1993, 1994; Exxon Company 1989a, 1992; Exxon Shipping 1988; Falk-Filipsson et al. 1993; Feuston et al. 1994; George et al. 2001; Gerde and Scholander 1987, 1989; Holland, Whitaker, and Gipson 1980; Gomer and Smith 1980; Howe et al. 1983; Johanson, Boman, and Dynesius 1988; Johanson and Fernstrom 1988; Kubaiewicz, Starzynski, and Symczak 1991; Larsen et al. 2000; Lyons et al. 1999; Morita et al. 1999; Orange-Sol 1987, 1991; Park and Holiday 1999; Rahimtula, O Brien, and Payne 1984; Rahimtula, Lee, and Silva 1987; Rolseth, Djurhuus, and Svardal 2002; Springborn Institute for Bioresearch 1985; Stubblefield et al. 1989; Sunshine Makers, Inc., 2002;Taneda et al. 2002; NIOSH, 1988; U.S. EPA 2003.

*These products (Corexit 9527, Inipol EAP 22, and Simple Green) contain 2-butoxyethanol, a known human health hazard. However, most of the studies on health effects were conducted with the chemical, 2-butoxyethanol, rather than with the products. The EPA lists 2-butoxyethanol as one of the ingredients to avoid, if possible, in products on its Janitorial Products Pollution Prevention Program. www.westp2net.org/janitorial/tools/haz2.htm. According to the EPA, Inipol is no longer being produced (as of January 1996) www.epa.gov/ceppo/ncp/inipolea.htm. However, Exxon still manufactures Corexit 9527 for use in oil spill response and Sunshine Makers, Inc., still manufactures Simple Green.

From this exercise, Miller and I concluded that a range of chronic symptoms *were possible* from overexposure to oil and chemicals present during the EVOS cleanup. The worker health survey described below indicates specific symptoms were reported in a subsample of the EVOS cleanup worker population.

Worker Health Survey

During summer 2002 Annie O'Neill, a graduate student at Yale Medical School's Department of Epidemiology and Public Health, conducted an internship with ACAT and AFER, the two nonprofit organizations investigating the health effects of the EVOS cleanup. She conducted independent research on Exxon's cleanup and an investigation of self-reported chronic health problems among EVOS cleanup workers for her master's thesis. Her mentor and key advisor was Dr. Mark R. Cullen of the Yale Occupational and Environmental Medicine Program. In 1987, Dr. Cullen edited and published, *Workers with Multiple Chemical Sensitivities, Occupational Medicine: State of the Art Reviews*, the first comprehensive collection of articles on this new disease.

O'Neill completed her thesis, "Self-Reported Exposures and Health Status among Workers from the *Exxon Valdez* Oil Spill Cleanup," in May 2003. Her study is the first attempt to assess chronic health problems reported among EVOS workers fourteen years after their employment on the cleanup. Her methods and findings are presented in brief with her permission; she plans to publish her study in a peer-reviewed scientific journal.

The specific aims of O'Neill's study were first, to assess the prevalence of chronic symptoms among former EVOS cleanup workers and secondly, to determine whether specific oil spill cleanup jobs and exposures are linked to higher reported prevalence of adverse health impacts. She designed her survey to test if workers who had the highest exposure to oil fumes and mist—those on the beaches and in the nearshore areas—would have higher incidences of self-reported chronic respiratory problems and neurological problems. She also designed her study to test if workers who used chemical solvents, specifically dispersants and Inipol, would report more symptoms of chemical sensitivity, neurological impairment, anemia, and liver disease than workers less exposed to these cleanup products.

She randomly selected workers for her study from public records obtained from the Alaska Department of Labor (ADOL) files, referrals from these workers, and community liaisons in Anchorage, Valdez, Homer, and Cordova who had kept private lists of cleanup workers. The majority (75 percent) of participants in her study were Caucasian, followed by Alaska Natives (14 percent).

O'Neill worked in collaboration with anthropologist Dr. Lorraine Eckstein to develop a questionnaire. They used the questionnaires in the Persian Gulf War Study and the Amchitka Workers Medical Surveillance Program as models for the EVOS worker survey. They found the study of Gulf War veterans, in particular, provided guidance for assessing respiratory impairment, chemical sensitivity, cognitive dysfunction (CNS problems), and general health. O'Neill hired a professional research firm, Anchorage-based Craciun Research Services, to conduct her telephone survey.

She analyzed data from 169 completed surveys and grouped workers into four oil exposure categories (no exposure, low, medium, and high), and three chemical exposure categories (no exposure, moderate, and high) on the basis of the jobs conducted during the cleanup. Jobs classified as risk of high exposure for oil included the use of pressurized hydraulic hoses and/or wands to spray beaches from onshore and offshore; crewing on skimmers, skiffs, and other boats used to deploy boom and/or pick up oil; and decontamination of equipment (boats) and gear (clothing). Jobs classified as potential risk of high chemical exposure included bioremediation and dispersant application; decontamination of equipment and gear; skiff and skimmer operators; garbage detail (hydrogen sulfide exposure); and wildlife care providers.

She identified and controlled for *"confounding variables,"* which might have similar impacts on health as the oil and chemical exposures and those which might bias results. These variables included smoking, alcohol consumption, age, gender, ethnicity, and previous or current employment involving exposure to oil and other hazardous chemicals. Workers were asked whether they *believed* the oil spill affected their health, which she factored into her results to reduce bias.

She included use of a respirator as a potentially confounding variable, because it would have reduced inhalation exposures. She found that 70 percent of her survey group reported that they were *not* provided with respirators.

She found EVOS workers in jobs with high oil exposure to oil fumes, mists, and aerosols have a greater prevalence of self-reported symptoms of respiratory problems, neurological problems, and chemical sensitivities than unexposed workers. Among workers with high oil exposure, nonsmokers reported a greater prevalence of symptoms of chronic bronchitis than smokers. Symptoms of chronic airway disease included sleep apnea, pneumonia, other lung conditions, chronic sinus and/or ear problems, asthma, persistent hoarseness. Symptoms of neurological impairment included amnesia or severe memory loss; difficulty thinking clearly or concentrating; trouble with speech; and general confusion or disorientation.

Similar to the oil exposure results, workers who had jobs with moderate chemical exposure reported a greater prevalence of symptoms of chronic airway disease than unexposed workers, while workers with high chemical exposure reported an insignificantly greater prevalence of symptoms of neurological impairment. (This may be an artifact of her job categories: "moderate exposure" jobs for her chemical analysis included workers who used pressurized hydraulic hoses and wands to spray beaches.) There were not enough workers in her sample to analyze for blood disorders or neurological impairment from Inipol or Customblen. She reported other "notable significant associations" between exposures and chronic symptoms. She did *not* find a correlation between cognitive dysfunction and oil or chemical exposure.

Exposure to the cleaning solvents De-Solv-It, Simple Green, and CitroKleen were all associated with increases in reported symptoms of neurological impairment, although the relation was not statistically significant. However, those who worked with De-Solv-It were more likely to report symptoms of chronic airway disease, chronic cognitive dysfunction, and chemical sensitivity than unexposed workers (71). She notes the latter finding is not supported by the available literature, but also notes there is little information available on health effects of this product.

O'Neill found several examples where the available scientific literature supported her study's results. For example, findings of increased symptoms of airway disease following oil mist exposure were reported previously, as were findings of increased prevalence of respiratory impairment and chronic neurological symptoms, following occupational exposure to volatile organic carbons (VOCs) and hydrogen sulfide.

O'Neill's findings support experiences of individuals such as Smith, La Joie, Moeller, Nagel, and others. Yet most oil spill-related toxic torts brought on symptoms of respiratory distress, CNS disorders, and chemical sensitivity, in particular, had not been successfully litigated (Chapter 9). Miller and I decided to review new developments in the medical and legal arenas to learn of any changes that might improve the odds for victims of such chemical-induced injuries.

The Petrochemical Problem: Gaining Recognition

Dr. Rea witnessed the slow emergence of chemical sensitivity as a new disease paradigm, and its slow acceptance within the scientific and medical communities, during his pioneering quarter century career in the field (Rea 2001). Dr. Rea and his colleagues had dubbed this "the petrochemical problem," because the increasing incidence of chemical sensitivity seemed to parallel the growth of the petrochemical industry and the increased use of synthetic products such as pesticides, plastics, food additives, synthetic textiles, and particleboard.

When energy conservation in the 1970s led to tighter building standards, fresh air intake in homes and offices was cut to a minimum. Levels of indoor air pollution from synthetic materials that offgas: gas furnaces, cigarette smoke, and other pollutants rose dramatically. As Americans spent more and more time indoors—on average 90 percent of the time by the 1990s—health complaints rose. The incidence of asthma, depression, and people reacting to low levels of everyday pollutants all surged upward. By 1987 the National Academy of Sciences estimated about 15 percent of the U.S. population was at risk from "increased allergic sensitivity" to chemicals, and signs of an epidemic were looming (Ashford and Miller 1998, 26, 233).

The federal government became undeniably aware of the health problems caused by solvents and crude oil, both petrochemicals, in two highly publicized incidents. In 1988 about two hundred EPA employees became sick with multiple symptoms after exposure to new carpet fumes—solvents—in the agency's Washington, DC, headquarters (ibid.). The ill workers represented about 10 percent of the workforce in the affected buildings and, for some of these victims, the acute exposure led to chemical sensitivity and permanent health problems. In an unrelated incident that same year, the Social Security Administration recognized chemical sensitivity in its manual for disability determinations.

Then, from 1992 onward, about 10 percent of the 700,000 returning Gulf War vets sought treatment for fatigue, depression, irritability, memory and concentration loss, muscle aches, shortness of breath, skin rashes, and other difficulties (Ashford and Miller 1998). Dr. Theron Randolph diagnosed some of the front-runners of this flood with petroleum poisoning (McGonigle and Timms 1992; *The Dallas Morning News* 1992). The federal government launched a pilot program to determine if the vets had been exposed to chemical weapons (Thompson 1993). When government doctors failed to recognize chemical sensitivity as a medical condition, frustrated vets and other victims turned to private doctors, legislators—and court, a route followed by an increasing number of people.

Claudia Miller, MD, a professor at the University of Texas Health Science Center in San Antonio, studied many of the vets (Ashford and Miller 1998). Gulf vets reported acute symptoms while in the war, from a host of chemicals including smoke from oil well fires and fuel in tent heaters, fuel vapors, diesel exhaust, and a carbamate drug used in pesticides and nerve agents. She noticed a striking similarity in chronic symptoms reported by Gulf War vets with those of civilians exposed to indoor pollutants in a new or remodeled building or to pesticides. Carpet glues, organic solvents, paints and lacquers, formaldehyde (in particleboard for example), pesticides, diesel and gasoline are now known to be associated with onset of chemical sensitivity. Dr. Miller found 78 percent of the vets she studied since the Gulf War reported onset of new chemical intolerances from food, medications, caffeine, alcohol, and smoking, among other things.

Dr. Miller collaborated with Nicholas Ashford, PhD, a professor of Technology and Policy at the Massachusetts Institute of Technology, where he teaches courses in environmental and occupational health law and policy. They published a groundbreaking book, *Chemical Exposures: Low Levels and High Stakes* (John Wiley & Sons: New York, 1991), which they revised in 1998. In it they explain their view that chemical sensitivity is not just a single syndrome, but rather a whole new class of disease, which they describe as "toxicant-induced loss of tolerance" or TILT. This phrase takes into account the fact that caffeine, alcoholic beverages, various drugs, and foods can trigger "chemical" sensitivity symptoms in individuals who already have lost their natural tolerance through an acute chemical exposure or other initiating event.

By the late 1990s it was understood that the initiating event for chemical sensitivity, or more broadly TILT, can occur either as an intense abrupt event like a pesticide spill or as an intermittent repetitive exposure (Wilkinson 1998). The EVOS cleanup workers had it both ways—the chemical exposure was intense, but also repetitive, intermittent for those who took breaks or were otherwise in and out of oiled areas, and prolonged for individuals who worked for months on the beaches. Those who studied TILT found that the nervous system, quite independent of the immune system, has the capacity to "amplify responses to stimuli that are perceived as dangerous to the organism" (ibid., 59). Once the stimulus is stopped, the nervous system initiates a process of amplification, so that the next time the person encounters that stimulus, or anything that can similarly trigger the nervous system, even at a much lower dose, there is an amplified or exaggerated response.

This process is known as *"limbic kindling,"* and it is the leading theory among environmental medicine doctors to explain the etiology or cause of chemical sensitivities and other TILT symptoms (Ashford and Miller 1998; Kilburn 1998; Rea 1995; Wilkinsen 1998). Limbic kindling is a type of epilepsy that involves abnormal firing of the limbic system—the part of the brain with a direct connection to the nose. The olfactory system is the normal pathway for airborne chemicals to interact with the brain; the limbic system is where the immune, nervous, and endocrine systems interact. Chemical-induced seizures cause the amygdala in the limbic system to misfire signals to the hypothalamus, which communicates with both the olfactory and limbic systems, regulating chemicals in the entire body.

The hypothalamus governs body temperature, reproductive urges and functions, metabolism, and even aggressive behavior. It also influences some immune system functions. Disrupting the hypothalamus—with any of a variety of chemicals once a person loses his or her initial tolerance—can create havoc in many different parts of the body and lead to the multiple system dysfunction experienced by people with severe chemical sensitivities such as La Joie.

Pesticides and solvent exposures are known to cause or facilitate limbic kindling. The EPA lists 2-butoxyethanol as one of the *pesticides* it has tested (CAS number 111-76-2) and it lists this chemical as one of the ingredients to avoid in its Janitorial Products Pollution Prevention Program. The EPA web page states products with the listed ingredients "pose very high risks to the janitor using the product, to building occupants, or to the environment." Comments under chronic effects for 2-butoxyethanol list reproductive and fetal damage, liver and kidney damage, and blood damage.

It is worth repeating that 2-butoxyethanol, a chemical of concern, is in Inipol, Corexit 9527, and Simple Green—all solvents used during the *Exxon Valdez* cleanup.[1]

Scientists and doctors are rapidly closing in on the mechanism of chemical sensitivity—that is, exactly how chemicals cause the varied symptoms—as lab tests and diagnostic procedures become more sophisticated. However, lack of a perfectly understood mechanism does not negate the existence of TILT diseases or attempts by doctors to treat people with symptoms. As Ashford and Miller point out in their book, "Useful interventions can precede a full understanding of mechanisms—that is, we do not have to know everything before we do anything" (1998, 311). The classic example they use to illustrate this point is the 1854 cholera epidemic (ibid., 51). A London physician noticed that people who developed cholera obtained their drinking water from the community well. He broke the pump handle and stopped the epidemic. It took thirty more years to discover the bacterium responsible for cholera epidemic.

There are also other developments in the public policy arena that show chemical sensitivity and other TILT symptoms are increasingly being recognized as a disabling disease (ibid.). For example, in 1991, the National Research Council held a workshop to determine research needs for chemical sensitivity and the Agency for Toxic Substances and Disease Registry (ATSDR) took the lead in fostering public research. In 1994, federal agencies involved in occupational and environmental health issues and the departments of defense and veterans affairs formed the Interagency Work Group on Chemical Sensitivity. Also in 1994, the first U.S. government-subsidized housing for people with chemical sensitivity opened in California. An increasing number of states regulate or call for regulation of chemicals such as pesticides, synthetic carpets, and carpet glues associated with chemical sensitivity and health problems. Canada and European countries recognize chemical sensitivity and implement policies to support people disabled by this disease.

"Chemical Injury and the Courts": Corporate Domain

Miller and I found during our investigation of sick EVOS cleanup workers' health claims that the scientific inroads towards understanding chemical sensitivity are not mirrored in the legal arena. In fact, just the opposite—we concluded that our nation's legal system has turned on its head to the point where it actually advocates protection of polluters at the expense of public health and victims' rights. How is this happening?

Sound public policies are based on sound science; the better the science admitted into the legal and regulatory arenas, the better the policies coming out of the system. We discovered that Exxon and other corporate polluters and manufacturers of dangerous products have devised ways to control the science being admitted in court and thus are able to control or bias court (judge or jury) decisions, which ultimately are used to shape and guide public policies. Exxon used three of these strategies to hide from the public the fact that its safety program failed to adequately protect workers from the harmful health effects of its cleanup.

Daubert

In June 2003, the Tellus Institute in Boston, Massachusetts, published a review called, "*Daubert*: The Most Influential Supreme Court Ruling You've Never Heard Of." The Tellus review was conducted by its Project on Scientific Knowledge and Public Policy and guided by a planning committee, which includes Dr. Eula Bingham who had previously advocated better health and safety protections for cleanup workers in 1989. This group encourages the understanding and use of the best available scientific evidence in policy decision-making—and they write critical and insightful reviews when something is amiss.

According to the Tellus report, "In June 28, 1993, the U.S. Supreme Court issued an opinion relating to how federal judges should decide whether to allow expert testimony into the courtroom." In *Daubert v. Merrell Dow Pharmaceuticals, Inc.*, the Supreme Court "instructed judges to examine the scientific method underlying expert evidence and to admit only evidence that was both 'relevant and reliable'." In other words, *Daubert* directs judges to act as "gatekeepers" of scientific evidence by allowing them to decide what does and does not get admitted as evidence to the jury (Tellus report 2003, 3).

Two later Supreme Court rulings considerably expanded *Daubert's* reach (ibid., 3). One made it virtually impossible to successfully appeal and overturn trial court decisions regarding admissibility of evidence (*General Electric v. Joiner* [1997]); the other ruled *Daubert* should be applied to *all expert testimony*, not just that based on scientific evidence (*Kumho Tire Co. v. Carmichael* [1999]).

The Tellus review (2003) found that, over the past ten years, *Daubert* has had unforeseen and very troubling consequences: "Polluters and manufacturers of dangerous products are successfully using *Daubert* to keep juries from hearing scientific or any other evidence against them" (3). The Tellus report explains, "Much of the evidence that forms the basis of a plaintiff's case, from the safety of drugs and consumer products to whether pollution has caused harm, is based on science. In many cases, pre-trial "*Daubert* hearings exclude so much of the evidence upon which plaintiffs intend to rely that a given case cannot proceed" (ibid., 3). The pre-trial hearings are argued in private, behind doors closed to public scrutiny. The Tellus review concludes that *Daubert* has become "the latest and most effect tool used by tort defendants to protect themselves from product liability and personal injury cases" (ibid., 3).

Daubert is one of the main reasons that most of the toxic tort cases filed by EVOS cleanup workers such as Smith and La Joie came to naught (Chapter 7). Toxic torts, in particular, rely on scientific evidence to demonstrate causality—that exposure to toxic substances, such as oil and solvent aerosols for example, will cause diseases, such as disabling chemical sensitivities, respiratory problems, and neurological damage in this example. The traditional western science-based medical community was—and still is to some extent—reluctant to accept chemical sensitivity as a new disease and has refused for years to publish scientific papers about chemical sensitivity in its "accepted" medical journals. This played right into the hands of corporate lawyers who used the very newness of the science, the innovative methods to diagnose chemical sensitivity, and the lack of acceptance by the scientific community to discredit the evidence under *Daubert*. This did not just happen to sick EVOS cleanup workers.

The Tellus report summary shows a battleground littered with plaintiffs injured by chemicals and faulty products. Following *Daubert*, the Tellus paper reports, "The percentage of expert testimony by scientists that was excluded from the courtroom rose significantly. This . . . has led to an increase in successful motions for summary judgment, since, without expert testimony, there is often little left with which to proceed. The percentage of summary judgments granted post-*Daubert* more than doubled. Over 90 percent of these judgments came down against plaintiffs" (ibid., 4).

The Tellus review warns, "Emboldened by their success in the courtroom, powerful interests are now trying to extend the reach of *Daubert*-like evidentiary standards to the regulatory arena, where they may affect the federal government's ability to understand and act to reduce risk from hazardous exposures" (ibid., 4). For example, according to the Tellus report, large corporations and their interests have challenged the EPA's right under the Data Quality Act (2000) to "draw conclusions about the potential ecological effects of [a] wide-ly-used herbicide . . . ." The Tellus review explains that the petition, backed by the chemical industry, "argued that [the] EPA could not include peer-reviewed academic studies documenting endocrine disruption effects in its assessment of risk from [the herbicide], because the EPA had not yet established testing protocols to characterize endocrine effects" (ibid., 15).

In this case, chemical industry interests argued that the EPA has the cart before the horse, but, as was demonstrated in the 1854 cholera epidemic and discussed for the chemical sensitivity epidemic, the lack of understanding of a known cause should not prevent recognition and treatment of the problem. A *Daubert*-like approach to scientific evidence severely undermines the ability of federal regulatory agencies to protect public health, because the agencies consider the totality of evidence in their decision-making process to assess risk, which is at odds with the piecemeal and fragmented approach encouraged by *Daubert* in court. The Tellus report (2003) warns, "*Daubert* and *Daubert*-like challenges threaten to paralyze the systems we use to protect public health and the environment" (ibid., 17).

Linda King's wonderfully informative little book designed to guide and encourage people suffering with chemical-induced injuries, *Chemical Injury and the Courts: A Litigation Guide for Clients and their Attorneys,* best sums up the state of affairs under *Daubert.* "Many [victims] feel they have been sold out three times— by the polluters, by the government, and again by the American legal system" (King 1999, 10).

Covenant of Silence

Unfortunately, it is not just *Daubert* that is limiting the public's right to know about hazardous chemicals and health outcomes of wrongful exposure. An investigative article in *The Washington Post* in 1989, "Toxic Waste, Court Secrets," reported on "how the American legal system covers up environmental hazards." The reporter, Benjamin Weiser, discusses the "growing use of secrecy procedures in the nation's civil courts and how that secrecy is hampering efforts by scientists and health officials to learn more about hazardous chemicals and their effects." The "covenant of silence" enforces secrecy through confidentiality in court proceedings or buys it through settlements—usually with injured parties desperate to pay medical bills.

Either way, chemical manufacturers and corporate polluters make sure that no one knows about the injuries that are caused and, in controlling access to the data on which scientific opinions are based, they influence the state of knowledge about the chemical of concern. Corporate interests argue that confidential settlements clear crowded court dockets, but this secrecy comes at the expense of public health. A *non*-confidential settlement would also clear court dockets, but the corporate interests insist on confidentiality as a condition of settlement.

Miller and I came up against the covenant of silence time after time in our investigation of the sick EVOS cleanup workers. All the civil settlements of the toxic tort cases we found are confidential and some impose gag orders to prevent workers from talking about their case (Chapter 9). Most all of the evidence in the single successful toxic tort we found, *Stubblefield v. Exxon* (1994), is protected by confidentiality. It was only through a series of unrelated and totally fortuitous events that we were able to obtain the evidence we did. In a case like *Stubblefield v. Exxon* (1994), humanity—not just the sick EVOS cleanup workers—would be greatly served by information on toxic effects of oil pollution and oil cleanup work; the documents should be unsealed for public review.

Occupational Laws

Dr. Nicholas Ashford, author of *Chemical Exposures* and other works on chemical sensitivity, told *Chemical and Engineering News*, "[A]s a rule, we have not factored neurotoxicity into the regulatory screening regime very much" (Wilkinson 1998, 58). He made this comment in reference to the federal process that screens chemicals for public health effects to determine whether to allow commercial production.

However, Miller and I learned that his comment also applies to occupational laws supposedly designed to protect workers from workplace hazards. Most of the workers' compensation claims for injuries incurred during Exxon's cleanup were filed in Alaska. Yet, as discussed in Chapter 8, Alaska's workers' compensation system flatly ignores chemical sensitivity claims—and most likely other claims of potential chemical-induced health problems (such as the headaches, nausea, and dizziness).

The sad irony is that until chemical-induced injuries are properly recognized—by state *and* federal systems designed to protect worker *and* public health, the injuries will continue to happen, injured workers will continue to be discarded by the system, and we will continue to drain our economy of skilled labor.

Conclusion: Long-Term Monitoring of EVOS Workers IS Warranted

Miller and I have concluded from our three-year investigation of EVOS workers that Exxon's cleanup was an occupational health disaster. Further, we conclude there are, unquestionably and undeniably, people who have died, and people who are suffering, from chronic health problems stemming from wrongful exposure during the cleanup.[2] There could be as many as two thousand or more former workers with chronic symptoms—and this estimate is likely conservative because it is based on Exxon's self-reported "peak" number of 11,000 cleanup workers, not the total number employed during the cleanup. However, scientists, state and federal oversight personnel such as ADEC and Coast Guard employees, and volunteers could be experiencing chronic health problems stemming from cleanup exposures as well.

We believe with proper medical attention from physicians who specialize in occupational chemical exposures that these symptoms can be relieved and some degree of health restored. Given the bright trail blazed by the Gulf War vets and the Vietnam vets (Agent Orange exposure), we also believe that some of the EVOS cleanup workers may prevail in a toxic court case. The injured workers' job would be made easier if documents in *Stubblefield v. Exxon* (1994) are made public.

But, most importantly, we believe that the public and *all* the EVOS workers, sick and healthy, have the right to know about the hazards and health effects of exposure to oil and solvents. To facilitate this end, we believe that the federal government (OSHA) should complete its unfinished tasks from 1989. First, it should subpoena all the medical records from the cleanup, which Exxon and VECO are required to keep for thirty years, and, with these records, it should authorize a full-blown epidemiological study with independent institutions on the EVOS cleanup workers. Secondly, the federal government should *require*—and fund—an independent long-term monitoring program to address health care problems and needs of the EVOS cleanup workers.

Additional lessons learned from our investigation and recommendations for change are discussed in Part 3, the final section of this book.

Part 2:
Sound Truth

"*There is something fascinating about science. One gets such a wholesale return of conjecture out of a trifling investment of fact.*"

Mark Twain

Pre-Spill Studies

Chapter 11.

1970s Science

(Author's note: The stories in Part 2 are framed on ordinary people doing extraordinary jobs—the popular definition of heroes. When Stanley "Jeep" Rice, PhD, discovered that his story led this section, he protested, "You can't start with me! I'm just a boring scientist!" As "boring" as chemistry and "toxicology" [the science of poisons] may be to some readers, the studies conducted by Rice and his team at the federal [NOAA] Auke Bay Lab are the mortar that glues together all of the wildlife and habitat studies. The composite yields a comprehensive new understanding of oil effects on marine life. Rice's story provides an insider's view on the process of developing water quality standards to protect aquatic life.)

Understanding Oil Toxicity

In 1971 a young scientist named Stanley Rice, PhD, was hired directly out of graduate school to start an oil pollution program at the NOAA National Marine Fisheries Service (NMFS) research lab in Auke Bay, Alaska (Rice 2001). Alaska had been a state for just twelve years, but it was very clear that oil was going to be a large part of the state's future. In 1967 oil had displaced fisheries as the leading source of income for the young state (Coates 1993). Discovery of the huge

Prudhoe Bay oilfield on the North Slope in 1968 had triggered
national debates over where to site the proposed pipeline to bring
Alaska's oil to thirsty domestic markets. The route favored by the
state and oil industry, and eventually authorized by Congress, brought
the oil down an 800-mile pipeline from the North Slope south to Port
Valdez. From there it went by tankers through Prince William Sound
and out Hinchinbrook Entrance to ports mostly on the West Coast of
the United States. Oil pollution from daily tanker operations and the
possibility of a catastrophic spill were acknowledged risks, considered remote by some, but very real to others. It was these risks that
brought Rice to Alaska.

"Jeep" Rice, as his friends call him, is a toxicologist, a scientist
who specializes in studying poisons and their effects on aquatic life.
After passage of the Clean Water Act in 1972, then the Trans-Alaska
Pipeline System (TAPS) Authorization Act in 1973, Rice was charged
with establishing "safe" levels of oil for aquatic life. Specifically, he
was to determine "safe" levels of oil that could be discharged from
the oily water treatment plant at the planned tanker terminal in Port
Valdez without harming the port's marine life. The Clean Water Act
required end-of-the-pipe standards to control pollution (Elder, Killam,
and Koberstein 1999). Each state is given responsibility to establish
its own water quality standards, so Rice and his team of federal scientists worked closely with state scientists to establish "safe" standards for oil in Alaska.

Determining safe levels of exposure to crude oil is challenging,
because each type of crude is a unique blend of hundreds of different hydrocarbons and other elements, a product of plant and animal matter cooked at different temperatures and compressed
through geologic time. Prudhoe Bay crude is chemically distinct
from crude oils from other areas in or outside Alaska. One crude oil
may be more or less toxic than another, depending on its unique
chemical makeup.

The chemical compounds in crude oil that pose a health threat
to aquatic (and human) life are *"aromatic hydrocarbons."* Aromatic
hydrocarbons contain one to five ring-shaped benzene molecules,
one of the oldest known human carcinogens. There are two types of
aromatic hydrocarbons: those that dissolve rapidly in water or air and
those that do not. Since the scientists were charged with establishing
water quality standards, they decided to simplify matters by focusing
on only those aromatic hydrocarbons that readily dissolved in water.
The *"water soluble fraction (WSF)"* of crude oil is mostly one- and

two-ring aromatic hydrocarbons. The scientists decided to largely ignore, for the time being, the *"polycyclic aromatic hydrocarbons (PAHs),"* the three- to five-ring aromatic hydrocarbons, which do not readily dissolve in water.

Scientists nationwide worked to develop standardized methods and tests that could be applied to all crude oils. To prepare crude oil for toxicity tests, scientists add crude oil to seawater in glass jugs and shake the mixture, using a paint shaker, for a set time at a set temperature. Then they allow the mixture to sit upright until black oil, containing PAHs and other oil compounds, rises to the surface and is thrown out. The remaining seawater contains dissolved crude oil fractions—the WSF—invisible to the eye, but deadly to aquatic life. Next, lab tests are conducted with living organisms to assess toxicity of the dissolved WSF, not the whole oil. These tests are called *"bioassays."* Dilute concentrations of the crude oil WSF are added to glass beakers along with test organisms such as juvenile fish or a variety of small invertebrates related to crabs and shrimp. The bioassays are terminated after forty-eight or ninety-six hours, which fit conveniently into a five-day workweek, and the number of organisms that died is recorded (U.S. EPA 1991).

Bioassays determine how many things *died* at different concentrations of WSF, ranging from zero to full strength WSF. From this range of concentrations, scientists calculate the concentration of WSF that kills 50 percent of the test organisms. The WSF is measured as *"parts per million"* or parts of WSF per million parts of water. To determine "safe" levels of exposure, the U.S. Environmental Protection Agency (EPA), the agency responsible for implementing the Clean Water Act, simply divides the level that kills 50 percent of the test organisms by one hundred (Rice et al. 2001). Rice (2001) called this "guess-work," but he explained that, at the time, scientists thought that one one-hundredth of the short-term toxicity level *ought to be safe* for fish and wildlife. The science of toxicology was still in its infancy in the 1970s and there was—and still is—a lot of room for improvement (Peterson et al. 2003).

During pipeline construction from 1974 to 1977, Rice and his team, including biologist John Karinen, chemist Jeff Short, and others, conducted standard bioassays to determine the toxicity of Prudhoe Bay crude to local pink salmon fry, shrimp, and other Alaskan species (Rice, Short, and Karinen 1977). Rice also directed his team to run thirty-day bioassays. They found that the longer creatures were exposed to the crude oil WSF, the less oil it took to kill them. This led them into a decade of research, during the 1980s, on "*sublethal*" effects of crude oil on pink salmon and herring (Karinen 1988; Moles, Babcock, and Rice 1987; Moles and Rice 1983; Rice et al. 1984; Rice et al. 1987a; Rice et al. 1987b).

Rice (2001) said, "We worried about long-term effects, but there really wasn't a way to measure more subtle effects." Rice suspected these subtle "*sublethal*" effects, such as reproductive damage or delayed growth, could also harm a species' ability to survive. In worrying about long-term effects of oil, Rice was ahead of his time, but he couldn't find funding to study these effects. This would all change twenty years later (Chapter 20).

Meanwhile, scientists across the country found that different crude oils were toxic to aquatic life at levels of 1 to 30 parts per million WSF. Dividing the higher number by the "safety factor" of 100, EPA established federal water quality criteria for some individual PAHs, but none for total PAHs. Instead EPA listed 300 parts per billion total PAHs as a guideline for states to use to develop standards (U.S. CFR45 79339 1980). (The guideline was considered to be the "Lowest Observed Effect Level" or LOEC.)

Rice and his team found that just below 1 part per million of crude oil WSF killed some of the more sensitive Alaskan species (Rice, Moles, and Karinen 1979). Applying the "safety factor," the state of Alaska adopted a total PAH criterion of 10 parts per billion and a total aromatic hydrocarbon criterion of 15 parts per billion. These criteria were more stringent than those established by the federal government.

The state water quality standards were established in 1979, less than one year after oil started flowing down the Trans-Alaska pipeline and the Valdez tanker terminal became operational. Alaska's standards are tough: they are the most stringent state standards for oil in the nation then and still now, twenty-five years later. Ten parts per billion of aromatic hydrocarbons is *less than one-hundredth* of a teaspoon (0.05 ml) of crude oil in an Olympic-size swimming pool.[1]

Baseline Studies

While Rice and his team were developing the state water quality standards for oil, they also started "*baseline*" studies to establish the types of sealife and background level of oil pollution in Port Valdez and Prince William Sound before oil started flowing down the pipeline (Karinen 1998). At the tanker terminal area in Port Valdez, they collected sediment samples and counted the number of animals and species present along the beaches and in the shallow nearshore area. Along beaches on either side of the proposed tanker lane in Prince William Sound, they collected intertidal sediments and mussels.

These baseline studies, conducted from 1977 to 1980, established that the tanker terminal area was relatively free of oil pollution *prior to startup* of oil shipping operations (Rice et al. 1981; see also U.S. Dept. of Commerce 1989). The studies also showed that the beaches of Prince William Sound near the tanker lane were generally free of oil pollution before the *Exxon Valdez* spill. Rice and his team found very low levels of hydrocarbons—PAHs—in the sediment at their two sites on either side of Hinchinbrook Entrance (Karinen and Babcock 1991).

Short analyzed the chemical composition of the Hinchinbrook hydrocarbons and determined the PAHs were from a natural source, but there were no such sources in the Sound. However, the North Gulf Coast from the eastern edge of the Copper River Delta and east to Yakutat Bay is peppered with small natural oil seeps and large coal fields. The powerful Alaska Coastal Current flows westward along this coast and straight into the Sound at Hinchinbrook Entrance (Royer 1982). The NMFS team reasoned this current could be transporting hydrocarbons into the Sound (Short 1998).

They searched upstream for possible oil sources, following the path of the Alaska Coastal Current. Short found the chemical composition of the Hinchinbrook hydrocarbons was consistent with the pattern for oil seeps near Katalla, seventy-odd miles to the east in the Copper River Delta. He realized that the source of the Hinchinbrook PAHs could also be coal fields in this area and to the east. Short (1998) explained that, given the state of analytical chemistry in the 1970s, coal and oil were difficult to distinguish on the basis of hydrocarbon composition patterns.

Wildlife, however, could distinguish between coal and oil. By the end of the 1970s, mussels were used worldwide to monitor oil pollution (National Research Council 1980). This shellfish readily soaks up the PAHs in oil and retains them in their tissues in a process called *"bioaccumulation."* Coal, however, is biologically inert: the PAHs are tightly locked into the solid matrix and cannot be absorbed by mussels or other wildlife (Short 1998). Short found that the mussels from the study sites in Hinchinbrook Entrance were free of hydrocarbons

(Short and Babcock 1996), which led him to suspect that the source of the Hinchinbrook hydrocarbons was coal.

At the time, the exact source of the Hinchinbrook PAHs seemed totally irrelevant—there were only very low levels of the PAHs, they were found in only one area, and they weren't a source of contamination for area wildlife. However, this minor detail would return over twenty years later as a roaring dragon, flamed by Exxon's strategic bellows (Chapter 21).

Alyeska's Studies: Deception by Design?

The TAPS owners did not object to Alaska's stringent water quality standards—the oil companies had promised there would be minimal oil pollution in Port Valdez (Redburn 1988; Townsend Environmental 1994).[2] The TAPS owners, including the big three—British Petroleum (BP), Atlantic Richfield Company (ARCO), and Exxon, which owned over 90 percent of the pipeline—formed a consortium, Alyeska ("al-e-ES-kah"), to operate and maintain the pipeline and the tanker terminal (Coates 1993). The Clean Water Act allows industry to monitor its own discharges for pollution, but unbeknownst to the public and government regulators, the TAPS owners and Alyeska cheated.

By the mid-1980s, Alyeska and its contractors had a full repertoire of manipulating and fabricating scientific data, as summarized by investigative reporter Charles McCoy of the *Wall Street Journal* after the spill (1989). Oil levels in the tanker terminal discharge regularly (and greatly) exceeded permitted levels. According to industry insiders, the oil levels "would have been off the scale" had the water samples been analyzed in Valdez. Alyeska scientists knew this, so they sent the samples—*unrefrigerated*—to Seattle. By the time the samples were tested, the toxic WSF had been partially broken down by bacteria in the warm temperatures or adsorbed, along with the PAHs, by the rubber test tube stoppers. The loss of the WSF and PAHs during transport and storage was usually sufficient to make the test results be within EPA's permitted discharge limits.

"Science" was easily bought by the powerful TAPS owners. Unfortunately, this track record of scientific abuse continued after the spill (Cohen 1992; U.S. EPA 1992).

Sound Truth

While the scientific and political controversy raged over how much or how little oil was in Port Valdez water and whether or not these levels were in compliance with permits (Epler 1985a, 1985b, 1988a; Pasztor and Taylor 1986; U.S. GAO 1987), the Sound presented her own "truth." The *Macoma* clam population in the mud flats near the tanker terminal crashed after TAPS became operational—85 percent of the clams vanished between 1978 and 1984 (Myren and Pella 1977; Myren, Perkins, and Merrell 1992).[3] These tiny clams filter water for food and are very sensitive to low levels of PAHs, which they concentrate in their tissues. Toxic hydrocarbons slipped undetected through the tanker terminal and slowly accumulated in Port Valdez sediments (Epler 1988b).

Early Oil Spill Studies
(1989 to 1992)

Chapter 12.

Tracking the Oil

In Seawater, Sediment, and Mussels

By Sunday, 26 March 1989, while NOAA's spill response team was still setting up computers at the command center in the Valdez Westmark hotel to track the oil spill, NMFS research scientists John Karinen and Malin Babcock were out in the Sound, collecting intertidal beach sediments and mussel samples from baseline monitoring sites (Karinen 1998; Karinen et al. 1993). These would be the only samples collected from these beaches before they were coated with oil during Sunday night's storm. In addition to the eight historical monitoring sites last sampled in 1980, they collected intertidal sediment and mussel samples from eleven new sites—seven before the beaches were plastered with *Exxon Valdez* oil and four outside the path of the oil. By hustling over Easter weekend to collect samples before the beaches were oiled, the scientists were able to establish that there were no changes at their sites despite twelve years of tanker traffic. The Sound was generally free of oil pollution immediately before the *Exxon Valdez* spill.

On Monday, 27 March, the NMFS Auke Bay Lab near Juneau held its annual three-day program review for the public and NOAA

191

administrators from Washington, DC. Jeff Short, the chemist who had worked with Rice on oil bioassays in the mid-1970s, was at the meetings, and he had a mission. He corralled some of the top administrators in NMFS and NOAA and asked for permission to measure oil concentrations in the water column under the surface slick.

Short (1998) argued that this oil, particularly the toxic aromatic hydrocarbons, could injure or kill larval, juvenile, and adult fish as well as copepods, which are food for everything from fish to humpback whales. He expected any oil dissolved or captured as whole droplets in the water column would be rapidly diluted and dispersed. He also knew the PAHs were the best indicator of the potential toxicity of spilled oil as they could be picked up directly by marine life or indirectly transferred through the food web. He needed to find out how long measurable amounts of oil—especially the persistent PAHs—were present in the water column to lay a chemical foundation to explain any biological effects on fish and wildlife.

Short knew he was asking his supervisors to make a commitment to document effects from the spill without guaranteed funding—there was no time for budget reviews or congressional appropriations. He waited anxiously for a response. A couple of hours later, Rice, now habitat program manager at the lab, told him, "Pack your toothbrush. You're leaving here in 48 hours and, before you leave, you've got to figure out how you're going to sample under an oil slick. Try not to spend more than $15,000" (Short 1998).

Panicked, Short explained the problem to old Norm Johnson who ran the maintenance and machine shop at the lab: how do you get a water sample under a giant oil slick from a boat in the middle of the slick without contaminating the sample by the slick itself?

Johnson figured it out. He bent stainless steel tubing, heat-sealed a Tygon (special plastic) tubing sleeve over one end, and connected the other end to compressed air. The end with the protective sleeve penetrated the slick and was deployed to a desired depth. Then the air compressor blew off the sleeve, which was connected by twine so it could be retrieved without littering. The air compressor was swapped with a suction pump that pulled the water sample onto the boat. The water sample was immediately treated with an organic solvent to separate any oil from the seawater and preserve the oil. This last step was extremely important. In unpreserved seawater samples, naturally-occurring bacteria would break down some of the oil before the sample could be analyzed in a lab, making there appear to be less oil than there really was. This would lead to incorrect results and misleading conclusions.

One week later Short and Pat Harris were in the Sound with a fishing boat charter arranged by CDFU. They witnessed what few others did—the huge slick moving through the Sound. Short (1998) recalled, "Northwest Bay (Eleanor Island) stank to high heaven—there was oil everywhere as far as you could see . . . We got to AFK [Armin F. Koernig] hatchery in Sawmill Bay in the middle of the night and all we could see were boats and booms and oil everywhere—it was a war zone . . . The north end of Smith Island was just devastated. There was oil everywhere, not just on the beaches, but up in the trees. But what bothered everybody was the deathly silence. That's when it really hit home. You know, usually when you're out on the water, you hear gulls, ducks, marine mammals, something. We'd seen a lot of ugly sites up until then—quite a number of dead sea otters and dead sea lions coated by oil and floating in the water, a still-born sea otter pup at Green Island. But the thing that really got to us was the absolute dead quiet at north Smith Island."

Short and Harris sampled thirty stations in four days, zigzagging their way around the Sound as they followed a cruise plan that alternated sampling between a clean site and a contaminated site to prove there was no carry-over contamination between sites. It meant a lot of traveling and basically no sleep. When they returned to Cordova, they rested for two days, and then they went back out to re-sample all thirty stations (Short and Harris 1996a).

While on the first seawater-sampling trip, Short and Harris shared a mutual concern. It was well known at the time of the *Exxon Valdez* spill that oil loss from a slick was rapid, involving initial losses of volatile hydrocarbons into the air and water column immediately after a spill. It was also well known that oil levels in the water column peaked immediately after a spill and then rapidly decreased as the oil became diluted and dispersed. They knew that once the giant slick passed through the Sound, oil would be increasingly difficult to detect in water samples. However, they suspected that oiled beaches would contaminate nearshore waters with low levels of PAHs that might be potentially harmful to sealife. Short and Harris had decided to see what oil they could find.

During their second seawater-sampling trip later in April, Short and Harris designed a caged mussel study while their skipper drove the boat to different stations. They knew a mussel filters about eight gallons of seawater daily and could accumulate PAHs in its fatty tissue.

They wanted to take mussels from a clean area, put them in cages, and suspend the cages in the water column at various depths near the shore. They figured mussels would soak up any oil in the surrounding water and capture "pulse events" or times when shots of oil might be mobilized from beaches such as after storms or during beach cleanup operations. Short and Harris thought caged mussels would be more reliable indicators than seawater of persistent low levels of oil. They planned to discontinue their seawater samples, rather than spend precious dollars collecting thousands of meaningless samples.

Their boss Rice agreed and helped them get permission from NOAA for the mussel study. During their last seawater sampling cruise in May, Short and Harris started the first of about 380 deployments of caged mussels at twenty-two stations in the Sound and sixteen locations outside the Sound at depths from three to eighty feet. They collected the mussels once a month from May through August, restocked the cages with fresh mussels, and reset them at the various depths. They continued this study through 1991 (Short and Harris 1996b).

The results of this study startled the Auke Bay Lab scientists. In the Sound at depths, where Exxon had found little or no oil in its seawater samples, the caged mussels revealed PAHs from *Exxon Valdez* oil were present in the water column through August, the last month sampled in 1989—*five months after the spill*. The path of submerged oil mirrored the path of the surface slick as it swept through the Sound and out into the Gulf of Alaska in 1989. The submerged oil was like a big underwater cloud that drifted through the water column. This cloud had the highest levels of PAHs in shallow seas adjacent to heavily oiled beaches. The PAH cloud dissipated over time, with depth, and with distance from oiled beaches so that by 1992, the last year of the study, it mostly persisted only off beaches that had been heavily oiled. Outside the Sound, the caged mussels told a different story. The PAH cloud had dissipated quickly and PAHs from the spill were found in caged mussels only sporadically in 1989 and were generally below detection limits in 1990 and 1991.

The caged mussel study proved that oiled beaches acted like oil reservoirs and continued to contaminate the adjacent shallow sea with toxic PAHs that could readily be picked up by local sealife. This study laid the chemical foundation for further examination of the biological effects of the *Exxon Valdez* spill.

* * *

While Short and Harris collected seawater samples and conducted the caged mussel study, their colleagues at the Auke Bay Lab collected sediment samples to track the movement of oil from the beaches to the seafloor. They focused their studies on Prince William Sound beaches where *nearly half* of the spilled oil stranded. They didn't have to look very far or very deep for oil. In 1989 they found the highest concentrations of PAHs immediately at the water's edge at the lowest reach of the tides, well below where the oil initially stranded (O'Clair, Short, and Rice 1996; Spies et al. 1996). *Exxon Valdez* oil was smeared across the shallow subtidal zone down to a depth of about 65 feet below the lowest tidal reach with the highest concentration of oil at the shallowest depths. They realized this oil probably was washed down the beach by the cleanup operations. At depths of over 130 feet, they lost the chemical trail of *Exxon Valdez* oil as it diluted and dispersed with depth.

Oil levels in sediments declined rapidly after 1989 and in 1991, the last year of the study, *Exxon Valdez* oil was positively identified in shallow subtidal sediments from only the most heavily oiled beaches. Lab tests showed that toxicity of intertidal sediments to standard lab test organisms also declined from 1989 to 1991 (Wolfe et al. 1996). The Auke Bay Lab scientists assumed that the oil had degraded and weathered on beaches with less initial oil. However, several years later, scientific advancements in understanding the oil "*weathering*" (degradation) process allowed positive identification of *Exxon Valdez* oil in archived samples from beaches where they had formerly found none (Short and Heintz 1997). The oil *was* still there; it was just a matter of refining the measuring tools to detect it. The scientists' struggle to detect low levels of lingering oil did not prevent this oil from harming sealife (Chapter 20).

Other federal scientists conducted standard bioassays with oiled beach sediments. The residual oil in intertidal sediments was potent enough to kill test animals for two years—1989 and 1990—but not thereafter. This lethal effect matched the area where the *Exxon Valdez* oil had caused major biological damage. Scientists wondered whether oil concentrations were high enough to cause subtler harm than outright death in fish and wildlife that used intertidal beaches.

Exxon's Studies: Deception by Design?

During their second seawater-sampling cruise in April, Auke Bay Lab scientists Short and Harris crossed paths with Exxon scientists who were also collecting water samples. Short talked briefly with the Exxon scientists and he observed that their seawater samples were stored directly in plastic containers (Neff and Stubblefield 1995, 147). Short immediately had doubts about their experimental design. He knew plastic containers would absorb PAHs, and bacteria in the seawater would break down whatever was left. Short figured, by the time the unpreserved seawater samples in their plastic containers were analyzed in the lab four to seven days later, there might not be any PAHs left. This trick of not properly storing oily seawater samples, which resulted in erroneously low hydrocarbon levels had been used before the spill by Alyeska's scientists (Chapter 11).

Exxon scientists used other strategies that reduced measurable levels of oil in the seawater samples. For example, during their first two seawater-sampling cruises, Exxon scientists used EPA's methods for preserving and analyzing samples of *drinking water* (Neff and Stubblefield 1995, 147). This method was totally inappropriate for seawater with its relatively high levels of bacteria. Exxon was well aware of the presence of these bacteria (Atlas, Boehm, and Calder 1981)—they were the basis of Exxon's much-touted bioremediation program for shoreline cleanup operations and of Alyeska's biological treatment system at the tanker terminal facilities. How could bacteria *not* eat oil in their seawater samples, yet still eat oil off the beaches and in the treatment ponds at the terminal? (Braddock et al. 1996; Leahy and Colwell 1990). It appears that Exxon was simply counting on people to not catch the deception. Incorrectly preserving seawater samples would have substantially *underestimated* the amount of oil in the samples.

By their third cruise in late April, Exxon scientists started to collect their seawater samples closer to the surface slick and to preserve the samples correctly, using the same EPA methods used by Short (Neff and Stubblefield 1995, 147). Not surprisingly, the levels of oil including PAHs in Exxon's seawater samples *increased*. Exxon scientists reported the highest level of oil occurred in late April, a month after the spill (Neff 1991; Neff and Stubblefield 1995, 158, Figure 8).

These results defied simple physics and known weathering patterns of oil spills, yet Exxon scientists ignored the critics and insisted their results were valid. However, Exxon scientists provided only a generalized summary of their data, which included data that were *averaged* across depths and even across areas. This made it impossible for others (including the author) to verify their results. They did the same—reported "averages" with no raw data—for their sediment samples and mussel tissue samples (Boehm et al. 1995, 392–393, Table A-1, and 396–397, Table A-3, respectively).

Huff, author of the popular book *How to Lie with Statistics* (1954), points out that such "an unqualified 'average' is virtually meaningless" (29). He warns, "Watch out for an average, variety unspecified, in any matter where mean and median might be expected to differ substantially" (127). This number game is similar to the situation in which a company has several high-paid corporate officers and a lot of lower paid workers. The average salary shows everyone has a pretty fair wage, however, the median—literally, the middle— shows that most folks are paid far less than the officers. Huff devoted an entire chapter to "the well-chosen average" in his book.

In Exxon's case, Exxon scientists amassed a huge data set with thousands of seawater and sediment samples and limited mussel tissue samples (Boehm et al. 1995) or surrogates for mussels (Shigenaka and Henry 1995). However, most of the samples had no *Exxon Valdez*-derived oil because they were collected from areas where there was not expected to be any such oil (too deep, out of oil path, too long after spill, etc.). This way Exxon's averages had little oil. These well chosen averages are the foundation for Exxon's claim that the oil spill has had only short-term effects on sealife.

For example, Exxon scientists argues that their average oil concentrations in seawater were less than the state's water quality standard in 1989 and, therefore, too low to outright kill sealife (Maki 1991; Neff and Stubblefield 1995, 170). Exxon scientists conducted bioassays to prove their point, using various non-Alaskan species and seawater stored in the plastic containers that Short had seen. Not surprisingly, whatever oil may have been left in the seawater by the time the bioassays were conducted was no longer toxic.

Exxon scientists also focused their sampling effort in deeper offshore areas of Prince William Sound and the Gulf of Alaska, well away from where most of the spilled oil stranded on the beaches (Page et al. 1995b). In these deep areas, below about 330 feet, Exxon discovered "oil," not *Exxon Valdez* oil, but petroleum hydrocarbons derived from the coalfields at the Bering River on the eastside of the Copper River Delta. The Auke Bay Lab scientists had also documented occurrence of these petroleum hydrocarbons at great depths, but they had ignored these hydrocarbons as their purpose was to document and track *Exxon Valdez* oil. However, Exxon scientists made such a huge ruckus about this source of contamination, and its implications for the ecological health of the Sound, that the NMFS scientists were eventually drawn into the fray to defend their own work. This controversy is covered separately (Chapter 21).

Long-Term Commitment

Exxon's spill offered a unique opportunity to advance scientific understanding of effects of oil in marine ecosystems, because long-term funding was finally available. From 1989 through 1991, public dollars were directed towards assessing injuries to fish, wildlife, and habitat from the spill. The 1991 civil settlement (Chapter 1, Sidebar 2, p. 9) provided $900 million over ten years to monitor, enhance, or restore wildlife and habitat injured by the spill. It established the *Exxon Valdez* Oil Spill Trustee Council (EVOS Trustee Council) to oversee expenditure of funds for this purpose.

"*Public-trust scientists*" were individuals who studied oil spill effects on the ecosystem with public dollars—through the EVOS Trustee Council, public agencies, or, in a few cases, with private (non-industry) funds. Some public-trust scientists, such as the NMFS Auke Bay Lab scientists discussed in this chapter, realized that the fate and effects of the spill should be crystal clear because of the unspoiled nature of the Sound before the spill. From the start, the Auke Bay Lab scientists committed a huge effort—about fifty people—to studying various long-term effects of the spill over the next decade. This included setting up a very expensive analytical facility from scratch for petroleum hydrocarbons. This was the first such lab in Alaska and it is still one of only a handful of such dedicated labs in the country.

Exxon also committed a huge amount of financial resources to studying the fate and effects of its spill, but for other reasons than those of the public-trust scientists. Exxon was liable for damages to both public resources *and* private individuals. While Exxon settled its damages for public resources in 1991, it did not even try to settle with private individuals. With the court trials looming in 1994 and billions of dollars of potential liability outstanding, Exxon scientists conducted studies to show that its spill had little effect on fish, wildlife, and other sealife. "*Exxon scientists*" were those funded by Exxon directly or indirectly through contributions to academic institutions.

Public-trust scientists conducted studies to quantify injuries and to monitor and understand recovery. Exxon scientists took the position that the levels of oil in the water column were too low to cause biological harm and they designed and conducted studies to validate their position. Both groups of scientists found what they were looking for. Huff (1954) warns, "It is dangerous to mention any subject having high emotional content without hastily saying whether you are for or agin it" (46). As the following chapters show, the public-trust scientists contend that Exxon shrouded its studies in an aura of scientific credibility and neutrality that the studies simply did not have, because Exxon was "agin" finding that its oil harmed Alaska's sealife.

These conflicting motivations set the stage for a bitter battle among scientists over the extent of harm to sealife injured by Exxon's spill and the prognosis for recovery. Eminent coastal ecologist Dr. Peterson noted, "[R]elatively simple decisions, assumptions, and conditions of the study design can . . . explain how conclusions may be dictated by choice of study design" (Peterson et al. 2001, 256). Huff calls this *"bias"*—whether conscious or unconscious (123). This chapter shows, for example, how simple choices in study design involving collection and care of seawater samples and use of mussels led to different results and polarized conclusions.

The following chapters present the public-trust scientists' journey of discovery—and expose some of Exxon's deceptions with the help of Huff's book. Huff's book has stood the test of time: it is still popular, still in print—and still true!—half a century after it was first published. Huff (1954) refers to the use of statistics to deceive as "tricks" (9). I adopted this term for Exxon's various statistical gyrations, intended to confuse, at best, or deceive, at worst.

Sound Truth

While the scientific controversy raged over how much or how little oil was in the Sound, and whether or not this oil was harmful to marine life, the Sound again presented her own "truth." Certain marine mammals, birds, and fish struggled to recover in areas of the Sound that had been most heavily oiled. A decade later, recovery in these areas still lagged well behind that in lightly oiled areas of the

Sound (EVOS Trustee Council 2000; Peterson et al. 2003).

In the next chapter, I start tracing the oil spill's effects on the Sound's ecosystem by first examining what happened to beaches and the sea plants and animals that live on the shores.

Chapter 13.

Coastal Ecology

Rocky Shores and Serendipity

Two weeks before the *Exxon Valdez* oil spill, Seattle-based coastal ecologist Jon Houghton, PhD, lost his job when his employer, a regional environmental consulting firm, found out about his plans to start his own business (Houghton 2000). Houghton immediately formed Pentec Environmental, Inc. He was working alone from his upstairs alcove and looking for business when he first heard news of the Alaska spill. Houghton called Exxon's senior science advisor Al Maki, PhD, who he had worked for previously on an Exxon project in Prudhoe Bay. He asked Maki if Exxon was interested in hiring a team of scientists to study oil spill effects on coastal ecology, the rich community of sea plants and animals that live on beaches. Maki hired him as a subcontractor through Houghton's former employer.

Houghton worked with a team of colleagues, including Dennis Lees and Bill Driskell as co-leaders. Lees and Driskell had done a considerable amount of work in Prince William Sound in the late 1970s, and Houghton and Lees had worked for the EPA in the 1980s, analyzing a very large collection of plants and animals from beaches in Puget Sound (Zeh, Houghton, and Lees 1981). As part of the latter study, Houghton and Lees had made recommendations for studying changes to *"community structure"* (assemblages of plants and animals) in the event of an oil spill. "So really what we did in Prince William Sound," Houghton (2000) said, "was implement those recommendations. They formed the basis of our study design."

Within five hectic days, the team was in the Sound, collecting samples. Their study focused on the *"intertidal zone"*—the beach exposed between high and low tides, and on the main habitat types in the Sound—rocky beaches, boulder-cobble beaches, and mixed sandy-muddy beaches.

Scientists estimated that about half of the spilled oil wound up on intertidal beaches, mostly in the Sound (Spies et al. 1996). These beaches and the adjacent shallow subtidal environment play a vital role in maintaining the overall health of the Sound's ecosystem. Intertidal rockweed and other algae, and subtidal eelgrass beds and kelp forests, provide food and shelter for a richly diverse community of animals. The small bottom-dwelling fish, mussels, snails, clams, marine worms, sea stars, sea urchins, small crabs, and other crustaceans such as tiny amphipods and copepods provide food for myriad larger predators. Migratory and resident waterfowl and shorebirds, sea otters and river otters, and even bald eagles, deer, and bears forage in this critical transitional habitat between sea and shore. In spring, the region serves as spawning ground for Pacific herring and feeding grounds for legions of migratory birds; in summer, a nursery for young salmon, herring, Dungeness crab, sea otters, and other young of the year; and in fall and winter, a life-sustaining foraging grounds for resident wildlife.

Houghton and his team were initially overwhelmed by the enormity of their task. All the beaches were remote and difficult to access. In their haste to sample beaches before the oil stranded, they didn't sample as systematically as scientists who initiated later coastal ecology studies. They couldn't: they had no idea what beaches would get oiled or treated.

After oil hit the beaches, Houghton and his team were again overwhelmed by the devastation, but heartened to find that much of the rockweed *Fucus*, a perennial brown seaweed that dominates the intertidal zone, survived the heavy oiling (see also Van Tamelen and Stekoll 1996a). Often the oil was not able to stick to these plants. When the oil stuck and dried, the plants shriveled and appeared dead, but in June, the team observed new growth on these plants.

Rockweed provides vital cover, protecting beach dwellers from predators, temperature extremes, and drying out between tides.

With the *Fucus* still largely intact, the team hoped the animal populations would recover rapidly in areas where many had been smothered or poisoned when the oil washed ashore. The oiled beaches reminded the team of a war zone where the buildings had not been destroyed—new colonists or survivors just had to move back to occupy the vacant spaces. Their hope was short-lived.

In April Exxon started its shoreline cleanup program. Houghton and his team sampled in the wake of the pressurized hot water wash and were sickened by the destruction it caused to the intertidal community. Nearly 90 percent of the rockweed was destroyed—blades scalded and holdfasts blasted from rocks. Up to 95 percent of the intertidal animals, and 100 percent of some species, that had survived the initial oiling were killed or displaced by the strong, prolonged, directed blasts of 140 degrees Fahrenheit (°F) water (Driskell et al. 1996; Houghton et al. 1996; Lees, Houghton, and Driskell 1996). Thermal limit of most intertidal organisms was well below 95°F. The creatures were literally cooked. Houghton and his team were horrified: the "cure" was worse than the original harm caused by the spill.

Exxon did not like the Houghton team's findings. According to Houghton, he and the others tried "real hard" to get Exxon to allow them to continue to document the biological effects of the cleanup operations. The scientists became so frustrated with Exxon's stonewalling that they prepared—unsolicited—a little paper about their field observations and impressions of the cleanup. Houghton (2000) said, "We drew analogies to the pacification of villages in Vietnam where, if you thought there were any Vietcong in the village, you napalmed it to cure it. If there was any oil on the beach, you hot water washed it to cure it—which had the same effect of killing everything that was there." Houghton delivered the paper to Exxon's Valdez headquarters as he left on a two-week break.

While Houghton was gone, Maki, who was Exxon's lead scientist, called in Houghton's colleagues Driskell and Lees to account for their paper. According to Driskell (2000), the meeting was very tense. Exxon was not at all happy with the evidence that its pressurized hot water wash was destroying what was left of the intertidal community. Despite a verbal grilling, they stood their ground. Adding to the drama, Driskell said, "There was one Exxon person sitting quietly in the corner behind the executive—an advisor, a tactician, probably a lawyer, we thought although we never found out, who just stoically glowered at us two intimidated biologists the whole time. His dark energy was stifling." Driskell and Lees were told there was no way Exxon could stop the cleanup or even talk about its damage; it would be "political suicide." Only if NOAA or the Coast Guard said to stop, could Exxon stop, they were told.

Despite early evidence of its biologically devastating effects from NOAA scientists (Whitney 1991), the Coast Guard never did shut down Exxon's pressurized hot water wash for reasons discussed in Part 1: Sick Workers. Hydraulic water washes became the primary beach treatment method in 1989.

Houghton and his colleagues were allowed to evaluate biological impacts of several alternative beach treatments, including three dispersants (Corexit 7664, Corexit 9580M2, and BP1100X) and one hydraulic-water washing method. They noted that Exxon's products, Corexit 7664 and 9580M2, appeared to be applied in dosages that "substantially exceed published LC50s" (lethal concentrations that kill half of the test organisms) for many intertidal animals (Lees, Houghton, and Driskell 1996, 329). These chemical products were eventually banned largely due to public opposition. The team was also allowed to conduct a brief study on biological effects of the cleanup in late summer.

That fall, team leaders from Exxon's oil spill program presented their findings separately to Exxon management and attorneys at Exxon headquarters in Houston. Houghton, Lees, and Driskell were told that they were not to discuss their findings on different programs with each other and they were not to attend meetings where other team leaders were presenting results. In their sequestered meeting, when Lees was reporting that kelp growth at oiled sites was several times lower than at reference sites, he heard one attorney murmur to another, "Are we shooting ourselves in the foot here?" (pers. comm.) Because of Exxon's legal strategy, Houghton, Lees, and Driskell were not allowed to interpret their results. The biologists had the distinct feeling that most of their work would never see the light of day. They were right.

In 1990 Exxon itself opted to discontinue its massive beach-washing operations. Exxon conducted some "spot washing" in 1990 and mostly discontinued even this scaled-down effort in 1991. Instead, Exxon's primary treatment methods became manual removal of oiled debris and application of chemicals. From 1989 to 1991, Prince William Sound beaches were doused with two Exxon products: 104,510 gallons of the liquid Inipol and sixty-five tons of the pellet Customblen. The human health costs of Inipol arguably exceeded any ecological gain (Chapters 1, 4, and 5).

Exxon also opted to discontinue another politically troublesome issue in 1990—the Houghton team's subcontract for another field season! Instead Exxon contracted directly with Houghton's former employer. However, serendipity intervened on Houghton's behalf. NOAA picked up the contract. Within one week of scrambling to get the new contract, Houghton and his team were back out in the Sound, resampling their old stations and adding new ones to strengthen their study design. Houghton and his team continued their study for the next seven years through NOAA.

In 1990 and 1991, Houghton and his team observed the first stage of recovery as waves of *"opportunistic species"*—ones that thrive in disturbed conditions—moved in to colonize oiled beaches.[1] They found sea plants and animals had more difficulty reestablishing on treated beaches than on *"set-aside beaches"*—oiled beaches that were spared the harsh cleanup. They reasoned this was because surviving individuals on set-aside beaches served as seed stock for new generations, while there were virtually no survivors on treated beaches. On these devastated shorelines, life had to start afresh with larval animals and plant stock carried from afar by drifting currents and tides.

In 1992 and 1993, Houghton and his team found that even the pressurized hot water washed beaches appeared to have recovered, at least superficially (Houghton et al. 1997). On some treated beaches, they found more rockweed than at the unoiled control sites, but this was because the population of grazers—sea urchins and others—had not yet returned to slow the seaweed's growth. They noted that the entire population of rockweed on treated beaches was of a single age, like a stand of new growth in a clear-cut forest, while on set-aside beaches, rockweed populations were of mixed ages. They were concerned that the single age stands might die off all at the same time. Houghton and his team continued their studies for another four years (Chapter 21).

In addition to the Houghton team's studies, a small legion of public-trust scientists conducted a huge coastal ecology assessment from 1989 to 1991 as part of the Natural Resource Damage Assessment (NRDA) process required by federal law (Sidebar 10, p. 206). Some of these scientists reported, "Injuries to the intertidal plant and invertebrate populations by the *Exxon Valdez* oil spill and subsequent cleanup treatment activities were documented over the entire area

Controversy, Confidentiality, and the Public Interest

In the wake of the *Exxon Valdez* oil spill, effective response and restoration were hampered by confidentiality restrictions, conflicting regulations, and lack of funds under the two federal laws that governed damage assessment in 1989.

The Clean Water Act (CWA) controls discharge and removal of oil affecting natural resources and is implemented by the U.S. Environmental Protection Agency (EPA). The EPA regulations cover immediate response studies—cleanup activities such as effectiveness of dispersant use or beach washing techniques, for example—and long-term restoration studies.

The Comprehensive Environmental Response, Compensation, and Liability Act (CERCLA, commonly known as "Superfund"), addresses recovery after a hazardous substance release and is implemented by the U.S. Department of Interior (DOI). The DOI regulations govern the Natural Resource Damage Assessment (NRDA; "ner-DAH") process, including the NRDA studies. Since these studies are designed for litigation, they are confidential until a settlement is reached. The DOI regulations also guide development of a restoration plan including a public review process.

The confidentiality of the NRDA studies hampered effective cleanup and restoration for three reasons (Cummings 1992). First, the NRDA studies could not be used to guide the cleanup to prevent more injury to oiled beaches. For example, NRDA scientists found fairly quickly that the pressurized hot water wash dramatically increased the injury to beaches, but this information could not be shared with the federal government scientists who worked on beach cleanup. The beach responders had to perform their own studies, separately from the NRDA studies (Wolforth 1990a, 1990b).

Second, the NRDA process forbade freedom of exchange among the NRDA scientists and between the NRDA and other scientists (Cummings 1992). No one knew for certain who was studying what, which made planning long-term studies difficult. No one could interpret his or her studies as part of the larger picture, which made it impossible to understand the full scope of injury to the ecosystem and to predict restoration needs. No one could refine and improve their studies because the normal critical review process was nonexistent.

Short-sighted litigation-driven science does not serve the public interest. For example, during one of the planning meetings for the coastal ecology study, a lawyer was overheard to comment, "We can't win a case with barnacles and mussels" (Highsmith 1999). Yet, ironically, over the course of the next decade, it was precisely the oiled "barnacles and mussels"—and the oil-soaked beaches—that led public-trust scientists to a new understanding of the persistent effects of oil to fish and wildlife.

Third, the NRDA process forbade freedom of exchange between the scientists and the public—the latter includes the spiller (Cummings 1992). This prevented the public from meaningfully participating in the restoration planning process required by CERCLA (ibid.).

Conflicts are compounded by the inadequacy of the DOI regulations for a spill the magnitude of the Exxon Valdez and by a spiller who can afford to outspend the federal government on "studying" spill effects. The NRDA process contained no dedicated funding mechanism for the damage assessment and restoration studies required by the law. When Exxon balked at paying, the federal government was vulnerable to pressure to prematurely settle claims to finance the assessment and restoration process (Collinsworth 1990; Harrison 1990; Stewart 1990; Suuberg 1990).

(continued)

The Oil Pollution Act of 1990 (OPA 90) was an attempt to fix the CERCLA by offering an alternative process to the confrontational NRDA process. This alternative, a cooperative planning process between the federal government and the spiller, introduced its own set of problems, which became apparent in light of the new scientific understanding of oil pollution.

The bottom line is that the federal laws were, and still are, simply not structured to deal with wealthy spillers who can afford to pay for the expensive damage assessment and restoration studies twice: once to reimburse the federal government for conducting studies and again to finance a second analysis that is more conducive to their interests.

Recommendations to strengthen OPA 90 from a public interest perspective are discussed in chapter 24.

---

of the spill, from Prince William Sound to the Alaska Peninsula," a distance spanning over 1,200 miles (Stekoll et al. 1996, 177). They found the sequence of events following the spill—initial impact, death, and recovery—was similar to that observed in other spills around the world such as the *Torrey Canyon* spill in England, the *Arrow* wreck in Nova Scotia, and the *Florida* grounding in Buzzard's Bay, Massachusetts. They predicted the rate of recovery would also be similar and take decades.

Like Houghton's team, the public-trust scientists also concluded that oiled beaches left untreated recovered faster than beaches that had been treated (Van Tamelen and Stekoll 1996a; Van Tamelen, Stekoll, and Deysher 1997). Others reported, "The main driving force causing differences between oiled and reference sites in this study was the creation of bare substrata [beach] as a result of the oiling and cleanup efforts. The rate of recovery will depend on the rate of recolonization of denuded areas with possible complications caused by any residual oiling effects" (Highsmith et al. 1996, 234).

Al Mearns, with NOAA's Hazardous Materials Response and Assessment Division, Biological Assessment Team, analyzed Exxon's various shoreline treatments in a review paper. He voiced strong concerns about the pressurized hot water wash and charged that "half of the 1989 shoreline operation was devoted to a shoreline treatment activity (Teal 1991) that left behind most of the stranded oil and killed half of all the biomass that was nominally injured by the entire event" (Mearns 1996, 323). According to Mearns, the loss of beach life was staggering—the cleanup alone killed as many plants and animals (measured by weight) as the original spill!

As a result of work of Houghton, Lees, Driskell, and others, mechanical and chemical treatments of beaches in response to oil spills have been largely stopped and the importance of conducting at least short-term post-spill monitoring has been institutionalized.

Exxon's Studies: Deception by Design?

Exxon scientists who replaced Houghton and his team also conducted a huge coastal ecology assessment program from 1990 to 1991, similar to the public-trust study. Faced with overwhelming evidence that the oil spill and cleanup caused severe and long-lasting harm to Alaska's coastal sealife, these Exxon scientists took extraordinary measures to hide the truth. They appear to have devoted a lot of time and effort into trying to show that the oil spill and cleanup had not caused lingering harm to Alaska's beaches. This was a crucial point, because recovery of the beaches basically set the pace for recovery of many injured species.

Exxon scientists obscured the obvious—their studies blurred differences between oiled and unoiled beaches. They used a number of tricks that hid the stark differences between oiled beaches, virtually wiped of life, and unoiled beaches, pulsing with life. How could this be done?

Detecting an Oil Signal

The challenge faced by all coastal ecologists who study community structure is that no two beaches are alike in the first place in terms of plant and animal species. Rocky wave-pounded beaches host collections of clingy creatures, ones that can hold firmly to rocks such as barnacles, mussels, periwinkles, chitons, limpets, and starfish. Gravel or sandy beaches host different communities of burrowing creatures such as clams, marine worms, and tiny pill bug-like crustaceans.

Even within one beach type, species and numbers of plants and animals vary, because of subtle variations in wave energy, slope, degree of ice scouring in the winter, and proximity of freshwater streams. For instance, exposed wide gravel beaches host species that are more tolerant of temperature and salinity extremes and more resistant to drying out between tides than species on narrow gravel beaches. And, as every beachcomber knows, beaches usually host patches of plants and animals along their lengths—things can look completely different ten steps further along the shore.

After an upset such as an oil spill, it is difficult, statistically, to pin a difference in community structure on oil alone, because of all these physical factors that also influence intertidal sealife. For example, if there are less of one species of periwinkle on one rocky beach than another, is it because the first beach is more exposed to battering from surf, or because the beach is more heavily oiled?

The ability of a study to detect differences from oil alone depends on its *"statistical power,"* that is, its ability to accurately detect differences from oil over and above all the differences from *"background noise"* or natural variability. This is like trying to find and listen to one radio signal above background static: the more carefully one tunes the radio, the better one hears the desired signal. Scientists determine the statistical power of their study through *conscious choices* made during study design. In the coastal ecology studies, public-trust scientists fine-tuned their radio to clearly "hear" the signal from oil, while Exxon scientists lost the oil signal by not carefully adjusting their studies to reduce the background noise. These simple choices in "tuning" their study design made huge differences in results.

Exxon scientists seem to have designed their coastal ecology studies to bury the stark effects of the spill in a jumble of statistics. Few scientists were brave enough to enter this statistical maze, but it hid the pins to burst the bubble of Exxon's claim of "ecological recovery" on beaches in Prince William Sound. Finally, over ten years after the spill, coastal ecologist Charles "Pete" Peterson, PhD, led an investigation to compare experimental designs and use of statistics between the public trust scientists' and Exxon scientists' coastal ecology studies.[2] Peterson and his team deciphered eighteen differences in design choices, which all dramatically biased Exxon's findings (Peterson et al. 2001). A closer look at just five examples will demonstrate how Exxon scientists failed to detect oil spill effects.

Threading the Statistical Maze

First, Exxon scientists built what Huff (1954) calls *"small sample bias"* into their study designs by using small samples. Exxon's beach samples were physically too small in size and there were not enough repeated samples to accurately represent the plant and animal species that lived on a particular beach. To explain the significance of small sample bias, I found it helpful to adapt an analogy used by Huff.

Imagine a barrel of beans. This barrel has lots of pinto and kidney beans, some garbanzo and black beans, and even a few lima beans. Each type of bean represents a different species—periwinkles, mussels, limpets, chitons, and so on—all the myriad life on intertidal beaches from common to rare. In their sampling procedure, Exxon scientists (Gilfillan et al. 1995a; Page et al. 1995b) grabbed only a few "beans" a few times compared to public-trust scientists (Highsmith et al. 1996; Houghton et al. 1996) who grabbed great handfuls of "beans" many times. The approach used by public-trust scientists increased their chances of accurately representing all the types and relative numbers of beans in the barrel (Sundberg et al. 1996).[3]

Peterson and his team (2001) demonstrated, in a separate test, that the small size and low level of replication of Exxon's samples did not accurately reflect the true numbers and diversity of species on the beach—the "beans" in the barrel. In Huff's bean analogy, the author notes, "If your sample is large enough and selected properly, it will represent the whole well enough for most purposes. If it is not, it may be far less accurate than an intelligent guess and have nothing to recommend it but a spurious air of scientific precision" (Huff 1954, 13). Small sample bias weakened the ability of Exxon's studies to detect oil effects and helped convince the media and public to believe something that wasn't true: that there were minimal oil spill effects on beaches.

The second example of tricks used by Exxon scientists was a creative use of what Huff calls *"the well chosen average"* (ibid., 27). Exxon scientists first *"pooled"* (combined) their data on individual species of invertebrates, then averaged the results. This obscured oil effects. Pooling is valid scientific approach in some cases, but not when it is used to obscure differences. In one instance, public-trust scientists (Highsmith et al. 1996) observed that two species of periwinkle responded differently to oil: both species decreased initially, then the more opportunistic species quickly increased to more than there were at the time of the spill. Exxon scientists (Gilfillan et al. 1995a) pooled their data for the two species—one high count and one low count—and the differences averaged out. *Voilà!* No more biological harm from oil—or at least ones that could be proven statistically. To demonstrate this misuse of pooling, public trust scientists combined their own data and found that *pooling hid 44 percent of the cases* where oil had affected the individual species of periwinkles (Highsmith et al. 1996, 234). The same thing happened with limpets and barnacles: when Exxon scientists pooled their data, the response signal from oil magically vanished (Gilfillan et al. 1995a,

415, Figure 3). Peterson's team (2001) noted simply, "[S]pecies composition matters" (275).

Beyond statistical games, species composition mattered for plant and animal communities on beaches, too. It makes a difference whether one has neighbors that are compatible or not. For example, public-trust scientists found that the opportunistic barnacle species *Chthamalus* reestablished more quickly after the cleanup than *Balnanus* (Highsmith et al. 1996). These barnacles differ in their ability to cling to rocks—*Balnanus* species have a much firmer grip. The rockweed *Fucus* that had attached to *Chthanalus* barnacles were lost when winter storm waves dislodged these barnacles from the rocks. This set back recovery of keystone *Fucus* species and further delayed recovery of the dependent intertidal community (Van Tamelen and Stekoll 1996b).

The third example of tricks used by Exxon scientists was the *error of omission*, where they selectively sampled and/or selectively reported their data. In the previous example, Exxon scientists failed to report their raw data on individual species; instead, they reported only pooled data to support their claim of no harm from the spill. In another example, Houghton's team (1996) and other public-trust scientists (Highsmith et al. 1996) observed one species of limpet *(Tectura persona)* that lived high in the intertidal zone was very sensitive to oil. Exxon scientists (Gilfillan et al. 1995a) selectively omitted the Houghton team's 1989 data for this limpet; the data clearly show this species vanished from beaches after the spill.

In fact, these Exxon scientists selectively omitted data from their own subsequent sampling program at Houghton's sites. They claimed Houghton's 1989 study focused on *"epifauna"*—animals that live on the beach surface, while Exxon's 1990 and 1991 studies focused on *"infauna"*—animals that live under the beach surface (Gilfillan et al. 1995a, 423). This was nonsense: Houghton and his colleagues thoroughly documented the effects of oiling and shoreline treatment on both groups of creatures from 1989 through 1992 (Driskell, Houghton, and Lees 1996; Houghton et al. 1996) and they found the cleanup, more than the oiling, had devastated the beach animals. Instead, in 1990 and 1991, Exxon scientists seemed to avoid making comparisons to the 1989 data set: Exxon scientists wrote, "Only chemistry samples from 1989 were included in the analyses for this program" (Gilfillan et al. 1995a, 423, Table 8, footnote).

Ironically, Exxon lawyers initially refused to give Houghton access to the 1989 data *that Houghton's team had collected* to prevent him from making obvious statistical comparisons, leading to findings that the pressurized hot water wash had been more deadly to clams and other infauna than the spill itself. Exxon lawyers were finally forced to release the data set to Houghton through legal proceedings in the litigation over damages to fishermen, Natives, communities, and others harmed by the spill.

By ignoring available data and limiting one of their studies to infauna, Exxon scientists appeared to hide the harmful oil effects. Peterson and his team commented that such omissions prevented "meaningful inferences" about community status and the recovery process (2001, 261). Huff refers to errors of omission as *"samples with built-in bias"* and he warns, "[T]he result of a sampling study is no better than the sample it is based on" (1954, 18).

The fourth example of tricks used by Exxon scientists was *improper choice of control beaches.* Scientific studies use controls to "control" the effect of natural environmental conditions that might influence the study outcome. Public-trust scientists paired oiled beaches with nearby unoiled control beaches that matched environmental conditions on oiled beaches—except for the presence/absence of oil. They did this specifically to reduce background noise—or, using the radio example, to tune out static from other signals so they could receive a crisp signal from oil differences alone.

Exxon scientists did not pair oiled sites with control sites. Rather, three of Exxon's four control sites in one study were from glacial bays in the southwest Sound (Gilfillan et al. 1995a; Page et al. 1995b). These beaches were untouched by oil, but they also had virtually no plant and animal life or, as Peterson and his team (2001) put it, "gross biotic impoverishment" (268). The cold, less saline, and more turbid glacial conditions did not support abundant intertidal sealife. When oiled beaches were compared to these bleak sites, the oiled beaches *appeared* statistically to be thriving with *relatively* more plants and animals.[4] Peterson's team pointed out in their paper that public-trust scientists had made this same error of using glacial sites for controls in 1989, but they had subsequently redesigned their study with the paired approach to match natural conditions to overcome this huge problem (Sundberg et al. 1996).

Improper choice of controls allowed Exxon scientists to maintain their credibility while making misleading statements. For example, Exxon scientists claimed "15 to 18 months after the spill, approximately 73 to 91% of the initially affected shorelines in Prince William Sound were statistically indistinguishable from unoiled reference

shorelines" (Gilfillan et al. 1995a). Careful readers will note that
Exxon scientists never said there were no oil spill effects to coastal
sealife—they only said they were not able to detect differences
between oiled and unoiled beaches. While nothing has been falsified,
the impression given does not square with the public-trust studies.

Improper choice of controls also allowed Exxon scientists to
imply that oil spills actually benefited coastal sealife: "In the shoreline
areas that were still different from reference, with few exceptions,
the communities present were more diverse, and had more species
and more individuals *than the corresponding reference sites*" (ibid.,
437, emphasis added).

When confronted with statements that defy common sense, Huff
encourages people to ask, "Does it make sense?" He wrote, "Many a
statistic is false on its face. It gets by only because the magic of numbers brings about a suspension of common sense" (1954, 137, 138).
Even if a person did not know anything about the nature of glacial
beaches or oiled beaches, Exxon's statement does not pass the common sense test and should not have been believed.

The fifth example of tricks used by Exxon scientists was "*incorrect inferences*," in which conclusions were not supported by the
study design. Public-trust scientists and Exxon scientists conducted
studies using a sampling design that specifically allowed them to predict the full extent of injury to shorelines and to estimate recovery tra-
jectories. Peterson and his team noted that public-trust scientists sampled the same beaches six years in one study and eleven years in
another. In contrast Exxon scientists sampled one year in one study
(Gilfillan et al. 1995a) and two years in another (Gilfillan et al. 1995b)
before they made sweeping conclusions that the Sound's beaches had
essentially recovered as early as 1990. These single snapshots in time
told an incomplete story and further, Peterson and his team observed
that Exxon scientists used different methods in their two-year study,
thus "preventing meaningful temporal contrasts" (2001, 264).

In cases with incorrect inferences, Huff warns, "reams of pages of
figures" and facts can lead to "a totally unwarranted conclusion"
(1954, 93). Conclusions regarding an evaluation of recovery *over
time* "cannot be evaluated by a single sampling," Peterson and his
team noted. They found that Exxon's study lacked "the ability to estimate the time course of recovery" (ibid., 265), the very purpose for
which Exxon's studies had supposedly been designed!

At the end of their lengthy technical review, Peterson and his colleagues concluded, "When any study combines several decisions that

fail to reduce error variance and fail to enhance power, the outcome of such multiple decisions is almost guaranteed to be a set of inconclusive results unable to detect even large impacts of environmental perturbation" (2001, 280)." In other words, Peterson and his colleagues found Exxon's studies appear to be *inconclusive by design*.

This type of bias creates what statisticians call "*uncontrolled Type 2 error*," which leads one to falsely conclude there are no differences when there are. In Exxon's studies, this type of statistical error allowed Exxon scientists to conclude there were no differences in beach life between oil and unoiled beaches when, in fact, there were huge differences. Huff calls uncontrolled Type 2 error simply "believing many things that are not so" (1954, 89). He said, "It is [a] sad truth that conclusions from such samples, biased or too small or both, lie behind much of what we read or think we know" (ibid., 13).

Exxon scientists appeared to have conducted credible studies, yet upon closer examination, these studies are fraught with what appear to be statistical tricks. Therefore, the conclusions drawn from Exxon's coastal ecology studies—that the oil and cleanup had little effect on beaches and that the beaches had recovered rapidly—are statements that are difficult to believe.[5]

Sound Truth

A toxic residual of Exxon's oil was buried under certain beaches in Prince William Sound. The short-term effects of this buried oil on marine mammals, birds, and fish are discussed in each of the next three chapters. The lingering effects of this buried oil on wildlife and fish are discussed in the subsection on Ecosystem Studies.

Chapter 14.

Marine Mammals

(Author's note: Sea otter biologist Lisa Rotterman declined to be interviewed for this book. While doing the research for this book, I encountered many people who still will not share their personal stories about the oil spill. Sea otters, however, are a public resource and people need to know what happened to these animals to try to avoid repeating similar mistakes in future spills. Rotterman and her husband paid dearly for trying to bring the sea otters' story to light. Their scientific papers left a trail for others to follow. This story is based solely upon these and other public records and publicly-available sources.)

Sea Otters

In 1984, Lisa Rotterman and her husband Chuck Monnett first flew into Cordova to study sea otters. They were graduate students at the University of Minnesota. Their work was funded by the Minerals Management Service (MMS) as the agency was interested in testing whether radio telemetry could be used to study adult female sea otters—their reproductive movements, behavior, and survival. MMS was concerned about potential impacts of a major oil spill on the small population of California sea otters and the agency wanted to work out methodology in Alaska before it started similar studies in California (Monnett and Rotterman 2000). Prince William Sound was the natural choice for studies as it had the highest concentration of sea otters in the world—believed to be between 5,000 and 10,000 animals (Burn 1994; Garrott, Eberhardt, and Burn 1993).

Rotterman and Monnett initiated the first long-term studies of adult female sea otters and, later, of dependent pups and weanlings. They also started a study on sea otter genetics, Rotterman's particular interest (Rotterman 1992). Like California, sea otters in the Sound had been hunted to near extinction. The Alaska animals recovered, but the remnant population in the western Sound was more established, while the eastern Sound was newly colonized. Understanding the genetic consequences of this history could perhaps help hasten sea otter recovery efforts in other areas.

The work was intense with long hours and exposure to wet, cold, and storms. They tracked their radio-tagged charges from an open skiff and learned when the animals had their pups, whether the pups survived, and where the pups moved as weanlings, then as adults. Like Dr. Jane Goodall, whose lifetime immersion with chimpanzees in Africa revolutionized understanding of these animals, Rotterman and Monnett were rewarded with a unique and intimate understanding of sea otters.

When the biologists first learned of the spill on 24 March, they were deeply concerned. USFWS was the lead agency responsible for conservation of sea otters in Alaska. Rotterman immediately drafted a proposal to the agency for comprehensive long-term studies to document effects of the spill on sea otters and their prey. This proposal was funded as a damage assessment (NRDA) study. During the next three years, Rotterman, Monnett, and their technicians monitored by plane and skiff 160 radio-tagged female and young sea otters in the Sound, as well as other sea otters (Monnett and Rotterman 1995a–c). This was a gargantuan effort; at the time, their study was the largest undertaking of radio-tagged sea otters in the world.

Before the team started their study, however, they were part of the effort to collect dead animals for the federal government. The federal government needed to quantify the number of sea otter deaths from the spill to leverage fines against Exxon (Ballachey, Bodkin, and DeGange 1994, 48). After the windstorm moved oil ashore and through the Sound, Rotterman and Monnett took out their whaler to recover carcasses. The bodies collected naturally with other flotsam and jetsam in the long wind-swept windrows that snaked their way through and out of the Sound. One day, fishermen Lynn Thorne and her crew on the *New Adventure* (Chapter 1) followed the biologists in their skiff around Knight Island Passage. Thorne filmed the team using dipnets to pull dead birds and sea otters up through the thick black goo (Weidman 2001). The biologists delivered their grisly cargo to a larger boat and then repeatedly went back into the mire to collect more carcasses.

In addition to the carcass collection effort, USFWS organized—and Exxon paid for—a highly publicized wildlife rescue operation, which started within a day of the spill (Batten 1990; Bayha 1990; Morris and Loughlin 1994). Instead of collecting dead animals caught in windrows, most volunteers and others worked near beaches where they could see carcasses and live animals. People seemed to think that sea otters and birds would be better off in the wildlife treatment centers than left alone in the Sound, because unoiled and lightly oiled animals were brought into the centers along with the heavily oiled animals (Ames 1990; Benji 1990; VanBlaricom 1990).

Pathology research under the NRDA studies suggested that many oiled sea otters died a horrible death (Lipscomb et al. 1993; Rebar et al. 1994, 1995):

Clinical, hematologic/blood chemistry, and postmortem findings, combined with previous research, suggest the following scenario. Oil-contaminated sea otters rapidly become hypothermic. They devote themselves to a life-or-death struggle to remove oil by grooming. Feeding is drastically curtailed, and energy stores are rapidly depleted. Grooming is marginally effective at best and results in ingestion of crude oil. By unknown mechanisms, exposure to oil causes interstitial pulmonary emphysema, which compromises respiration. Their desperate situation causes a powerful stress reaction. Gastric erosions (ulcers) form as the physiologic effects of stress reach a critical level. Hemorrhage into the gut begins. The combined effects of these factors overwhelm the otters; shock ensues, followed by death. Some sea otters succumb to hypothermia rapidly, and no lesions form. Others live long enough to develop some or all of the morphological markers that characterize this syndrome: interstitial pulmonary emphysema, gastric erosion and hemorrhage, hepatic and renal lipidosis, and centrilobular hepatic necrosis. Otters that are captured and taken

to rehabilitation centers are subjected to additional stressors
... (Lipscomb et al. 1994, 277 [quoted with permission]).

The final body count on sea otters collected from the entire spill region in the six months following the spill was 994. Of these, 493 were from the Sound (Ballachey, Bodkin, and DeGange 1994; Hofman 1994). Based on available carcasses, visual surveys, and limited modeling, federal researchers estimated some 3,500 to 5,500 sea otters died after the spill from an estimated population of 30,000 throughout the entire spill region (Bodkin and Udevitz 1994; Burn 1994; DeGange et al. 1990; DeGange and Williams 1990). Because there was no large-scale effort to collect carcasses in offshore windrows, it is likely that these estimates grossly understate the number of sea otters that died.

By mid-summer 1989, Exxon wanted to release its "rescued" sea otters back into the Sound. There was intense political pressure to support this undertaking (Rappoport, Hogan, and Bayha 1990). Exxon had poured out millions of dollars in a highly publicized effort to save a couple of hundred animals. Of the total 343 captured sea otters, 123—over one-third—had died in the treatment centers (Hofman 1994). Exxon wanted to release the survivors, perhaps to trumpet its success and bolster its public image (Ames 1990, 140; Batten 1990).

Rotterman and Monnett, among other scientists (Spraker 1990), were concerned that the captive sea otters might have picked up diseases from domestic animals and that releasing diseased animals could infect wild stocks (Monnett and Rotterman 1993). A herpes-like virus had already been identified in sea otters at the Seward treatment center (Haebler, Wilson, and McCormick 1990; Harris et al. 1990).

In July and August, 196 animals were released into the eastern Sound—right into the stocks of animals that Rotterman and Monnett had studied prior to the spill. Immediately after the releases, a virulent herpes virus swept through the stocks, killing a large number of prime age breeding adults (Monnett and Rotterman 1995b). The virus that devastated the stocks occurs naturally in wild populations of sea otters from the Aleutians down to Southeast Alaska (Monnett et al. 1990, 406), but it seemed that introduction of the captive animals into the wild population had somehow triggered a disease outbreak.

Rotterman and Monnett had radio-tagged forty-five of the healthiest captive otters prior to release (Monnett et al. 1990). The team tracked these animals over-winter and found that nearly half of them died (ibid). The surviving females did not reproduce nearly as well as other radio-tagged otters. Instead of supporting this telling work, USFWS greatly reduced the scope of the final data analysis for this study (Monnett and Rotterman 1995c).[1] The record suggests that decision-makers within the agency were influenced by the politics of the pro-oil Bush administration of 1992, which may not have wanted to know the truth about the effects of Exxon's spill and its wildlife rescue effort on sea otters.

Ironically, it was later reported that the thirty-seven sea otters, including thirteen orphaned pups that had been treated and released to aquariums, also died at a higher rate than other captive sea otters (Gruber and Hogan 1990; Lipscomb et al. 1994). Necropsies performed on these animals found organ damage consistent with oil exposure in adults and with shock and general poor health in pups. For pups, treatment centers had served as a poor surrogate for their natural mothers.[2]

A year passed. In the fall of 1990, all of the field scientists presented the findings from their summer research to the funding agencies and ever-present lawyers. The results of the weanling study conducted by Rotterman and Monnett were very unexpected. The team had planned to start this study in 1989 (Rotterman and Monnett 1995), but the radio-tag order was delayed. When the radio-tags did arrive, there were far fewer than expected and needed for a sample size large enough to detect an effect—or so the scientists had thought.[3] But it turned out the effects of oil were so profound that the results were statistically significant—because almost all of the weanlings died in the western Sound! (Rotterman and Monnett 1993, 1995).

This was of interest to government lawyers who were looking for ways to leverage a higher fine against Exxon for natural resource damages (Ballachey, Bodkin, and DeGange 1994). However, the critical weanling study was not repeated until 1992—and the work was done by others, not Rotterman and Monnett. Even so, results still found more weanlings had died in oiled areas than in unoiled areas, although to a lesser extent than in 1990 (ibid.).[4] USFWS did not publish its findings until over ten years later (Ballachey et al. 2003).

Rotterman and Monnett knew that young sea otters foraged in shallow subtidal seas, while older animals, ones with larger lung capacity, dove deeper and longer and fed more offshore (Monnett and Rotterman 1995a; Rotterman and Monnett 1995). Mussels and clams in shallow seas off oiled beaches were contaminated with oil, as documented by NOAA Auke Bay Lab studies (Chapter 12); it seemed possible that this oil may have compromised the health of young sea otters (Doroff and Bodkin 1994). Studies to determine effects of oiled habitat and oiled food on sea otter health, using parameters such as body condition, growth rates, blood chemistry, and organ function were conducted by other researchers several years later as related in Chapter 18).

In late 1991, USFWS notified Rotterman and Monnett that funding for all their studies had been cut completely. USFWS had funded the bulk of the oil spill damage assessment (NRDA) studies on sea otters and the lion's share of this funding had gone to Rotterman and Monnett (J. Bodkin, Alaska Science Center, USGS [formerly USFWS], pers. comm., 27 February and 4 March 2003). After the October 1991 civil settlement, the NRDA funding was discontinued as the focus shifted to restoration studies. According to USGS employees, the sea otter budget dropped from a million dollars a year to the pre-spill level of about $250,000 a year (ibid.). The damage assessment studies of Rotterman and Monnett were expensive, in part because of the large sample sizes and amount of replication—all necessary to get reliable data and draw meaningful conclusions.

However, as revealed in the NRDA study reports discussed and cited above, Rotterman and Monnett had encountered difficulty funding and conducting their research well before the 1991 budget cuts. Their problem possibly was largely political: the biologists had found out that the oil spill caused long-term damage to sea otters—information that the federal agency did not seem to want made public at the time. Without funding, Rotterman and Monnett were unable to publish much of their oil spill work or to continue working in their chosen field of sea otter research. Yet their study on sea otter weanlings was—and still remains—the most profound and strongest direct evidence of chronic impacts of the *Exxon Valdez* oil spill on marine mammals at the population level. Because of the secrecy order imposed by the government lawyers on the public-trust scientists, few people knew that many of the sea otter pups in oiled areas continued to die for several years after the spill.

Further, Rotterman and Monnett had found that Exxon's highly touted wildlife rescue effort had not gone well for the sea otters—either treated, released animals or wild stocks. Their study, along with the pathology work on the captive, treated sea otters call into question the value *to the animal* of wildlife rescue centers. During the course of my research, I found that many of the public-trust sea otter biologists appeared to have thought that it may be more humane to euthanatize heavily oiled animals and allow nature to take its course with the rest rather than to attempt to rescue these wild creatures (Hofman 1994).

Orcas

Craig Matkin's first encounter with orcas—the sleek black and white "killer whales"— changed his life. In summer 1976 he was kayaking alone in the entrance of Eschamy Lagoon in Prince William Sound, when suddenly he found himself in the midst of a group of orcas chasing salmon. Fascinated and afraid to move, he watched the whales long into that magical evening. At the time almost nothing was known about orcas in the Sound—how many there were, their social structure, their habitat range, or preferences. That evening, he committed to study these fascinating creatures (Matkin 1994).

Matkin earned a masters of science in marine mammals from the University of Alaska. In 1978 he, along with a handful of other researchers and their mentor Pete Isleib, organized the nonprofit North Gulf Oceanic Society (NGOS). Pete Isleib was a Cordova fisherman and eminent bird biologist who motivated many young Alaskans to pursue their dreams to study the abundant wildlife in the Sound and the Copper River Delta. Matkin (1998) followed Pete Isleib's example by combining fishing and research into a lifestyle. By 1983, after five years of sporadic whale research and scrimping money from crew jobs, he had his own fishing boat, a salmon fishing permit, and a research contract from the Alaska Council for Science and Technology to conduct the first photo identification of orcas in the Sound. Other funding followed for the next five years.

By photographing orcas consistently from the left side when they surfaced to breathe, Matkin and his crew learned to recognize individual whales. Orca biologists in British Columbia had discovered the shape of an individual whale's pronounced dorsal fin and the markings and form of its gray saddle-patch directly behind the dorsal fin were unique—the equivalent of a human thumbprint (Morton 1990). Watching whales and working with photos of 320 living animals, the NGOS team saw that different whales consistently grouped together, and they began to piece together genealogy, social structure, and behavior of the various groups (Bigg et al. 1990).

Matkin learned there were two distinct types of orcas based on social structure and feeding habits (Matkin 1994).The different types avoided contact. One type, the *"resident orcas,"* traveled in large, tightly bonded groups or *"pods"* of maternally related animals. Seven resident pods frequently visited the Sound.They were all highly vocal using trills, high whistles, deep wavering groans, and sharp grunts to echolocate salmon or keep in contact with each other in the dimly lit waters. Matkin learned resident pods were bonded so tightly that, when a whale disappeared from its pod and was not seen for two consecutive years, he could count that whale as dead.The resident AB pod became the NGOS team's favorite: it was the most commonly sighted group in the Sound and its members were unusually friendly, often approaching and following the researchers' boats (Matkin and Saulitis 1997).

The other type of orcas, the *"transients,"* traveled in small groups and hunted marine mammals, usually Steller sea lions, harbor seals, and Dall porpoise.Transient orcas emitted quiet rapid clicks like muffled machine gun fire to echolocate as they navigated near craggy shorelines. Matkin learned the social structure of transient orcas was very loose and groupings changed frequently, making it difficult to confirm deaths.The NGOS team sighted only one group of transients every year, the AT1 group.This group was unusual, as it was a genetically distinct group of twenty-one orcas (Saulitis 1993).

In 1985, the AB pod learned to raid the black cod catch of long-line fishermen. There were brief and intense skirmishes between whales and humans as the fishermen desperately tried to scare away the whales with explosives and gunfire (Matkin, Steiner, and Ellis 1986). Six whales died. By 1987, changes to the Marine Mammal Protection Act and to the long-line fishery had alleviated the conflict. The AB pod recovered rapidly from its losses. Five new calves were born in 1988, which brought the total number of whales in the pod to thirty-six.

On 24 March 1989 Matkin was in Seward getting his seine boat, the *Lucky Star*, ready to go herring fishing when he heard on the radio that there had been a big oil spill. He quit wrenching on his diesel engine and waited through thirty seconds of ads to see, as he said, "what country had been screwed."What he heard next, he didn't believe: Prince William Sound? "No way!" He dropped everything and rushed to the Coast Guard station. "There must be some mistake," he told them. Still disbelieving, he called his friend Rick Steiner, a politically savvy Cordova fisherman who also ran the University of Alaska's Sea Grant Program in Prince William Sound. Matkin caught Steiner moments after he returned from an over-flight of the spill. As he listened to his friend's dazed and numb voice, Matkin said, "That's when I realized, 'I ain't going herring fishing'."

Five days later he pulled into Sawmill Bay with part of the NGOS team—his pregnant wife Olga, their toddler daughter Elle, and research aid Eva Saulitis. The fishermen were frantically booming off the bay entrance to protect the San Juan hatchery. The *Lucky Star* joined them for a while but the next day, as oil advanced down Knight Island Passage, Matkin heard of orca sightings at the southern end of Knight Island. They left immediately and intersected the resident AB pod at the south end of Applegate Rock.

Hearts pounding, the team did a quick tally. Seven animals were already missing—three adult females and four juveniles. Two of the females had left offspring less than four years of age. Abandoning calves was unprecedented behavior for resident whales—Matkin and the others realized that these animals had probably died. Shocked by the sudden losses, just six days into the spill, they only slowly became aware that the pod was behaving strangely. Matkin said, "The pod milled in circles. It was intense—they wanted to go north, but not through the oil. Finally, the whales went right up through it, along the eastside of the island." The *Lucky Star* followed.

Matkin and the NGOS team monitored the Sound's whales all summer. For the first time, other researchers actively sought them out, because the NGOS held the trump card—all the pre-spill baseline data. They turned down research contracts. The first was an offer of a $280,000 check from Exxon. He asked his friend, Toby Rilling, who had arranged the contract from Exxon without his knowledge, to tear up the check. Rilling didn't believe him. Matkin explained gently, "If I take that check, all my work from now on will be paid off. I won't be able to do my own work ever again."

Exxon was furious that Matkin had not taken the bait, and Exxon tried unsuccessfully to hire each member of the NGOS team before they finally left the NGOS team alone. Different government people also offered Matkin lucrative contracts or just came to check out his operations. The visiting scientists didn't impress him. "They had no field experience and no feel for what was going on," he said.

However, after the field season that fall when he went to negotiate multi-year contracts with the NOAA National Marine Mammal Laboratory, Matkin discovered that some of the federal scientists were trying to take over his project. His blue eyes blazed as he said, "They expected to take our field data, write the papers, take the credit, and have full control over future projects." He had to get into "intense negotiations" even to get credit for his past work. He persisted and by spring 1990, the NGOS had a two-year federal contract. The relief at being funded and back out in the field was bittersweet. The NGOS team found that six more members of the AB pod were missing and presumed dead—one female that left a surviving calf, four more juveniles, and one maturing male (Matkin et al. 1994). The team noted that this mortality rate—about 20 percent for both 1989 and 1990—was about ten times higher than the pre-spill rate for other resident pods in the Sound. The AT1 group of transient orcas was also missing nine whales by spring 1990 and two more subsequently disappeared. Three of the missing AT1 whales had been observed near the *Exxon Valdez* after the spill.

In 1991, the NGOS team found that the AB pod was missing five more whales—all three orphaned calves and the two males (Dahlheim and Matkin 1994). The males had collapsed dorsal fins. The fins had started to collapse in 1989 after the spill. The NGOS team had interpreted this as a sign of failing health. The fins had completely flattened against the orcas' backs before the whales disappeared.

Matkin was fairly certain that the cause of death for the missing orcas was Exxon's oil, but there were no whale carcasses in the USFWS freezer vans full of birds and sea otters. He drew evidence to support his theory from the pathology studies on oiled sea otters and harbor seals. These studies reported inhalation of oil vapors caused inflammation of mucous membranes and lung congestion (Frost, Manen, and Wade 1994). The volatile hydrocarbons travel rapidly from the lungs to the blood and accumulate in the brain and liver, causing lesions, neurological damage, and death.

Seal biologists Kathy Frost and Lloyd Lowry observed that harbor seals, the primary prey for the AT1 pod, were extremely disoriented and lethargic immediately after the spill. In one study, nine of fourteen oil-drenched, dizzy seals collected and examined had mild to severe lesions characteristic of oil poisoning in their brains' thalamus (Lowry, Frost, and Pitcher 1994). The thalamus acts as a central control switchboard, relaying messages from sensory systems to other parts of the brain to coordinate movement. The seal biologists thought that damage to this central switchboard garbled messages to other parts of the body and caused the normally fluid, quick, and graceful seals to stagger, tip, and reel and have difficulty breathing, swimming, feeding, and diving (Spraker, Lowry, and Frost 1994).

Matkin reasoned orcas that inhaled oil vapors would suffer similar organ damage and fate, however, Matkin's federal "co-workers" were edgy about legal liability and reluctant to clearly state that the oil spill caused the orcas to die, although they all agreed that oil *could* have caused the deaths. Matkin brought all the lines of evidence together that pointed to oil as the culprit, but his federal colleagues and grant administrators would not draw the inevitable conclusion. The acting executive director of the NRDA team, Dave Gibbons, said this was a position the federal government was not willing to take. It was like an unfinished game of "Clue." Matkin had missing bodies, the murder weapon, the criminal, and the room, but he was not allowed to state that Exxon's oil caused the death of orcas in the Sound.

The EVOS Trustee Council funded Matkin's research in 1992—$50,000 for two projects, one in the Sound and another in southeast Alaska. Then Matkin noticed another budget item of $100,000 for federal "oversight" of his two projects! Matkin showed the budget to his friend and NGOS member Rick Steiner, a leading advocate of public participation in the EVOS Trustee Council process. According to Matkin, "Rick had a fit." Furious about the excessive padding of agency pockets instead of responsible stewardship of public monies, Rick Steiner shared his concerns with Congressman George Miller (D-7th CA) who requested a Government Accounting Office investigation into the matter (U.S. GAO 1993). The whole deal erupted in public the following year (Chapter 19).

Harbor Seals

From 1975 until the spill in 1989, ADFG marine mammal biologists Kathy Frost (2003) and her husband Lloyd Lowry studied seals—ringed, bearded, spotted, and ribbon seals—that lived on or near pack ice in the far north. They focused on *"feeding ecology,"* exploring relationships between predator and prey to understand the role of seals and other animals, such as bowhead whales, in the Arctic Ocean food web. During summers, they explored Prince William Sound, boating and fishing in remote areas. They were ADFG's obvious choice to study harbor seals in the Sound after the spill. When the phone call came to their Fairbanks home on 24 March, Lowry told their boss that they were in the middle of the Tok champion sled dog race and they would call back first thing Monday. Frost realized, "The minute we went to Valdez, we wouldn't be home for a long, long time."

Harbor seals are one of the most abundant marine mammals in the Sound with population estimates of several thousand (Frost 1997). The seals frequent over fifty haul-out sites—intertidal reefs, rocky shores, sand and gravel beaches, and floating glacial ice—where they feed and rest year round and give birth, care for their young, and molt during the long Alaska summers.

In lieu of actual animal counts—which are hopelessly inaccurate because of all the Sound's nooks and crannies, ADFG biologists began to monitor population *trends* in the Sound in 1983 by aerial surveys of key haul-out sites. From these surveys, Frost and Lowry knew the harbor seal population in the Sound had declined by about 40 percent between 1983 and 1988, the years immediately before the spill. They gathered earlier unpublished surveys of marine mammals, including harbor seals, and found populations of fish-eating marine mammals—harbor seals, Steller sea lions, Dall porpoise and minke whales—all had dramatically declined by as much as 70 percent between the 1970s and the 1980s (Hill et al. 1996). The harbor seal population in the Sound had plummeted by nearly two-thirds during this time.

This complicated their job. They realized they had to determine oil spill effects on harbor seals and recovery rates, while finding the cause of the ongoing population decline—and they had to keep these two events separate. "So," Frost said, "we sat down like good basic biologists and made a shopping list of what could be affecting the ongoing decline in harbor seals—things like disease, contaminants, mortality by predation or hunting, and changes in food abundance or distribution." The EVOS Trustee Council funded their various investigations for the next decade.

To monitor population trends and effects of the oil spill, Frost conducted aerial survey trend-counts annually with Cordova pilot Steve Ranney. The surveys included seven sites that had been oiled and eighteen that were not. Frost and ADFG *"biometricians"* (biologists who specialized in mathematics and statistics) refined their methods over the years until they knew the best window of time—which summer month and what time of day, relative to low tide—to conduct surveys to obtain the highest counts of animals. Then they developed a sophisticated statistical model to correct past trend-counts, some of which had actually *underestimated* counts and masked the extent of the pre-spill decline.

From the surveys, Frost and Lowry estimated about 300 seals died in 1989 due to the spill, and pup production was about 26 percent lower that year than previous (or subsequent) years (Frost et al. 1994).The high loss of animals in 1989 appeared to be limited to one year.The scientists realized that the ongoing decline complicated and delayed recovery of harbor seals from the spill, because the animals could not replace those lost during the spill.

Their aerial surveys showed that the decline continued during the 1990s, but at a much slower rate—2 to 6 percent compared to 40 percent during the 1980s (Frost, Lowry and Ver Hoef 1999). They realized mortality from orca predation and limited hunting pressure could account for the low but persistent annual decline during the 1990s. Subsistence hunting had not caused the major, pre-spill decline "by any stretch of the imagination," Frost insisted, "but once you have a population that's declining, every animal that's removed contributes to the decline."The oil spill had removed a huge chunk of this population.

To test whether food might be limiting survival—and recovery— and to prove missing seals from 1989 had not simply moved away, Frost and Lowry needed to better understand harbor seal habits. Starting in 1991, they began gluing satellite-linked depth recorder tags to the backs of wild harbor seals. Since the satellite tags were shed during the annual molt, they also attached small flipper tags to each animal they handled. By tracking tagged animals, the scientists found the vast majority of seals fed locally within fifteen miles of their haul-outs and almost all fed within thirty-one miles of their haul-outs (Frost, Manen, and Wade 1994). Occasionally seals ranged more widely, presumably for food, but perhaps to explore new sites. Frost referred to these adventurers as "Columbus seals." Other researchers found substantial genetic differences in harbor seals among regions within Prince William Sound and between the Sound and other areas, confirming that harbor seals were essentially "homebodies." These findings suggested that missing seals from 1989 had most likely died rather than moved away.

Frost and Lowry realized that if something was limiting food, it had to be a region-wide problem to consistently affect seals at all the haul-outs. The couple had "lively intellectual discussions" about whether food was limiting seal survival and recovery. Frost described her husband as "wisely skeptical and not at all ready to believe" that limited food availability had caused the precipitous pre-spill decline in population.

They hired a post-doctoral student, Bob Small, to model possible causes for the decline.After factoring in orca predation, reproductive rates, hunting pressure, the spill—"and anything else we could imagine happening to them," as Frost said, the bottom line was that only a hypothetical 40 percent reduction in their food could readily explain the observed 40 percent decline in their population. The model convinced Lowry that food was the key. But what was affecting their food?

They needed to obtain scientific data to support their hypothesis. Faced with having to analyze harbor seal diet using crude methods with inherent biases, Frost and Lowry searched for scientists with new, improved techniques. They found what they were looking for in 1993 (Chapter 19).

Exxon's Studies: Deception by Design?
Sea Otters

Exxon scientists (Johnson and Garshelis 1995) monitored sea otters in 1990 and 1991 and reported that "1–2 years after the spill, otters remained abundant in the oil-affected area of PWS and showed no apparent spill-related effects on their distribution or pup production" (925). Exxon scientists did not monitor the weanlings to see if they survived. (Rotterman and Monnett's weanling study [1995] showed that most did not survive in heavily oiled areas.) Unfettered by such troublesome details, Exxon reported in 1991 in glossy brochures that "Sea otters thrive in Prince William Sound." This was Exxon's story and the company stuck with it, despite plenty of evidence to the contrary.

In his book *How to Lie with Statistics,* Huff (1954) encourages people to ask, "What's missing?" The author notes, "Sometimes what is missing is the factor that caused a change to occur" (127). In Exxon's case, the weanling study was missing and, with it, all damning evidence that the oil spill continued to kill sea otter pups in heavily oiled areas for years after the spill.

Orcas and Harbor Seals

Exxon scientists did not conduct any studies on orcas or harbors seals from 1989 to 1992, the time period covered in this section.

Sound Truth

Sea otter weanlings continued to die at higher rates in oiled areas through at least 1992 and prime age adults well beyond 1992 (Chapter 18).

The resident orca AB pod lost thirteen of its thirty-six members in the first two years following the oil spill. The transient AT1 orcas lost eleven of its twenty-one members in the first two years following the oil spill. No new calves were born into either of these pods in 1989 to 1992, the time period covered in this section.

Chapter 15.

Marine Birds

Murres and the Gulf of Alaska

When seabird biologist John Piatt, PhD (2000), heard of the spill on 24 March, he felt sickened. Piatt was a research biologist, working for the USFWS in Anchorage. He knew the northern Gulf of Alaska, including Prince William Sound and Cook Inlet, hosted some of the largest populations of seabirds in North America. Millions of *"pelagic"* (oceanic) seabirds, including petrels, fulmars, cormorants, kittiwakes, and Piatt's personal favorites the *"alcids"*—murres, guillemots, and puffins, breed on isolated islands and rocky outcrops in the Gulf of Alaska. Each summer, millions of short-tailed and sooty shearwaters migrate through the Gulf of Alaska. Hundreds of thousands of coastal seabirds such as loons, grebes, murrelets, and sea ducks—scoters, eiders, and oldsquaws (now called long-tailed ducks)—over-winter in the region's bays and the Sound's protected inland seas.

Piatt knew it only took "a dime-sized spot of oil" to soil feathers enough to cause hypothermia and death. Before moving to Alaska in the early 1980s, he had spent seven years in Newfoundland, Canada, studying (among other things) the effects of oil pollution on seabirds. He knew that alcids, loons, grebes, and sea ducks are the

most vulnerable to oil, because they spend most of their time swimming on the sea surface and often raft together in dense flocks. Each
year in March, murres aggregate in the Gulf of Alaska near breeding
colonies. He hoped some of the other alcids had not yet returned.

Piatt took with him to Valdez his well-worn copy of *Oil in the
Sea: Inputs, Fates and Effects*. This National Research Council tome,
published only four years earlier in 1985, provided the first and last
words on oil spills. Piatt thought the book "was a gold mine. It had
everything you could imagine that you'd want to do in response to a
spill. It talked about petroleum chemistry, toxicity, collecting samples, dispersants, and legal issues. There was a whole chapter on legal
issues." Piatt wondered, "Why is this in a volume about scientific
effects of spills?" He didn't have a clue how dominant and pervasive
the legal side would become in assessing damages. He found a whole
chapter on birds and another on marine mammals. It was all based
on thirty years of research and experience.

Unfortunately, Piatt couldn't get anyone to even crack the cover
of this 'gold mine.' Instead of the coordinated response described in
his book, he saw "nothing but chaos. There was no leadership. It was
just a mish-mash of people all doing their own thing as best they
could, but that really didn't make for a very good response program."

Frustrated, Piatt drafted and delivered a game plan to USFWS four
days after the spill. It was based on published studies and his own
years of experience measuring the impact of oil pollution on
seabirds in Newfoundland. He proposed three steps be taken immediately, assuming oil would pass through the Sound and soon endanger seabirds in the Gulf of Alaska. First, he advised that beach surveys
be conducted both before and after the oil hit the beaches to get an
idea of how many birds were killed. Second, he suggested setting up
a processing center—a wildlife morgue—as a way to handle all the
bodies once they started coming in to Valdez. Third, he stressed that
carcass drift experiments were needed to establish the proportion of
dead birds that drifted ashore in different areas so that the total number of birds killed could be estimated.

Most carcasses of seabirds killed by oil at sea, sink at sea and, of
those that drift ashore, many are scavenged by predators or buried in
debris. Historical studies, conducted worldwide, report that anywhere between 0 to 59 percent of birds killed in a spill drifted
ashore, depending on currents, winds, tides, and proximity of beaches, while the rest were lost at sea (Piatt and Ford 1996). Piatt calculated a middle-of-the-road median recovery rate of about 10 percent

for seabird carcasses from these fifteen historical studies, but he knew recovery rates were needed for the Sound and Gulf of Alaska in order to estimate how many seabirds died from the spill.

No one paid him any attention initially because during the first few days after the spill, no one brought any dead birds to Valdez. A week passed, ten days: still nothing. But there were plenty of dead birds—Piatt saw them the few times he was able to catch over-flights of the Sound, and he heard about them from other agency people who walked the beaches.

USFWS was slow to react because the Sound has no wildlife refuges, management of which is the agency's responsibility. A bigger part of the problem was that the spill was never "federalized," so USFWS and NMFS, the federal agencies responsible for collecting dead wildlife to assess *"damages,"* (legal term, here, monetary value of losses) had no funding initially to carry out their charge. The agencies literally had to wait until Exxon promised to make response funds available, which required cooperative agreements, lawyers, and time (Zimmerman, Gorbics, and Lowry 1994). It was to Exxon's advantage to control the collection of incriminating evidence and the longer the company stalled, the less evidence there would be.

This all changed because of Kathy Frost, an ADFG marine mammal biologist. Piatt was at a press conference in Valdez, fuming as he listened to a federal on-scene coordinator tell national media and spill responders, "We're not seeing anything. We don't know that there's a real problem yet." Then he heard, "Wait a minute!" from the hallway and Frost marched up to the podium in her orange survival suit drenched in oil. She announced, "I just came back from the beaches and I'm here to tell you there's dead animals everywhere!" This fortuitous timing and Frost's pluck turned into a defining moment.

USFWS came out of its stupor of disbelief. The on-scene coordinator was replaced and more staff were sent to assist with the response. Cal Lensink, a retired USFWS research biologist with over thirty years experience and a commanding presence, rolled into Valdez in his VW van to volunteer his services. In their shared hotel room, he and Piatt developed a response strategy. Lensink immediately organized USFWS staff to build receiving and processing stations for wildlife carcasses. Piatt walked beaches in the Sound, collecting and examining dead animals, and distributing collection and documentation kits—bags, data forms, and instructions—to a small army of untrained crews who had been hired by Exxon to collect dead and dying animals.

When the first dead sea otters arrived in Valdez, Lensink yanked teeth to determine age, took measurements, and determined the animals' sex (DeGange and Lensink 1990). When the first dead birds came in, Lensink organized people with manuals to properly identify the birds. Soon, bags of dead animals came in with important documentation inside revealing the "who, what, where, and when" details of collection. He and Piatt set up a computer data entry program to record this relevant data along with other biological information. Eventually USFWS set up similar morgues in Seward, Homer, and Kodiak. These morgues processed nearly 37,000 dead birds in 1989.

In the second week following the spill, Piatt helped colleagues at the USFWS initiate some carcass drift experiments in the Sound, but Anchorage bureaucrats squashed the project, citing concern about "bad PR" (public relations). They worried that people might be angry with the government for returning carcasses to the sea that had been collected as evidence of spill damages.

A few weeks later, Piatt led a team of researchers on the USFWS research vessel R/V *Tiglax* to conduct bird surveys and collect carcasses from beaches along the Alaska Peninsula, retracing the path of the oil. Inside the Sound, the oil was relatively fresh, black and liquid. It formed sheens that sometimes were difficult to distinguish from the water. On slippery oil-stained beaches, Piatt collected black slimy soft lumps, which he discovered were mostly marbled murrelets, loons, bald eagles, and sea ducks—scoters, oldsquaws (long-tailed ducks), buffleheads, and goldeneyes.

Outside the Sound, wind and waves had churned the black crude with seawater and air, creating a thick, sticky, chocolate-brown mousse that completely coated entire beaches in a 100- to 200-yard wide swath of slime. "It was extraordinary," Piatt said. "We couldn't really tell if there were any birds at first. Then we found them, rolled in a thick case of mousse. We had to pick off inches and inches of this muck before you could even see what bird it was. It was really quite gross." They picked up many, many more dead birds, mostly murres. They also picked up bodies of other colonial seabirds—cormorants, shearwaters, guillemots, and puffins, but fortunately many of these species returned to colonies in the spill zone later than murres and so largely avoided the wave of black death that washed out of the Sound.

In Puale Bay on the Alaska Peninsula (across from Kodiak on Shelikof Straits), Piatt and the *Tiglax* team made a grisly discovery. Underneath layers of muck, he found murres, thousands of dead murres. Murres look and act somewhat like penguins except they are strong fliers. Like penguins, murres eat small fish and invertebrates. Their almost erect stance and smart uniforms of black hooded capes with white fronts reminded Piatt of little eighteen-inch soldiers. He had expected to find common murres and thick-billed murres—they were some of the most numerous seabirds in Alaska. But Piatt wondered where they came from, because the local colony of murres in Puale Bay appeared to be thriving and hardly oiled. He suspected the carcasses had drifted from the huge murre colony at the Barren Islands north of the Kodiak archipelago. The Barren Islands hosted one of the largest and most diverse seabird colonies in the Gulf of Alaska and it was directly in the path of the oil.

With hundreds of relatively fresh carcasses at hand, Piatt and his team decided to conduct a very small-scale carcass drift study in order to determine the origin of the birds. They took one hundred dead soldiers—all murres—from Puale Bay, tagged them, brought them up to the Barren Islands, and dumped them out in the middle of the night. Then Piatt waited while various beach survey teams patrolled the beaches of the Alaska Peninsula. Sure enough, three murre carcasses drifted about 200 miles back to the beach at Puale Bay. Now Piatt had a recovery rate and he could estimate the total number of birds killed—once they were done collecting bodies.

In the fall, Piatt, Lensink, and others worked up the data from the wildlife morgues. Of the nearly 37,000 total carcasses, Piatt and the others estimated that 7,000 had died from causes other than oil. There had been a "*wreck*," a large-scale natural die-off, of puffins and shearwaters in Shelikof Straits during the fall migration. Piatt explained, "Massive die-offs are not uncommon for seabirds. Some species are high-density animals like caribou that feed in large aggregations and they require high-density foods. Once in a while, when that food's not around, what happens? Large numbers die *en masse*."

Of the 30,000 seabirds killed by oil, nearly 90 percent were collected outside the Sound; 74 percent of the total were murres (Piatt and Ford 1996). Piatt and his co-workers thought the 3 percent recovery rate from the Barren Islands carcass drift study was probably low for the Sound where there were more islands and more people walking the beaches. Based on the median recovery rate from the worldwide drift studies and considering the oil spill region, they reasoned that a 10 percent area-wide recovery rate would yield a conservative estimate of the spill's toll on murres. Using this figure, they estimated the oil spill killed between 100,000 and 300,000 murres.

This was by far the largest number ever documented of seabirds killed by an oil spill. Piatt and the others thought it could represent about 10 percent of the existing population of common murres in the Gulf of Alaska and probably more than half of the breeding colony at the Barren Islands. They speculated there might be long-term effects of the spill on survivors as well. Murres are long-lived birds that lay only one egg each year. It was reasonable to anticipate delayed recovery from reduced egg production and fewer young birds. Based on a model of murre population dynamics published by other seabird biologists *before Exxon hired some of them* (Ford, Page, and Carter 1987), Piatt estimated it could take twenty to seventy years for murres to recover.

Piatt and the others decided that it was their job as public-trust scientists to share their information with the public. Over the objections of federal lawyers, they published their initial findings in a 1989 issue of *Nature* and more complete findings in a 1990 issue of *The Auk* (Piatt and Lensink 1989; Piatt et al. 1990)—three years before most of the other public-trust scientists pierced the veil of secrecy imposed by federal attorneys.

Exxon immediately attacked the report as preposterous. Exxon scientists claimed, "The only things that are certain are that approximately 30,000 oiled carcasses were retrieved and that this number represents some unknown fraction of the total number of birds killed by the spill" (Wiens 1996, 596). Exxon scientists steadfastly ignored the obvious—that the 30,000 dead birds *collected* were only a small fraction of the *total killed*. Piatt's papers increased public furor over spill damages and public pressure for a high monetary settlement (Medred 1989).

In 1990 Piatt returned to his normal duties, conducting research on auklets in the Shumagin Islands, some 300 miles southwest of the Kodiak archipelago. That summer, he learned that the federal lawyers, now in complete control of damage assessment projects, ironically were paying a million dollars for a carcass drift study! Just as the *Oil in the Sea* book had predicted, the federal lawyers eventually needed to quantify the number of bird bodies that had been recovered in a way that was defensible in court. The NRDA process revolved around proving *exact* damages—in this case, the exact number of birds killed by oil. More birds equaled more dollars in compensation.

Finally, the federal lawyers realized that the ratio of birds killed to those actually found on beaches was of major economic importance.

Contractors doing the study needed 300 dead oiled birds and the only way to get them, in 1990, was to shoot live birds and oil them prior to releasing the oiled carcasses for the drift study. For political reasons the researchers weren't allowed to shoot birds in the Sound, so, in another ironic twist, Piatt ended up "collecting" many birds for them in the Shumagin Islands. When people found out about this study after the fact, they were furious, but the federal lawyers got their numbers: 43 percent recovery in the Sound, 13 percent along the Kenai Peninsula, 6 percent around Kodiak, and 2 percent along the Alaska Peninsula (Ecological Consulting, Inc. 1991).

When the contractor cranked through the numbers, their best estimate of total bird mortality was 375,000 with a range of 300,000 to 645,000 (Ford et al. 1996). Despite the fact that the carcasses were subjected to different winds, currents, and weather patterns than in 1989, Piatt said the U. S. Department of Justice study "will probably stand forever as the most thorough examination of carcass drift recovery that has ever been done." He also was pleased that the average estimate of mortality in the new study not only corroborated his earlier estimate, but made it look conservative!

In the meantime, other USFWS scientists (Nysewander et al. 1993) conducted extensive surveys from 1989 to 1991 and found that five murre colonies in the path of the spill had 40 to 60 percent lower population counts compared to pre-spill counts than two colonies outside the spill path. They also found delayed effects as Piatt had predicted—murres from oiled colonies did not produce as many offspring as murres from unoiled colonies. Scientists thought the oil spill had disrupted critical social behavior (Phillips 1992). Murres time egg-laying so that it is both synchronized, to overwhelm predators and reduce predation, and early—to fledge young before fall storms. With experienced breeders nearly wiped out by the spill, scientists measured nearly total reproductive failures at every colony they monitored within the spill path and reproduction remained below normal throughout the study. There was some evidence of recovery by 1992.

Even though Piatt was no longer assigned to oil spill research, he continued to dabble on the side. Initially, he blamed the murre population declines, disrupted and delayed egg-laying, and reduced productivity on the oil spill. But when murre productivity at oiled colonies stayed suppressed for several years after the spill, Piatt began to have second thoughts.

As he dug deeper into the published literature, going back in time, he found there were serious declines in a number of murre colonies *outside of the spill path*—declines of 30, 40, 50, and 80 percent—in the two decades preceding the spill. Besides the evidence of spill mortality, here was evidence of sustained changes in productivity for years before the spill. What could account for that kind of a large-scale phenomenon, he wondered?

Piatt had done his doctoral work on predation by common murres and Atlantic puffins on capelin, a high-calorie forage fish. When he first came to Alaska, he found seabirds in the Gulf of Alaska were eating mostly juvenile pollock, which were much lower in calories and lipids (fats) than capelin, but he had dismissed this as a difference in diet preference between Atlantic and Pacific seabirds. Now he wondered if capelin in Alaska had died off for some reason, forcing murres to feed on lean cuisine—the pollock.

As luck would have it, he started to correspond with a NMFS shrimp biologist in Kodiak, Paul Anderson, who, Piatt described, "had just been out there plugging away for twenty-five years doing small-mesh trawl surveys every year in the Gulf of Alaska." Over the decades, NMFS had done hundreds of small-mesh trawl surveys in the northern Gulf of Alaska to monitor stocks of commercial species. The small-mesh gear also captured small forage fish such as capelin, a noncommercial species. These unpublished surveys showed that shrimp and capelin had dominated catches in the 1960s and early 1970s, but they had virtually disappeared in trawl catches in the late 1970s. They had been replaced with walleye pollock, cod, and flatfish.

Intrigued, Piatt dug out more historic data sets (Sanger 1986). "Lo and behold!" he said. "It was like an epiphany." He found the most important food, by far, during the mid-1970s for five of the most abundant species of seabirds in the northwest Gulf of Alaska was capelin. A decade later, these seabirds were eating sand lance, pollock, and squid. Piatt found similar evidence for Steller sea lions, harbor seals, and northern fur seals. During the early 1970s, the diet of these marine mammals was rich in capelin and herring, but a decade later—well before the oil spill—these mammals were eating far fewer high-calorie fish and far more pollock (Hansen 1997). Populations of fish-eating *"apex"* (top-level) predators, including common murres, black-legged kittiwakes, sea lions, and harbor seals, crashed coincident with the loss of shrimp and capelin (Merrick, Loughlin, and Calkins 1987).

Piatt suspected that changes in the marine climate were ultimately controlling population size in these apex predators. He looked up oceanographer Tom Royer's studies (1993) and found seawater temperature in the Gulf of Alaska had abruptly warmed during the late 1970s and warm conditions had persisted through the 1980s, which favored arctic cod, pollock, and flatfish. Piatt realized for murres, the shift in marine climate had acted like a slow-moving wreck, decimating the Gulf of Alaska population over time. The oil spill had worsened an already bad situation for murres and other seabirds in the spill path; it was an environment where rapid recovery was not likely.

He shared his theory and evidence with scientist Bruce Wright who was managing NOAA's damage assessment and restoration program. Bruce Wright (1999) encouraged Piatt to publish his findings in the proceedings of the first public symposium on oil spill science that had been held in January 1993. Piatt eagerly accepted. He and NMFS biologist Paul Anderson crafted their paper as a NRDA study on common murres and this paper served as a "heads up" to their colleagues that the dramatic declines in seabird populations, first noticed after the spill, might be caused by more than just oil (Piatt and Andersen 1996).

Their effort eventually led to a suite of studies on apex predators—common murres, black-legged kittiwakes, pigeon guillemots, and harbor seals, among others—to better understand the link between biological productivity and marine climate (Chapter 19).

Harlequin Ducks and the Sound

In May 1975 when Sam Patten, PhD (2003), and his wife first set foot in Cordova, they were warmly greeted by Alaska's legendary birder Pete Isleib. "I don't know how he found out who we were," Patten said, "but he gave us all kinds of advice and support." Patten had a NOAA contract to study gull colonies along the south central coast of the Gulf of Alaska. He found Isleib already knew a great deal about the gulls in this remote area. Isleib advised him to check out the gull colony at the mouth of the Alsek River near Yakutat, well to the east of his camp at Egg Island. There, Patten discovered a hybrid gull

colony, which led to a study of gull evolution and a doctorate in public health from Johns Hopkins University in 1980. He said, "I will
always be incredibly indebted to Isleib."

When Patten first learned of the oil spill, he was living in western Alaska in Bethel and working for ADFG as Area Biologist for the
Yukon-Kuskokwim Delta District. He immediately offered to help the
state study effects of the spill on wildlife. The state moved his family
to Fairbanks and that summer his ten- and twelve-year old sons
camped with him on Egg Island where he repeated his study on gulls
to look for any oil spill damage. He found none—gull productivity
remained unchanged from 1975 and 1976. The NRDA lawyers
announced, "Wonderful study, Sam. Good clear results. We're cancel-
ing it because you found no effects. Go find some damage."

It was fall 1989 and Patten didn't have much time to prepare a
new research proposal. He did some quick thinking. He knew most
of the oil was stranded on the shoreline in western Prince William
Sound and he figured wildlife that depended on this area for food,
shelter, and rearing young would most likely show oil spill effects.
From Isleib's work (Isleib and Kessler 1973), Patten knew that up to
several hundred thousand sea ducks over-winter in the Sound and
protected waters of Kodiak and Shelikof Straits. He also knew that
the Sound was an important stopover area on the migration route of
sea ducks in the spring and fall—the ducks built up their fat reserves
by feeding on plentiful supplies of blue mussels, chitons, snails, and
other invertebrates.

Patten wrote a proposal to study oil spill effects on six species of
sea ducks—three species of scoters, two species of goldeneyes, and
harlequins. With these six species, he figured he had the oiled shoreline completely covered. Harlequins fed at or near the water's edge
in the intertidal zone; the goldeneyes and two species of scoters dove
to feed in the lower intertidal and shallow subtidal areas; and white-
winged scoters fed further offshore, diving over ninety feet for scallops and clams. Patten was set to 'find some damage.' His proposal
was funded.

After studying his charges during the winter of 1989 to 1990,
Patten and his crew realized that scoters and goldeneyes are not residents of the Sound, but some harlequins are. As Patten said, "That's the
first thing that hits you in the head. Some harlequins are here all the
time—they're feeding in the intertidal and they breed locally. It was
an 'ah-ha' experience. Our study rapidly evolved from a study of scoters, goldeneyes, and harlequins into a focused study on harlequins."

When Patten started his study, he knew only basic harlequin ecology. They are the smallest of the sea ducks. They earned the name by their striking streaks and patches of white and deep russet highlights splashed across the males' dusky blue bodies. Pairs of colorful males and drab females gather near stream mouths in late May and fly upriver to nest in suitable holes—tree cavities, root wads, stumps, or stream banks. For reasons known only to harlequins, they prefer to nest along wildly rushing, turbulent, white-water streams. Once females begin incubating eggs, males return to the Sound and congregate in saltwater bays and offshore rocks. When the ducklings are about two weeks old, they grow restless and bravely launch themselves into the stream torrent and cascade haphazardly down to the sea. A surprising number of them survive this perilous journey. Females raise their young in shallow estuaries with boulder-strewn shoals and rocky islets. By October, migrant harlequins that breed further inland join resident harlequins to over-winter in the Sound, swelling the Sound's population to an estimated 10,000 of these ducks until May when the cycle begins anew.

During the summer of 1990, Patten and his crews conducted a preliminary survey of breeding harlequins in the western and eastern Sound. They enlisted the help of dozens of ADFG fish technicians who volunteered to record sightings of harlequins and broods along streams while conducting their anadromous fish surveys. Streams were carefully traversed three times from mid-May through September.

Patten fully expected to find breeding harlequins everywhere: Isleib himself, ever helpful and supportive, told Patten he had seen breeding harlequins throughout the Sound before the spill. The best estimates of breeding populations of harlequins in the Sound, pre-spill, ranged from *minimum* numbers of 2,600 in the early 1970s survey to 5,500 during surveys in 1984 and 1985 (unpublished USFWS reports in Klosiewski and Laing 1994). Patten and his crews found breeding harlequins in the eastern Sound, but the numbers of breeding harlequins in the western Sound were, according to Patten, "vanishingly low."

He had found injury, but he had not yet pinned down how the injury occurred. He and his crews also found that mussel beds in the western Sound were "sopping with oil." They were dumbfounded. Most scientists were still operating under the old assumption from the 1970s that oil remaining on beaches years after a spill was not harmful to wildlife. Patten said, "The mussel beds were just loaded with oil. And what did we find most common in the birds' gullets? Small blue mussels. Then things really started going off in my mind."

He wondered if the harlequins might be having reproductive problems from eating oiled mussels. This was a natural cognitive leap based on his training in physiology and public health. In fall 1990, Patten revamped his harlequin study to include a field study of the breeding habitat of ducks in the unoiled eastern Sound as a control— and a review of the history of oiling of harlequin duck habitat. The government lawyers tallying natural resource damages told him, "Sam, your study is worth $25 million!" Everything was funded.

Patten supervised the three-year study in the eastern Sound. He said, "Dave Crowley did the work and he got his Master's out of it. He did a nice job. Basically, he found that harlequins prefer certain kinds of habitat and they were reproducing normally."

From 1991 to 1993, Crowley compared the streams and general basin topography—the lay of the land—in twenty-four areas where harlequins were breeding to twenty-four areas were they were not (Crowley and Patten 1996). He found the ducks selected larger salmon streams, but chose small, steep streams if larger ones were not available. He described the ducks as having "an ecological dependency" on the rocky intertidal area where streams met the sea—activities such as feeding, resting, courtship, and brood-rearing mostly occurred near the stream mouth. He also found the harlequins in the eastern Sound returned to the same nest sites and streams each year. Such *"site fidelity"* among harlequins had been reported elsewhere.

Meanwhile in the western Sound in 1991, Patten and his crew found so few harlequins near stream mouths during spring surveys they had to guess which streams the ducks would have used for breeding had the ducks been present (Patten et al. 2000a). They chose twelve streams to monitor for breeding activity based on similarity to known breeding streams in the unoiled eastern Sound (Patten 1993). During the long Alaska twilight—when reclusive females were most likely to leave nests to forage—Patten and his crews manned mist nests along the stream banks.

In 132 hours of effort, they captured *not one* harlequin duck. They didn't even see any ducks flying along the stream corridors. In 1992, they more than doubled their efforts in the western Sound. They selected thirty-nine streams and worked with mist nets for over 384 hours during twilight hours. They captured two females, only one of which was incubating eggs along an unoiled stream in the oiled region. In comparison, over the two-year study in the eastern

Sound, Patten's crew captured sixty-five breeding harlequins in 522 hours along sixteen streams.

Patten and his crew also mapped the distribution of the summer molting population of harlequins. In late July and August, adult harlequins gather at certain sites to molt. During this time, the birds are flightless and so are forced to feed in a relatively small area. Ducks prefer molt sites with protected offshore rocky islets and easy access to plentiful food. Harlequins have a strong affinity for their molting area and return to the same safe spot year after year.

In 1991 and 1992, Patten discovered about 20 percent of the molting ducks in the western Sound gathered near mussel beds (Patten et al. 2000a). In 1991, all of the mussel beds used by the molting ducks were oiled; in 1992, nearly 60 percent were oiled. He also detected a disturbing trend—the harlequin population in the western Sound was dwindling. Far fewer molting ducks returned to the western Sound the second year of his study and the rate of decline was nearly three times greater than that in the eastern Sound.

In fall 1991, Patten and his colleagues (2000b) searched the state's oil spill files to map the distribution of oiling to compare it to their map of habitat use (Supplement 1, p. xviii). They found extensive documentation of oiling to known harlequin habitat in the western Sound (Supplement 2, p. xx). They also discovered that many of these same areas, especially the mussel beds, had been treated extensively with Inipol (which was known to kill mussel and oyster larvae) and other chemicals in 1989 through 1991 (Supplement 3, p. xxviii). They found that ADEC had issued a citation when Inipol was applied in excess of levels thought to be safe for fish and wildlife (CFS 1989 2[34]).[1]

In summer 1992, armed with their maps, he and his crew surveyed as many as possible of the 130 oiled sites used by molting harlequins in the western Sound. They dug in suspect beaches with trowels and clam shovels, sketched maps of their finds, and carefully took samples, which they sent to the Auke Bay Lab for analysis.

Their labor-intensive work led to discovery of nearly fifty mussel bed sites in the western Sound still heavily contaminated with *Exxon Valdez* oil (Patten et al. 2000a, Appendix 3). The oil was virtually fresh crude, trapped beneath thick mats of mussels in *"anoxic conditions."* Without oxygen, the oil had not weathered and it still contained measurable levels of PAHs—over 4 parts per million in the mussel tissues and ten times that in the underlying sediments. Patten suspected these levels were high enough to pose a threat to harlequins.

By 1992, the Sound provided its own collaborating evidence: other biologists observed reproductive failure among black oyster-catchers and high death rates among young sea otters and river otters in areas of the Sound that had been most heavily oiled (Andres 1997, 1998, 1999; Bowyer et al. 1994, 1995; Rotterman and Monnett 1993; Sharp and Cody 1993). All of these predators ate mussels.

Alarmed state and federal agencies conducted a pilot study in 1991, then a much larger study in 1992 and 1993, to determine the extent of the oiled mussel bed problem (Babcock et al. 1996). During the larger study, Auke Bay Lab chemists measured extraordinarily high levels of PAHs in sediments collected from oiled mussel beds—thirty-one of seventy sites in the Sound had concentrations greater than 10,000 parts per million PAHs! This was the equivalent of *one percent oil*—and it meant these sites were "sopping with oil," as Patten had described earlier. Mussels from these same beds contained levels of up to 8 parts per million PAHs, which validated Patten's findings. When this chemistry study along with the growing biological evidence became known in 1993, public pressure forced government agencies to try to restore the oiled mussel beds (Chapter 21).

In fall 1992, Patten hit the paper trail again, looking for evidence to support his theory that the lingering oil was still toxic to harlequins. This research was tough for him. He was a field biologist and he had struggled with organic chemistry in school. Gamely he persevered through stacks of technical reports.

His literature review of effects of low levels of oil exposure on seabirds and waterfowl yielded evidence to support his theory (Patten et al. 2000a, Appendix 1; Varanasi 1990; Varanasi et al. 1993). He found metabolic changes and poor health from oil ingestion and reproductive problems, such as "complete cessation of reproductive activity for long periods of time" and lifeless eggs (1-1). He also found other supporting evidence of low levels of oil in mussel beds in 1989 from a NOAA study on subsistence seafood—shellfish and fish (Hom et al. 1996). This study reported that intertidal shellfish (mussels, clams, chitons, and snails) had levels as high as 12 to 18 parts per million—right in the ballpark of the levels he had found in the mussel tissue samples from 1992 (Patten et al. 2000a, Appendix 5).

Ironically, the closer Patten got to pinpointing oil as the cause of the harlequins' demise, the harder it became for him to do his work. By late 1991 the shifting political winds finally caught up with Patten. New political appointees under the Walter Hickel state administration began to make his life miserable. His supervisor from the

ADFG Oil Spill Division had come directly from BP Exploration in Anchorage and he controlled oil spill research funding. Patten had to rewrite his proposal for his 1992 fieldwork *twenty-eight times* before he secured funding where before, under the previous (Steve Cowper) administration, he had had to make only one or two revisions. Isleib encouraged Patten to stick with his work and told him, according to Patten, "As the chips come down, I can help you out. I've got the data set. I'll back you to the hilt."

Patten managed to keep the harlequin duck study going until 1993. He described these years as "a period of hell" where everything that he did and had done was challenged. Patten said, "It came down to scientific assassination—and character assassination." There was turmoil, conflict, and struggle at every turn. Patten felt as though he was "a rat on an electric grid" where he was shocked by everything he touched. One of Patten's loyal field biologists told him, "Do you know what's happened to you? You've become politicized!"

Of this time Patten said, "The biggest, the most hurtful thing of all was being separated from my wife and children." Patten was not allowed, as other researchers were, to conduct his oil spill work from Fairbanks, so he could be with his family. He paid, out-of-pocket, for his weekend commute from Anchorage. To protect his marriage, he didn't take his troubles home. His blood-pressure sky-rocketed.

When he submitted the first draft of his final report in early 1993, his supervisors objected to everything—the design, the methodology, the results . . . Patten said, they even accused him of sloppy science. Patten wrote and rewrote his report. In June 1993 supervision of the politically troublesome harlequin project was transferred to the ADFG Waterfowl Program. The director there, Tom Rothe, was sensitive to politics—he wanted more revisions. Then, tragically, in July, Isleib was killed in a freak accident. Patten wept for his friend when he heard the news. Stubbornly, Patten refused to back down on his findings even though he no longer had access to Isleib's vast knowledge of harlequin pre-spill habits.

His new supervisor figured out another way to handle the situation. In May 1994, a waterfowl management job suddenly opened in Fairbanks and Patten was offered "the bait." Patten said he was "superbly overqualified" for the job, but he took it. He said, "It was the correct decision because I moved back to my wife and I resumed being a father again."

The final report of Patten's harlequin study—published by ADFG

seven years later in 2000—contains the concluding chapter of this
saga.
   *"By June 1994, all of the original staff, including the principal
investigator, had left the project after submission of a revised draft
of the final report. Since then, this final report was extensively edited for format and style, verification and expansion of original data
presentations, and additional statistical analyses by the ADFG
Waterfowl Coordinator. Moreover, important points of discussion
and initial conclusions have been critically reviewed and edited for
prudent interpretation of findings" (Patten et al. 2000a, ii).*

Patten noted the "Faustian allegories" in the final drama of his study.
He said simply, "These people sold out." However, others validated
Patten's work before his final report was even released (Chapter 18).

Exxon's Studies: Deception by Design?
Murres

Under pressure from the federal studies, media reports, and growing
public concern about murres, Exxon scientists repeatedly attacked
the public-trust studies, vehemently objecting to the published estimates of 375,000 to 645,000 dead seabirds. It was well known that
carcasses found on beaches represented a mere fraction of the total
killed, but Exxon scientists avoided mention of the fifteen worldwide
studies on carcass drift recovery rates. Instead they poked fun at the
high-end estimate, claiming that it was enough to "kill the highest
estimate of the entire spill-path murre population more than twice
over" (Parrish and Boersma 1995a, 113). Exxon scientists used the
colony-count of breeding adults in the oiled area, which represented
only a fraction of the total population. Exxon's studies and letters created controversy and public confusion which Exxon lawyers were
capable of exploiting to great advantage (Chapter 23).
   Exxon scientists also conducted two studies on murres, allegedly to determine spill effects. In one study, Exxon scientists
(Boersma, Parrish, and Kettle 1995) measured breeding success at
the largest colony of murres in the spill path at East Amatuli Island
in the Barren Islands. Swept by strong winds and currents, the
Barren Islands are bathed in turbulent nutrient-rich waters, which

provide a local environment that is rich in forage fish preferred by breeding seabirds. The Barren Islands are a regional "hot spot" for murres that could be expected to attract new recruits quickly, following losses of breeding adults. Because of this, the Barren Islands' murre colonies were the least likely, among the dozens of colonies in the spill path, to show a harmful effect from oil.

Predictably, Exxon scientists found no dramatic changes in the number of chicks fledged by murres during their three-year study (1990 to 1992) compared to murre productivity in the 1970s. They concluded the lack of a clear pre- to post-spill pattern was due to high natural variability and since they couldn't detect any trends, neither could anyone else—specifically the public-trust studies, which they called "not scientifically supportable" (Boersma, Parrish, and Kettle 1995, 847).

In the second Exxon study, a scientist surveyed "*attendance*" (number of birds) at thirty-two of the thirty-six murre colonies in the spill path in 1991 and compared the 1991 attendance with historical data (Erickson 1995). The Exxon scientist concluded that the spill had "no major apparent effect" on colony attendance (ibid., 809), although he admitted this was probably because birds from large at-sea rafts filled the vacancies at the colonies (ibid., 809). He also found colony-attendance levels in his 1991 survey was not "drastically lower than historical levels" (ibid., 811), but this was largely because of his choice of historical data used to evaluate trends. In other words, this Exxon scientist's survey *completely missed*—was unable to detect— the large-scale population declines, largely from natural causes.

This is a critical point. Wildlife populations cycle high and low through good and bad years—even decades. Species have evolved various strategies to ensure a breeding population survives the bad years. In murres, the survival strategy involves a buffer of large numbers of surplus adults in at-sea rafts. The oil spill wiped out hundreds of thousands of murres, but there was enough buffer to withstand these losses. However, the huge declines from natural causes before and after the spill put the population at extreme risk of an all out crash if exposed to another stressor—such as the spill. The Gulf of Alaska murres managed to survive these double insults—this time, but the threat of extinction from combinations of man-made and natural disasters is very real.

Exxon's studies had a very limited ability to detect effects from the spill or natural causes. Exxon scientists seem to have used a number of statistical tricks of the nature described by Huff (1954). For example, Exxon scientists built into their survey design *"small sample bias"* by focusing on a single study plot in one study or on a single season two years after the spill in the other study. They presented data on a per colony basis. With several small colonies increasing and several very large colonies decreasing, there were no statistical trends, however, there was a pronounced decrease in the overall numbers of murres attending the colonies in 1991 compared to historical data. Exxon scientists failed to mention there were fewer birds—*"an error of omission."* In their thicket of statistical biases, inferences, and omissions, Exxon scientists failed to detect not only the spill effect on murre populations, but also the almost unmistakable imprint from natural causes.

Harlequin Ducks

Exxon scientists (Day et al. 1995) found initial injury to harlequins, but claimed the population had recovered within two years and no long-term harm was expected. Problems with their study design and conclusions are discussed elsewhere (Chapter 18).

Sound Truth

Murre populations did not recover in oiled areas of the Gulf of Alaska and harlequin duck populations continued to decline in oiled areas of the Sound during the time period covered by this section (1989 to 1992).

This led public-trust scientists to further investigate the cause of the lack of recovery (Ecosystem Studies: Chapters 17 to 21).

Chapter 16.

Fish

Pink Salmon and an Ace in the Hole

On 24 March 1989 Sam Sharr (2001) started his day as he had for the past five years by jogging around Cordova in the pre-dawn quiet. In the Prince William Sound and Copper River fisheries (designated by the state as management "Area E"), Sharr was the lead biologist for salmon and herring research for the Alaska Department of Fish & Game (ADFG). He had come to Cordova in 1981 to start a "catch stock identification" program to help fishery managers prevent over-fishing of any one stream or area. He stayed on to help rebuild the catch and escapement sampling programs for salmon and herring as these programs had atrophied during state budget crunches in the 1970s. He had been hired for his current position when his predecessor retired. He loved his job and he loved his community. Unpretentious and quick to smile, Sharr was an ambassador of good will for the local ADFG office.

Eventually he swung by the home of his boss James Brady, the Area E fisheries manager, where Brady's wife, Nancy, joined him for a few more laps around town. When the joggers returned to the Brady home, Brady told them quietly that he had learned of "a sizeable spill

Contrasting Life Histories of Pink Salmon and Pacific Herring

Wild Pink Salmon

Wild pink salmon in Prince William Sound are unique in that they spawn in streams on intertidal beaches where the stream meets the sea (Cooney et al. 2001a). This was not always so; the 1964 earthquake uplifted land in the Sound and made many streams too steep for salmon to ascend. Pink salmon adapted by becoming intertidal spawners, which makes them vulnerable to oil spills. The fish spawn in "redds" or nests dug in the stream gravel from July through mid-September in some 800 short streams in the Sound and they die shortly after spawning.

Depending on the water temperature in the fall, the bright orange eggs hatch in October or November. The "alevins"—larval salmon with big yolk sac bellies—work their way deeper into the streambed for the winter. They emerge as "fry" in about six months with their yolk sac absorbed or "buttoned up." The skinny hungry fry salmon "outmigrate" or drift to the sea from late March through early June (or later if the water temperatures are colder than average). They spend the summer as juveniles in nearshore nursery areas, feeding and growing rapidly. Many fall prey to other fish and seabirds. By August or September, surviving fish, now adults, move outside the Sound and feed in the open ocean of the North Pacific until they return to spawn a year later.

Hatchery Pink Salmon

Prince William Sound hatchery pink salmon go through the same life stages as wild pink salmon, but spend their early embryonic phases at one of five hatcheries in the Sound (www.ctcak.net/~pwsac/). Hatchery fish return to spawn at the hatcheries where they were released as fry. The eggs and developing alevins incubate in controlled laboratory conditions. Fry are held in saltwater net pens and fed until the spring phytoplankton bloom peaks. The fry are released into this rich food source to increase the odds of their survival. Once released, the juveniles spend the summer feeding and growing in nearshore bays with their counterparts. During this time, they could be exposed to oil contamination in nearshore waters. They move outside the Sound in fall as adults and swim in the open ocean of the North Pacific until they return to the hatcheries to spawn a year later. Since the fish are not held in captivity their entire lives, this form of fish farming is called, "salmon ranching."

Pacific Herring

Prince William Sound herring arrive at their natal beaches in late March (Cooney et al. 2001a). When water temperature warms to about 40°F, they spawn in shallow waters near shore and in the intertidal zone of beaches, blanketing sea plants and the seafloor with their sticky milky white eggs. Many eggs are lost to seabirds, fish, invertebrates, and pounding by waves. Surviving eggs hatch in two to three weeks and release tiny yolk-sac larvae to drift in the water column.

Herring larvae absorb their yolk-sacs within days and the weak swimmers must learn to feed on passively drifting "microplankton" (microscopic sea plants). During the extended larval drift, many herring are lost to "zooplankon" (tiny drifting animals) or currents that carry them out to sea.

Survivors are swept into nearshore bays where they safely metamorphose to juveniles in late July or August. Chances of survival greatly improve after the first critical winter—a four-month period of semi-fasting.

(continued)

Juvenile herring spend the next two to three years in the protected bays, feeding and growing. They join the adult fish as four-year-olds and for the rest of their lives, eight years or so, they return to shallow water to spawn every spring and move to deeper offshore water to overwinter. They do not die after they spawn like salmon.

going on" in the Sound. With a sickening feeling, Sharr raced home to change clothes and join his boss for an over-flight at first light.

The biologists were stunned by what they saw. Oil was roiling out of the side of the stricken tanker and boiling into the water. Already the slick was huge. Filled with foreboding, the two realized that the spill could not have happened at a worse time of year for fish. Adult herring were already returning to protected beaches to "stage" or just hang out for a couple weeks before spawning *en masse* on seaweed (Sidebar 11, p. 250). After incubating for just over three weeks, the herring eggs would hatch and the pelagic larvae would drift for a couple months at the whim of currents and winds. Lucky ones would be carried into protected bays where they would spend a full year feeding and growing. Pink salmon also depended on clean intertidal streams and protected beaches. Hundreds of millions of pink salmon fry, the 1988 brood year, were still over-wintering in gravel redds (spawning beds) in intertidal streams throughout the Sound. In about a month, when the bulging yolk sacs would be nearly absorbed, the salmon fry would wriggle up out of the gravel to drift downstream to the sea where they would spend their summer months in nearshore nursery areas.

The dazed biologists wondered: how many of these critical nursery beaches would be oiled? What effect would the oil have on the Sound's wild herring and pink salmon populations? Usually by now, Sharr was out in the field collecting samples of young pink salmon fry before they emerged from their gravel redds in the spring. But the spring fry survey had been delayed because streams were still frozen. Suddenly Sharr realized he had to beat the oil to the beaches.

Four days later, Sharr and his field crew headed out to the Sound on ADFG's cutter, the R/V *Montague*, commanded by skipper Jim Vansant to dig fry samples from intertidal streams in a spring ritual that predated statehood in 1958. ADFG used the data to predict salmon run strength and set harvest and escapement levels. Over the years, some streams in this standardized survey had been dropped and others added so that the so-called *"index streams"* always represented a good cross section of wild pink salmon populations throughout the Sound.

Sharr realized that many of his usual index streams in the southwest district were going to get oiled because the dominant circulation flow exited the Sound through Montague Strait. He had brought along on the *R/V Montague* all the historic reports, stuffed in boxes, so he could pick additional index streams from the database, including ones he anticipated would not be oiled.

Racing ahead of the oily plume, Sharr and his crew collected fry—twice—at their still unoiled, selected sites by digging nearly fifty shallow holes in the streambed at various tidal heights. First, they used carefully cleaned clam rakes to collect fry samples for tissue analysis of oil content. They took great care to preserve the fry samples for these oil studies, following the procedure hastily outlined by Seattle-based NMFS chemist Bob Clark before they left Cordova. Next they placed a gas-powered pump in the holes to collect fry samples for forecasting adult returns, using the same method that had been used historically (Pirtle and McCurdy 1977). They recorded the number and condition—live or dead—of fry carefully, just as they had done in preceding years.

Sample collection was painstaking work made even more excruciating by frigid howling winds that plummeted air temperature to a wind chill of fifty degrees below zero and froze fish scoop nets solid within seconds. "It was something else," recalled Sharr of the 1989 spring fry survey. But they got their precious fry samples.

Meanwhile, Brady had been very busy. During the first week of the spill, he invited all the fish experts and oil experts in the state to meet in Cordova with his team of biologists. In order to attend the four-day meeting, Sharr flew back and forth to Cordova from the field between fry digs, while Vansant motored the R/V *Montague* to different streams. The gathering in ADFG's storage garage, hastily (and permanently) converted into a research space, included chief fisheries scientist John Clark, who had created the Fish Stock Biology Group in the 1970s to save the state's prized historic fisheries management program from budget cuts. Mixed with the biologists were NMFS oil specialists Bob Clark, Rice, Short, and others. John Clark told the gathering, "Write me an integrated program and an operational plan to determine the effects of the spill on herring and pink salmon."

This team effort resulted in the first and most comprehensive NRDA program that relied on the historic database as a solid foundation against which to measure and compare effects from the spill. Sharr said, "The amazing pre-spill time series of historic data became the backbone of the NRDA studies for pink salmon—at all life stages. Residents of the Sound owe a very, very large debt of gratitude to the pioneering biologists (Ralph Pirtle, Wally

Noerenberg, and Mike McCurdy, among others) for that legacy."

The NRDA program for pink salmon had six elements, which were assigned to different resource managers. ADFG's work included spring fry and fall egg surveys; summer aerial monitoring of adult salmon; summer foot surveys of adult salmon to calibrate the aerial monitoring; and a coded-wire tag program to monitor juvenile growth and adult returns. The federal work included monitoring growth, feeding, and survival of juvenile salmon and measuring level of oil contamination in tissue.

Field studies documenting effects of an oil spill on pink salmon and herring had never been done. No one really knew what to expect, not even the experts. The integrated team was determined to find answers to questions that no one had asked during the past two decades of oil research. Their success depended on close coordination, good communication with open sharing of data, and solid commitment to the multi-year program.

Back out in the field, Sharr was completing the fry survey and planning his next studies when news came from town: a lab technician had inadvertently left the precious fry samples for the oil studies out of the freezer and they had thawed. They were useless for the oil studies—the fry survey had to be repeated! Sharr and Bob Clark spent two frantic days in a helicopter, racing around the Sound to collect new samples to replace those lost.

Meanwhile, the R/V *Montague* and its crew resupplied in Cordova and headed back to the Sound to collect the second batch of fry samples for the short-term oil effects study (Wiedmer et al. 1996). This time oily sludge lay thick in the northwest-facing bays. Sharr recalled the "absolute dead quiet, so quiet it was startling. Just when the Sound was typically alive, there was absolutely nothing."

Sharr and skipper Vansant realized it was a mistake to treat streams in the unoiled eastern side of the Sound as controls, because physical differences like geography, tidal exposure, and volume of freshwater discharge, among others, could overwhelm and hide any differences due to oil alone. Instead Sharr wanted to pair streams—one oiled and one unoiled—in close proximity to each other, matching as many physical variables as possible. Reviewing the historical studies, the team found there were limited data from a lot of streams on the most heavily oiled sides of Knight, Chenega, Evans, Bainbridge, and Latouche islands and some of the mainland watersheds in southwest Prince William Sound.

In a normal year, critical decisions were based, in part, on the annual fry surveys and subsequent forecasts: whether fishery managers opened or closed fisheries; what price processors' paid fishermen for their catch; whether fishermen upgraded boats and equipment; whether banks extended large (or any) loans to fishermen. All these decisions and more depended on the biologists' forecast. And this was not a normal year—the forecasts would also be used in court to show how many salmon should have returned, but didn't because of the spill. Millions of dollars hinged on Sharr's forecast.

Determined to collect the best data in this intense situation, Sharr called Brian Bue in Anchorage and asked him to join them in the Sound. Bue was a biometrician with ADFG's central office. He took numbers from sampling efforts like Sharr's fry survey and applied different statistical tests to interpret results. Bue also helped biologists design their studies to ensure precise results and insightful conclusions. With Bue's guidance, Sharr picked ten oiled streams and fifteen unoiled streams—all index streams with historical data. Again, the crew collected fry samples for the oil effects study.

When finished with the fry survey and back in Cordova, Sharr supervised preparations for the coded-wire tag study (Willette 1996), which was a massive undertaking. Each spring for three years, hatchery crews carefully inserted coded-wire tags—nearly microscopic slivers of wire with identifying binary codes—into tiny snouts of a million juvenile pink salmon. The original intent of this program was to enable fisheries managers to distinguish between hatchery fish and wild fish in commercial harvests in order to estimate the proportion of wild fish in the harvest and prevent over-fishing of these stocks. Each fall lab technicians collected fish heads from processing plants and hatcheries to recover coded-wire tags from the previous year's release and make their catch estimates.

After the spill, the coded-wire tag program took on new importance because wild fish—with eggs incubating in oiled streams—were more likely to be harmed than hatchery fish, which would first encounter oil as juveniles in nearshore bays. To estimate loss of wild fish from the spill, it was critical to determine what percentage of the harvest was wild fish. The coded-wire tag program also allowed fisheries scientists to monitor growth of juvenile salmon. In salmon, as in many marine species, juvenile growth was an indicator of adult survival: the faster the growth, the less chance of being eaten, and the higher the survival. Anticipating delayed growth after the spill, both ADFG and NMFS sent crews to collect juvenile fish from nearshore areas throughout several summers. The coded-wire tag program also allowed fish crews to measure amount of *"straying;"* that is, adults returning to somewhere other than their natal stream or hatchery. Biologists suspected oil might impair the renowned homing abilities of salmon and result in increased levels of straying.[1]

By July, Sharr and his fish crews were back out in the field conducting "foot surveys" to monitor returns of wild pink salmon by walking along stream banks and counting salmon in the stream. Foot surveys were originally started to improve the accuracy of the historical aerial survey program. These improved estimates were essential for calculating losses of wild pinks after the spill. Sharr stationed foot survey crews on the R/V *Montague*, which made ten-day circuits of the Sound, stopping to survey fifty streams. Sharr also stationed crews at fish camps along index streams throughout the Sound. He visited the different crews to keep on eye on things.

"For sheer physical exertion," Sharr said, "There was nothing like the foot surveys from the fish camps." Each day at 4:00 A.M. Sharr and his crew jumped into small whalers and raced out of camp just ahead of two other crews. They ran up streams in full rain gear regalia, calibrating the extent of tidal reach with survey equipment, while the other crews followed, counting salmon in the stream. The crews also had to be wary of feeding bears. One of Sharr's crew lost twenty pounds in the first round of foot surveys. At one fish camp where crews monitored a weir and did daily surveys of several creeks, the dispatcher seriously underestimated the caloric needs of the crew. Sharr sent two planeloads of food to the famished technicians whom he described as looking "somewhat skeletal."[2]

In September and October, Sharr and his crew were back out on the R/V *Montague*, collecting pink salmon eggs for the fall egg surveys from the same index streams they had sampled earlier during the fry surveys (Sharr, Bue, and Moffitt 1990a). The streams were a little more crowded this time—they crossed paths with ADFG habitat biologist Mike Wiedmer and his crews and with Exxon's fish crews, all out collecting eggs and sediment samples to assess the level of oil contamination. In 1990 the various fish crews repeated their entire sampling schedules.

In spring 1991 the public-trust fisheries scientists attended a high-stakes meeting to compete for funding against all the other groups of public-trust scientists working on NRDA studies. Ironically, Sharr and the state fisheries biologists found themselves competing against Rice and his federal scientists for funding different parts of their comprehensive NRDA program! Federal lawyers in charge of funding projects were ruthlessly eliminating all studies that did not bolster their legal arguments for a big bucks settlement with Exxon. The fisheries scientists did not know what would become of their comprehensive program. In fact, they didn't even know how it was going because the lawyers had imposed secrecy orders after startup, which isolated the ADFG biologists from the federal scientists. The fisheries scientists even had to present their pieces of the comprehensive program separately to the lawyers.

The lawyers learned from ADFG biologists that weathered crude oil was not environmentally benign as Exxon claimed. Chemically, this oil was mostly PAHs, the aromatic hydrocarbons largely ignored since the 1970s. In response to this toxic threat, the lawyers were told that eggs died (Wiedmer et al. 1996). Further, surviving alevins, incubating in gravel streambeds, soaked up the oil residues (PAHs) sponge-like through lipid-rich eggs and tissues. Alevins increase production of a specific enzyme, known as cytochrome P450-1A, to metabolize or breakdown toxic PAHs.

ADFG and Auke Bay Lab scientists (Carls et al. 1996a) reported elevated levels of cytochrome P450-1A in tissues *from nine different organ systems*, including the liver, gills, peritoneal (abdomen) connective tissue, spinal cord cartilage, pharyangeal epithelium (throat lining), and the *"vascular endothelium"* (blood-rich lining) of the kidney, brain, heart, and intestine. Alevins from oiled streams were saturated with PAHs. Tissues of alevins from oiled streams also had more lesions or abnormalities than fish from unoiled streams. Thus, scientists pointed out, the PAHs were still inducing measurable biological harm in wild salmon two years after the spill. *This was an unprecedented finding.*

The lawyers learned from both state and federal scientists that surviving juveniles had markedly reduced growth in oiled areas in 1989. Juvenile pink salmon, feeding in dense schools near the water surface, physically gulped whole oil droplets and sheen floating on the surface (Carls et al. 1996b; Celewycz and Wertheimer 1996; Sturdevant, Wertheimer, and Lum 1996; Wertheimer and Celewycz 1996). Chemically, this oil was mostly PAHs, the multiple benzene ring compounds largely ignored since the 1970s. The Auke Bay Lab scientists believed this was the first field evidence confirming that oil exposure was linked with reduced growth. Again, *this was an unprecedented finding.* They suspected that young fish exposed to oil were forced to shunt some of their energy reserves to break down and eliminate the poison instead of using the energy for growth.

Lawyers were told that the spill had population-level effects on pink salmon (Geiger et al. 1996). Both state and federal scientists reported that surviving adults had high levels of straying—up to 50 percent of some pink salmon stocks failed to return to their natal hatcheries and wandered instead into streams to spawn (Sharp, Sharr, and Peckham 1994; see also Wertheimer et al. 2000). Further, the oil actually resulted in fewer salmon returning to streams and hatcheries. Bue and other ADFG biologists calculated that nearly two million wild pink salmon—over a quarter of the potential wild stock production from the heavily oiled southwest Sound — had failed to return in 1990 (Bue et al. 1996). Hatchery pink salmon from the oiled southwest Sound had about half the survival rate of pink salmon from unoiled hatcheries, a potential loss of nearly seven million fish in 1990. The primary cause of the population-level reduction in pink salmon survival was reduced growth at the early marine stage from exposure to oil—*another unprecedented finding*.

The government lawyers were unimpressed by the fisheries scientists' presentations. So was Bob Spies, the scientist hired by the EVOS Trustee Council to head the NRDA studies. The problem was that the record harvests of both pink salmon and herring in the Sound in 1990 undermined the scientists' arguments for continuing their studies with NRDA funding. The comprehensive fisheries program, as originally designed in the ADFG garage, was far too sophisticated for the lawyers who preferred obvious, blunt tools such as sea otter and bald eagle carcasses. As Auke Bay Lab scientist Rice explained, "Pink salmon were not a good bet for bringing in large settlement dollars at that time."

It was Sharr who saved the comprehensive fisheries program. He had worked nearly sixty hours straight preparing his talk. He argued it was too early to detect the full effects of the spill and that it would waste money to discontinue promising research. He advocated continuing basic fisheries research in order to understand the full range of effects from the spill. He argued convincingly that pink salmon were a keystone species in the Sound and a crucial indicator of the ecosystem health. Sharr said, "That speech was the one instant in time that I was proud of my career."

Other fisheries scientists listened in amazement to their inspired compatriot. Bue dubbed Sharr's talk, the "Patrick Henry speech"— "Give me salmon, or give me death!" In a break after his talk, a secretary whispered to Sharr, "That's the first time anybody here's said anything that's made sense." Sharr's speech, along with Bue's skilled technical analysis and some effective lobbying by Auke Bay Lab scientist Rice and others, tipped the balance. The comprehensive fisheries program was funded. According to Sharr, "The public never knew how close we came to losing everything."

In summer 1991, there was another record return of hatchery pink salmon, although the commercial fisheries harvest was a bust due to unusual behavior and disrupted run timing. Many of these fish had likely been exposed to oil either in the streams (wild stocks) in 1989 or in the nearshore areas (wild and hatchery stocks) in 1990. The adult salmon returned to the Sound, but instead of returning to their birth streams or hatcheries over a two-month period, the fish milled and ripened in deep water beyond the reach of fishermen's nets. Then, in a two-week period in August, millions of pink salmon bolted for streams and hatcheries. This unprecedented behavior overwhelmed the capacity of the Sound's fishing industry to catch, transport, and process the fish. At the end of the disastrous 1991 season, three of the five salmon processing plants in Cordova declared bankruptcy (Adams 1991a–g).

Meanwhile, Exxon scientists used the record returns of hatchery pinks in 1990 and 1991 to validate their claim that the spill had no effect on pink salmon. The public-trust scientists, however, knew the record returns were a result of favorable ocean conditions and that many millions more salmon would have returned had it not been for the oil spill. The public-trust scientists could not talk to the media because of the secrecy order (Sidebars 2, p. 9, and 10, p. 206; Cummings 1992). In the absence of other information, the national media picked up on Exxon's story. This left most of the public with the impressions that pink salmon were not affected by the spill; that the Sound was recovering; and that fishermen were getting rich from the record salmon returns!

Then in 1991 Sharr drew the ace in the hole during the fall egg survey. Ironically, fall egg surveys had been discontinued after budget cuts in the mid-1980s, but they had been reinstated as part of the comprehensive fisheries program at the insistence of Bue. During the two previous years, the highest egg mortality had matched the pattern of oiling, moving from the lower and mid-level intertidal zones in 1989 to the high intertidal zone in 1990 (Bue et al. 1996; Sharr, Bue, and Moffitt 1990a, 1990b). Then in 1991, everything changed. Sharr found 40 to 50 percent egg mortality in oiled streams, nearly double that of unoiled streams and much larger than egg mortalities during the previous years (Sharr, Bue, and Moffitt 1991). He also found that high egg mortalities were mismatched with beach oil—dead eggs occurred across the intertidal zone *and above it in areas that had never been physically oiled.* Sharr's initial reaction to his discovery was, "Oh, this really messes things up!"

But the more Sharr thought about it, the more he realized there were only two ways to explain these findings. One was to assume that the egg mortalities weren't "real" but rather an artifact caused by physical differences in the environment between oiled and unoiled streams. Sharr, Bue, and ADFG geneticist Jim Seeb worked cooperatively with the hatchery biologists two years later to incubate eggs from different streams in exactly identical laboratory conditions (Bue, Sharr, and Seeb 1998). Eggs from oiled streams had persistently higher mortality than eggs from unoiled areas. This finding proved the egg mortalities in oiled streams were real.

The only other explanation for the persistent egg mortality was that oil caused long-term heritable effects, which, if true, had somehow eluded scientists for decades. Sharr reasoned there were two mechanisms to explain his findings. One was that the entire parent stock of the 1991 year-class had incurred reproductive damage when they had incubated as eggs and alevins in oiled streams in 1989 and 1990, and this damage manifested as an inability to produce viable offspring—the dead eggs Sharr found in 1991. The only other way to explain the high number of dead eggs was to assume that the oil itself was somehow still directly killing the eggs.

Sharr realized that the implications of his egg survey data were profound—and completely outside his job description as an ADFG research biologist. That fall, he shared his findings and thoughts with NMFS researcher Rice and his team at the Auke Bay Lab. The federal scientists decided to further explore Sharr's extraordinary findings. Nearly a decade passed before public-trust scientists more fully understood the ecology of young salmon and the effects of oil on salmon life history (Chapters 17 and 20, respectively).

Pacific Herring and Political Suicide

On the Good Friday holiday in 1989, herring biologist Evelyn Biggs (1999, 2001) was surprised to get an early morning telephone call from her boss Brady, telling her to be at work by 7:00 A.M. She arrived to discover the ADFG office in an uproar and everyone in shock from the spill. Biggs was the fisheries research scientist in charge of stream surveys and the herring spawn deposition study for ADFG. It was her dream job, one that she had persistently pursued after finishing her masters in fisheries biology in 1980 at Corvallis, Oregon. She had clung to her dream through nearly a decade checkered with jobs commercial fishing, chartering sail boats, doing seasonal fisheries research in the Pacific Northwest and Alaska, and starting a family. Now her dream job had turned into a nightmare.

For Biggs and other staff biologists at the ADFG office, March 24 was the beginning of fourteen-hour workdays, seven days a week, with no relief until September. A single mother, she entrusted her three-year old son to daycare centers, grandparents, and friends. "It was really hard," she said, but quitting her job was unthinkable. She plowed her considerable energy into her herring research. She explained at the time of the spill, "Huge masses of herring were moving through the southwest Sound toward Naked Island. We hadn't even flown our first aerial surveys yet to monitor pre-spawning aggregations of herring. At that point, I knew zip about oil exposure and zip about oil toxicity. But it was pretty clear oil was going to intersect herring—adults, eggs, and larvae. I was panicking."

At the gathering of fish biologists and oil experts in the ADFG garage during the first week of the spill, Biggs found the help she needed to devise an integrated program to determine oil spill effects on herring. The program focused on early life stages—embryo and larval growth and development—because these stages were thought to be the most sensitive to oil effects. The herring team planned parallel lab experiments with oil exposures to help interpret and validate any field observations. They included sampling programs to determine distribution of the drifting pelagic larvae and to document oil exposure through Short's caged mussel studies.

On 31 March, three days after the violent spring storm churned the sea surface and mixed oil down into the water column, the herring started to spawn. For reasons known only to herring, they choose to spawn mostly in four areas: the northern ends of the Naked Island archipelago and Montague Island in the oil spill path and Fairmount Bay and Tatitlek Narrows outside the spill trajectory. For three weeks, thousands upon thousands of fish swarm in seaweed forests near these beaches and released sticky eggs and milt. Biggs monitored the activity from a spotter plane, watching the dark swarms of fish move through milky white clouds of spawn. By 20 April, the herring were done and there were 106 linear miles of spawn, a huge amount for Prince William Sound herring.

During her over-flights, Biggs observed that herring spawned on very few heavily oiled beaches, but there was considerable spawn along lightly and moderately oiled beaches. She saw visible oil sheens in spawning areas only a few times. Mostly the surface oil sheens moved and swirled ominously further offshore. Under the sea surface, however, it was a completely different story. Chemist Short found *Exxon Valdez* oil in all his caged mussels within the oil spill trajectory (Short and Harris 1996b). It was if an invisible plume of dissolved oil and oil droplets and particulates swept through the Sound under the sea surface. Based on Short's work, Biggs estimated that 40 to 53 percent of the total egg deposition—basically all the spawn at the Naked Island archipelago and Montague Island—was exposed to this invisible subsurface oil plume (Brown [formerly Biggs] et al. 1996).

The herring eggs incubated for about three weeks. During this time, Biggs dove along beaches with herring spawn to monitor incubating eggs and collect samples for lab studies planned by team member Mike McGurk, PhD. During her dives, Biggs relaxed and the spill trauma receded, pushed to the back of her mind by the stunning beauty of the underwater seascape. Filtered sunlight danced with waving seaweed fronds, freshly coated white by billions of tiny eggs. Rainbow-colored *"nudibranchs"* (shell-less snails) flapped gracefully along like slow-motion butterflies as they moved among the sea plants to graze on herring eggs. Other fish and tiny crabs nibbled on herring eggs, too, working their way slowly along the fronds. The entire scene just pulsed with vibrant life—or so it seemed.

Biggs and McGurk first noticed problems when the herring eggs should have hatched to release pelagic larvae. Along the oiled beaches, millions of larvae died immediately after hatch. Biggs called it a "high instantaneous mortality rate," and she calculated *a loss of 99.9 percent of herring larvae from oiled beaches*. Because about half of the herring eggs were deposited at beaches in the oil path, according to Short's chemistry data, this meant that half of the young-of-the-year were dead! McGurk's lab studies confirmed her field observations (McGurk and Brown 1996). They estimated that several million herring larvae hatched successfully from oiled beaches compared to several billion from unoiled sites.

The surviving herring larvae drifted at the mercy of currents and tides, which carried the young fish along with the oil through the western and southwestern Sound (Norcross and Frandsen 1996). During spring and summer, an unusual weather condition prevailed and subsurface water was retained in the Sound much longer than normal (Chapter 17). Virtually all the young herring, captured by McGurk and his crew, were within the upper eighty feet of the water column (Brown et al. 1996). So was the dispersed oil, according to Short's caged mussel data. Biggs's team estimated that at least 85 percent of all young herring were exposed to oil during summer 1989. Most larvae had twisted skeletons, missing lower jaws, and tissue lesions (Marty et al. 1997a; Norcross et al. 1996). Even the normal-looking young herring grew more slowly—at about half of the normal growth rate (Marty et al. 1997a; Norcross et al. 1996).

These lines of evidence, or "threads" as Biggs called them, all strongly suggested that oil had devastated the 1989 herring year class. But the government lawyers thought her evidence was inconclusive and they wanted to discontinue funding her work (Brown 1999). Bob Spies, the lead scientist for the *ad hoc* pre-EVOS Trustee Council, gave Biggs a 1987 review of biological effects of oil by Judith Capuzzo. For Biggs, Capuzzo's review "became somewhat of a bible in terms of predicting long-term effects" such as genetic damage, reduced reproductive potential, and disease. By reading scientific literature on oil effects and interacting with colleagues in other disciplines, she gained valuable insight, which helped her anticipate research needs and long-term oil spill effects. Instead of shutting down her program, Biggs decided she needed to expand her research to include a fish disease program and a genetic study and she fought ferociously for funding.

No one thought Biggs had a chance of proving any direct damage from oil because there were so many uncontrolled variables: for example, were the effects Biggs reported due to oil or differences in temperature, salinity, predation, or chance that some larvae drifted into patches of food while others starved? But Biggs's tenacity surprised everyone. Sharr (2001) said, "Nobody ever really gave her credit at the time, but man, she had tremendous intuition and she was enough of a forceful personality that she hung in there and made that herring program work." Biggs's request was funded.

Biggs added Jo Ellen Hose, PhD, Dick Kocan, PhD, and Gary Marty, PhD, to her team. Hose was an expert on injury caused by toxic chemicals to embryo and larval stages of fish and on general *"ecotoxicity,"* or the nature of pollutants acting on biological systems. Kocan's specialty was fish diseases and *"genotoxicity,"* or the nature of pollutants and viruses acting on fishes' genetic material. Marty was a budding fish pathologist—as a doctoral student, he studied fish tissues and organs to understand how diseases and pollutants caused physical abnormalities, such as skeletal bends and jaw deformities, and how these abnormalities affected the whole organism. Biggs used Capuzzo's review to guide and coordinate her expanded herring program.

As Biggs's expanded team conducted studies from 1990 through 1992, a very clear understanding of oil spill effects on herring emerged. At the cellular level, virtually all the herring captured in May 1989 near the Naked Island archipelago, within the oil spill trajectory, had deformed cells and genetic damage, while most larvae captured near unoiled spawning beaches did not (Hose et al. 1996). Larvae from oiled areas also had much higher rates of internal tissue and organ damage compared to larvae from unoiled areas (Marty et al. 1997a). Preserved samples from the 1989 lab incubation tests confirmed these findings, pointing to oil as the cause of the effects (ibid.). The researchers found that the skeletal deformities and cellular damage in larvae from oiled areas diminished rapidly over the next couple years along with the oil, according to Short's mussel tissue data.

The tissue lesions and organ damage had never before been documented in fish larvae following an oil spill. To validate their fieldwork, Biggs collected egg masses from an unoiled area for team member Kocan who exposed the eggs and larvae to low levels of oil in a laboratory setting, simulating conditions in the Sound. They found that the exposed herring developed skeletal deformities, cell and genetic damage, and tissue lesions similar to those observed in their wild counterparts (Kocan et al. 1996a). They also found that these abnormalities increased with increasing concentration of oil, specifically PAHs. In fact, one particular type of genetic damage, *"anaphase aberrations,"* was so closely correlated with PAHs that the researchers recommended using it to assess injuries during future oil spills. Hose, working with NMFS researchers, found abnormalities occurred at lab exposures as low as *one-third of one part per billion PAHs* (Hose et al. 1996). This was well within the range of oil known to be in the Sound in April and May of 1989 when the herring larvae hatched.

Biggs had anticipated that the oil spill might affect different age classes of herring as well. She suspected that herring exposed to oil as one-year old fish in 1989 might have suffered genetic or reproductive damage. Since young herring stayed in protected bays or nearshore areas for the first two years of their lives, she figured the

year old fish would have been exposed to subsurface oil in 1989 just
like the larvae. Since over 90 percent of adult herring returned to
their natal beaches, Biggs's team decided to evaluate the reproductive potential of these juveniles when they matured and returned as
four-year old adults in 1992.

Her suspicions were confirmed. In a combination of lab and field
studies, her team found that four-year old herring returning to oiled
sites to spawn had fewer embryos hatch and more abnormal larvae
than four-year old herring returning to unoiled sites (Kocan et al.
1996a). The four-year old fish with the worst internal tissue lesions
produced the highest number of abnormal larvae—four out of every
five fish.

Biggs also had concerns about the health of herring that had
returned as adults to spawn in 1989. The Auke Bay Lab team had
found liver lesions and high levels of PAHs in tissue of adult herring
from oiled sites in the Naked Island group and Rocky Bay in 1989
(Moles, Rice, and Okihiro 1993). The liver lesions and PAHs in tissues
were not present in fish from unoiled sites or in any fish in 1990 or
1991. (These findings were later confirmed in lab tests; Carls, Rice,
and Hose 1999.) By now she knew that exposure to oil stressed fishes' immune system and compromised health, which could lead to disease outbreaks. Would the herring population crash at some point in
the future like some scientists speculated?

By fall 1992 Biggs faced a wall of resistance to funding her herring program. The herring population was at an all time high and
the fish appeared healthy. She appealed to ADFG Commissioner
Carl Rosier and Jerome Montague, Chief of Restoration in ADFG's
Habitat Division, for support. She argued that long-term injury was
already occurring in herring and she warned that disease outbreaks
might occur. Her stalwart supporter Sharr said, "Evelyn was down to
being one person against the whole damn process." Her appeal fell
on deaf ears. All funding was cut for the herring program.

Biggs felt the main reason for the abrupt demise of her herring
program was that her team was among the first to suggest that oil,
specifically PAHs, was incredibly more toxic than previously thought.
Public support for such research was political suicide—a career-
breaker—under the pro-oil Hickel and Bush administrations. Other
studies that found long-term damage from oil, such as the juvenile sea
otter studies and the harlequin duck research, had run into similar
problems with funding cuts (Chapters 14 and 15).

Half a year later in April 1993, the same people who cut her program faced an irate public—and the possibility that Biggs was right. During the winter when Biggs and her team were closing out their research program and writing papers for publication, a disease outbreak swept through the herring stocks, decimating the population. The first anyone knew of it was in April when the few surviving fish, weak and riddled with lesions, straggled into the nearshore areas to spawn (Brown et al. 1996). Over winter the adult herring population had crashed from a twenty-year high to a twenty-year low. Biggs said, "All of a sudden, ADFG wanted 'magic' answers after ignoring our plea to follow the thread of evidence."

ADFG sent Biggs out immediately to collect herring from the Sound. She was shocked. She described the fish as "looking horrible. They swam in circles and spirals or lay at the surface belly up. They were covered with lesions—open bloody sores—and, when we opened up the females, we found solid white clumps of undeveloped eggs—the fish were reabsorbing them." Biggs called Alaska's fish disease expert who determined the disease was an outbreak of a "viral hemorrhagic septicemia" (VHS) virus. Beyond that initial identification, she was not impressed with the state's fish pathologist. She explained, "He wanted us to find fish kills. Well, three-quarters of the stock were missing, but there were no dead bodies. Looking for fish kills is a waste of time—they're eaten. We never see them."

Biggs recommended to Jim Ayers, the executive director of the EVOS Trustee Council, that he get a herring disease study back in the Sound and that he get a project leader, but one year passed before herring were collected again in the Sound. During this gap, Biggs's thread of evidence between the spill and the disease outbreak was irretrievably broken. She said, "A concrete link between oil and the disease outbreak was never really made. Most people still believe there was a connection—we simply could not prove it."

Meanwhile, despite Biggs's frustration with being marginalized by the EVOS Trustee Council and ADFG, she continued to believe that the herring research was critical to understanding the long-term ecosystem effects of the spill. She took time off work to help strengthen the consolidated class action brought by fishermen, Natives, and others damaged by the spill. She spent most of fall 1993 working with the new ADFG herring scientist, researching and outlining herring studies for the new Sound Ecosystem Assessment (SEA) program funded by the EVOS Trustee Council.

Finally, in 1994 Biggs decided to move to Fairbanks to pursue her doctorate in herring ecology at the Institute of Marine Sciences. She realized that she was totally stressed out and this hampered her effectiveness. Biggs had another reason for moving as well. In 1991 she had remarried and had her second child. Her family needed her as much as she needed them. For the next ten years Biggs (now Brown) studied the lives of juvenile herring for the SEA (Sound Ecosystem Assessment) project and forage fish for the APEX (Alaska Predator Ecosystem Experiment) project to fill a void in both oil spill research and fisheries management. This was work she had tried to do through ADFG. Her work became an integral part of the SEA and APEX programs (Chapters 17 and 19).

In reflecting on his wild ride after the oil spill, Sharr (2000) said, "We were extremely lucky to have a group of young, energetic, conscientious, and ethical biologists at ADFG. They tackled an immense problem without huge amounts of education or exhaustive resumes. They took on the extra work with little funding and no support from Anchorage or Juneau, where supervisors resented the oil research for competing with the normal management duties. What they accomplished was extraordinary—and largely unrecognized by the public. Their dedication salvaged the Prince William Sound research projects."

Exxon's Studies: Deception by Design?
Pink Salmon

Exxon scientists also studied the effects of the oil spill on all life phases of pink salmon from eggs to adults. However, careful examination shows that Exxon's studies appear to be loaded with statistical tricks that reduced the odds of finding harmful effects from oil, as was the case with Exxon's coastal ecology studies. For example in the fall egg surveys, the sample size was too small to detect harm; the signal from oil—the poor egg survival in oiled streams—was overwhelmed by natural physical differences among the streams. Further, there were not enough streams sampled or number of samples per stream to allow scientists to detect an oil signal; that is, to tune out background noise and statistically find differences in survival between oiled and unoiled streams.[3]

Another example of bias was use of what Huff calls "*the well-chosen average*." In the spring fry surveys, both public-trust scientists and Exxon scientists found that survival and number of normally-developing fish was far less in oiled streams at the lowest tide level in 1990 than in 1991. Exxon scientists combined and averaged their data across all tidal zones—which they themselves noted was "not valid" (Brannon et al. 1995, 567)—and the oil effect vanished. Exxon scientists commonly averaged their data, which had the net result of obscuring oil spill effects (Chapters 12 to 15).[4]

One classic example of confusing results was Exxon scientists' use of what Huff calls "*the semi-attached figure*." Huff wrote, "If you can't prove what you want to prove, demonstrate something else and pretend that they are the same thing" (1954, 74). Exxon scientists (Maki et al. 1995) designed one study to determine whether adult salmon returned to oiled and unoiled streams, but *not* whether fewer fish returned to oiled streams. Exxon scientists then pretended that the mere presence of fish in streams was the same thing as no effects of the oil spill on adult pink salmon. In this example, Exxon scientists attached their conclusion of "no effects of the spill" to something completely different—the fact that salmon return to streams to spawn!

With their arguably biased study designs, Exxon scientists consistently found they could detect no differences between various life stages of salmon from oiled and unoiled streams. They interpreted this to mean there were no measurable effects from the oil spill. The salmon studies were prime examples of Exxon scientists' mastery of deception—or as Huff wrote, "Counting up something and then reporting it as something else" (1954, 80). "No measurable effects" was a lot different from "no effects"—it meant only that Exxon scientists did not measure any effects—but few people other than scientists understood the creative semantics.

Exxon also played up the record returns of hatchery salmon in 1990 and 1991 with the national press, which had the effect of hiding the truth about oil spill effects. Unfortunately, this was relatively easy for Exxon to do, because the public-trust scientists were still under gag orders. Public-trust scientists knew the record returns were related to favorable environmental conditions and record crops of plankton—and that the returns would have been even better without the spill.

Pacific Herring

Exxon scientists steadfastly maintained that the spill did not harm herring. However, Exxon's herring studies, like the salmon studies, also appear to be designed to obscure and minimize oil spill effects (Pearson, Moksness, and Skalski 1995). For example, Exxon scientists ignored the extensive subsurface oiling of herring spawning beaches, eggs, and larvae. Instead, they based their study on *visibly* oiled beaches and they estimated only about 4 percent of the herring spawn occurred along *visibly* oiled shores in 1989 (ibid., 626). This strategy increased the chance that eggs and embryos exposed only to subsurface oil would be classified as "unoiled." Exxon scientists essentially compared oiled fish with oiled fish and looked for differences in effects. Not surprisingly, they found few differences. This strategy discounted a lot of oil spill effects, such as the dead eggs and grossly deformed herring larvae, observed by Biggs and her colleagues.

It appears that to further obscure oil effects, Exxon scientists collected one water sample and an unreported number of sediment samples at each site where they collected herring spawn (Pearson, Moksness, and Skalski 1995, 633, 637). The water and sediment samples were supposedly indicative of the oil in the environment (instead of caged mussels). However, only *average* PAH concentrations *by bay* were reported for the water samples—with no confidence intervals (ibid., 646, Figure 9a) and the sediment sample results were not reported at all. The sediment samples may, or may not, have revealed the truth about the extent of subsurface oiling to beaches, eggs, and larvae and the fallacy of relying on visible oil to classify herring as oiled or not.

Exxon scientists then looked for oil effects in herring larvae against this backdrop of averaged PAH concentrations—and, not surprisingly, found few effects that could be pinned on oil in 1989 and 1990 (ibid., 627). Unlike the public-trust scientists, Exxon scientists did no further oil toxicity studies to determine if the 1993 population crash, which wiped out literally over 100,000 tons of herring, was related to the spill. Instead, ten years later, despite substantial evidence to the contrary, Exxon scientists still claimed that the spill had not harmed herring (Chapter 20).

Sound Truth

Pink salmon embryos continued to die at higher rates in oiled streams through at least 1993. In 1992 and 1993, pink salmon stocks collapsed in the Sound.

In 1993, Prince William Sound herring stocks collapsed from a viral disease outbreak.

These unexpected fisheries collapses had devastating consequences for the commercial fishing industry, individual fishermen and their families, and the Cordova community. The fisheries collapses and the resulting socio-economic upheaval with political consequences led public-trust scientists to further investigate the causes of the lack of recovery and disease problems (Chapters 17, 20, and 21).

Ecosystem Studies
(1993 to 2003)

Chapter 17.

Sound Ecosystem Assessment (SEA) Program

Introduction to the SEA Program

In September 1993, Ted Cooney, PhD, (1999) a research professor with the Institute of Marine Science at the University of Alaska Fairbanks, was in Cordova on a mini-sabbatical. He was wrapping up some work he had started when the oil spill hit in 1989. It was a small cooperative study with the Prince William Sound Aquaculture Corporation (PWSAC) and ADFG to examine relationships among sea temperature, growth of juvenile salmon, and juvenile salmon survival.

At the time, fisheries biologists knew that ocean survival was of utmost importance to controlling and predicting salmon survival and run strengths, but they didn't know where in the ocean this occurred, nearshore or far off at sea. The quest to understand this relationship was akin to the quest for Holy Grail and many fisheries biologists, including Cooney, devoted their careers in its pursuit.

Cooney's approach to this quest was through the base of the marine food web—the plankton—the community of tiny sea plants

("phytoplankton") and animals *("zooplankton")*, including fish larvae, that largely drift with currents and tides. He had started his
career in Prince William Sound in the mid-1970s, working with the
hatcheries to determine if there would be enough plankton to feed
millions of hungry hatchery-reared salmon fry when they were
released *en masse* into the Sound. He followed his passion, studying
mostly zooplankton communities in the Sound and Gulf of Alaska
and effects of climate and season on these small creatures. He also
studied growth and feeding behavior of hatchery-reared salmon on
zooplankton to determine relationships that might help more accurately predict adult salmon returns. This work eventually had led him
to initiate the cooperative study, which was put on hold after the spill
when Cooney had been swept up in NRDA studies, looking at spill
effects on plankton.

In September 1993, Cooney was finally getting caught up with
the life he had dropped in 1989. He had only been in Cordova a few
days, he said, "when word came down that, as a result of the blockade at the terminal and pressure by the fishing community, the EVOS
Trustee Council was going to take a new approach at trying to understand what was going on in the Sound" (Sidebar 12, p. 275). Cooney
was called to attend an emergency teleconference meeting with Jim
Ayers, EVOS Trustee Council executive director, Wright, the NOAA
liaison to the council, and Gary Thomas, the Prince William Sound
(PWS) Science Center executive director. The meeting stretched past
midnight and resulted in money awarded to the PWS Science Center
to plan a comprehensive study of pink salmon and herring.

Cooney was already involved with a group of local fishermen
and scientists, known as the PWS Fisheries Ecosystem Research
Planning Group, that had been guiding fisheries management decisions since the spill. This group swung into action. Cooney said, "The
upshot was that we all worked together long, long hours for nearly
three months that fall to create this plan."

Cooney and other scientists were selected to present the
group's plan, dubbed the Sound Ecosystem Assessment or SEA program, to a panel of international scientists—and to the community—
on a dreary cold December day. They described the SEA program as
a 5-year integrated study of physics and biology affecting survival of
juvenile life-stages of pink salmon and herring in the Sound. The plan
called for *"bottom-up"* investigations, looking at effects of varying
marine conditions and plankton on fish survival, and *"top-down"*
investigations, looking at what might influence predation on salmon

and herring (Pearcy 2001). The goal was to understand the various processes that affected loss of these juvenile fish so mathematical models could be refined to more accurately predict adult returns, among other things.

It was a masterful plan, drawing on skills from different disciplines to pierce the mystery shrouding the lives of young salmon and herring. Nothing quite like it had ever been tried for lack of funds or opportunity or both. Cooney said, "It was quite daunting and a little bit scary to look across the table at notable oceanographers and fisheries scientists and answer questions about why we

---

Changing the Course of History: Fishermen's Blockade and Ecosystem Studies

In the wee morning hours of 20 August 1993, the Prince William Sound fishermen seine fleet—some 100 boats strong—battled through a fierce fall storm and converged in Valdez Narrows, the geologic bottleneck to Port Valdez. There, the fishermen formed a blockade with their boats, effectively halting all tanker traffic for three days (Bickert 1993a). The Cordova community supported the blockade by sending out food, banners, and townspeople on the smaller gillnet boats.

This act of community civil disobedience was an act of desperation. The Sound's pink salmon fishery had collapsed the year before (1992) and by August 1993, it was obvious the pink salmon fishery was a bust again. Further, the Sound's lucrative herring fishery had collapsed in spring 1993 when over 100,000 tons of an estimated 120,000-ton stock failed to return to spawn. Faced with financial ruin and bankruptcy, the fishermen decided to protest to focus attention on persistent spill-related problems in the Sound (Wuerth 1993a, 1993b).

Interior Secretary Bruce Babbitt detoured from Anchorage to meet with protest leaders in Valdez as did Alaska Governor Walter Hickel who was overheard to say, "Well, if I was a fisherman, I'd probably go out there" (Steiner and Grimes 1999). The fishermen asked for relief from the state on boat and permit loans for one year; a promise they would not be fined for civil disobedience; and "ecosystem studies" to determine the cause of the Sound's ailments and a realistic time to recovery. Once their demands were met, the fishermen quietly disbanded the blockade. No one was fined (Bickert 1993b).

As a result of this civil action (*New York Times* 1993), the newly-appointed EVOS Trustee Council under the Clinton Administration pledged over $5 million to start the Sound Ecosystem Assessment program, the first of three ecosystem studies. These massive, multi-year studies, and the complementary research at the NOAA/NMFS Auke Bay Lab to verify the field work on oil effects to pink salmon and herring, laid the foundation for a new understanding of oil toxicity, discussed in Part 2 of this book. By observing the Prince William Sound ecosystem as a whole, rather than as individual disconnected parts, scientists ultimately discovered oil was much more toxic than previously thought (Peterson et al. 2003). The ramifications of this profound discovery have yet to be fully realized and are discussed in Chapter 24.

should get $5 million to begin this project." But the international panel seized the moment and approved the conceptual approach with some refinements.

In January 1994, Cooney and others with the fisheries planning group presented the SEA project to the EVOS Trustee Council—where it was nearly summarily dismissed as being too grandiose, too risky, and too expensive. U.S. Forest Service (USFS) trustee George Frampton finally suggested that the SEA plan should at least be forwarded to Dr. Spies, the lead scientist of the EVOS Trustee Council, for formal review. Frampton and others persuaded the members of the EVOS Trustee Council of the merits for funding a major ecosystem study. In mid-April 1994, Cooney got a phone call from Jim Ayers who said the Council was going to fund the SEA project at $5.1 million for the first year—and they wanted him to be the lead scientist. Cooney said, "There went my life again."

For the next six years, he ran "flat-out," managing, in total, a $22 million budget and juggling the four main components of SEA—oceanography, plankton, pink salmon and herring projects, each with their separate studies—while actively participating in some of the studies himself. He also expended "quite a bit of effort," as he said, keeping people on track and focused without getting diverted down intriguing branches off the main stem of the SEA program. In the end, it was Cooney's dedication and drive that guided all the principal investigators—the leaders of the separate studies—through a successful, massive closeout with a complete write up and synthesis of the SEA program's findings (Cooney et al. 2001a; Pearcy 2001).

Bottom-Up Investigations
Physical Oceanography

Oceanographers Shari Vaughan, PhD, Shelton Gay, and others at the PWS Science Center looked at marine life through the lens of large-scale physical processes. They tracked sea temperature changes and movement of nutrient-rich water masses, which drove production of the ocean's tiny sea plants and animals. These cycles of plankton abundance, in turn, influenced growth and survival of juvenile fish such as pink salmon and Pacific herring. Abundance of forage fish, in turn, influenced growth and survival of fish-eating seabirds, marine mammals, and even other fish. Physical oceanography was a study of

The crippled *Exxon Valdez*—with 8 of 11 cargo tanks ripped open—was towed to Outside Bay, Naked Island, Prince William Sound (PWS). Repaired and renamed, the tanker plied the foreign trade while Exxon tried unsuccessfully to lift the ban on its return to the Sound (Associated Press 2002). In 2002, the 16-year old tanker was retired in an undisclosed foreign port. (Little 2002).

Cleanup workers used a pressurized hot water wash to push oil into the sea where crews with booms attempted to collect the oil. Exposure to oil mist and oil aerosols (PAHs or polycyclic aromatic hydrocarbons) posed a health risk for cleanup workers. Many still report ill health. Smith Island, May 1989.

Cleanup workers who handled oily trash (in orange bags) were at risk of exposure to hydrogen sulfide. Smith Island, PWS, May 1989.

Some cleanup workers and skiff crews were housed on the Alaska State ferry. Volatile oil compounds evaporated from the surface sheen and posed a health risk to everyone exposed, not just the paid crews. Oil compounds also dissolved in the water column under the sheen and posed a similar health threat to young salmon, herring, and other sealife. PWS, May 1989.

Barge-mounted cranes directed pressurized hot water sprays at oil on rocky cliffs and harder-to-access beaches. Barge crews were exposed to oil mists and PAH aerosols. Perry Island, PWS, 10 June 1989.

Cleanup workers used solvents and pressurized hot water wash to clean miles of oily booms. These workers were exposed to oil mists, PAH aerosols, and other chemicals. Herring Bay, Knight Island, PWS, 11 September 1989.

A cleanup worker without a respirator sprays Exxon's product, Corexit 9580M2, on Quayle Beach, named after the vice president who visited it. Many cleanup workers who sprayed this product on Disk Island were hospitalized with respiratory problems. After this incident, Exxon paid cleanup workers to sign an indemnity form to protect the company. Ironically, Corexit 9580M2 was eventually banned due to concerns for wildlife, not human, health. Smith Island, PWS, 8 August 1989.

Cleanup workers guide a pressurized hose across a beach, while the worker on the end sprays Inipol EAP 22, which contains an industrial solvent, 2-butoxyethanol, a known human health hazard. In 1990, Exxon provided cleanup workers with more protection against chemical exposures. However, health problems possibly linked to use of this product surfaced after the cleanup. Inipol EAP 22 is no longer manufactured. Northeast end of Green Island, PWS, 1990.

Cleanup workers such as Phyllis "Dolly" La Joie washed workers' personal clothing with Tide, Simple Green (shown), and other solvents to remove oil stains. They washed the raingear and outer garments with pressurized hot water wash, which exposed the decontamination workers to PAH aerosols. La Joie is suffering multiple health issues such as respiratory problems, memory loss and other central nervous system problems, and chemical sensitivities, among other problems. She claims her health problems stem from her cleanup work. Simple Green contains 2-butoxyethanol, a human health hazard. The now discontinued Inipol EAP 22 and some of Exxon's Corexit products also contain 2-butoxyethanol.

According to public-trust scientists, nearly half of the total beaches where herring spawned in 1989 were oiled by either visible surface sheen or dissolved oil: 99.9 percent of the herring embryos exposed to oil died immediately upon hatching in 1989.

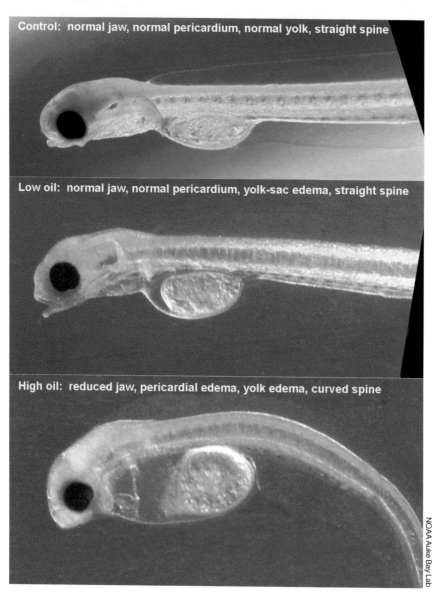

Control: normal jaw, normal pericardium, normal yolk, straight spine

Low oil: normal jaw, normal pericardium, yolk-sac edema, straight spine

High oil: reduced jaw, pericardial edema, yolk edema, curved spine

NOAA Auke Bay Lab

Federal scientists at the NOAA Auke Bay Lab discovered that very low levels of oil-in-water maim and kill both herring and salmon embryos. Herring embryos (shown) were harmed by PAH levels in the water as low as 9 parts per billion; exposure to higher levels (86 parts per billion) caused more harm. Weathered oil, similar to the oil buried under beaches in PWS, harm fish embryos at PAH levels in the *parts per trillion*. All these levels are below outdated federal water quality guidelines for PAHs (300 parts per billion), once thought adequate to protect aquatic life. The federal guideline is not a legally enforceable standard.

In 2001–2003, federal scientists at the NOAA Auke Bay Lab documented extensive buried oil in the biologically rich "green zone" near the low tide line on PWS beaches. Although the beach surface appeared clean, residue of Exxon's oil welled into pits (right) dug on the beaches. Scientists found invertebrates soaked up this oil. Scientists linked exposure to low levels of buried oil (PAHS) and oil-contaminated food with delayed recovery in sea otters, harlequin ducks, and other predators that eat invertebrates. Lingering harm to wildlife from persistent, bioavailable, and toxic oil was unanticipated in 1989. By 1999, the U.S. EPA had identified PAHs as a human health hazard.

Prince William Sound fishermen, facing financial ruin after collapses of herring and pink salmon populations, blockaded Valdez Narrows from August 20 to 23, 1993, to focus attention on the ailing Sound. As a result of this blockade, scientists funded through the *Exxon Valdez* Oil Spill Trustee Council conducted three seminal ecosystem studies and determined that very low levels of oil were much more toxic to fish and wildlife than previously thought. Similarly, medical doctors have found oil (PAHs) also poses human health risks.

factors that influenced the ocean's biological abundance literally from the bottom of the food web up to the apex predators.

Vaughan and Gay also studied the physical processes that influenced temperature changes and nutrient supply such as the exchange of water between the Gulf of Alaska and the Sound; mixing of large water masses in the Sound caused by seasonal variations in temperature and salt content; currents produced by tides and winds; and yearly differences in "climate" as energy exchanges between the sea surface and overlying air masses. They produced hydrographic maps, which were similar to topographic maps, but instead of showing elevation gradients of landmasses, their maps showed temperature, salinity, and density gradients of water bodies at different depths and seasons.

The oceanography team was tasked with mapping the seasonal physical conditions in Prince William Sound during the SEA investigation. Specifically, Vaughan and others were to map the central Sound, basically extending in time the work of earlier oceanographers, while Gay focused on surveys of deep fjords and shallow bays used as nurseries by juvenile herring. They hoped their map collections would define distinct circulation patterns so that another team member, mathematician wizard Vince Patrick, could construct computer models to predict how the circulation might vary under different climatic conditions. Their goal was to have their hydrographic maps and computer model help fisheries biologists, starting with the SEA members, understand what was driving survival of young salmon and herring in order to more accurately forecast adult returns.

Vaughan and Gay conducted twenty hydrographic cruises over four years. They augmented the hydrographic cruises with data from an acoustic Doppler current profiler, which they moored in the central Sound. This instrument recorded changes in flow of water masses as reflected "pings" in much the same way as a radar detector measures the speed of an automobile. They released satellite-linked drifting buoys, equipped with "holey-sock" drogues, to track speed and direction of currents in the upper water layer. They accessed meteorological data—wind speed and direction, wave height, barometric pressure, and air and water temperature. These data were recorded every thirty minutes from buoys that they moored in mid-Sound and at Seal Rocks in Hinchinbrook Entrance. The data were relayed directly to their computers in Cordova from a communications repeater station, which the oceanography team installed on a remote mountaintop in the Sound. Their computer technology at the PWS Science Center was so high-tech that it impressed even visiting Alyeska oilmen, who use complex computer programs to operate the Trans-Alaska Pipeline System.

In the 1970s and 1980s oceanographers had charted the circulation within and immediately outside of the Sound (Royer 1998). They had found the circulation is driven largely by yearly and seasonal variation in the transport of Gulf of Alaska water into the Sound by the Alaska Coastal Current (Royer 1981a). This powerful boundary current, one-fifth again as large by volume as the Mississippi River, flows westward along the North Gulf coast. It is fueled by the massive outflow of freshwater from glacial ice and snow fields, located along the coastal arc from British Columbia to the Kenai Fjords (Royer 1979, 1981b). Near Prince William Sound, the Alaska Coastal Current swells during the summer and early fall when high rainfall coincided with glacial melt, and it shrinks during the winter as the coastal rivers freeze up. It is generally believed that this current flows along the coast and loops into the Sound through Hinchinbrook Entrance, swooshing around the central Sound in a counter-clockwise gyre and mixing with the Sound's water before exiting through Montague Straits (Royer and Emery 1987; Royer, Hansen, and Pashinski 1979).

The oceanography team found striking differences in this pattern during the SEA years (Vaughn, Mooers, and Gay 2001). During the winter-spring transition, they observed that, as expected, the exchange of water between the Gulf of Alaska and the Sound at Hinchinbrook Entrance was uniform, orderly, and largely responsive to prevailing winds. During the summer-fall transition, they found, to their surprise, that circulation split into horizontal layers and the direction of flow within these layers varied between years. For example, in 1995, the upper 500 feet of water reversed and flowed *out* of the Sound, while the deeper layer flowed *in*. In 1996 and 1997, this pattern reversed. They found strong bursts of inflow always occur within the upper layer in response to strong winds.

Vaughan and Gay found the winter flow is driven by frequent and gale- and storm-force winds and it snakes through the Sound like a strong river. In contrast, the summer flow does not have the same powerful pizzazz and the water mills, creating "lake-like" conditions in the Sound of a seemingly indecisive mind, sometimes eddying in a weak clockwise gyre, sometimes counterclockwise. By fall, the circulation of the central gyre would sort itself into the usual counterclockwise motion, picking up speed as the Alaska Coastal

Current quickened and swelled with freshwater glacial melt. During winters, the circulation pattern matched that observed by the early oceanographers.

Strong winds, channeled by the mountains, and the high volume of freshwater runoff from streams and glacial melt create a circulation pattern in the Sound that is unique among inland seas around the world. Vaughan and Gay found this general circulation pattern varied, sometimes annually, but in ways that could be predicted based on factors like wind-strength, volume of freshwater discharged by the glaciers and rivers, and upwelling of deep ocean water flowing across the continental shelf into the Sound.

From spring to fall, they found the Sound is a rich composite of water masses, each with distinct characteristics. Deep-ocean water is cold and briny, the Alaska Coastal Current is warmer and less salty, and water within most fjords is brackish from freshwater runoff and glacial melt. All these water masses form layers with the lightest, freshest, and warmest ones at the surface and the heaviest, coolest, and saltiest ones on the bottom. These water masses are unstable and in near constant motion.

The sun creates a seasonal heating that stirs and mixes the Sound's water masses. In spring, increasing sunlight heats up the Sound from the top down. Small fjords fringing the Sound heat and freshen the quickest, but at different rates, depending on their depths, rate of freshwater input, and amount of mixing due to wind and tidal currents. The array of distinct water masses are set apart by density fronts. Density-driven currents flow between water masses, creating unstable eddies, which loop and spin through the Sound. In fall, the surface waters cool and gradually currents and winds mix the water layers from the surface to the bottom, creating a uniform water mass again until spring when the solar heating cycle begins anew.

Gay found that the circulation and hydrography within the small fjords turned out to be extremely variable, but, in certain cases, not impossible to predict (Gay and Vaughn 2001). While the fjords were all unique, they could be grouped into broad categories by presence or absence of glaciers; inner lagoons or lakes; and *"sills,"* underwater glacial moraines across the fjord mouths. Glaciers and lagoons influenced circulation in the upper water layers through freshwater input and delayed timing of freshening and heating events, respectively. Sills influenced circulation of the deep water by acting like speed bumps across fjord entrances. Gay's system allowed him to predict hydrographic patterns in the spring and summer, but not so much in the winter when strong winds and storms drastically altered circulation patterns.

The hydrographic maps produced by the oceanography team of these bays explained a great deal of the variability in zooplankton abundance, which was measured simultaneous with collection of the oceanographic data. Along with Patrick's three-dimensional computer model (Wang et al. 2001), the maps distilled order from seeming chaos. The maps clearly illustrated, in a predictable fashion, why the different fjords in the Sound were so variable in terms of plankton abundance and why this abundance varied so drastically in the same fjord between years. These watery 'road maps' assisted the biologists with their investigations.

Plankton Communities in the Sound

For about two decades prior to the SEA project, plankton communities in the Sound had been studied extensively by many researchers including Cooney. It was known that the zooplankton community is a mixture of coastal species (mostly small copepods) and oceanic species (mostly large copepods). The latter waft into the Sound from the Gulf of Alaska. Oceanic copepods take refuge in deep trenches in the Sound, some of which lie nearly half a mile below the water surface. There, they reproduce and die each spring, while their offspring enter the surface waters in April. Researchers had noted that the annual crop of tiny drifting plants appear to be nitrogen limited during each spring because the bloom is usually short-lived, but not always so.

The plankton community varies tremendously between seasons and years, mocking efforts of fisheries managers to predict the strength of pink salmon returns based on food availability. No one had figured out the relationship between the plankton communities and the physical environment of the Sound—this was the prize sought by researchers to take some of the guesswork out of forecasting salmon returns.

As the SEA project was being developed in 1993, Cooney made a startling discovery: the abundance of zooplankton in surface waters in April and May near the pink salmon hatchery in the southwest Sound was tightly linked with something called the *"Bakun upwelling index"* (Eslinger et al. 2001). This upwelling index was thought to reflect the intensity of the onshore spread of surface water from beyond the continental shelf into the Sound each spring. For the decade preceding the SEA program, the upwelling index accounted for fully 74 percent of the variability in zooplankton abundance between years.

For a biological oceanographer, this was an astonishing correlation. Cooney theorized that when the Aleutian low-pressure system was strong, this surface water flow would be strong, like a river, and it would flush out resident zooplankton in the surface layers in the Sound like a good spring housecleaning. Conversely, when the Aleutian low-pressure system was weak, the flow of Gulf of Alaska water into the Sound would also be weak. Under this scenario, the Sound would act like a lake and would retain its zooplankton community.

Cooney structured the plankton component of the SEA program to investigate his "river/lake" theory, controlled by horizontal mixing, as well as the traditional nutrient-limiting theory, controlled vertical mixing. His goal was to explain the origin of the seasonal and annual variability in the springtime zooplankton blooms so that his team could develop a predictive model to improve forecasts of pink salmon returns.

Cooney's team included other plankton specialists, oceanographers, chemists, and modelers with the PWS Science Center and the Fairbanks-based Institute of Marine Science and the International Arctic Research Center. The team conducted monthly ten-day sampling cruises from March through July, with opportunistic cruises in the interim from 1994 to 1997. They also obtained data on plankton abundance and weather conditions from moored buoys, rigged with various sampling devices, and from staff at the Armin F. Koerning salmon hatchery in the southwest Sound.

Cooney and his team (Eslinger et al. 2001) found that everything boiled down to a critical window in spring—as short as two weeks—when cold air and strong winds combine to determine the degree of mixing of water masses, which sets the stage for the spring plankton bloom. In a calm warm spring, the upper water layer warms quickly and remains separated from the cooler deeper waters. Tiny sea plants bloom in profusion in the warm surface layer, quickly deplete the nutrients in this layer, and die, sinking to depths inaccessible to surface-loving zooplankton. The population of these small grazers swells initially in response to an abundance of food, but before they can reach huge swarms, the plant population has died back. Cooney and his team realized that warm calm springs yield mega-blooms of phytoplankton, but only a small return on zoo-plankton, which are not able to take full advantage of their rich food source before it vanishes.

But during a cool stormy spring, phytoplankton bloom in bursts in the calm between storms. Turbulent winds and waves mix the warm upper water layer with cool deeper waters and temporarily limit phytoplankton growth while at the same time replenishing nutrients in the upper layer to support the next burst of plant growth. Cool stormy springs extend the phytoplankton bloom in time, which in turn supports dense swarms of animal grazers. Phytoplankton continue to bloom until nutrients become depleted as the upper water layer stabilizes when spring storms taper off. Contrary to prevailing opinion, Cooney found cool stormy springs with a small standing plant crop actually result in an abundance of zooplankton.

After five years of data collection and analysis, Cooney realized variability in the standing stock of zooplankton during the SEA years could be mostly explained by the traditional nutrient-limiting theory, not his "river/lake" theory. He was discouraged, but then it occurred to him to apply the spiffy SEA computer model *retroactively* to the decade prior to startup of the SEA program. He found that prior to 1992 the computer simulation was completely out of synch with the field data—when the model predicted high zooplankton abundance, the field data showed low stocks and vice versa. Cooney knew the model did not account for horizontal movement of water masses: it only simulated vertical mixing. He suddenly realized the model was actually showing that the vertical mixing *was not controlling* nutrients or plankton abundance prior to 1992, something else was—like perhaps the river/lake notion.

Cooney knew horizontal transport of Gulf of Alaska water into the Sound occurred. As part of the SEA program, chemist Tom Kline at the PWS Science Center measured carbon isotopes in the large oceanic copepods and other zooplankton. Kline traced the origin of at least half of the Sound's population of these animals (most of it in some years) to the Gulf of Alaska. Cooney also knew that the 1980s were a time of great zooplankton abundance in the Gulf of Alaska, which some researchers linked to a shift to warmer sea temperatures during this time (Chapter 19). He figured some of this abundance must have been swept into the Sound. Cranking through the available data, he determined zooplankton stocks in the Sound were twice as high during the 1980s compared to stocks in the 1990s.

For Cooney, the retrospective analysis was bittersweet. It proved he was at least half-right, and possibly 100 percent right, with his river/lake theory. During the SEA years, many "river-like" spring times prevailed in the Sound and zooplankton stocks were harnessed to physical processes that controlled vertical mixing and replenished nutrients in the upper water layer. Prior to the SEA years, historical evidence suggested that "lake-like" conditions often prevailed and controlled zooplankton abundance.

Lake-like conditions may have even intensified the effects of the oil spill. During 1989, the freshwater discharge from the Alaska Coastal Current was the lowest in its fifty-nine-year recorded history and the north winds, which had blown strong from 1976, relaxed. The resulting lake-like conditions created minimal flushing action and the Sound retained oil—especially oil dissolved or entrained in the water column, which proved so deadly to fish eggs and embryos.

The problem was timing. Cooney (1999) said, "Plankton cycles have a temporal component. Mother Nature welcomes us into her laboratory, and we go and listen to her lecture. However, unless we stay in the lab, listening long enough, we don't get the major lessons and we don't fully understand how the Sound functions." The oceanographers with the SEA program had detected a *"decadal"* (varying on the order of decades) pulse to the circulation pattern, alternating between horizontal and vertical mixing processes with the strength and position of the Aleutian low-pressure system. Cooney realized his river/lake theory required "listening" for more than five years to fully understand it. This mystery would remain to be unraveled by future researchers.

Besides the annual, and possibly decadal, rhythm of the physical processes, Cooney and his co-workers also detected a seasonal pulse within the plankton community itself (Cooney et al. 2001b). In March young stages of the large deep-dwelling oceanic copepods ascend to the near surface waters where they drift and grow, providing critical high-fat food for young pink salmon, herring, pollock, and other fish. These oceanic copepods comprise the largest chunk of zooplankton biomass in the upper layers until they complete their growth phase in June and sink slowly back to deeper water. Smaller coastal copepods dominate the near surface waters in all months, but during May, June, and July other major groups appear in short-lived bursts like annuals in a flower garden.

The names of these tiny animals are disproportionate to their size—*"pteropods"* (swimming molluscs also called sea butterflies), *"larvaceans"* (tunicates with clear jelly-like bodies), barnacle *"nauplii"* (babies or larvae), krill-like *"euphausids"*—and dozens of others. Collectively, the assemblage looked like miniature space aliens with their various adaptations to stay afloat and alive.

Cooney and his SEA colleagues found this seething soup of odd life-forms, especially the larger copepods, provides food for young pink salmon and some relief from larger predators, which, under some conditions, are more content eating zooplankton than small fishes.

Top-Down Investigations: Fish, Part 2—Ecology
Ecology of Young Pink Salmon

Studying the food web or *"trophic"* relationships of juvenile pink salmon—what salmon eat and what eat salmon—is akin to opening the proverbial Pandora's Box. Young salmon are linked in some way with nearly everything else in the food web. Simplifying these interactions *and* the key driving physical forces into a predictive model seemed an impossible task to many. SEA project leader Cooney said, "There were many who said, 'you guys will never figure this out, as far as salmon are concerned—it's just too complicated.' But mathematician Vince Patrick didn't pay a lot of attention to the nay-sayers. As a result, one of the coolest things that happened in the SEA program was a success that occurred because of a novel collaboration."

That collaboration was primarily between two young scientists—the energetic Vince Patrick, PhD, and Mark Willette, a quiet unassuming ADFG fisheries biologist who led the SEA pink salmon team. When this team gathered to discuss their Herculean task, they decided to focus on understanding what influenced death in young fish. If they could successfully model these factors, then they could improve salmon forecasting. They started with one idea to test and they developed and tested two others during the course of the SEA program.

The first idea stemmed from earlier observations by Cooney, Willette, and others that wild pink salmon in the Sound have evolved so that their fry reach the sea exactly when their main food—young of the large deep-dwelling oceanic copepods—reaches the upper water layer. Originally, the basis for this remarkable co-occurrence seemed straightforward—a simple predator-prey relationship that optimizes growth of young salmon. But then Willette's team realized another advantage to salmon fry of this timing. By entering the sea coincident with a huge swarm of young copepods, the salmon fry lessen their own odds of being eaten by larger fish and fish fry-eating seabirds. Willette's team reasoned that the young copepods might afford shelter to salmon fry from predation, because the predators could switch from salmon fry to copepods in the presence of such abundance. This was dubbed "the prey-switching hypothesis."

The two other ideas also stemmed from breaking conventional molds and thinking more like a salmon fry than a scientist. When the swarm of large copepods thins out, usually by mid-June, hungry juvenile salmon are forced to leave protected refuges to forage over deeper waters where they are more likely to get eaten. Bingo—Willette's team realized the timing of migration is critical for young fish: too early and there might not yet be other zooplankton to fill hungry salmon stomachs; too late and they become prey for herring and pollock that invaded nearshore refuges.

More mulling produced another idea. Since smaller fish are more likely to get eaten, the team realized that size of fry when the fish left their nearshore refuge is critical to survival. If pink fry are blessed with abundant food early during their marine residency, they grow quickly and are relatively large when they migrate to deeper water. Larger fry are less likely to get eaten—they are more agile swimmers and physically too big to fit in mouths of young herring and pollock. Willette's team saw where physical processes could intersect and influence both the timing of migration and fry size when the fish leave their shelter.

Willette and his team had their work cut out for them and they plunged eagerly into several dozen overlapping studies. Five years and much ado later, the upshot of the SEA program was surprisingly "simple-minded," as Cooney said, but the team had documented—and modeled—something that had never been done before (Willette et al. 2001). Cooney explained, "Our work confirmed the notion that when there's a lot of plankton in the Sound, almost everybody that *can* eat plankton is eating plankton and nobody's eating little fish. The opposite is also true: when plankton is gone, then bigger fish eat little fish and little fish have problems surviving."

What had really helped nail down the story of young salmon ecology was the unique coded-wire tag system used by the hatcheries in the Sound (Chapter 16). Each hatchery marked up to twelve groups of salmon fry every year with coded-wire tags and these individual

groups were released at different dates and places during the zooplankton bloom. By tracking average growth rates of the various
groups of coded-wire tag fish (Willette 2001), Willette's team learned
a great deal about how prey and predators affected survival as the
salmon fry grew up. In addition to salmon size, growth, and timing of
migration, the team also found salmon school density and predator
size at time of migration to deeper water were important influences
of overall salmon survival. The scientists were startled to discover that
up to 75 percent of pink salmon fry could be eaten within their first
forty-five to sixty days at sea! That meant that it was local Prince
William Sound predators that were eating the juvenile salmon.

This understanding was a huge step forward in the field of
salmon ecology (Cooney et al. 2001a). For Prince William Sound
pink salmon at least, Cooney and his team had found the highly
sought Holy Grail. They had discovered that the first two months of
marine survival of young pink salmon determined adult survival.
Further, they determined that young salmon survival resulted from a
complex mix of predator-prey interactions and zooplankton abundance and timing, all of which could be predicted based on oceanographic conditions.

This drama was modeled successfully by Patrick to be predictive
for hatchery salmon. One of the most important aspects of Patrick's
model was that it helped the biologists explore things that could not
be directly observed. For example, Willette could measure the size of
salmon fry, prey, and predators or the timing of the zooplankton
bloom, but he could only guess how these things fit together. Cooney
explained, with a wry smile, that "Vince was able to understand the
interactions, because the system boiled down to a few key dynamics.
He aggregated half a dozen key prey species that were common to
pink salmon, herring, and pollock. He worked with six to twelve
groups of coded-wire tag salmon from two hatcheries for two years.
This way he successfully coupled oceanography with ecology."

The model's ability to reproduce survival patterns from multiple
coded-wire tag groups of salmon gave the biologists confidence that
it was accurately mirroring how the system functioned. Cooney
described Patrick's effort to push ahead with his novel cutting-edge
model as "heroic and ground breaking. Truly remarkable."

In summing up the SEA program's effort, Cooney said, "The real
lesson from this kind of work is that, if you want to break a paradigm
and learn more about a system, then it might be the novel collabora-
tions between people who normally don't work together that are

important. Taking advice from the inner circle all the time is maybe not the best thing for advancing science." The unlikely collaboration between a mathematician and a biologist—and the hard work of their dedicated staff—revolutionized our understanding of the role of pink salmon in the Sound's ecosystem. However, the pink salmon model needed to be further validated and used before it could improve the accuracy of forecasting returns. The EVOS Trustee Council continued this work as part of the Gulf Ecosystem Monitoring (GEM) program (Chapter 23).

Ecology of Young Herring

Evelyn Brown (formerly Biggs) watched the herring controversy spin itself in knots from the sanctuary afforded by distance and her dissertation work in Fairbanks (Chapter 16). The move away from the pressure and politics in Cordova worked wonders on her psyche and she was able, once again, to plough her energies into her true passion—understanding the ecology of young herring.

The northern location of Prince William Sound with its extreme seasonal light and temperature cycles, together with its glacier-carved bays and fjords, created a unique nursery area for juvenile herring (Norcross et al. 2001). Over thousands of years, these isolated and harsh conditions forged herring stocks that were genetically distinct among all the world's herring—a treasure trove of tough genetic material found only in Prince William Sound. Yet, oddly, at the time of the spill, the location of the Sound's herring nursery areas was not known. No one really understood how currents, temperature, nutrients, plankton, and herring predators all combined to yield these slim silver fish. Brown's role in the larger SEA program was to help solve these mysteries by studying growth and survival of young herring.

As with the complementary team on young salmon, the herring team first mulled over possibilities. Herring have a different life cycle from salmon; young herring over-winter in nearshore nurseries for the first two years of life, while young pink salmon spend just one summer in the Sound. The herring team realized that nursery areas—the nooks and crannies forming the edge zone of the Sound—varied in terms of quality of available nutrients and zooplankton. The quality of the nursery areas likely affects the nutritional status and overall survival of the young fish, because during the winter plankton stocks are at their lowest levels. Cooney (1999) explained, "We had some inkling that herring store energy to get through that period of time, but we had no idea what caused winter-kill. Cold temperatures? Starvation? We thought all these ideas were important to study."

The SEA team's main focus was factors affecting survival of juvenile young-of-the-year and one-year old herring. Supplemental studies on herring eggs and larvae found these pre-juveniles had a rough early life (Bishop and Green 2001). Spawning herring attract large flocks of glaucous-winged gulls, mew gulls, surf scoters, surfbirds, and black turnstones, among others, which gobble up one-third to three-quarters of the herring spawn! A circulation model predicted surviving larvae would be wafted into nearshore bays on currents circulating through the Sound, while some—the unlucky ones— would be flushed out of the Sound to their death (Wang et al. 2001). Although the model details were sketchy, the SEA team realized that *any* protected nearshore bay might serve as a potential nursery for juvenile herring. They studied the same two deep fjords and two shallow bays selected by the oceanography team.

Brown's studies—and half a dozen others—began in August when herring larvae drift into these bays and transform into tiny inch-long silver juveniles. Two-year old fish move out of the bays as new young drift in to join the one-year olds (Stokesbury et al. 2000). Summer fare is a diverse diet of fish eggs, barnacle nauplii, and small and large copepods; autumn fare shifts slightly to other available species (Foy and Norcross 1999). Young herring grow rapidly over the next couple months, storing much of their food as high-energy reserves—in the form of fats. As fall approaches, young herring glut themselves on the late plankton bloom while one-year old fish target prey with high-energy content (Sturdevant, Brase, and Hulbert 2001).

From December through early March, all the juveniles enter a period of semi-fasting, drawing down their energy reserves to bridge these lean months (Foy and Paul 1999). Young herring that enter this period of semi-fasting at lengths shorter than three inches simply vanish during the winter, victims of insufficient energy reserves to buffer against starvation, disease, and predators. There is little margin for error. Herring that survive the winter find themselves swimming in a sea of plenty in March when offspring of large oceanic copepods surface from the depths.

Nursery areas play a crucial role in nourishing young herring (Stokesbury, Foy, and Norcross 1999). In some bays, fresh water, Sound water, and shots of water from the Gulf of Alaska combine and mix in just the right way to sustain a rich soup of phyto- and zooplankton. In these bays, young herring grow long and fat. But in other bays, physical conditions do not support a good crop of zooplankton and pickings are slim for young herring. No one bay is consistently optimal. It is more like an annual lottery. But if too many fish draw the winning bay, fortunes could reverse and the winners could become losers. By depleting limited food resources, young fish grow more slowly, imperiling their own winter survival.

Working as a team, the SEA herring scientists had discovered the key to forecasting adult survival, another huge advance in fisheries ecology (Cooney et al. 2001a; Stokesbury et al. 2002). Brown (1999) explained, "There are four critical times in the lives of these fish that determine how many will survive. First is egg mortality, which can easily be a steep loss—80 percent of the hatch or more. Then as larvae, they've got to intercept a patch of food within ten days and stay with it or they starve. This is purely a chance thing, because larvae drift passively with the currents. Next, they've got to get to an appropriate nearshore bay—another chance thing—by the time they're ready to metamorphose into juveniles. If they drift out into the Gulf of Alaska, they're toast. And finally, they have to grow fast enough to reach a critical mass that will sustain them through their first winter. Young fish that make it through their first winter are pretty much home free. The mortality really tapers off at that point"—absent other stressors such as oil spills.

Exxon's Response

Exxon scientists did not conduct any comprehensive studies comparable to the SEA program.

They did not look for long-term effects of the oil spill, because they claimed there would be none. They based this claim, in part, on the record harvests of pink salmon in the years after the spill. The record harvests were due to the PWS Aquaculture Corporation's hatchery system reaching full production, after twenty years of planning and hard work by fishermen and others (Sidebar 13, p. 290). Exxon scientists refused to acknowledge that low levels of oil could harm egg and embryonic pink salmon and herring and that harm to these early life stages could affect adult returns, as suggested by the 1992 and 1993 collapses of the fish stocks. Rather than doing their own studies, Exxon scientists attacked the work of those who found low levels of oil caused harm to young fish (Chapter 20).

Synthesis of SEA Program

The SEA program developed a refined holistic picture of the ecological processes affecting juvenile salmon and herring populations in Prince William Sound. This picture is actually a series of intricate overlays of oceanographic conditions, crops of phyto- and zooplankton, juvenile salmon and herring stocks, and key predators. These overlays can be viewed from the bottom-up or the top-down—the picture is the same either way—because they were created simultaneously through a tightly-coordinated program.

For juvenile pink salmon, the picture shows how the intensity of spring storms and the amount of Gulf of Alaska water flowing into the Sound, (the latter a product of interacting oceanic and atmospheric conditions), modifies predation losses during a critical window each spring. For young herring, the picture shows that winter-kill during

---

Interpreting Oil Spill Effects on Wild and Hatchery Pink Salmon

The ADFG has monitored production of hatchery and wild pink salmon ever since the PWS Aquaculture Corporation first began to produce pink salmon in 1977. These records show that production of hatchery pink salmon rose steeply from 28,000 (rounded) in 1977 to 17 million in 1989 (ADFG 2002, Appendix F.9).[a] As planned, full production of 31 million pink salmon was achieved in 1990 and has been maintained ever since with a range from 15 to 39 million fish.[b] This planned and anticipated wealth of pink salmon from hatchery production was the basis of Exxon's claims that pink salmon had recovered from the spill by 1990.

ADFG's records also show the stark drop in both hatchery and wild pink salmon returns in 1992 and 1993, which now, in the hindsight of ten more years of data, stand out as an anomaly. These are the lowest years of wild stock production in all the years of monitoring returns continuously from 1977. These are also the lowest years of hatchery returns since the hatcheries reached full production in 1990. Exxon claims that these two years of low production had nothing to do with the *Exxon Valdez* oil spill. However, Cordova fishermen have reached a different conclusion, based on public-trust science. Fishermen point out that pink salmon exposed to oil as juveniles in 1989 returned successfully as adults in 1990, but these fish failed to produce viable offspring—as witnessed by the population crash in 1992. Similarly, pink salmon exposed to oil as eggs (wild fish) in 1989 and juveniles in 1990, returned as adults in 1991, but failed to produce viable offspring—as witnessed by the population crash in 1993. Pink salmon subsequently recovered from any oil effects as observed by increased production from 1994 onward.

[a]These numbers are conservative as they have not been adjusted based on coded-wire tag data, which was not available prior to 1989.
[b]These numbers are conservative as they have not been adjusted based on coded-wire tag data, which has not been available since 1998.

the first year is critical to forecasting adult survival and that loss from winter-kill can be predicted based on the pre-winter condition of the young fish, not winter duration or temperature. The picture also shows that the Sound is often enriched with nutrients, large copepods, and other zooplankton, entering from the Gulf of Alaska.

In reflecting on their findings, Cooney and his team of lead scientists advocated a more comprehensive approach to management of fish and wildlife populations—a holistic approach that would encompass the ecological processes influencing species survival. Ecologists have advocated such ecosystem-based management for decades. Cooney and his team agreed with earlier researchers who wrote, "It is time to manage and protect whole ecosystems. This will not be a linear extension of single-species thinking . . . as generally envisioned by contemporary resource managers." Indeed, the EVOS Trustee Council elected to use the SEA program as a foundation to continue and expand ecosystem studies outside the Sound in the adjacent Gulf of Alaska (Chapter 22).

Cooney retired after completion of the SEA program. During the final days of his tenure in 1999, he warned, "The job is not yet finished. Because pink salmon and herring depend on the edge zone of the Sound—they will remain at risk for as long as there is toxicity from oil in that region" (Cooney 1999). Studies to understand lingering harm to herring from oil were continued by others (Chapter 20). The political ramifications of this lingering harm are discussed in the final section of this book (Part 3: The Legacy).

Chapter 18.

Nearshore Vertebrate Predator (NVP) Project

Introduction to the NVP Project

In 1992 Leslie Holland-Bartels, PhD (1999), accepted a job to lead the newly reorganized marine mammal and fisheries program at the USFWS Alaska Fish and Wildlife Research Center. A portion of her job included the EVOS studies, which involved mainly sea otters. Holland-Bartels was a strategic planner with a gift for seeing the big picture. She loved tackling complex problems. Her work was guided by a promise she had once made to her grandfather, a commercial fisherman in Lake Michigan who had been upset about the way state biologists regulated the doomed lake trout fishery without input from any villages. He had told his granddaughter, "When you become a biologist, understand what the problem is and make them do it right."

Based on this promise, Holland-Bartels had developed a unique style of facilitating problem solving among scientists, public agencies, and private groups whose lives intersected with fish and wildlife. When the call came from Alaska, Holland-Bartels immediately recognized, as she said, "a science planning and strategy challenge" far exceeding the scale of anything she had done in the past. She and her husband packed up their cat, dog, toddler, "and stuff" and moved from Atlanta, Georgia, to Anchorage, Alaska.

When Holland-Bartels took stock of the EVOS research in her new domain, she realized that a fresh approach was desperately needed—not just within the Center's research program, but within the EVOS Trustee Council's approach to studies. She explained, "We were finishing up a number of studies designed and controlled under that damage assessment mind set. We needed an opportunity to look at the question of 'what are the long-term consequences of this spill on the ecosystem?' instead of 'how many animals died or were damaged?' The old avenues of research weren't getting us anywhere in terms of long-term interest." Holland-Bartels felt strongly that a cross-disciplinary approach—teaming up fish, seabird, and mammal researchers—to focus on key species in the nearshore area might provide insight on long-term oil effects and recovery estimates.

Within a year, Holland-Bartels vision was realized. In 1993 federal auditors soundly criticized the EVOS Trustee Council for not including the public in their decision-making process as ordered in the 1991-settlement agreement—and as repeatedly requested by the public. People were also furious about the 1993 collapse of the Sound's fisheries and they unleashed their frustration upon the EVOS Trustee Council. People were fed up with the single species studies—they wanted to know what was wrong with the whole system and the prognosis for recovery. At the federal level, Clinton was now president and his administration was friendly to the environment. Holland-Bartels didn't pay much attention to politics, but she was heartened by the change in management philosophy within the EVOS Trustee Council.

In response to public criticism and public pressure, the EVOS Trustee Council sponsored a workshop in April 1994 to look at different ways of doing business. It turned into a watershed event. In an effort guided by the Council's executive director, Jim Ayers, the Council adopted new principles to guide "Science for the Restoration Process," as the resulting publication was called, and they put these new rules into action. They started a "Public Advisory Group" to look at the science questions from a public perspective and an external peer review process to tighten up the science. They also encouraged cross-disciplinary research to frame an ecosystem approach, and public-private partnerships to break agency dominance.

This was exactly what Holland-Bartels had been waiting for and she wasted no time in convening a working group—across twenty-five disciplines initially—to put together a research proposal under these new rules. The group agreed to actively seek public input as part of the decision-making process. They worked with the new Public Advisory Group and they went out to Native communities to gain insight into everyone's perspective on what was going on in the Sound. Holland-Bartels said, "Every time my staff came back, they had learned something that set their thinking a little bit differently." In her heart, she knew her group was trying to 'do it right' as she had promised her grandfather long ago. She ended up, one year and numerous drafts later, with a core of fifteen lead scientists from eight different research organizations including two private groups. The trustees approved the mammoth six-year, $6.5 million proposal in March 1995 and the Nearshore Vertebrate Predator (NVP) project was born.

The NVP project focused on species that apparently continued to be affected by lingering oil hidden in the beaches of Prince William Sound. The project was eloquently simple in its design (Holland-Bartels 1998). It focused on four species that fed in nearshore areas. Two of species, sea otters and harlequin ducks, fed on nearshore invertebrates; the other two, river otters and pigeon guillemots, fed mostly on fish. The scientists focused on these two food pathways, because invertebrates soak up and store oil contaminants in their bodies, while fish metabolize and excrete pollutants. The NVP group reasoned that predators that feed heavily on oiled invertebrates might show persistent harm from oil exposure, while predators that feed on fish might not. Holland-Bartels explained, "We wanted to balance different habitat use to gain insight into what was going on." They regarded these four species as *"sentinel species,"* ones that are sensitive to pollutants and so are broad indicators of ecosystem health—the canaries of the Sound.

The NVP group then developed studies for each species to address the common question, "Was there a problem with recovery or not?" This question would be addressed through *"demographic studies,"* which measure a species' vital statistics—population size, density, distribution, female survival and reproduction, and survival of young, among other things—to determine a population's capacity for expansion or decline. If populations were depressed in oiled areas relative to unoiled areas, the scientists agreed to test each of three common questions to determine what was holding back recovery. "Was it habitat?" "Was it food?" Or, "Was it oil?" To attain the statistical power to resolve these questions, the NVP group decided to focus on two previously heavily oiled sites—Naked and Knight Islands, and two mostly unoiled or lightly oiled sites—Montague Island and Jack Pot Bay. These choices represented worst and best case scenarios, respectively, within the Sound and they provided the best chance of finding the cause-and-effect type answers the scientists sought.

The first two questions about habitat and food involved relatively straight-forward studies, but the question "Was it oil?" involved assessing health through body condition and testing blood chemistry and liver tissue to measure levels of specific *"biomarkers"*—biological red flags—indicating exposure to aromatic hydrocarbons.

Vertebrates (whether fish, birds, mammals, or humans) have an enzyme system that breaks down aromatic hydrocarbons. This enzyme system does double duty. It cleans house by metabolizing the body's own complex organic compounds that are no longer needed, and it wages active defense by recognizing pollutants as biologically dangerous chemicals and breaking them down. Specific enzymes are produced to break down specific groups of compounds. The specific enzyme that breaks down PAHs also breaks down PCBs (polychlorinated biphenyls) and is called cytochrome P450-1A. However, the defense system is imperfect: when PAHs are oxidized through the break down process, some of the intermediate compounds are highly reactive and more toxic than the original poison. These intermediates can trigger cancer or DNA mutations before they are fully broken down. Thus, elevated levels of cytochrome P450-1A and related enzymes are biomarkers; they indicate exposure to PAHs or PCBs and also the potential for harmful health effects.

Measuring activity of cytochrome P450-1A and related enzymes was tricky, especially given the NVP group's desire to "do no more harm" to injured species; that is, they did not want to sacrifice any animals in the name of science, a practice which was (and still is) common. Holland-Bartels said, "We refused to do surgical work on sea otters and river otters because of the risk to the animals." Instead they worked with scientists at Purdue University to develop a blood assay method for mammals that allowed them to collect cytochrome P450-1A data from a simple blood test instead of an invasive, risky liver biopsy (Ballachey et al. 2000).

However, this method did not work for bird blood, which Holland-Bartels described as "very different from mammalian blood."

So they perfected safe surgical techniques that could be performed quickly in the field on live adult birds—and then finally, in the last year of their study, they successfully performed liver biopsies on live fledgling pigeon guillemots. With this technique, they were able to measure levels of a liver enzyme EROD (ethoxyresorufin-O-deethylase), which was directly related to production of cytochrome P450-1A (Seiser et al. 2000). Holland-Bartels said, "It took a lot of effort, and a lot of experimental design work, and a lot of work from our veterinarians and a whole bunch of folks to develop safe sampling procedures, especially for the pigeon guillemot chicks." But this effort paid off in ground-breaking field research—the scientists tracked oil exposure seven, eight, and nine years after the spill. The NVP group found PCB exposure as well, but after accounting for its confounding effects, enzyme activity was still higher in oiled than unoiled areas. Holland-Bartels was rightfully proud of the forensic blood chemistry and tissue work done by her group of detective-biologists.

The NVP project, guided by Holland-Bartels's steady hand and her adaptive, open management style, produced pioneering results and untangled much of the mystery surrounding the full effects of oil spill on these four species—and more broadly, on the Sound.

Harlequin Ducks

U.S. Geologic Survey (USGS) bird biologist Dan Esler (1999) described Prince William Sound in winter as "a big duck pond." Many seabirds such as puffins, kittiwakes, and murres spend the winter farther offshore. However, the nearshore waters of the Sound are prime wintering grounds for sea ducks, a group that includes scoters, goldeneyes, buffleheads, mergansers, long-tailed ducks, and the diminutive harlequin duck. In March 1989, as in March of every year, thousands of wintering harlequin ducks decorated the shorelines of the Sound. Natives refer to them as "rock ducks" for the ducks' love of exposed rocky beaches. When oil washed ashore those fateful days in 1989, it claimed the lives of an estimated 1,000 harlequins from the Sound. This was roughly 7 percent of some 14,000 ducks present, but the oil claimed a much higher proportion in the western Sound.

As part of the NVP project, Esler found that harlequin densities were lower in oiled areas than unoiled areas during winters of 1996 and 1997 (Esler et al. 2002). This corroborated findings of other public-trust scientists (Rosenberg and Petrula 1998) and pointed to a problem with recovery in oiled areas. His challenge was to figure out whether it was habitat, food, oil, or some other factor that accounted for the delayed recovery. When Esler looked for differences in habitat that could account for the lower densities of harlequins in oiled areas, he found that harlequins definitely prefer certain areas—rocky beaches near offshore reefs and stream mouths and the more exposure to crashing surf and wild winds, the better. But after accounting for these habitat relationships and differences between areas, Esler still found oiled areas had fewer ducks (Esler et al. 2000a).

To determine if food influenced density patterns, Esler had to first find out if the same ducks returned to the same sites to feed every year. During the late summers of 1995, 1996, and 1997, when the harlequins gathered to molt, Esler and his crew rounded up the now-flightless ducks with sea kayaks and herded them into net pens so they could be banded and radio-tagged (Esler et al. 2000b). Fully 96 percent of the banded adult females homed with unerring accuracy to the exact shoreline from the previous year, while the other 4 percent were found on adjacent beaches less than a mile away from their original point of capture. Juveniles and males strayed a bit more than females, but in all cases the male and young ducks were found within twelve miles of their original capture point. Most radio-tagged ducks wintered near their molt sites.

Once Esler established that the same ducks fed repeatedly in oiled areas, he then could determine if there was less food in oiled areas and, if so, whether this was limiting the ducks' survival. Prince William Sound is one of the farthest north wintering ranges for harlequins, because the ducks are visual feeders—they need to be able to see to eat. During the brief mid-winter days, harlequins feed almost constantly to meet their energetic needs for survival. Esler calls them *"trophic generalists,"* referring to their habit of gobbling up anything they could find among the beach rocks—snails, amphipods, chitons, limpets, mussels, and other tidbits. The ducks can't afford to be picky eaters. They need all the calories they can get to survive during the long months of lean daylight hours. Esler's colleagues scraped and collected invertebrates from several beach plots in oiled and unoiled sites. They found more than enough food for the ducks at all sites—but they also found, from the radio-tagged ducks, that winter survival of harlequins was lower in oiled areas than in unoiled areas.

These findings pointed to oil as the culprit responsible for the lower survival in oiled areas. Esler found ducks from oiled areas had much higher P450-1A enzyme activity, indicative of toxic exposure, than ducks from unoiled areas (Trust et al. 2000). He also found that heavier ducks had less enzyme activity, indicating less oil exposure. Esler knew heavier ducks were more likely to survive because they had more stored energy reserves. This combination of information suggested that harlequins were suffering physiological consequences—health problems—from oil exposure that could be related to poor survival.

Esler mulled over his findings. He thought that the Sound was not, and never really was, prime breeding habitat for harlequins. In fact, he thought of it as prime *nonbreeding* habitat. He knew some ducks breed locally in the coastal forests, especially in the eastern and northern Sound, but many of the wintering ducks leave the Sound to breed elsewhere in Alaska. The Natives did not recall people gathering harlequin eggs or seeing young ducklings. Esler saw the delayed recovery more as a failure for wintering ducks to survive rather than as a failure of ducks to breed in the western Sound. Surveys conducted by ADFG supported Esler's findings and showed declines in wintering populations of harlequins in oiled areas during 1995 to 1997.

Esler realized, even if all of the negative effects from oil didn't exist, the length of time to full recovery—replacing ducks lost by the spill—still boiled down to a matter of basic bird biology. Harlequins have long life spans and low reproductive rates. Juvenile ducks follow their mothers to wintering sites and then most return to those same sites, or nearby areas, for the rest of their lives. There was some immigration from other areas—just enough to keep the gene pool well-mixed, but not enough to replace catastrophic losses from the oil spill in 1989 (Lanctot et al. 1999).

In piecing together data from five years of studies, Esler concluded that harlequin ducks still had not recovered and that the primary reason for this lack of recovery was most likely continued oil exposure (Esler et al. 2002). Harlequins had what Esler called, "an unfortunate combination of characteristics" (ibid., 283) that made them particularly vulnerable to oil spills and chronic oil pollution. The ducks' preference for nearshore beaches, diet of invertebrates, and loyalty to particular sites put them at high risk of being contaminated by oil. Their life history—longevity and low reproductive rates—guarantees a lengthy recovery period from oil spills or chronic oil poisoning. Finally, harlequins in Prince William Sound live at a precarious energetic threshold that can't tolerate increased metabolic costs. Even the smallest additional challenge can upset this balance and plunge the ducks' population into decline. And the metabolic cost of dealing with toxic oil is no small challenge.

Esler concluded that full recovery would be delayed until oil remaining in the Sound reached low enough levels that it no longer caused biological harm. He felt there was nothing anyone could do at this point to hasten recovery. It was too late for prevention and the remaining oiled areas were too large for intensive restoration. Removal of the remaining oil would be expensive and intrusive. He reasoned recovery was best left up to Mother Nature. Esler thought full population recovery of harlequin ducks could take decades—far longer than scientists had imagined.

Sea Otters

In 1992 USFWS marine mammal research biologist Jim Bodkin (2000) started to develop a technique to use aerial surveys to count sea otters in the Sound. Bodkin had helped with some of the earlier skiff surveys conducted in 1984 and 1985 and, more recently, with the annual spring *"beach cast"* (carcass) collection, which USFWS staff had conducted from 1976 to 1985, and since 1989. Bodkin felt the spring counts accurately portrayed age distribution of sea otter deaths, but he was not happy with the accuracy of counts of *live* sea otters. He explained, "There were two problems with these skiff surveys. They didn't count animals offshore or ones that were underwater." Bodkin received funding from the EVOS Trustee Council to perfect an aerial survey technique. He and a colleague started with the basic design used for decades by ADFG to survey moose, and then they tinkered with the details.

In 1993 Bodkin conducted his first annual aerial survey (Bodkin et al 2002). The data revealed a problem. He explained, "In some of the places where mortality was highest and oiling was most extensive, sea otter populations had pretty much failed to recover by 1993." At northern Knight Island between Herring Bay and Bay of Isles, Bodkin counted only 75 animals in both years where he expected to find a *minimum* of 165 sea otters based on pre-spill counts. Bodkin stressed that a lot of sea otters in the oiled western Sound—roughly 2,000 or 66 percent—survived the spill. But there were pockets like heavily oiled Herring Bay where Bodkin estimated nearly 90 percent of the sea otters had died. From his aerial surveys, he felt recovery was lagging in Herring Bay compared to other areas in the western Sound. In 1994 Bodkin still counted only seventy-five animals in northern Knight Island.

Bodkin reported his findings at the group workshop organized by Holland-Bartels. He proved why Exxon's claim of sea otter recovery within two years was "demographically impossible," as he said. He argued that, based on a 9 percent annual replacement rate (the average rate for this region) and the number of surviving animals, it would take *at least* four years to replace the *minimum* number of 1,000 animals lost in the spill. Factoring in age-specific reproduction and survival rates resulted in projected recovery times of ten to twenty-three years—and that was if nothing was interfering with recovery.

But something clearly was interfering with recovery—at least at northern Knight Island. Other studies, including Exxon's, had found that reproduction was occurring, yet there was a zero growth rate among sea otters populations in the northern Knight Island area. Bodkin also pointed out that the beach cast surveys had found a disproportionately high number of prime age adult carcasses in oiled areas, relative to pre-spill surveys, indicating that spill effects on annual survival *were increasing* rather than dissipating (Monson et al. 2000a).

Bodkin's work on sea otters was expanded to become one of the four cornerstones of the NVP project. In addition to the annual summer surveys, Bodkin also conducted intensive aerial surveys of the NVP sites each summer from 1995 to 2000. Of all his varied jobs, Bodkin liked flying aerial surveys the best. Oiled or not, Prince William Sound was still one of the most beautiful job sites anyone could imagine. Bodkin counted sea otters from 300 feet above the sea surface, first along the twisting coastline, and then offshore, as his pilot followed invisible transect lines marked by a global positioning system. Within each transect, the plane banked into tight circles five times so he could estimate animals not counted during the first pass. Bodkin sometimes wondered what an observer might think of the purple, green, and white Bellanca Scout flying back and forth, back and forth, then around and around in circles. Between the annual surveys and the intensive surveys, which were repeated up to six times each summer, Bodkin logged nearly 6,000 miles each year and he developed a keen awareness of where one might find the territorial sea otters during the summer months.

His primary finding from eleven years of summer surveys (1993 to 2003) was that sea otters were recovering in most of the western

Sound. A surge of growth between 1996 and 1998 had added about 600 animals to the population (Bodkin et al. 2003). While this was still well below the numbers needed to replace animals lost in 1989, the trend was promising.

Meanwhile the northern Knight Island area portrayed an entirely different story. In bays once filled with Exxon's oil, there was zero population growth from 1993 to 2001; the population flat-lined at an annual average of 79 sea otters. This was roughly half the pre-spill population count. Then in 2002 and 2003, the population plunged to 38 and 26 animals, respectively. Bodkin is still sorting out the reasons for the drastic decline; lingering harm from oiled habitat is a prime suspect (pers. comm., January 2004).

Bodkin knew that the sea otter surveys didn't tell the whole story. Young adult sea otter males were the first to colonize new territories—or newly vacated areas in the case of the oil spill. If there were sufficient food resources, young females, then breeding females and males moved into the area, pushing the young males out. In an area occupied by sea otters for a long time, breeding females outnumbered males. Sea otters had occupied the western Sound for three-quarters of a century before the spill. Surveys conducted in the 1970s and 1980s reported 62 to 87 percent females in the population. Thus, the spill had wiped out a large portion of breeding female sea otters—and their pups—as the spill occurred at a time when many females were near term in their pregnancies. Replacing lost animals meant waiting for the slow process of colonization to play out.

Bodkin witnessed this colonization process in action. He captured and tagged a higher percentage of young adult males and non-breeding females in the oiled area than in the unoiled area where he found more breeding females. He figured the surge in population growth in the western Sound between 1996 and 1998 occurred after a critical mass of breeding females finally moved into the area. But that colonization process had stalled in the northern Knight Island area.

The most obvious question to ask was, "Was it food?" that was causing the delay in recovery of the northern Knight Island population. When young male sea otters moved into an area, they were looking for food. Good sea otter habitat meant good food—a seafloor rich with a variety of clams, crabs, mussels, sea urchins, sea stars, octopi, and other delicacies. To answer this question, biologists used SCUBA equipment to dive and collect what they observed sea otters eating, they measured the caloric content of this food, and they captured sea otters to measure *"condition"* or weight per unit length (Dean et al.

2002). While they found no *statistical* differences in any of these parameters between the study areas, they did observe more high-energy food, larger sea urchins, and heavier sea otters, especially young females, in the oiled area. While Bodkin's colleagues tested whether prey populations were thriving in response to fewer sea otters at the oiled sites, Bodkin found that sea otters at the oiled site had relatively easy pickings of a rich food supply (Dean et al. 2000). Otters ate significantly more calories per hour and spent significantly less time foraging than those at the unoiled site. Food differences were clearly not the problem constraining sea otter recovery in northern Knight Island.

Neither was reproduction. The few breeding females at northern Knight Island produced pups at the same rate as females at the control site. However, data from the beach cast surveys, which included ten years of pre-spill data and ten years of post-spill data, clearly showed that otters born after the spill had lower survival rates. The spill also took a toll on prime breeding age animals and older animals for years after the accident. Similar findings were reported by aquariums and other public display facilities that received captive sea otters from Exxon's treatment centers in 1989—these animals died at a higher rate than other captive sea otters (Rebar et al. 1995). In the western Sound the effects of the spill on the population grew less distinct with time as new animals replaced ones damaged by the spill. The divergent population trends at northern Knight Island indicated that spill effects on survival lingered longer where oil impacts were greatest.

Bodkin's group (2002) tested 157 sea otters for continued exposure to residual oil. They found levels of the biomarker cytochrome P450-1A were significantly higher in sea otters at oiled sites than unoiled sites. They also found significantly higher levels of the serum enzyme GGT (gamma-glutamyl transferase) in sea otters from Knight Island. Elevated GGT levels indicated liver disease or injury such as had been observed in captive minks exposed to oil. The levels of these two enzymes were highly variable among individuals. Bodkin explained that sea otters were "basically diggers." They excavated huge amounts of sediment—up to five cubic meters a day if they were feeding solely in soft substrate. The oil was not evenly distributed across the subtidal seafloor—it collected in pools and patches. Bodkin's group reasoned that some sea otters foraged in oil pockets more often than others and the variable exposure was reflected in the individual animal's blood and liver chemistry.

This persistent harm in wild sea otters from exposure to residual oil is a thread of evidence that trails back to 1989 (Chapter 14). Autopsies on sea otter carcasses immediately after the spill found lung damage (interstitial pulmonary emphysema), liver and brain damage (lesions), neurological damage, stomach ulcers, and other damage to internal organs from acute oil exposure. Treated captive sea otters were sickly; autopsies on ones that survived Exxon's treatment program (and died in aquariums) found chronic liver and lung diseases associated with acute exposure to oil.

Bodkin and his group concluded that sea otters had not fully recovered from the spill and that oil was the most likely culprit causing delayed recovery in localized areas. Putting the pieces together from their combined studies, they realized there was only one possible explanation for their findings of injury to animals that had not even been exposed to the initial spill. The animals' health was compromised from continued exposure to oil either through digging or diet or both. Sea otters had evolved a survival strategy similar to harlequin ducks: they were long-lived and had low reproductive rates. This survival strategy left them unprepared, as a species, to deal with catastrophic loss, especially from something such as oil that continued to poison their environment.

Pigeon Guillemots

When USFWS bird biologist Greg Golet (1999) took the lead on the pigeon guillemot study for the NVP project, he inherited a wealth of pre-spill data on these small diving seabirds—and some curious tidbits from NRDA studies on seabirds in general. At the 1994 workshop organized by Holland-Bartels, Golet learned that populations of many *"piscivorous"* or fish-eating marine birds had sharply declined in Prince William Sound between the early 1970s and the 1990s (Agler et al. 1999). The numbers were grim: fourteen of seventeen species including loons, cormorants, mergansers, certain gulls, black-legged kittiwakes, some murrelets, terns, puffins, and guillemots had plummeted an average of 65 percent—in some cases up to 95 percent. In contrast, many marine birds that fed on invertebrates either had not declined at all or had experienced an initial decline, but then recovered. Harlequins were one of the notable exceptions.

Golet found the studies on pigeon guillemots by two USFWS staff, Karen Oakley and Kathy Kuletz, particularly useful. Initially, the scientists conducted studies at Naked Island from 1978 through 1981. These studies were more detailed than most baseline studies conducted in the advent of oil tanker traffic through the Sound, because they used their research to complete master's degrees. The *Exxon Valdez* grounded about twenty miles from Naked Island and oil coated parts of the island. From 1989 to 1992, Oakley and Kuletz (1996) repeated their earlier studies.

The later surveys found fewer guillemots—the post-spill population was 43 percent smaller—and the relative declines in population were greater along oiled shorelines. The biologists found no obvious disruptions to breeding from the spill as they had expected—egg laying, hatching success, chick growth were not significantly different from pre-spill. However, they noticed a change in diet—parent birds fed chicks more sand lance pre-spill and more cod after the spill. They calculated that the highest growth rate of chicks was in 1979 when chicks ate the most sand lance—60 percent of their diet. They concluded that the spill contributed to the decline in the population of pigeon guillemots, but was not solely responsible for it.

Golet's NVP project picked up where the NRDA study left off. He was charged with determining what role oil had played in the guillemots' decline from 15,000 in 1972 and 1973, as reported by Pete Isleib and others, to just over 3,000 in 1993 (Hayes and Kuletz 1997). This promised to be a more complex task than the companion studies on sea otters and harlequin ducks, because of evidence that changes in food availability also influenced the decline.

Most of the piscivorous birds required abundant sources of high-quality lipid-rich schooling *"forage"*(food) fish such as capelin, sand lance, and juvenile herring to successfully raise young (Springer and Speckman 1997). Scientists suspected that the population of forage fish had shifted from one dominated by energy-rich herring and capelin in the 1970s to one dominated by lower-energy cods in the 1980s in response to an abrupt warming of sea-surface temperature. Investigating the roles of climate change, forage fish, and the oil spill on seabird populations became the subject of the APEX project, the primary source of Golet's funding (Chapter 19).

The life history of pigeon guillemots made them a prime candidate for study under both the NVP and APEX projects. The feeding habits of pigeon guillemots helped Golet and his colleagues detect differences in oil effects and food effects on population dynamics (Golet et al. 2002). Guillemots are unique among alcids because they feed nearshore in small home ranges, while their cousins, such as murres and puffins, range widely offshore in pursuit of schooling fish. Seabirds that forage far and wide for pelagic schooling fish criss-crossed between oiled and unoiled areas, which blurred oil effects. Pigeon guillemots forage in much smaller areas close to their colonies, which were usually distinctly oiled or not. This smaller geographic scale made it easier for Golet to distinguish between effects of food and oil on recovery.

Pigeon guillemots' diverse diet also gave Golet and his team clues to help distinguish between oil effects and climate effects (Golet et al. 2000). Guillemots forage for invertebrates in shallow waters and dive up to 165 feet in pursuit of fish—mostly demersal sculpins, blennies, greenlings, flatfish, and rockfish, but also schooling fishes such as herring, capelin, and sand lance when available. Guillemots are unique among seabirds in that they had hedged their bets for survival in favor of dietary diversity rather than efficiency. Schooling fish have a higher energy content (Anthony and Roby 1997), but they are also ephemeral, always on the move, and harder to locate than the predictable fish that hang out along rocky subtidal shores. Guillemots, by virtue of their flexible diet, are partially buffered from shifts in forage fish populations, so the guillemots' population does not fluctuate as wildly as some other seabirds.

To determine to what extent food differences influenced survival at the oiled and unoiled sites, Golet and his co-workers circled entire islands in skiffs to census guillemots during early morning high tides before adults left their nests to forage. Using SCUBA equipment, they dove at guillemot foraging areas to determine the types and abundance of demersal species. They peered for hours through binoculars and spotting scopes to identify the fish held crossways in bills of adults that returned to their nests to feed their chicks. The researchers measured growth of chicks every five days from hatching until fledging. They confirmed that, before the spill, adult pigeon guillemots were heavier, in better condition, and they produced more offspring and chicks grew more quickly and had higher survival rates. After the spill, neither chicks nor adults survived as well in oiled sites, but for different reasons, as Golet found out.

For the chicks, the evidence pointed to differences in diet, likely caused indirectly by the spill. The reference colonies in Jackpot Bay were located next to a nursery area for herring, a lipid-rich fish that made up 45 percent of the chicks' diet (Golet et al. 2000). In the Naked Island colonies before the spill, adult guillemots preferred to feed their offspring sand lance, another high-lipid forage fish. These

fish had made up the single largest component of the chicks' diet—
42 percent. However after the spill, sand lance accounted for only 13
percent of the chicks' diet. Sand lance were known to be sensitive to
oil—they didn't avoid oiled sand, which made them vulnerable to
injury from oil exposure (Pearson, Woodruff, and Sugarman 1984).
Golet took a closer look at diet and reproductive success as part of
the APEX studies (Chapter 19).

For the adults, the evidence pointed to a strong oil spill effect. To
ferret out oil effects, Golet and his team drew blood samples and
took liver biopsies of adults and chicks without killing the birds or
the biologists—which was no easy thing (Seiser et al. 2000). Adults
were wary of decoys set on their roost rocks to lure them in for capture and they easily avoided rocket nets and noose snares. Biologists
had to dangle from ropes or cling to wobbly extension ladders and
dip-net the birds lacrosse-style out of the air as they flushed from cliff
nests. Since adults abandoned nests after such treatment, Golet and
his team made a point to capture and take samples from both mates
to reduce the number of nests disturbed. They also carefully collected abandoned eggs, which were airlifted to the Alaska SeaLife Center
where they were hatched and the chicks used in laboratory tests to
help interpret findings from their and other field studies (Prichard et
al. 1997; Roby and Hovey 2002).

Golet's hard-won data proved worth the effort. He found that
pigeon guillemot populations in the Sound were definitely lower
than pre-spill levels and that the decline was greater—and still continuing downward—at oiled colonies. At unoiled sites, the population
actually increased between the survey years of 1993 and 1998. Ten
years after the spill, the blood chemistry and liver tissues of adult
guillemots from oiled areas reflected continued oil exposure and
liver damage. Other blood parameters such as elevated levels of cor-
ticosterone and glucose—in birds with measurable P450-1A enzyme
levels—indicated the birds were stressed as their bodies mobilized
energy reserves in response to a toxic threat.

The levels of cytochrome P450-1A in adults, while higher in oiled
areas, were still low and variable, consistent with what one might
expect from individual birds foraging in a patchy mosaic of oil along
the seabed. Other NVP scientists confirmed the presence of subtidal
oil—through elevated levels of cytochrome P450-1A and related
enzymes—in two species of fish, masked greenling and crescent gun-
nel, up to ten years after the spill (Jewett et al. 2002). These fish were
collected from at least one site near the guillemot colonies studied

by Golet and others. Golet found no evidence of oil exposure in chicks. He knew chicks were fed exclusively fish, while adults ate invertebrates as well, so he figured the adults were picking up oil from eating contaminated invertebrates and foraging in oiled sediments. Golet reasoned that ongoing oil exposure and organ damage in adults could have contributed to reduced survival and delayed recovery in oiled areas.

Golet and his team (2002) concluded that pigeon guillemots in the Sound had not recovered from the spill and that the delayed recovery was most likely due to significant and lingering harm to adults still foraging in oiled habitat and eating oiled prey. They realized that *additional* harm from oil spills and pollution could be the last straw for wildlife populations already in decline from unfavorable natural conditions such as climate changes.

River Otters

At the initial meetings organized by Holland-Bartels, scientists interested in river otter studies brought a wealth of information on biomarkers that became central to the NVP project. Use of biomarkers to chart chronic injury and monitor recovery of wildlife populations exposed to oil spills was cutting-edge research in 1995. In 1989 it was pioneering work and the fact that it had been done at all was a testament to three stubborn biologists with a bold vision.

After the oil spill in 1989, Governor Cowper called the director of the Institute of Arctic Biology at the University of Alaska in Fairbanks (UAF) and told him in no uncertain terms to get involved with the oil spill research. Under orders from the director to write proposals, mammal biologist Professor Terry Bowyer, PhD, (2003) submitted a study on river otters to the federal lawyers in charge of damage assessment. Then he discovered that an ADFG staff had submitted a nearly identical proposal. The two biologists discussed their predicament and decided to collaborate rather than compete. Thus began a very productive relationship and some of the earliest work to investigate chronic effects of the oil spill on river otter populations.

In the Sound and throughout coastal areas of the Gulf of Alaska, "river" otters spend much of their time feeding and swimming in the ocean. No one knew how many river otters were in the Sound in

1989, but biologists later estimated there were about eighty animals for every sixty miles of good habitat, which meant they were very common (Testa et al. 1994). They live in old-growth forests mostly at the edge of the sea. Individual home ranges are lengths of shoreline, a narrow strip of habitat along the coastline and they rarely use inland habitat away from the sea. River otters feed mostly on a diverse array of marine fish in inshore waters. They also take invertebrates from the intertidal zone, which makes them extremely vulnerable to oil spills and lingering effects from oil exposure. The animals are very secretive, but conveniently for researchers, river otters keep a record of what they eat preserved in *"scat"* (feces) at community latrines. The record is imperfect, as some easily digestible foods such as soft parts of clams leave no trace, but much can be learned about river otters' diets from the bits of bone, fish scales, shell fragments, and fur and feathers left at the latrines.

Bowyer and ADFG biologist Jim Faro faced an uphill battle to convince the NRDA lawyers to fund their proposal. Only twelve river otter carcasses were found after the spill, which didn't impress the federal lawyers who refused to list river otters as an injured resource. (The biologists suspected, and later confirmed, that sick animals crawled into dense underbrush or dens to die and so carcasses were nearly impossible for beach crews to find.) The biologists successfully argued that river otters were ideal sentinel species—other researchers used these animals during the decade prior to the spill to study biological effects of heavy metals, pesticides, and PCBs. The biologists proposed to study the traditional indicators of pollutant effects—diet and prey availability, habitat use, and demographics. Bowyer's colleague, Professor Larry Duffy, PhD, joined the team to study biomarkers to measure stress and overall health. The lawyers balked at funding the blood chemistry work. It was just too new of a technique (for wildlife) to be used in court to prove damages.

Bowyer's team was insistent. They strongly suspected that critical clues to solving the mystery of oil effects on river otters were contained in the animals' blood. They argued that in humans, it is well documented that exposure to aromatic hydrocarbons or stress triggers an immediate defense response, which ultimately involves the immune system. Immune system activity can be monitored through various blood components including enzymes and plasma proteins. For example, elevated levels of two small proteins, *"interleukin-6"* and *"haptoglobin"* indicate a toxic effect or trauma in humans (Heinrich, Castell, and Anders 1990).

Bowyer's group believed that a similar response in river otters exposed to oil could have disastrous consequences, because hapto-globin binds "free" or excess hemoglobin in the plasma, preventing it from transporting or storing oxygen. The biologists realized that by limiting oxygen supply, this cellular trauma could limit dive time and the ability of river otters to catch fish, possibly leading to starvation and death.

The lawyers remained unconvinced that the biomarker study was worthwhile. Bowyer and Larry decided to fund that portion of their proposal out of their summer salaries. Grudgingly, the lawyers reimbursed them, but each year for four years from 1989 to 1992, Bowyer said, "We fought bloody battles" to preserve the biomarker research and keep their project intact.

Bowyer's team chose two study sites, each with fifty miles of sim-ilar habitat and number of latrines, an indicator of river otter density (Bowyer et al. 2003). They hustled to identify their sites in April, because they thought the number of latrines would reflect river otter density before the oil spill impacted the population. They figured cor-rectly. Study sites at oiled northern Knight Island and unoiled Ester Passage were far enough apart so even the notoriously mobile male river otters would not cross the twenty-five miles of open sea between the two areas.

The researchers live-captured twelve river otters at Knight Island and ten at Ester Passage, outfitted the animals with radio-transmitters, took blood samples and body measurement, then released the animals and tracked them. The biologists also tramped through the woods and located all the latrines within each study area—over 110 latrines at each site, which they proceeded to "clean" on a regular basis to gath-er data on diet. The initial collection of scat in April 1989 provided a good record of what the animals ate the winter before the spill.

One of the first things Bowyer's team noticed was home ranges of river otters were about twice as large in oiled areas than at unoiled areas (Bowyer, Testa, and Faro 1995). River otters at Knight Island also frequented beaches with steep tidal slopes and large rocks, while ani-mals at the unoiled sites chose areas with gently sloping beaches. The biologists thought river otters at Knight Island might be actively try-ing to avoid beaches where oiling was most severe and more likely to pool and persist—the flatter beaches—in favor of exposed beach-es with more wave action.

By summer 1990 the biologists noticed a startling shift in diet in river otters in the oiled area, while the diet of river otters at unoiled sites remained essentially unchanged (Bowyer et al. 1994). At oiled sites, scat samples indicated river otters switched to slower moving prey such as crabs and flatfish—rather than the swifter sand lance, gunnels, and ronquils. River otters at oiled sites also ate far fewer species than animals at unoiled sites. The year lag time initially puzzled the biologists, but in speaking with other scientists, they realized the delay probably was related to the movement of oil from intertidal areas in 1989 to subtidal habitats in 1990.

Since fish are a large portion of a healthy river otter's diet—80 percent they found out—the biologists figured an abrupt change to a diet with far fewer fish might have huge consequences for these animals. Sure enough, male river otters in oiled areas were found to weigh significantly less than animals from Ester Passage. Male river otters foraged more widely than females even at unoiled sites, but the lower body weights meant the males were starving. They were obviously looking for food, as evidenced by the larger home ranges, but they either weren't finding it or couldn't catch it.

The biologists wondered if this was because their preferred prey were less abundant or because river otters at oiled sites were less successful at catching the faster fish. They couldn't fully answer this question, because funds to study prey availability had been cut. However, river otters at oiled sites had elevated levels of blood proteins, indicating that they were suffering a toxic effect from oil exposure (Duffy et al. 1993). The biologists also found elevated levels of *"porphyrins"* in scat (Bowyer et al. 2003). Porphyrins are involved in synthesis of heme, the critical component of hemoglobin involved in oxygen transport, and this biosynthetic pathway is readily disrupted by oil exposure. The biologists interpreted elevated levels of porphyrins to mean that river otters in oiled areas were oxygen-stressed because of oil exposure and the animals were trying to compensate by producing more hemoglobin. Oxygen-deprivation could explain why otters in oiled areas had switched to slow-moving prey.

The evidence was mounting: elevated levels of blood proteins and porphyrins, lower body mass, larger home ranges, and change in diet (Duffy et al. 1994a). This added up to what Bowyer called, "an overwhelming indication" of chronic effects from the oil spill. Still the lawyers and political scientists with the EVOS trustee process refused to list river otters as an injured resource. Ironically, by 1992 Bowyer's team even found evidence of recovery—levels of blood haptoglobin and interleukin-6 were no longer significantly different (although the biologists cautioned the sample size was too small to

be reliable) and average body mass of animals was nearly identical
between oiled and unoiled areas (Duffy et al. 1994b).

Finally, in 1993, river otters were officially listed as an injured
species by the EVOS Trustee Council (Bowyer et al. 1993). Then all
funding for their study was cut while Faro was in the field and
Bowyer was out-of-state. Under Governor Hickel, funding for oil spill
studies was at the mercy of Mark Fraker, a staunch oil industry advocate and former BP employee who showed little regard for studies
that could potentially damage this industry. Fraker wielded a lot of
influence among the three state trustees and all it took was one vote
to terminate a study. Bowyer's group was livid, but there was nothing
they could do to reinstate their funding. The cuts so demoralized the
biologists that Bowyer said he never wanted to hear of river otters
after that. Faro eventually retired from ADFG and moved to Sitka.
Studies on river otters languished.

In 1995 University of Alaska Fairbanks Professor Dan Roby, PhD, and
graduate student Gail Blundell picked up the river otter studies
where Bowyer's team had left off. As part of the NVP project, they
repeated the earlier studies, adding a study on prey availability and
refining the blood chemistry work. Buoyed by their enthusiasm and
the support from the EVOS Trustee Council, Bowyer served as project advisor and Duffy continued to do the chemical analyses. This
time they used Jackpot Bay as a control, which proved to be prob-
lematic. Some of the male river otters commuted between unoiled
and oiled sites—a round trip of about seventy miles—by swimming
short distances and island-hopping. Bowyer said, "We had no clue the
animals moved that much!" Some of the animals in Jackpot Bay also
forsook their marine life and reverted to being "river" otters, feeding
and living in the large freshwater system in the area. (Neither these
animals nor the commuting males were considered as part of the
NVP study.)

After four more years of work, the scientists found strong evidence of recovery despite evidence of continued oil exposure
(Bowyer et al. 2003). Oil on river otters' fur, collected by wiping their
coats with gauze soaked in solvent, and elevated levels of
cytochrome P450-1A indicated the animals were still exposed to oil
present in the Sound (Duffy et al. 1999). However, the levels of oil
did not appear to be high enough to create measurable health problems. Male river otters continued to gain weight in oiled areas until

eventually there were no differences in body mass between oiled and unoiled sites. All the blood biomarkers, haptoglobin, interleukin-6, and others, and the corresponding levels of porphyrin in scat were greatly reduced from the earlier studies, although the levels were still slightly elevated at oiled sites. Home ranges gradually shrunk back to normal sizes in oiled areas and the animals no longer avoided oiled shores, but used the same types of gently sloping beaches. Eventually no differences in diet were found and the inshore waters were well stocked with diverse fishes in both areas.

In a lengthy monograph (2003), Bowyer and others concluded that river otters had suffered population-level injuries from the oil spill, but had largely recovered by 1998. He used laboratory studies on river otters conducted by Merav Ben-David at the Alaska SeaLife Center in 1998 to 1999 to link his early and later field studies (Ben-David, Williams, and Ormseth 2000). Ben-David found that oil exposure resulted in health costs, particularly reduced hemoglobin levels, which made diving and moving on land more exhausting. Indeed, animals with lower levels of hemoglobin perished soon after release and more of these experimental animals died of starvation than wild otters during a time of food shortages (Ben-David, Blundell, and Blake 2002). She also found a correlation between malnutrition and porphyrins: starving animals produce more porphyrins (Ben-David, Bowyer, and Duffy 2001). Bowyer realized that the change in diet he had observed—from fish to slower moving prey—with resulting weight loss and increased level of porphyrins meant otters in oiled areas had had trouble catching prey and had lost weight as a result of exposure to residual oil.

By 1997 most of the river otters alive in 1989 had been replaced by younger animals and there were no differences in survival between oiled and unoiled areas. In the end, Bowyer concluded that integrating individual-based studies on biomarkers and population-level studies on demographics, diet, and habitat proved essential to understanding the effects of oil on river otters—just as he, Duffy, and Faro had first envisioned in 1989.

Exxon's Response

Exxon scientists did not conduct any comprehensive studies comparable to the NVP project. They did not look for long-term effects of the oil spill, because they claimed there would be none.

They based this claim, in part, on a "battery of laboratory toxicity tests" conducted "in compliance with standard and published test procedures . . ." (Stubblefield et al. 1995, 665). They claimed the 'battery of tests' was "thorough, equaling or exceeding baseline testing requirements necessary for pesticide registration in the United States" (ibid., 688). These tests were all short-term (two weeks or less) and used relatively large doses of oil, fed to European ferrets and mallard ducks as surrogate species for wildlife in Prince William Sound. Exxon scientists concluded, "WEVC [weathered *Exxon Valdez* oil] would be considered practically nontoxic using the EPA criterion . . ." (ibid., 686), considering that in the Sound, potential exposures after 1989 were well below those shown to cause harm *in their tests* (ibid., 689). The problem was that *their tests* were a poor substitute for going out into the Sound and looking for oil effects in the actual wildlife at risk over a period of years, not merely days.

Exxon scientists also based their anticipation of no long-term spill effects on their study of habitat use in seabirds, conducted from 1989 to 1991. Exxon scientists (Day et al. 1995) found that "most marine-oriented bird species in PWS and along the Kenai [Peninsula] were using habitats without regard to the initial oiling that the bays had experienced . . ." (755). From this they concluded "that the negative effects of the *Exxon Valdez* oil spill on bird habitats had largely dissipated by that time" (ibid., 755). Whether or not birds *used* oiled habitat was different from whether or not birds *survived and successfully fledged young* in oiled habitat. Huff (1954), author of *How to Lie with Statistics*, points out, "There are many forms of counting up something and then reporting it as something else" (80). Huff calls this form of deception, "*the semi-attached figure,*" which I discussed earlier (Chapter 16).

For it to work in this case, *the number* of birds had to be the same in oiled and unoiled areas *and* the same before and after the spill. Exxon scientists found this largely to be so (with some exceptions): "[T]he overall abundance of most species in these bays had not changed significantly since 1984–1985" (Day et al. 1995, 754). This was, in fact, not so, but Exxon's study appeared to be designed so carefully to avoid detecting oil effects that it also failed to detect dramatic declines from natural causes (Chapter 19). For example, during the three-year study, there was only one survey conducted during the mid-winter period critical for harlequin duck survival; the other ten surveys were conducted from late March through October.

Huff observes that "[T]he result of a sampling study is no better than the sample it is based on" (18). In this case, according to USFWS biologist Esler (1999), the summer surveys yielded results with "limited relevance for understanding dynamics of wintering populations," which he considered central to population recovery of harlequins. In his understated way, Esler (1999) dismissed Exxon's habitat use study by simply saying, "Our approach [the NVP project] had greater resolution and power to determine habitat affiliations and evaluate oil spill effects" (see also Esler et al. 2000c, 846).

Sound Truth—NVP project Synthesis

The task of synthesizing the voluminous information from the NVP project fell to Holland-Bartels (2002). Using what she called "a collective weight of evidence approach," she skillfully wove the information from the four vertebrate predators studied, whose recovery status was uncertain at the beginning of the NVP project, into a compelling story of delayed recovery from persistent oil spill effects. With deft, confident strokes, she broadly painted the following big picture in the nearshore environment ten years after the spill.

Sporadic releases of residual oil were occurring and bottom-dwelling species, primarily invertebrates, exposed to this oil soaked it up and passed it on to wildlife that ate contaminated invertebrates. In areas that had been heavily oiled, the continued exposure to residual oil, even though patchy and irregular, was sufficient to cause increased levels of mortality and delayed recovery of localized stocks of sea otters and harlequin ducks.

In contrast, for wildlife that ate primarily fish, this was not the case. Piscivorous river otters were recovering from initial oil effects, even in heavily oiled areas. The story was a little more complicated for pigeon guillemots. Young guillemots ate only fish and they showed no oil-related impacts. However, adult guillemots supplemented their fish diet with some invertebrates. The continued lack of recovery of pigeon guillemot colonies seemed to be caused by sporadic oil exposure in adults, foraging for invertebrates and eating oiled prey, as well as fewer forage fish, because of the spill *and* natural climate change (Chapter 19).

Holland-Bartels (2002) concluded, "The collective weight of evidence from this project indicates lack of full recovery of the nearshore ecosystem from the *Exxon Valdez* oil spill nearly a decade

Chapter 19.

Apex Predators

The Alaska Predator Ecosystem Experiment (APEX)

Introduction

Bruce Wright's official position, since 1991, was chief of NOAA's Office of Oil Spill Damage Assessment and Restoration. He referred to his job as being a "gofer biologist" to the EVOS Trustee Council, but he actually held a prestigious and pivotal position with a lot of control and input as to which projects were funded (Wright 1999, 2003). He filled a critical role, relaying information from field biologists to resource managers and the trustees. In 1994, after startup of the SEA and NVP projects, Wright realized that the EVOS Trustee Council needed to do some sort of ecosystem study on seabirds and harbor seals. After all, the EVOS Trustee Council (1994) had listed seven species of seabirds as injured resources either as "not recovering" (three species of cormorants, pigeon guillemots, and common loons) or "recovery unknown" (common murres and marbled murrelets). The EVOS Trustee Council had also listed harbor seals as an injured resource "not recovering" from the spill. (It was difficult to detect oil effects on some species such as harbor porpoise, Dall porpoise, and Steller sea lions.)

Wright and many other scientists believed that food was somehow limiting recovery of marine mammals and seabirds, independent of the oil spill (Alaska Sea Grant College Program 1993; Hansen 1997). There were a variety of ongoing, independent projects on forage fish and apex predators such as seabirds and harbor seals. USFWS seabird researcher Dave Irons became the spark for a more coordinated approach. He had actually run boat transects for a couple years looking for forage fish with a simple fish finder, because he suspected there was a link between forage fish and black-legged kittiwakes in the Sound.

Frustrated with these unsophisticated methods, Irons called Wright and proposed a project with USFWS studying seabirds and NOAA studying forage fish using sophisticated hydroacoustic equipment and fish sampling techniques. Wright got the EVOS Trustee Council to fund a workshop to integrate all the various studies on apex predators into an ecosystem program, but the marine mammal scientists were unable to participate. Wright presented the workshop results to the EVOS Trustee Council as the "Seabird-Forage Fish Project." It was funded and started in August 1994 as pilot project in Prince William Sound.

The project evolved rapidly. John Piatt (Chapter 15) was the first to realize that the Sound did not have enough diversity of forage fish and productivity among seabird colonies for comparisons to resolve the food question. Piatt (2003) convinced his bosses at the USGS and Wright to expand the Seabird-Forage Fish Project to lower Cook Inlet. He explained, "We've got the perfect system in Cook Inlet. We've got the seabird colonies most affected by the oil spill—the Barren Islands colonies where murre populations are relatively large and, in recent years at least, relatively stable. We've got the Chisik Island colony where populations of murres and kittiwakes have declined by 40 to 80 percent over the past twenty years. And we've got the Gull Island colony in Kachemak Bay where populations of murres and kittiwakes have increased by the same amount over the same timeframe."

Piatt argued that targeting these extremes—stable, thriving, and failing seabird colonies—could lead to understanding what was driving differences in survival. He pointed out that over 90 percent of the seabirds killed by the oil spill were killed in lower Cook Inlet and this was where recovery efforts on seabirds should be focused, not in Prince William Sound. Piatt considered the Sound, "a backwater to the Gulf of Alaska in terms of seabird productivity." Wright said Piatt was "very persuasive"—the Cook Inlet studies were funded, starting in 1995. The entire collage of seabird studies was renamed the "Alaska Predator Ecosystem Experiment" or APEX.

Wright and others realized they could have resolved the "Is it food?" question (Alaska Sea Grant College Program 1993) just by studying Cook Inlet, but the growing collection of studies under Wright's watchful eye included support studies for the SEA program and the NVP project in the Sound. Wright was overwhelmed. As he explained, "This was not my job." He was already managing a four million-dollar program for the EVOS Trustee Council. The APEX studies foundered for lack of clear direction.

A crucial reorganization of APEX occurred in late 1995. Wright brought in Dave Duffy with The Nature Conservancy to relieve him as project leader and he hired Dan Roby, a professor at University of Alaska Fairbanks to develop models of energy flow among seabirds and forage fish. Several other agency scientists and contractors were added until the APEX project consisted of twenty integrated components

Wright described the trio of Duffy, Roby, and Piatt as "the scientific powerhouses of APEX." He said, "They understood the big picture better than anybody else. They were very knowledgeable of bird dynamics and ecology and they were key in making the project work." Wright's unflagging support for the program helped deliver the funding through the EVOS Trustee Council. APEX was fully funded at $10.8 million as a five-year ecosystem project, starting in 1996.

With as many as one hundred scientists collecting data and monitoring field activities during the field season, APEX was a logistical nightmare. Just producing data sets that could be compared across the large and varied study area required a year-long effort by Duffy and Irons to coordinate new sampling and data-sharing procedures and figure out project protocols. Another problem was data interpretation. When APEX started, data collection technology outpaced data analysis capabilities. Sophisticated hydroacoustic equipment was available to detect and measure density of schools of fish, but the mathematical procedures to analyze the data were not available as promised by the contractor. Complicated mathematical algorithms had to be worked out before the scientists could interpret their own data!

Then there was the issue of cross-checking the tricky hydroacoustic data on forage fish biomass and availability. USFWS Dave
Roseneau developed a simple and cost-effective method of monitoring stomach contents of halibut and lingcod caught by sports fishermen (Roseneau and Byrd 1997). He found the diet of these predatory
fish mirrored the availability of forage fish in lower Cook Inlet. His
program was immensely popular with the public. Herring biologist
Evelyn Brown and others (1999) pioneered the use of different techniques to photograph forage fish schools during aerial surveys for the
SEA program (Chapter 17). One of the most technical and successful
methods was lidar, which used pulsed laser light to penetrate into
shallow water even when the water was not clear. During APEX,
Brown advanced and refined use of lidar as a practical remote sensing
tool to assess forage fish populations (Brown et al. 2002).

APEX, with its component projects in Cook Inlet and Prince
William Sound, was one of the first attempts anywhere to integrate
effects of marine climate and ocean conditions on populations of forage fish and seabird predators.

Gulf of Alaska Studies—The 'Big Picture'

The APEX scientists were tasked with understanding the effects of
changes in forage species, particularly energy-rich forage fish. Wright
was intrigued by Piatt's theory about changes in ocean temperatures
and Paul Anderson's long-term small-mesh trawl surveys, which he
had kept running on a tiny budget (Chapter 15). Piatt and Anderson
reasoned that warmer ocean and air temperatures were functions of
storms and atmospheric pressure differences over the Gulf of Alaska.
They were searching for what was driving the shifts in marine climate change and they thought this might be related to shifts in
atmospheric pressure. Wright secured funding for Piatt and Anderson
to expand their historical analysis of small-mesh trawl surveys and to
gather historical satellite data on sea temperatures and atmospheric
conditions in the northern Gulf of Alaska.

While on sabbatical at the University of Washington, Piatt attended seminars with scientists from the University of Washington and
elsewhere who were just beginning to report on abrupt shifts in sea
temperature in the northeast Pacific Ocean, including the Gulf of
Alaska. These scientists found the shifts usually occurred in cycles,
generally over the space of a few years during every other decade or

so (Francis and Hare 1994). They dubbed this climate regime shift, the *"Pacific Decadal Oscillation."*

They described the resulting biological regime shifts, the effects, of these physical flip-flops between warm and cool climate regimes as "abrupt sledgehammer blows" to wildlife populations (Francis and Hare 1994, 281). They found salmon thrived in the Gulf of Alaska during warm regimes from the early 1920s to the late 1940s and early 1950s, and again from the mid-1970s to 1992, the end of their study. Salmon stocks plummeted during the twenty-five-year cool period in between the warm eras. They concluded there were "very significant and coherent linkages" between climate regime shifts and salmon production in the northeast Pacific Ocean (ibid., 287).

Following clues from these earlier researchers, Piatt and Anderson took a hard look at the basic physical oceanography of the Gulf of Alaska (Wright et al. 2000). Seaward of the arc of mountains rimming the coastline of the Sound, Cook Inlet, and the Alaska Peninsula, the narrow continental shelf drops abruptly from 550 feet to the seafloor at about 13,720 feet. The shelf break acts like a giant underwater riverbank, channeling the flow of the swift and narrow Alaska Stream westward into the huge North Pacific gyre. This counter-clockwise circulating water mass provides the oceanic link between the Gulf of Alaska continental shelf and the Pacific Ocean. The gyre, in turn, is closely coupled to the Aleutian low-pressure system.

Fierce winter storms, spawned by the Aleutian low-pressure system, are pushed either south or north across the Pacific by slow-motion atmospheric flexes in a high pressure ridge usually centered over Siberia. When the storms track more southerly across the Pacific, cold northwest winds tug at nearshore surface waters and pull cold waters up from great depths, causing *"upwelling"* along the shelf break. When the storms track more northerly, there is *"downwelling"* where warm waters along the continental shelf sink at the shelf break. According to 1,500 years of climate data recorded in bristle cone pine tree-rings, the two climate regimes—cold and warm— oscillate back and forth an average of every fifteen years or so with one regime lasting anywhere from one to three decades before reversal—the Pacific Decadal Oscillation.

Meanwhile in Kodiak, it took Anderson several years to collect and sort through the data from nearly 9,000 trawl surveys, collected since 1953 in the western Gulf of Alaska, while Piatt was compiling four climate indices for the same area and time from colleagues at the University of Washington. When they coupled the physical and biological data sets, they found that when the Aleutian low was weak, coastal water temperatures were unusually low and shrimp and capelin dominated trawl catches (Anderson and Piatt 1999). During periods with a strong Aleutian low-pressure system, coastal waters were warm and cod, pollock, and flatfish dominated trawl catches. All the climate indices pointed to an abrupt change from cool to warm around 1977. Anderson and Piatt analyzed trawl data for thirty-six individual species and determined that the average total catch of biomass dropped by over half from 1972 to 1981, mostly due to a crash in catch of forage species such as shrimp and capelin.

Anderson and Piatt (1999) reasoned that starvation and predation were driving the changes in the ecosystem. (Many of the thirty-six species they looked at, such as capelin, were not fished commercially in the Gulf of Alaska.) Blooms of the sea's microscopic plant life (phytoplankton) occur in response to warm sea temperatures and are closely followed by an increase in numbers of tiny animal grazers (zooplankton). Change in the timing of peak abundance of this rich food source could wreak havoc on young fish, shrimp, and crab. For example, other researchers found that populations of one main food of forage species, a *Neocalanus* copepod, peaked one to two months later during a cold period (ibid., 121). Warm regimes, with an earlier abundance of these copepods, favored the young of species that spawned in early spring, such as pollock. In warm regimes, young of species that spawned in late spring, such as capelin, might completely miss their main meal ticket and starve. Anderson and Piatt also believed that the abrupt rise in cod, pollock, arrowtooth flounder, and halibut had hastened the demise of shrimp and capelin as the groundfish were all voracious predators of these forage species.

Piatt and Anderson concluded that these large-scale decadal fluctuations in sea temperature had an indirect impact on seabirds and marine mammals that subsisted on forage species and juvenile groundfish. During the cold regime prior to 1977, seabirds and marine mammals relied on fatty forage species such as capelin. With the onset of the warm regime, capelin crashed and seabirds and marine mammals replaced capelin in their diets with juvenile pollock (Anderson, Blackburn, and Johnson 1997). Then several seabird and marine mammal populations crashed. Juvenile pollock are lean fish without the high energy content of fatty forage species such as capelin. The transition from cold-to-warm regimes took about fifteen to twenty years to ripple through the ecosystem and alter community structure of the apex predators—the seabirds and mammals.Anderson and Piatt concluded that short-lived shrimp and capelin reacted more swiftly— within a couple years—to temperature shifts and so were good indicators of decadal-scale changes in northern marine ecosystems.

Wright described Piatt and Anderson's work as "the clearest picture of a biological regime shift anywhere on the planet." Their work established a framework to understand effects of climate and food on seabird populations.

Lower Cook Inlet Studies

Piatt (2003) viewed Cook Inlet as a large inland extension of the continental shelf in the northern Gulf of Alaska (Drew and Piatt 2002). It was about the same length as Chesapeake Bay. He figured patterns of seabird productivity in the Inlet would mirror the same physical influences at play in the Gulf of Alaska. So from 1995 to 1999 under the APEX program, he mustered a small army of colleagues, graduate students, and support crew to collect the same types of physical and biological data that others were collecting in Prince William Sound. It was a monumental undertaking.

Sea-surface temperature images from orbiting NOAA satellites and temperature-salinity-depth profiles from boat transects yield a revealing profile of the Inlet's physical oceanography. Upwelling of cold, nutrient-rich water at the shelf break near the Barren Islands is carried into Cook Inlet by the Alaska Coastal Current and strong tidal currents, driven by the second highest tidal exchange in North America.This plume of cold Gulf of Alaska water flows up the eastside of the Inlet and mixes with warmer freshwater from coastal rivers at the head of the shallow estuary.The warmer estuarine water then flows down the westside of the Inlet, collecting silty, freshwater glacial melt before it empties into the Gulf of Alaska. A distinct boundary forms between the two water masses as a mid-channel tidal rip, visible as a collection of kelp, logs, and debris.

The stable Barren Islands seabird colony near the shelf break is bathed in cool, well-mixed, oceanic waters from the upwelling. The thriving Gull Island colony in Kachemak Bay on the eastside of the Inlet is situated in a plume of oceanic water mixed with freshwater runoff, and the struggling Chisik Island colony across the Inlet to the west is surrounded by warmer and turbid estuarine water.

Piatt and his team found these oceanographic processes ultimately accounted for the pronounced differences in seabird productivity at the three colonies. To determine this, they started at the bottom of the food web and worked their way up. They found standing
crops of phytoplankton and zooplankton were always much higher
near the Barren Islands and on the eastside of Cook Inlet (Drew
2002). In these two areas, the tiny plants grow in profusion in the
cold, nutrient-rich water, while the surface water near Chisik Island
is a virtual desert—it supported about one-tenth of the drifting plant
life as the water in Kachemak Bay.

At the Barrens Islands, forage fish school in the nutrient-rich
waters and consume abundant plankton (Robards et al. 1999b). From
hundreds of beach seines and mid-water trawls, and analysis of hundreds of thousands of fish, Piatt's team from the Alaska Maritime
National Wildlife Refuge in Homer found far more fish, especially the
high lipid or fatty fish preferred by seabirds, in nearshore waters at
the Barren Islands. There, energy-packed capelin and sand lance
thrive and comprise over 99 percent of the total seine catch by number. In comparison, there were one-tenth as many fish in Kachemak
Bay, where sand lance represent 71 percent of the seine catches, and
one-hundredth as many fish at Chisik Island, where sand lance
account for only 24 percent of the seine catches. Trawl surveys
revealed that walleye pollock and capelin dominate the deeper shelf
waters near the Barren Islands, while sand lance dominate deeper
waters in Kachemak Bay (Abookire, Piatt, and Robards 2000; see also
Speckman and Piatt 2002).

Piatt and his team found that seabird breeding success was closely related to an abundance of high lipid forage fish within a short distance of the colony. The team discovered that adult murres and black-
legged kittiwakes consistently fed their chicks mostly high lipid forage fish. Murres provisioned their single chick with capelin at the
Barren Islands colony, sand lance at Gull Island, and smelt at Chisik
Island (van Pelt and Shultz 2002). Kittiwakes fed their one or two
chicks primarily sand lance at all three colonies. Pigeon guillemots in
the eastern Inlet and horned puffins at Chisik Island also fed their
chicks mostly sand lance, caught in the inner reaches of Kachemak
Bay where the fish thrived in conditions that favored explosive
growth of plankton (Litzow et al. 2000).

With sand lance prominent in so many seabird diets, Piatt and his
team conducted separate studies on this fish to learn about its habits.
From a literature review, they found these fish are a keystone species,

filling a critical ecological role as prey for nearly one hundred species of fish, birds, and mammals. From fieldwork, they discovered that sand lance live nearly year-round in the shallow subtidal zone near beaches so they are almost always accessible as prey. Unlike herring and other forage fish, sand lance spawn in October, instead of spring, so they have high-energy reserves in summer when other adult forage fish have low fat reserves (Robards et al. 1999a). Sand lance provide seabirds with a shot of lipid-rich food when it was most needed to feed hungry chicks.[1]

With the seabirds at the different colonies all feeding their young high-energy food, Piatt wondered, "What could account for the differences in chick survival?" He and his team found part of the answer was how often adults fed their young. Frequency of feeding depended on how easily the fish could be caught. Kittiwakes had to work much harder than murres to feed their young (Zador and Piatt 1999). Piatt and his team discovered this by accounting for "loafing time," as Piatt called it, when both parents were present at a nest. Adults can only afford such a luxury when there is ample food to satiate hungry young; otherwise one parent hunts for fish while the other guards the young from predators. Piatt realized that loafing time or adult attendance was actually a very sensitive measure of forage fish abundance—as well as a very effective buffer against food shortages.

Piatt and his team found kittiwakes had no loafing time—at least one parent was usually hunting even at Gull Island where plentiful food was a short distance from the colony (Shultz 2002). He explained, "Kittiwakes pretty much wear their hearts on their sleeves when it comes to food supply and breeding success." When food is abundant and nearby, like at Gull Island, they successfully raise two chicks. But when food is scarce and there is no buffer, every hardship lowers chick survival. At the Barren Islands, where kittiwakes had to forage farther to find surface-schooling fish, chick survival dropped. At Chisik Island, few birds ever fledged chicks, because adults could seldom find enough food. Piatt said, "God knows why the birds are even there."

In contrast, murres have a lot of loafing time at the Barren Islands and Gull Island, but not at Chisik Island (Zador and Piatt 1999). Murres switch from leisure time to hunting fish when food supplies drop, so they are able to successfully raise their single chick across an incredibly wide range of food abundance from scarce to plentiful. It helps that murres can dive over 600 feet in search of prey and are not restricted to surface feeding like kittiwakes. As proof of the success of this approach to feeding, murre fledgling weights and body condition are remarkably consistent among all the colonies, despite huge differences in food supply. But there is a threshold—a critical limit of prey below which even murres, constantly hunting, can not find enough food for their young.

Piatt and his team witnessed a reproductive failure of murres at Chisik Island in 1998 (Piatt el al. 1999). Prior to this, an *El Niño* event occurred in 1997 in the Gulf of Alaska. Super warm air from the equator lofted into the atmosphere, causing unusually warm sea surface temperatures in the Gulf of Alaska beginning in June 1997. There was a large *"wreck"* or natural die-off of young starving shearwaters and some kittiwakes in the southern Bering Sea and Aleutians during summer 1997. By October 1997, water temperatures in Cook Inlet rose and remained unusually warm throughout the whole water column until May 1998. There was a moderate wreck of young starving murres in lower Cook Inlet during spring 1998. After a tough winter, seabirds were in poor condition for the 1998-breeding season. Murres at Chisik Island had erratic attendance (absent much of time), delayed breeding, and higher levels of stress hormones (corticoid steroids) than in 1997. Most failed to produce eggs. Kittiwakes, with their lower tolerance for food stress, failed to reproduce at both Chisik and the Barrens in 1998 (Shultz and Harding 2002), and horned puffins at the Chisik colony had the slowest chick growth in all five years of APEX studies (Harding 2002; Harding, Piatt, and Hamer 2003).

Predicting Murre Recovery

After five years of intensive studies across a range of disciplines, combined with extensive analysis of a montage of historical data, Piatt finally felt he could piece together enough information to answer the critical question: when would murres recover from the oil spill? (Piatt 2002) He concluded that the shift from a cool-to-warm regime in the late 1970s and the persistence of this warm period reduced food availability of preferred high lipid (fat) forage fish to seabirds through the 1980s and 1990s (Piatt and Roseneau 1999). This resulted in widespread population declines, lower breeding success, and mass mortality events both before and after the oil spill. He had documented this climate shift and its effect on forage fish and seabirds on a large scale in the Gulf of Alaska and on a regional scale in lower Cook Inlet.

Against this larger drama, Piatt demonstrated that it was possible to measure effects of smaller-scale events, such as intermittent *El Niños* or random oil spills, on breeding success and population, but it was tricky. He knew strict census data can be misleading, for example, when nesting sites are saturated with breeding birds or when all suitable breeding habitats at a colony are filled. Both may have been the case for kittiwakes at Gull Island and murres at the Barren Islands prior to the spill: there was no room at the inn and there were no more inns in town. However, after the spill, there likely was a lot of vacant breeding habitat that suddenly opened at the Barren Islands due to the high oil-kill of murres. Piatt believed the thousands of murres that had died at the Barren Islands in 1989 may have been largely replaced by surplus birds from the reservoir of at-sea rafts, rather than replaced through reproduction. Piatt also now knew that murre egg productivity was not a good predictor of population trends. Breeding success was very similar at the Barrens, Gull Island, and Chisik Island—the birds all *produced* eggs, however not all the chicks *survived* as evidenced by pronounced differences in population trends.

Without historical data on murres at the Barrens and records of more subtle indicators of breeding success than productivity, Piatt knew that it was impossible to precisely determine the spill's effect on this population and the time to recovery. However, from a collection of post-spill studies from 1989 to 1999, it appeared that murres were increasing at the Barren Islands by just over 4 percent annually. Piatt concluded that current ecological conditions were adequate to sustain a stable population at the Barrens and support this modest population growth. This supported his earlier speculation that it would take twenty to seventy years for murre populations to recover to a stable age distribution at colonies that had suffered the most from the oil spill (Chapter 15).

Murre recovery also depended on favorable environmental conditions. But seabirds were struggling to survive in the longest warm regime on record. The good news was that Piatt and others anticipated a return to a cold regime—and more favorable conditions for murres and other seabirds—in the early 2000s. (Subsequent research proved this to be true.)

Prince William Sound Studies

The APEX studies in Prince William Sound grew out of a swirl of controversy and some seemingly disjointed clues that pointed in directions of oil and changes in food as the causes of shifts in seabird populations. USFWS seabird researcher David Irons and others dug out Sound-wide surveys of marine birds conducted from 1989 to 1993 and compared these data to Pete Isleib's pre-spill surveys in 1972 and 1973. They found populations of many species had declined—quite drastically in several cases (Agler et al. 1999). Most of the species that had declined were fish-eaters—loons, cormorants, mergansers, certain colonies of black-legged kittiwakes and some other gulls, Arctic terns, murrelets, puffins, and parakeet auklets. Most of the handful of species that had increased during the roughly twenty-year period—goldeneyes, harlequin ducks, black oystercatchers, and scoters—ate shellfish and other invertebrates. Several species of the shellfish-eaters declined after the spill, but the populations had fared well up until then. The reason for the decline of piscivorous seabirds was not clear from the surveys, but the evidence pointed to forage fish as a good place to start looking for answers.

APEX studies in the Sound were a collage of work by seabird research biologists, forage fish assessment teams, ecologists, and modelers. These scientists set out to find the reasons for the dramatic declines in populations of fish-eating seabirds in the Sound. Unlike the APEX studies in the Gulf of Alaska and lower Cook Inlet, where the scientists focused on physical oceanography and worked their way up to the apex predators, the Prince William Sound studies focused on forage fish (capelin, sand lance, herring, and juvenile pollock) and seabirds. The APEX studies in the Sound depended on the SEA program to provide physical and biological oceanography data (Chapter 17).

Prince William Sound is a complex estuary with deep fjords carved by moving rivers of ice—the glaciers still present and active in the area. The SEA investigations (Chapter 17) revealed, in general, there are three broad regions distinguished by different water masses. The dynamic central and southwest region is influenced by Gulf of Alaska waters, carried by a spur of the Alaska Coastal Current that sweeps into the Sound through Hinchinbrook Entrance. The current spins through the central Sound in a counterclockwise gyre, exiting through Montague Strait. The northeast and east region is usually warmer but still influenced by the nutrients and plankton in the central gyre and inflow from the Gulf of Alaska. The north and northwest region has the most stable water masses and is more influenced by outflow from glaciers and streams than by Gulf of Alaska waters.

Large-scale atmospheric processes affect the Gulf of Alaska and create differences from year to year, and even from season to season, in the amount of Gulf of Alaska water flowing into the Sound and in the warmth and degree of mixing of the Sound's waters. This, in turn, drives production and distribution of plankton (Chapter 17). Concentrated plankton blooms in warmer bays attract young herring and sand lance to surface waters where the fish become available to surface-feeding seabirds.

Black-legged kittiwakes became a central focus of the APEX program in the Sound. These small white gulls with inky black wing tips and legs and yellow bills are the most abundant colonial seabirds nesting in the Sound (Irons1996). Some 20,000 pairs nest at twenty-seven colonies located throughout the Sound. Irons, Rob Suryan, and others conducted a suite of studies during the post-spill decade and found that three-quarters of the kittiwakes were concentrated in two large colonies in northern Sound and the rest were spread out in several small-to-medium colonies in the rest of the Sound. Sound-wide, the kittiwake population has remained stable since the 1970s; however the *distribution* of kittiwakes has gradually shifted over time from south to north (Suryan and Irons 2001). In 1972 Isleib found most kittiwakes were in the central and southwest region. By the mid-1980s, the population was distributed equally between north and south, and by the 1990s, 70 percent of the kittiwake population was in the north, split between two colonies located far away from each other in the extreme northwest and northeast corners of the Sound.

Kittiwakes are visual feeders, spotting schools of herring and sand lance from the air and foraging almost exclusively in shallow waters close to shore. The foraging grounds for the various colonies usually did not overlap and this held the key to understanding the shift in distribution of the population. At the Shoup Bay colony in the northeast Sound, chicks thrived on juvenile herring, which had increased dramatically during the 1980s. The increase in herring had supported an increasing number of birds and this colony had swelled in size. At the failing colonies in the central and southern Sound, chicks were fed mostly sand lance and capelin, but when these colonies were thriving in the 1970s, adults had fed their chicks herring and eulachon (Kuletz et al. 1997). During the 1980s, there were far fewer capelin and eulachon in the central Sound, reflecting climate changes in the greater Gulf of Alaska waters, and these kittiwake colonies had declined.

Chick survival is closely tied to a high-energy diet. One-year old herring are packed with energy-rich lipids, more so than sand lance, capelin, and young-of-the-year herring (Anthony and Roby 1997). The oil spill nearly wiped out young-of-the-year herring in 1989, resulting in a scarcity of surviving one-year old herring in 1990 (Suryan, Irons, and Benson 2000). Adult herring also had low rates of reproduction during the post-spill decade, which contributed to wildly fluctuating abundance of one-year old herring during this time (Chapter 20).

Irons and Rob Suryan found that kittiwakes responded different-ly to food stress in the two colonies they studied—Shoup Bay and Eleanor Island—from 1996 to 1999 (Ainley et al. 2003). At the Shoup Bay colony, there was no alternative to one-year old herring readily available and adult kittiwakes were hard-pressed to find enough sand lance and younger herring to feed their chicks to compensate for this shortfall. By radio-tagging some birds, researchers determined that, in years when herring were extremely scarce, it took longer for adults to find fish and the increased time between feedings resulted in high-er chick mortality (Suryan et al. 2002). At the Eleanor Island colonies, sand lance, and sometimes capelin, were usually available. When one-year old herring were scarce, these kittiwakes switched to alternate fatty forage fish and younger herring.

Kittiwakes, unlike murres, have a very limited capacity to buffer their young from the vagaries of prey abundance (Kitaysky, Piat, and Wingfield 1999; Kitaysky, Wingfield, and Piatt 1999). At the Shoup Bay colony, their limits were exceeded at least three times during the post-spill decade: in 1990, most likely due to direct oil spill effects, and in 1997 and 1998, due to generally poor forage fish abundance from the 1997 *El Niño*. Stress of raising chicks depleted limited ener-gy reserves of adult birds and reduced life expectancy and subse-quent productivity (Golet and Irons 1999; Golet, Irons, and Costa 2000; Golet, Irons, and Estes 1998). These studies concluded that the overall reduction in herring in the Sound as a result of the *Exxon Valdez* oil spill left a lasting impact on these kittiwake colonies.

Studies on pigeon guillemots supported the kittiwake findings. USFWS biologists Kuletz, Golet, and others teamed up to continue the earlier studies on pigeon guillemots at the Naked Island colonies in the central Sound. Roby measured the fat content of forage fish and determined that sand lance, herring, and capelin had the highest energy density, while young pink salmon and pollock had the lowest (Anthony and Roby 1997). Kuletz and Golet found that the diet fed to young guillemot chicks gradually changed from the 1970s to the

1990s (Hayes and Kuletz 1997). In the 1970s, guillemots fed their chicks a diet rich in fatty fish; during the 1980s, the diet contained more lean forage fish; and during the 1990s, the diet consisted mostly of lean fish. While guillemots successfully produced a steady stream of chicks during these decades, the population plunged. By tracking individual pairs of adult birds, Kuletz and Golet found chicks fed fatty fish grew faster and survival was much better than chicks fed a leaner, more diverse diet (Golet et al. 2000). They concluded pigeon guillemots need fatty forage fish to maintain healthy productive colonies.

Irons updated his Sound-wide marine bird survey in 1996, 1998, and 2000 (Irons et al. 2000). Of the fourteen species he studied, he found populations of nine species still had not recovered—to varying degrees—from oil spill effects. Fish-eating cormorants, mergansers, murres, and pigeon guillemots showed the strongest evidence of long-term effects eleven years after the spill. Harlequin ducks, goldeneyes, and black oystercatchers were greatly impacted initially, but at least oystercatchers appeared to be recovering in later years. Black-legged kittiwake populations remained mostly stable, but with a redistribution of birds from southern, oiled colonies to northern, unoiled colonies. Loons showed only weak negative effects from the spill.

Irons had not expected to find these persistent negative effects and his surveys were not designed to elucidate the reasons for the declines. However, he believed part of the reason for lower counts of marine birds in 1993 and 1998 was probably the *El Niño* events in these years. Another reason for the decline was revealed from the NVP studies, which provided convincing evidence of continued exposure to oil of some marine birds and sea otters during the years of Irons's surveys (Chapter 18). Based on corroborating evidence, he concluded persistent oil in the environment and reduced prey abundance were affecting the recovery of marine birds in Prince William Sound.

Prince William Sound Marine Mammal Studies
Harbor Seals

ADFG biologists Kathy Frost (2003) and Lloyd Lowry searched diligently for alternative ways to analyze harbor seal diet, other than through scat or stomach content, which had inherent biases and were difficult to collect. They discovered a Canadian scientist, Sara Iverson,

PhD, who was pioneering the use of fatty acids to study marine food webs and diets of *"pinnipeds"* (seals, walruses, and sea lions).

Operating under the 'we are what we eat' principle, Iverson had discovered that fatty acids in prey species are deposited directly into blubber of pinnipeds, such that blubber predictably mirrors current diet. Pinnipeds go through annual cycles of feast and famine, and during the 'feast' phase, Iverson found small blubber biopsies taken from live animals contained a record of recent diet transcribed in a code of fatty acid *"signatures."* These signatures, or chemical profiles, were unique to different prey like PAH signatures were unique to different crude oils. Fatty acids are the largest constituents of lipids (such as blubber). Iverson was developing methods to translate the fatty acid code by using specific lipids as natural biomarkers and certain fatty acids as indicators of different prey.

When Iverson and Frost proposed using fatty acid signatures to study harbor seal diet to the EVOS Trustee Council in 1993, the method was mostly promise and potential (Iverson et al. 2004). Frost said, "It was a real credit to the EVOS Trustee Council for funding it."

Starting in 1994, Iverson, Frost, and Lowry collected blubber biopsies from yearling, juvenile, and adult harbor seals at various haul-outs throughout the Sound when the animals were captured for tagging (Frost et al. 1999). In 1997, they added weaned pups to their study after advancements in technology created reliable satellite-tags for small animals. They collected prey species—capelin, cod, herring, eulachon, salmon, sole, smelt, sand lance, sculpins, squid, shrimp, octopus, pollock, and rockfish—from trawls near the haul-outs, and they built a "prey library" of fatty acid composition by analyzing thousands of individual specimens (Iverson, Frost, and Lowry 1997). They found they could identify prey with an average of 95 percent accuracy by its fatty acid signature—essentially a chemical profile or fingerprint. With the same accuracy, they could determine what harbor seals had eaten.

The three scientists found large differences in diet, based on fatty acid signatures, in harbor seals from Prince William Sound, Southeast Alaska, and Kodiak—areas separated by 250 to 500 miles. They could distinguish seals from different regions within the Sound at geographic scales of 50 miles or more, simply based on fatty acid signatures reflecting differences in abundance and distribution of forage species near the haul-outs. They found many animals stayed close to their haul-out, diving and feeding nearby (Frost, Simpson, and Lowry 2001; Lowry et al. 2001). They could even identify seals from specific haul-outs within six to ten miles of one another—but with less accuracy. Inaccurate identifications turned out to be the more adventurous "Columbus seals" that ate a wider variety of prey in their broader range of travels.

The team noted dietary differences by age, between years, and most significantly, between decades (Frost et al. 1999; Iverson, Frost, and Lang 1999). Archived blubber samples revealed that diets of harbor seals from the Sound were generally lower in flatfish and higher in pink salmon in the 1970s, while diets of harbor seals from Kodiak were generally higher in sand lance and much more diverse compared to the 1990s.

A breakthrough occurred in 1997 when the team began using a laboratory technique called *"isotope dilution"* to measure body composition of fat, protein, and water in pups, yearlings, and juveniles in Prince William Sound (Iverson et al. 1998). To their surprise, they discovered these young seals were heavier and had more blubber than any other young harbor seals that anyone had ever measured! (Iverson, Frost, and Lang 2003) As Frost (2003) succinctly stated, "To come up with young harbor seals that are *twice* as fat as young harbor seals elsewhere on the planet suggests that food is not currently the problem." However, she added, "It *could* have been a problem during the 1970s." They found body fat content is strongly correlated with diet diversity. In other words, juveniles with more diverse diets are in better condition. The team also found that some high-fat species such as eulachon, which averaged less than 5 percent of the total diet, contributed up to 30 percent of the fat in blubber and, thus, are more critical than other food in building fat reserves.

Frost and Lowry retired from ADFG in 2000. Although their work contributed to understanding the relationship between food and past declines of harbor seals in the Sound, the question "Is it food?" was still not fully resolved. Frost said, "The best scenario I can give is that something happened in the late 1970s and early 1980s. There apparently was an ecosystem change that altered the composition of the prey base—the food—but we may never be able to pinpoint the problem, because we don't have the baseline data."

Iverson continued her, by now, widely accepted and widely used work on fatty acid signatures.[2] Frost felt that Iverson's work was "one of the real payoffs of the restoration program." She said, "Now fatty acid signatures are routinely included in studies. Nobody thinks of them as something unusual."

Studies on apex predators spawned a flurry of investigations on the role of diet diversity and quality of lipids on body condition, survival of young, and population growth and size in both seabirds and marine mammals. Such research may help solve the mystery of the population declines and explain why food quality may be as important as food quantity for species success. This research continues beyond the scope of this book.

Orcas

In 1993 Craig Matkin's NGOS team lost its funding for orca research. Instead the EVOS Trustee Council contract went to federal scientists; however, the project was set up so that whoever did the work would have to turn over all the data to the federal "principal investigators" for publication. Matkin (1998) viewed this as a blatant attempt to take over his work, because the research proposal was virtually identical to his past work—there was no independent or divergent line of research. This contract clearly put the federal scientists in control, something they had been trying to achieve since 1989. Matkin applied for and received separate private funding for his orca research during the 1993 season.

That fall, Matkin raised "holy hell" at the EVOS Trustee Council meetings. He brought with him the U.S. Government Accountability Office's report (1993), which found the EVOS Trustee Council process suffered from a lack of competitive bidding, conflicts of interest, and excessive administrative costs. He accused the federal agency scientists of using the huge pot of restoration money as an opportunity for "their mega-research dreams and budgets and getting all the credit, so they could march on with their careers and power-building." Matkin insisted, "Our observations and research are important, too." He argued—adamantly and successfully—for adoption of a policy stating that federal and state research would not displace private researchers from the work they were doing (EVOS Trustee Council 1994). He was well aware of what had happened to sea otter biologists Lisa Rotterman and Chuck Monnett. "Now I'm being displaced," he announced, "*and this is not going to work.*"

But Matkin might have suffered the fate of the sea otter biologists had it not been for another voice crying foul. Homer fisherman Jim Diehl was a member of the new EVOS Trustee Council Public

Advisory Group. A public advisory group was required under the civil settlement, but was not integrated into the bureaucratic-heavy EVOS Trustee Council's decision-making process until 1992. Creation of this group had been forced by public outcry—championed by Cordova fisherman Rick Steiner and others.

Diehl felt strongly that the EVOS Trustee Council should encourage private research, especially people like Matkin. Diehl didn't know Matkin personally, but he knew of Matkin's track record from long before the spill. Jim started asking questions of EVOS Trustee Council members. When he didn't get satisfactory answers, he called the Department of Commerce in Washington, D.C. By the fall 1993 EVOS Trustee Council meeting, Diehl had sixty-five letters of support for the NGOS research. Matkin said, "the EVOS Trustee Council had to take me seriously after that."

The federal researchers offered a compromise. The EVOS Trustee Council funded the NGOS research, but the new $60,000 study to look for contaminants in whale blubber was awarded to federal researchers who relied on the NGOS team to provide them with samples. For the contaminant study, the federal scientists *assumed without pretesting* that the sampling and analysis techniques developed by Frost for harbor seals would work for whale blubber. But this was not the case. They found out that these methods did not work for orcas. The $60,000 NOAA fatty acid study failed. The EVOS Trustee Council, now under public scrutiny, could not afford this kind of waste and poor science. Eventually the scientists who were most troublesome to Matkin were phased out of oil spill studies and out of the Sound. Matkin and the NGOS team have been awarded annual contracts, including the contaminant study, since 1994.

In 1999 Matkin summarized his team's post-spill findings since 1989 (Matkin et al. 1999; see also Scheel, Matkin, and Saulitis 2001). He reported the resident AB pod still had not recovered from its loss of thirteen members. Seven calves had been born into the pod since the spill, but in 1993 one matrilineal group split off to travel with another resident pod. This was unprecedented behavior for the tightly-bonded resident orcas. Matkin believed that the social disintegration and persistent high death rate in the AB pod were triggered by dramatic changes in social structure after the spill, following the sudden deaths of several mature females. The other six resident pods commonly seen in the Sound increased in numbers since the spill.

Matkin reported that the transient AT1 group of orcas also had not recovered from its losses since the spill, but for these whales, the oil spill complicated an already complex situation. Matkin and the NGOS team discovered no calves had been born into AT1 group since 1984. Matkin figured one reason might be that their primary prey, harbor seals, were becoming scarce (Saulitis et al. 2000). From the mid-1970s to the mid-1990s, harbor seal populations in the Sound and Gulf of Alaska declined precipitously on the order of 60 to 90 percent. However, Matkin also suspected environmental pollution was contributing to the lack of births.

From 1994 through 1999, Matkin conducted contaminant studies on orcas through the EVOS Trustee Council. His group was stunned to find that Alaska transient killer whales contained some of the highest levels of organochlorines (DDT, DDT breakdown products, and PCBs) in the eastern North Pacific—comparable to orcas found in more polluted waters (Ylitalo et al. 2001). Blubber biopsies were used for contaminants and genetics studies. The results from these studies revealed an astonishing story: organochlorines increase as young whales sexually matured, then drop dramatically as adult females transfer the bulk of their contaminants to their first born offspring. Subsequent offspring receive lower and lower amounts of contaminants from their mother. Adult males, on the other hand, continue to accumulate organochlorines throughout their lives.

Over 90 percent of the Alaska transient orcas contain organochlorine levels found to cause reproductive dysfunction and immune system failure in other marine mammals (ringed seals, harbor seals, and sea otters; see, for example, Helle, Olsson, and Jensen 1976). The resident orcas have elevated contaminant levels too, but not the high life-threatening levels found in transient orcas. The difference, of course, is diet: transient whales pick up the organochlorines from their main prey, harbor seals, while resident whales eat fish, prey with much lower contaminant levels. Studies to determine how these pollutants affect reproduction in Alaskan orcas continue and are beyond the scope of this book.

Exxon's Response

Exxon scientists did not conduct any comprehensive studies comparable to the public-trust studies on seabirds and marine mammals, but this did not prevent Exxon scientists from refuting the findings of the public trust studies.

Seabird Studies

Exxon scientists spent a couple years studying the effects of the spill on murres and other seabirds—and then over a decade generating papers to attempt to discredit public-trust research on seabirds.[3] In their initial spill studies on murres, Exxon scientists focused on censuses and productivity as indicators of recovery. However, Exxon studies of productivity and recovery were limited to the Barren Islands (Boersma, Parrish, and Kettle 1995), where historical data were sketchy and findings lent themselves to differing interpretations. Census data collected by Exxon scientists at other colonies were similar to counts by public-trust scientists, but Exxon scientists appear to have limited their use of historical data in order to indicate rapid recovery from the spill (Erikson 1995). Public-trust scientists showed that population recovery or growth could not be projected accurately from these measures, but Exxon scientists made it seem like they could (Chapter 15).

Exxon scientists projected an annual growth rate of 25 percent at the Barren Islands murre colonies after the spill (Boersma, Parrish, and Kettle 1995). Piatt had demonstrated that ecological conditions at the Barren Islands could *not* support such an optimistic growth rate; it was six times higher than what he had observed, based on a diverse array of measurements. Some of these same Exxon scientists had studied murres *before* the spill. Their model of murre population dynamics—created in 1982 *before* they were under Exxon's hire—demonstrated that murre colonies could take decades to recover from large losses of prime breeding adults, the very study upon which Piatt based his projections in the 1990s (Ford, Page, and Carter 1987).

Exxon scientists conducted post-spill surveys of marine birds in limited areas of the Sound in 1989 to 1991 and compared their findings with USFWS surveys from 1984 and 1985 (Murphy et al. 1997). However, instead of looking for general trends in abundance, Exxon scientists looked for oil effects, based on habitat use (Day et al. 1997). They reported most of the species they studied either increased or did not change over time (i.e., habitat use was basically unchanged)—which they interpreted as no oil effects. However, Exxon scientists violated a basic assumption of their own study design. They *assumed* birds would choose *not* to return to heavily oiled colonies and beaches when, in fact, many birds did. Violating this basic assumption made their conclusions invalid.[4]

In their apparent obsession to disprove seabird populations had declined as a result of the oil spill, Exxon scientists employed intricate statistics to normalize data (Wiens et al. 1996). Exxon's statistical mill was so effective it even ground up and hid the huge population crashes from natural causes. Artificial data manipulation create what author Huff calls "*bias.*" Huff (1954) wrote, "By the time data have been filtered through layers of statistical manipulation . . . , the result begins to take on an aura of conviction that a closer look at the sampling would deny" (ibid., 18). The flaws in Exxon's surveys became more obvious with time as the public-trust scientists documented huge declines in bird abundance from natural causes.

To this author's knowledge, Exxon scientists never updated their post-spill surveys, but this did not stop them from attacking Irons' more recent surveys (Wiens et al. 2001). In response, Irons and his colleagues (2001) pointed out that Exxon scientists failed to shed any new light on these old issues and "the issues that they continue to debate cannot be resolved with certainty using available data" (ibid., 893). This was a polite way of saying Exxon scientists sounded like broken records, repeatedly rehashing old arguments with old evidence. Irons stuck by his earlier conclusions.

Harbor Seal Studies

Exxon chose initially not to study harbor seals, a decision Frost described as "conspicuous." Exxon hired Frost and Lowry's former boss at ADFG, who had retired in 1988, to do some population counts in the Sound (Burns 1993). Their former boss and other Exxon hires started showing up at meetings to challenge Frost and Lowry's position that the spill had killed an estimated 300 harbor seals. In a paper published in 2001, Exxon scientists offered an alternative interpretation of oil spill impacts on harbor seals. They argued, "Under adverse conditions, movement of seals away from impact areas is the simplest explanation to account for missing seals" (Hoover-Miller et al. 2001, 131).[5]

In response to this statement, which was published after Frost and Lowry retired, Frost said simply that in 1989 she and Lowry had spent "hundreds and hundreds and hundreds of hours" in the Sound in a seventeen-foot Boston whaler, counting seals and noting degree of oiling and behavior. She and Lowry never saw any oiled seals in unoiled areas. She said, "I don't understand where all those oiled seals went when no one saw them."

Exxon scientists also largely ignored or misinterpreted the collaborating evidence of oil damage from the tagging studies, the fatty acid signature analyses, and the genetic studies. Such selective interpretation of data is similar to a trick commonly used by people employed to "prove" points and suppress unfavorable data, according to Huff (1954, 75, 123). Exxon scientists also looked at population trends while ignoring absolute counts. Exxon's seabird scientists had taken a similar approach with surveys of marine birds. As discussed above, this approach was unable to detect huge declines in seabird and harbor seal populations from natural causes and so could not be expected to detect more subtle spill-related losses.

Orcas

Exxon scientists did not conduct any studies on Prince William Sound's orcas.

Synthesis of Studies on Apex Predators

I asked Wright, now with the Conservation Science Institute in Santa Cruz, California, to help summarize the massive and complex studies on apex predators. For nearly a decade, Wright had relayed information between the seabird and harbor seal biologists, because he believed the same factors were ultimately controlling populations of these apex predators. He was also eager to help bring closure to this important suite of studies, which had never been formerly synthesized like the SEA and NVP projects. In a way, studies on apex predators ended like they had started—a set of seemingly independent disjointed studies. Only those involved knew of the connections.

The studies on apex predators showed that survival and productivity of wildlife in the North Pacific is dependent upon food quality and availability, predation, change or loss of habitat, disease, and effects from contaminants, including oil and other pollutants. The studies also found that effects of these factors are very much intermingled and even connected by the heavy hand of human impacts at many levels.

Studies on seabirds and harbor seals demonstrated that the North Gulf coast ecosystem, from forage fish up to apex predators, shifts in response to small changes in ocean surface temperature.

Climate regime shifts—the flip-flops between warm and cool sea temperatures and back—were found by other researchers to be part of the naturally-occurring Pacific Decadal Oscillation.

Wright pointed out the implications of this temperature-sensitivity of the vast North Gulf coast ecosystem suggest that human impacts that affect sea temperatures are likely to elicit a similar response in the marine ecosystem. He observed that loss of habitat is usually thought to be more important in areas with high human populations, but a loss of habitat can also occur in response to human-induced warming of the ocean surface temperature. Thus, phenomena such as global warming and climate change from increased greenhouse gas emissions may have devastating effects on sealife in the North Pacific Ocean.

The APEX and marine mammal studies found that climate regime shifts affect both prey abundance and quality, which in turn affects survival and productivity of apex predators. Analyses of historical data shows that prior to the late 1970s, sea temperatures were cool and capelin were plentiful in the Gulf of Alaska and the Sound. The apex predators—orcas, marine birds, sea lions, seals, and predatory fish—thrived under these conditions. But when sea conditions warmed up and the forage fish disappeared, apex predators were forced to eat something different or starve. Predators that had found it easy to find capelin were now out searching for alternative prey. Available forage fish such as juvenile pollock and young pink salmon are not energy-rich morsels like capelin. The loss of food quality affects seabird health and productivity, at a minimum, and probably survival as it is harder for less fit individuals to dodge predators and chase down food. Populations of seabirds, harbor seals, and other marine mammals crashed, squeezed out by a triple whammy of loss of forage fish, a low-energy diet, and heavy predation.

The studies on apex predators show how natural cycles and contaminants can interact to affect wildlife populations. While murre *colony counts* appear to have bounced back to near normal within ten years of the spill, the estimate of time to stabilize the *population* and replace birds lost by the spill was twenty to seventy years. The lower estimate for recovery depended on favorable environmental conditions; that is, a shift to a cooler climate regime. Other studies found Prince William Sound kittiwakes were stressed by the loss of herring after the spill. The population dipped in 1990 due to lack of young herring to feed offspring, and it dipped again in the *El Niño* years of 1997 and 1998 from a combination of poor environmental conditions and food shortages.

For orcas, the effects of the oil spill and change in food availability were compounded by the increases in persistent organic pollutants such as DDT and PCBs. Orca groups that lost the most members after the spill have not recovered from oil spill losses. Changes in social behavior and size of the fish-eating resident AB pod from the Sound may be directly related to the spill, because in comparison, the overall population of resident orcas from the Gulf of Alaska has increased since the spill. In contrast, the mammal-eating transient AT1 pod has failed to produce calves for the past twenty years. This is most likely due to a combination of declining prey abundance, from natural causes, and high levels of organochlorine pollutants in the whales' blubber. The future of this genetically-distinct stock of whales is uncertain. Lack of reproduction and loss of whales after the spill led a coalition of Alaska environmental and Native groups to petition the National Marine Fisheries Service in fall 2003 to have this stock declared depleted under the Endangered Species Act (O'Harra 2003).

Finally, Wright concluded that, like the SEA program, this collection of studies demonstrated the importance of a consistent and comprehensive long-term approach to monitoring wildlife populations and ecological processes. With such an approach it was possible to measure effects of human disturbances such as contaminants, oil spills, and global warming against the natural ebb and flow of abundance in wildlife populations. Studies on apex predators, prey, and marine climate became part of the large-scale GEM Project, that continue and are beyond the scope of this book (Chapter 22).

Chapter 20.

Fish and Oil Toxicity

Pink Salmon and Persistent Oil Effects

The large number of dead pink salmon eggs in oiled streams in the Sound, discovered by ADFG biologist Sam Sharr in fall 1991, had the state biologists stumped (Chapter 16). In January 1992, Auke Bay Lab oil pollution division director Stanley "Jeep" Rice (2001) invited Sharr and other state biologists to the Auke Bay Lab for a strategy session with federal scientists. After the 1991 civil settlement for damages to fish and wildlife, state and federal scientists were finally free to share their data and findings. The gathered scientists decided to test whether they could repeat in the lab what Sam had discovered in the Sound—persistent harm from residual oil. This became known as "the Bue effect" after the scientist, Brian Bue, who had insisted on reinstating the fall egg surveys. They also decided to test what effect oil exposure of eggs might have on surviving fish and to see if they could verify some of the other earlier findings such as stunted growth and reduced adult survival.

Other studies reported that breakdown of polycyclic aromatic hydrocarbons (PAHs) by the enzyme cytochrome P450-1A produced intermediate compounds that caused genetic damage and cancer in some species. But oil doses had been high and the length of exposure usually short in the previous studies. Only low concentrations of hydrocarbons had been found in salmon redds after the spill, so scientists at the meeting assumed that whatever had caused the high egg mortalities was doing it at very low levels, even near or below detection limits of analytical equipment. They decided to culture two separate lines of pink salmon. One line would descend from fish exposed to very low levels of oil during the eight-month embryonic development. The other line would descend from fish not exposed to oil (controls) (Heintz, Short, and Rice 1999).

This historic experiment started in September 1992 at the Little Port Walter hatchery on the southeast tip of Baranoff Island. Like Prince William Sound before the spill, this largely wilderness area has very low levels of background pollution. The salmon stock chosen for the test was local, from Sashin Creek, as these fish spawn in the intertidal zone like pink salmon from the Sound. The hatchery has served since 1938 as a federal research station where some of the seminal work on pink salmon biology has been conducted. Federal biologist Ron Heintz was assigned project leader for his experience culturing salmon. Quiet and unassuming, Heintz (1998) had no idea at the time that this assignment would shift his life's work—and that he would still be researching chronic effects of oil on pink salmon over a decade later.

Conceptually, the experimental design was simple: incubation chambers were set up to mimic conditions in Prince William Sound. Different amounts of weathered North Slope crude oil were sprayed in a one-time application onto gravel, which was flushed with flowing water (for forty-eight hours) prior to adding the salmon eggs. Initial water concentrations of total PAHs ranged from 2 to 51 parts per billion. Within two months, PAH levels were barely detectable at 100 parts per *trillion* using sophisticated equipment (gas chromatography/mass spectrometry). At fry emergence eight months later, PAH levels had slipped well below the limits of detection even with state-of-the-art equipment. Chemist Short was disappointed in the low PAH levels because studies from the 1970s suggested levels 1,000 to 100,000 times higher were needed to cause biological harm. Both Short and his boss Rice thought the whole experiment was a long shot.

Even though the idea was simple, the experimental set up was complex and there was a lot of room for things to go wrong. In March 1993 when the first fish tissues were analyzed, a lot of abnormalities were found. Upon further investigation, it was discovered that just after the eggs were fertilized, someone had turned on a saltwater pump. Salmon eggs are sensitive to saltwater before they are water-hardened and nearly three-quarters of the incubators had to be discarded. In a typical understatement, Heintz said, "It's a slow painful process to get something new up and running."The team decided to salvage what they could and to restart the experiment the next fall.

Results from the first experiment were startling. First, the lab setting allowed the scientists to closely examine the miraculous biological changes that a baby salmon undergoes as it develops from an egg to an emergent fry (Sidebar 11, p. 250; Marty, Heintz, and Hinton 1997). Just like with the lab set up, the biological changes are complex and there is a lot of room for things to go wrong.When the Auke Bay Lab team examined the fish that had been exposed to oil, they found a number of delayed developmental problems, production of energy-expensive P450-1A enzymes, and increased deaths (Marty et al. 1997b).The scientists suspected that some of the developmental changes would affect the fishes' ability to survive. For example, the abnormal bulging of the gonadal cells could be related to reproductive impairment in adults, such as had been documented in field studies through 1993 (Chapters 16 and above). The scientists realized that the egg stage was more sensitive to oil exposure than other life stages, as they had concluded twenty years earlier (Rice, Moles, and Short 1975).

Second, the chemistry samples showed that not all the PAHs coating the gravel dissolved into the water, but a subset did. PAHs in fish tissue mirrored only those PAHs present in water, not the total suite of hydrocarbons that coated the gravel (Marty et al. 1997b).This meant pink salmon larvae had picked up PAHs from water flowing through the incubators and gravel, not from direct contact with oiled sediment as previously thought. But the real surprise was that the subtle but significant effects of oil exposure in young salmon occurred at levels as low as about 5 parts per billion. *This is well below the state's water quality standard of 10 parts per billion PAHs and sixty times lower than the federal guideline of 300 parts per billion PAHs!* (State of Alaska Water Quality Standard Regulations 18 AAC 70; U.S. CFR 45 79339 1980).

The Auke Bay Lab team realized they had entered a brave new world where the old thinking based on short-term oil effects no longer applied. Of his earlier work in the 1970s to develop state water quality standards for PAHs, Rice said, "Fifteen years of worthless research! How depressing!" Determined to figure out the extent of chronic oil effects and how oil caused these effects, the Auke Bay Lab team added two studies to the initial experimental design. When they restarted the main experiment in fall 1993, they suspended another batch of eggs *above the gravel* to test if biological harm occurred from contact with just oiled water as earlier results suggested. They also incubated another batch of eggs in oiled gravel that had been weathered a full year—the rocks left over from the first experiment. This was to test for "the Bue effect" to see if the larger, more toxic PAHs were lost from residual oil in the gravel at a slower rate than the smaller molecules and, if so, whether this could account for persistent toxic effects like Sharr had discovered in the Sound.

While the second tests were running, Short and Heintz worked feverishly to understand how low levels of oil could cause these persistent biological effects. They studied data on weathering of Prudhoe Bay crude oil from lab experiments that measured reductions in individual hydrocarbons with time. They compared these results with literally thousands of samples taken from the Sound over the years since the spill. They boiled all this information down into a new oil-weathering theory—a model—that predicted, once oil spilled, nearly all the changes observed over time for PAHs were the result of simple physics (Short and Heintz 1997).

Heintz explained, "In the 1970s people used to think about oil toxicity in terms of solubility. Since not much oil dissolved in water, people thought oil must not be really bad for aquatic organisms. On the other hand, if you think of solubility in terms of kinetics or *the rate of chemical reactions*, the molecules are all jiggling and those near the surface occasionally jiggle right out of the oil film. Oil molecules don't like water, so if they happen to encounter another oil or lipid, they're going to stick to it. In the Sound, you have oil upstream on rocks and it's constantly oozing these PAHs—that's the weathering process. Any little lipid packets downstream, like salmon eggs, are going to soak up PAHs as they wash by."

The bottom line was that the 1970s understanding of oil toxicity, related biological harm to how much oil dissolved in water *quickly*; this oil, mostly the WSF, was the basis of the water quality standards. The emerging understanding related biological harm to how much oil, measured as PAHs, dissolved *over time*. The new model explained how a residual patch of oil could retain its toxicity and even be more toxic over time—"the Bue effect." The larger PAH molecules are released more slowly than the WSF, because it is less likely that the larger molecules would jiggle loose at all. The larger PAHs become more concentrated in the residual oil as the WSF dissolves out of the oil. As the residual oil begins leaching the more persistent and less easily dissolvable PAHs, the toxicity of the residual oil increases and persists until most of the PAHs are removed.

Their emerging understanding explained the increase in dead salmon eggs and deformed embryos observed by Sharr in oiled streams in the Sound in 1991. The Auke Bay Lab team realized that, eventually, the residual oil would lose enough PAHs where it no longer caused biological harm, but they didn't yet know when that would occur in the Sound. They thought it "could persist on the time scale of years" (Heintz, Short, and Rice 1999, 501).

These different fractions of oil also have different biological effects. In animals, the WSF acts mostly on the cell membrane and causes a quick-acting narcotic effect, which acts on the central nervous system. This explains things such as the dizzy, disoriented harbor seals and the brain lesions in sea otters and seals immediately after the spill. PAHs act *within* the cell, disrupting the basic functions of life over time. The new understanding of oil toxicity could account for the chronic maladies observed in the Sound during the years following the spill; the 1970s understanding does not. For example, deformed salmon embryos and stunted juvenile growth can result from lingering PAH exposure (Heintz et al. 2000; Heintz, Short, and Rice 1999; Marty et al. 1997b).

Heintz and his team were anxious to see if their new oil-weathering model was consistent with their test results. In spring 1994, they discovered PAHs in the tissue of eggs incubated in or suspended above oiled gravel mirrored PAHs in the water (Heintz, Short, and Rice 1999). They detected higher mortality, metabolic problems, and deformities in embryos exposed to initial levels of PAHs as low as 1 part per billion or one-tenth of the state's water quality standard (ibid.). And, they found higher egg mortality in tests with the more weathered oil (ibid.). All these findings validated results of their first experiment and were consistent with their emerging understanding.

The Auke Bay Lab team puzzled over how their laboratory findings related to what was going on in Prince William Sound. When the spill first occurred, many scientists thought the salmon streambeds would be protected from direct oiling by the fresh water outflow, which pushed the oil away from the stream mouth. But the new findings proved that oil did not have to be directly in the gravel redds to harm developing embryos—*it could be on nearby stream banks or even upstream.* Rice pondered the problem for weeks before he figured out that when the tide rose, seawater flooded upstream past buried pockets of oil along the stream banks. When the tide fell, oil-contaminated water flowed back through the gravel, this time draining to the lowest point in the stream—the channel in which the eggs were buried (Heintz, Short, and Rice 1999).

To prove this theory, Rice and his team needed samples, lots of samples, from streambeds and stream banks from the early years of the spill. This seemed like wishful thinking. Then serendipity struck.

In November 1994, Short received a relayed a message from ADFG biologist Mike Wiedmer about "a bunch" of sediment samples collected from stream deltas in 1989, 1990, and 1991. ADFG staff was cleaning out their freezers and the samples were literally headed to the dump. Did Short want them? *Hell, yes!* Short and Rice never even knew these samples had been collected, because of the secrecy orders imposed by the government lawyers. Wiedmer rushed to the city dump and retrieved the sediment samples—a half-ton of them—sitting outside, still frozen with their chain of custody seals intact. Short caught the next plane to Anchorage and arranged for their shipment down to the Auke Bay Lab near Juneau. Rice said, "I know those samples were the missing link we needed to support our theory."

The sediment samples were a gold mine of information. In 1989 ADFG had sampled 172 streams at least once in the intertidal area near the stream channel to document oil contamination. Some streams had been sampled extensively from streambed to stream bank to map the distribution of oil in 1989 and again in 1990 and 1991. There were over 300 samples in total. Rice said, "There was the meticulous documentation of the sample locations with maps, field notes, photographs, and videotape." Short analyzed the samples and found the chemical fingerprint of oil contamination were consistent with weathered *Exxon Valdez* oil as predicted by the new oil-weathering model.

Using all the documentation, Auke Bay Lab scientist Mike Murphy re-sampled twelve of the streams in 1995; Short found that residual oil consisted mostly of the large toxic PAHs, consistent with their new model. By extrapolating, Short and the others determined that PAHs in streams had, for the most part, dropped below the lowest level found to kill salmon eggs in lab tests at five years after the spill. This was consistent with Sharr's work—he had found high pink salmon egg mortality in oiled streams up to four years after the spill, but not thereafter (Murphy et al. 1999). The new theory perfectly explained the field observations!

Their data and model even explained how oil could be both available to sealife and long lasting, a conundrum that had stumped the Exxon scientists operating under the 1970s paradigm. The key was uneven or patchy distribution of oil. Where oil pooled in mats and pockets under surface sediments, it weathered slowly until it spread into thinner layers by storms, tides, or foraging wildlife. Every time the residual oil was redistributed, a new cycle of weathering and killing began as PAHs were released and this cycle continued until the concentration of PAHs fell below that causing biological harm. Rice referred to these patches as little timed toxic land mines, poised to destroy salmon, harlequin ducks, sea otters, or whatever sealife might disturb them.

While Short fine-tuned the oil-weathering model, Heintz and his team continued the incubation studies as life-history experiments, which ended up spanning years and generations. Surviving fry of the 1993 experiment were marked with coded-wire tags and released to monitor growth and adult survival. Because of the low number of surviving fry, Heintz repeated his experiment in 1994 and 1995, working with both odd- and even-year genetic stocks of pink salmon to verify and validate his initial findings. Each time over 200,000 fry were marked with coded-wire tags and released, and both first- and second-generation offspring were monitored for persistent effects of oil exposure.

In continuing these incubation studies, the Auke Bay Lab team made important discoveries. They found delayed growth in juvenile salmon from embryos exposed to initial PAH levels in the water of 18 parts per billion compared to unoiled stocks (Heintz et al. 2000). They knew that slower growth meant more losses to predation in the marine environment (Chapter 17). Still, they were astonished to discover, consistently, a 40 percent reduction of adult returns of embryos exposed during incubation to initial PAH levels in the water of 15–18 parts per billion—repeated with three different brood years. They also found a 20 percent reduction in adult returns from embryos exposed to about 5 ppb PAHs. Further, the harm inflicted on the parent stocks from exposure during the incubation stage caused a reduction in the adult return of the second generation, the offspring. The biological harm incurred from low-level oil exposure to pink salmon eggs and embryos in the lab reduced individual health and resulted in population-level harm at PAH levels *well below* those thought to be "safe" for fish and wildlife![1]

These findings, among others (Bue, Sharr, and Seeb 1998), suggest that the pink salmon run collapses in 1992 and 1993 may have been related to exposure of the *parent stocks* to oil as eggs, embryos, and juveniles. For example, the parent stocks of the even-year class (born in 1988) were exposed to oil in the nearshore nursery areas as juvenile fish in 1989. This year class returned as adults (record run of 1990); however, they could not produce viable offspring as evidenced by the even-year run collapse in 1992. The parent stocks of the odd-year class (born in 1989) were exposed to oil in the intertidal streams as eggs and fry in the winter of 1989–1990 and as juveniles in the nearshore nursery areas in 1990. This year class returned as adults (record run of 1991); however, they could not produce viable offspring as evidence by the odd-year run collapse in 1993. It is significant that only the herring stocks and the pink salmon stocks in Prince William Sound collapsed during these years, while other stocks in the state, which were not oiled, were not affected. At a minimum, this evidence suggests that complex mechanisms were operating, and we may never know all the causes and linkages.

As foreseen by Sharr, the implications of his egg survey data—"the Bue effect"—were enormous, but Sam wasn't around to see the resolution of his findings. Sam left Alaska in January 1995. In parting, Sharr (2001) noted one of the more depressing aspects of the oil spill research was that "chunks of the scientific community were vulnerable to being bought off." It was these "chunks"—the Exxon scientists—that continued to oppose, at every opportunity, the emerging understanding of persistent oil effects. It was an argument Exxon could not afford to lose, because it meant oil was, in fact, much more toxic than previously thought. The full scientific and political ramifications of the new oil toxicity paradigm—the central legacy of the *Exxon Valdez* oil spill—are discussed in the final section of this book.

Pacific Herring: Low Levels and High Stakes

The dramatic crash of the Prince William Sound herring stock in 1993 provided the political push to fund more herring research (Chapter 16). After Brown left, her team splintered as former members pursued different research angles. For the next several years, two scientists, Dick Kocan and Gary Marty, separately focused on adult herring and sought to understand the disease outbreak and its effect on reproduction. At the Auke Bay Lab, Rice reassigned researcher Mark Carls from pink salmon to herring. Carls (2001) had over a decade of experience with oil toxicity studies involving early life stages of fish—cod and mackerel from his work in the early 1970s on Atlantic species, and pollock, herring, and salmon in the 1980s after he joined Auke Bay lab (Carls 1987). Carls focused on adult herring and disease for one year, then he returned to his chief interest, starting what became a seminal study on oil exposure to herring embryos.

In 1994 Carls and his co-workers found several clues to help explain the disease outbreak. Adult herring, ripe with eggs and milt and exposed to weathered oil at post-spill levels in the Sound, developed a VHS virus outbreak (Chapter 16). The relationship was straightforward: the more PAHs in the water, the more fish had open skin lesions and internal damage to tissues and organs (Carls et al. 1998). At very low PAH levels, Carls and his colleagues found "subclinical viral infection" or tissue damage that preceded a full disease outbreak and indicated the immune system was weakened. They also found ripe herring exposed to oil did not break down the harmful PAHs nearly as efficiently as fish did after they spawned. Carls concluded that ripe or pre-spawning herring were vulnerable to oil exposure, because oil weakened the fishes' immune systems, which could lead to disease outbreaks.

As Carls became more involved in the herring research, he became aware of the heated controversy between public-trust scientists and Exxon scientists over whether the oil spill had damaged herring or not. This controversy was worrisome to Carls. One of the most exciting aspects of science for him was the opportunity to watch developing life with the thought that his research would help protect it. "The astonishment of watching a single fertile egg develop cells, form an embryo, and eventually hatch never gets old," he said. He remembered his excitement upon learning that developing embryos were ultra-sensitive to contaminants—he knew then he had found his life's work. He knew there should be a straightforward answer to the question of what level of oil was "safe" for young herring and he set out to find it.

In April 1995 Carls and his team captured adult herring, just ripening to spawn, from pristine areas in Southeast Alaska. They artificially spawned these fish at the Auke Bay lab. The fertilized eggs were exposed to water contaminated with oil as it flowed over rocks coated with weathered crude—the same technique developed by chemist Short for the pink salmon studies and at the same low PAH levels, well below those thought to be "safe" for aquatic life. He needed large numbers for his experimental design, especially to detect small differences between the control fish and the oiled fish. So he decided to collect more adult herring and run a duplicate experiment, just in case. Then, Carls said, "Serendipity struck."

When he called Short for advice, the chemist suggested that he *not* re-coat the rocks with oil. Instead, Short suggested that he just expose the second batch of herring eggs to water flowing over the older, more weathered oil from gravel that had already been flushed with water during the pervious year's experiment. Short told Carls that the pink salmon researchers were finding effects on embryos around one part per billion PAHs—and that the larger PAHs, the ones that weathered out last into the water, were more harmful to salmon embryos than the aromatic hydrocarbons in the WSF. Carls decided to take a chance. He exposed the second batch of artificially fertilized herring eggs to initial PAH levels in the water of *less than one part per billion*, a fraction of the level supposedly "safe" for marine life.

The data were crystal clear—and astonishing. Larvae exposed to oil from the first experiment had twisted spines, misshapen jaws, and other skeletal deformities as well as genetic damage (Carls, Rice, and Hose 1999). The tiny fish had metabolic problems and tissue damage as well, a frequent problem being severe *"ascites"* or swollen bellies caused by retained water. The balloon-bellies restricted blood flow to tissues and organs, stunting growth and development. Larvae exposed to oil had trouble swimming, they were a smaller size because of premature hatch, and many more died than larvae not exposed to any oil. Carls detected harmful effects in larvae exposed to initial PAH levels that were *30 times lower* than the federal water quality standard. Results from the second batch of eggs were generally identical, but more frightening. The more weathered oil was much more toxic—larvae suffered harmful effects at initial PAH levels that were *750 times lower* than the federal standards. The Bue effect!

Carls realized that state and federal laws regulating oil pollution are not at all protective of aquatic life *because the laws are based on the wrong oil fraction!* The Auke Bay Lab fish research proved large PAHs are much more deadly to precious fish embryos than smaller aromatic hydrocarbons in the WSF; however, the laws based on 1970s research treat the large PAHs as if they are harmless. Carls realized that to adequately protect aquatic life, the water quality standards needed to be revised and strengthened. This simple solution, however, was trapped in a quagmire of scientific controversy, which effectively stalled such public policy changes. It was very clear to Carls that the water quality standards would not change until the controversy was resolved.

A decade after the spill, he set out, along with Marty and Jo Ellen Hose, to reconcile divergent government and industry conclusions over oil effects on herring much like Peterson had done with the coastal ecology studies (Chapter 13). Like detectives, these scientists retraced all the steps in the two herring data sets, looking for clues that had polarized Exxon and government researchers. It was slow and tedious work, but they found what they were looking for (Carls. Marty, and Hose 2002).

Starting with initial spill impacts, Carls and his colleagues first carefully examined the ability of the different analytical methods developed by Exxon scientists and federal scientists to correctly identify Prudhoe Bay crude oil. They tested and compared the methods in different situations, using fresh and weathered oil and various fish tissues. Exxon's model was structured so that an oil source was always suggested, but ambiguous results were always classified as something other than Prudhoe Bay crude. Carls and his team found that results from Exxon's model were routinely misinterpreted to suggest that the hydrocarbons present originated from sources other than Prudhoe Bay crude. They found Exxon's method correctly identified Prudhoe Bay crude oil in only *4 percent of the herring eggs and 0 percent of the herring tissue!*

If a student was wrong on 96 and 100 percent of all tests, that student would flunk. But Exxon had already published its findings as "science," because no one had challenged them—until Carls and his team did, ten years after the fact in a technical paper.

Continuing their investigation, Carls and his co-workers next used the best method—the government's method—to reanalyze hundreds of sediment, seawater, mussel tissue, and herring egg samples. Exxon refused to give its raw data to the federal government, so Carls and his team made do with Exxon's published papers (Pearson, Moksness, and Shalski 1995; Pearson et al. 1995). The two data sets agreed that only 4 to 6 percent of the herring spawn was visibly contaminated with oil, but the government's caged mussel data showed that subsurface oil could be taken up directly from the water. This proved, for mussels at least, that visible sheen was *not* a reliable indicator of oil exposure, because mussels did not have to be *visibly oiled* to soak up PAHs. When Carls and his team tested the validity of using mussels from many sites as surrogates to indicate oil exposure to herring eggs, they found the level and composition of oil in mussel tissue almost perfectly mirrored that in herring eggs across all tide zones. Based on this re-analysis, Carls and his team essentially "flunked" Exxon's claim that oil was not bioavailable.

Next Carls and his team combed through both data sets, matching evidence of PAH levels in the water with PAH exposure to herring eggs and condition of oiled larvae. They found that PAH levels in mussel tissue increased after the spill, peaking in April—during the time herring spawned and embryos hatched. PAH levels *in the water* during the peak in mussel tissue ranged from about 2 to 8 parts per billion. In comparing lab and field data, they found that PAHs in herring tissue from lab tests with the most weathered oil best mirrored PAHs in herring tissue from the Sound. Egg and larvae deaths, larvae deformities, and other fitness problems increased with increasing levels of large PAHs in the water. The team determined that PAH levels were actually high enough to harm herring embryos at 83 percent of the sites that Exxon scientists had sampled and declared unaffected.

These findings debunked Exxon's argument that its oil did not harm herring, because total PAHs never exceeded water quality standards. Quite the reverse—these findings demonstrated that the water quality standards are not protective of marine life! After Carls and his co-workers untangled Exxon's garbled statistics, they had to conclude that major philosophical differences underpinned the polarized conclusions. They wrote, "The philosophy behind the science was apparently different—industry initially found minor effects, stopped testing, and concluded the spill caused little damage, but the NRDA [public-trust] group was not convinced that initial equivocal results could be interpreted as not harmful, continued testing, and ultimately concluded that there were major spill effects" (Carls, Marty, and Hose 2002, 165).

In other words, Exxon scientists set the bar for "no effects" at the state's water quality standards and proclaimed that since the PAH levels in the water were *below* this bar, the oil caused no harm.

In contrast, the public-trust scientists made no initial assumptions, determined there was harm, and *then* figured out that the bar itself—the water quality standard for PAHs—was not high enough to protect herring. Why? Because the public-trust scientists are responsible for

setting the bar to protect fish and wildlife from oil and other pollutants and if the bar isn't protective, it is up to these scientists to sound
the alarm and sound it they did (Carls et al. 2001b; Rice and Heintz
2000; Rice, Short, and Heintz 2001; Rice et al. 2000). The ramifications
of their bell-ringing are discussed in the final chapter of this book.

Exxon's Response

Exxon scientists wrote scathing and overly dramatic reviews of the
NOAA Auke Bay Lab research. Significantly, to my knowledge, Exxon
scientists never published any technical papers to show that they
duplicated any of the Auke Bay Lab studies on pink salmon or herring
to verify or discredit the results themselves. This is the usual scientific route of challenging other scientists' work. Exxon scientists just
created a paper controversy. The most common deception was built-
in bias through careful selection of data. Readers were never told the
whole story. Instead they were presented with only a few discon-
nected bits and then led to believe Exxon's gospel.

Pink Salmon

Some Exxon scientists referred to Auke Bay Lab's findings as "alleged
injury to pink salmon in PWS from the *Exxon Valdez* oil spill"
(Brannon et al. 2001, 572). In this review, these Exxon scientists
attacked three central findings of the new oil-weathering model and
its theories of biological harm. First, they claimed that the stunted
growth detected in pink salmon fry from oiled areas in 1989 was
"hopelessly confounded" (ibid., 573) by differences in temperature
and age of fry between oiled and unoiled areas. However, they did
not mention that the sample size in the public-trust studies was large
enough—with a million coded-wire tagged fish—to detect an oil signal over the background noise. They also claimed it was "highly
improbable that hydrocarbon concentrations in the water column
were high enough to affect growth" (ibid., 573). However, they based
this claim on the state's water quality standards for acute toxicity and
ignored the public-trust data showing harmful effects of PAHs, well
below the supposedly "safe" standards.

Second, Exxon scientists presented several alternative explanations for the damning evidence of persistent harm from Sharr's fall

egg surveys. Exxon scientists claimed Sharr's data were "completely confounded" (ibid., 573) by not accounting for deaths from the sampling procedure itself. This nonsense was later discredited by the Auke Bay Lab studies, which validated Sharr's fieldwork. Exxon scientists did not mention these later studies when attacking Sharr's early work. Exxon scientists referred to alevins as the "most vulnerable stage to oil toxicity" (ibid., 573), citing papers from the 1970s where the exposure was short-term. However, they failed to mention that the authors of these papers, Rice and his Auke Bay Lab team, now considered eggs to be the most vulnerable stage when exposed for long time periods. Exxon scientists also claimed that the high survival of alevins in 1989 and 1990 proved that oil had not harmed eggs, because "oil was at its highest concentration" (ibid., 573) immediately after the spill. They ignored two of the principle findings of the Auke Bay Lab studies. First, very low levels of oil exposure to embryos affected embryonic development, juvenile growth, and adult survival. Second, oil effects persisted after the spill for at least four years (two generations) due to increased concentrations of PAHs in the incubation waters of oiled spawning streams from residual pockets of oil (the Bue effect).

Third, Exxon scientists attacked the egg incubation studies. They claimed that the level of PAHs to kill eggs was "many times greater than what was measured *in the actual incubation substrate (0.5 ppb to 267 ppb mean PAH)*" (ibid., 571, emphasis added) in oiled streams after the spill. The central finding of the incubation studies, which the Exxon scientists did not mention, was that the concentration of PAHs in the gravel was irrelevant: it was the concentration of PAHs *in the water* that mattered. They failed to note that PAH levels *in the water* of the incubation studies were estimated to have been very similar to PAH levels *in the water* of oiled streams for several years after the spill. Exxon scientists also kept harping on the point that the PAH concentrations were "well below toxic levels" (ibid., 573) based on the water quality standards. They ignored the Auke Bay Lab findings that PAH levels *lower than those previously thought to be "safe"* reduced individual fitness, delayed recovery, and caused population-level harm.

Exxon's "reviews" of the public-trust studies on pink salmon seem to be nothing more than manipulations of selected data to minimize oil spill effects on these fish.

Pacific Herring

Regarding possible persistent oil spill impacts, Exxon scientists were the first to review the different theories and available evidence to determine the most likely cause of the 1993 and 1994 herring population crashes in the Sound (Pearson et al. 1999). They conducted their review well before the Auke Bay Lab team had even started their seminal study on oil exposure to herring embryos. Exxon scientists argued that the cause of the 1993 collapse could not be oil when oil had not caused any effect in 1989. They based their claim of no initial harm on 1970s science and the water quality standards, support they clung to like drowning men clinging to life rings. They steadfastly ignored—or tried to discredit—all the early evidence that low levels of PAHs harmed herring embryos. Predictably, Exxon scientists concluded, "The 1989 oil spill did not contribute to the 1993 decline" (ibid., 711). Instead they claimed the herring crashes were caused solely by a dangerously high population and resulting disease outbreaks.

In contrast, public-trust scientist Carls and his colleagues reviewed all the evidence for the herring crashes with the advantage of a more complete understanding of persistent oil effects. From this perspective, they ruled out a weakened immune system due to exposure to oil in 1989, because subsequent studies had found this condition was reversible in surviving fish. They ruled out delayed reproductive effects from oil exposure in the 1988- and 1989-year classes, because they documented that reproductive effects were undetectable by 1990 or 1991. And, they ruled out large-scale environmental shifts, because the dramatic collapse occurred only in the Sound and not in other herring stocks around Alaska.

Finally, Carls and his colleagues examined the relationship between density and disease just as Exxon scientists had done. In other herring populations, disease outbreaks and population crashes always occurred when the population was at or near *"carrying capacity,"* or the ability of the environment to support the population. They found evidence that the record high herring population in the Sound might have been near the mythical number for carrying capacity. In the two years preceding the crash, herring in the Sound had low annual growth and were small for their age, both signs that too many fish were competing for too little food and the limited food was limiting growth. Other researchers had found the VHS virus was common to herring stocks, but only in the Sound was it associated with collapse of the population.

Carls and his team (2002) concluded that, while too many fish and disease were the most likely causes of the 1993 herring crash, they could not rule out a combination of disease, environmental stress, and indirect oil effects. At a minimum, the 1989 closure of the Sound's herring fisheries—due to the spill—added more fish to an already huge population. This could have tipped the balance to exceed the Sound's carrying capacity and triggered the disease outbreak.

Subsequent research by public-trust scientist Kocan found that the youngest herring in a spawning biomass were the most susceptible to the VHS virus and that infections could spread rapidly to older fish during spawning aggregations (Kocan and Hershberger 2001). His study suggests that when survivors of the ill-fated 1989-year class of herring first joined the adult population in 1993, these young fish may have been more vulnerable to viral infection and infected youngsters may have contaminated the older fish. However, it was impossible to say with certainty that this—or any of the other hindsight theories[2]—had caused the 1993 herring crash.

Sound Truth

Overall, pink salmon appear to be recovering from the oil spill. However, there is still concern about egg mortality from residual oil in or near streams used by spawning salmon, especially in areas that were heavily oiled in 1989. Winter storms or other disturbances could redistribute this oil, resulting in potentially harmful exposure to salmon eggs. Toxic effects of this oil are expected to diminish with time.

Since 1993, Prince William Sound herring have been plagued by disease outbreaks. The definitive cause of these disease outbreaks has so far eluded public-trust researchers. At the time of publication of this book—fifteen years after the spill, Prince William Sound herring have failed to recover from the spill and they have failed to produce a single large year-class of offspring.

Chapter 21.

Habitat is Where It's At!

Coastal Ecology—Two Steps Forward, One Backward . . .

In 1994 and 1995, Jon Houghton and his team confirmed that the superficial recovery they had observed on treated beaches the previous two years was just that—superficial (Driskell et al. 2001). They found the single age stands of rockweed, which had colonized the bare rock left in the wake of Exxon's harsh cleanup, had simultaneously died of old age over large areas. This die-off triggered another wave of colonization and succession of rockweed and beach creatures, which could not survive without the sea plants. The second wave of succession carried into 1996 when another die-off of rockweed triggered a third wave of succession. Each wave was less violent than its predecessor as single-age stands of rockweed slowly diversified to multiple-age stands.

This pattern of recovery with two steps forward and one backward slowed the pace of recovery on treated beaches. These crash and restart cycles were not dependent on the presence or absence of oil; rather, they were caused by the initial wipeout of huge swaths of sea plants and animals by the pressurized hot water wash. The cycles did not happen on 'set-aside' beaches, the oiled beaches left untreated for Mother Nature to heal. On these beaches, Houghton and his team found that opportunistic species steadily gave way to the more competitive pre-spill species (Houghton et al. 1997). On untreated beaches, recovery progressed relatively quickly and in a straight-forward manner, unlike the fits and spurts on the treated beaches.

In 1996 Houghton's team concluded that, given enough time, succession would advance to restore the natural balance among predators and prey at all the oiled beaches. They reasoned that the crash/restart cycles on the treated beaches would eventually dampen until they blended in with the natural rhythm of the Sound. Houghton and his team estimated that the time to full recovery on most *set aside* beaches might take thirteen years or so. Intertidal communities at similar latitudes around the world had taken ten to fifteen years to recover after other oil spills.

The outlook for beaches treated with the pressurized hot water wash was grimer. Team member Dennis Lees noted that it might take clams and other *"infauna"* (the animals that live under the surface of beaches) decades or longer to recover on these beaches (Lees, Driskell, and Houghton 1999). Lees and his colleagues attributed this to loss of adult burrowers and changes in sediment structure caused by the washing process of the cleanup. The fine sediments and organic matter, normally trapped under the beach surface, had been flushed away by the pressurized water washes and these materials took time to replace. Young burrowing animals could not settle in such habitats until there was adequate food and protection.

What Houghton and his NOAA colleagues witnessed on the Sound's shorelines was widely known among ecologists as *"convergence."* Peterson (2000), now one of the leading coastal ecologists in the country, explained, "Convergence is a way of defining recovery as converging to a condition in species abundance and diversity, and community structure, that you would have expected to exist without the disturbance." Convergence was the notion of getting back to the way things would have been without the disturbance, in this case the oil spill, and it was the standard indication of ecosystem recovery.

In 1997, after seven years of work for NOAA, Houghton and his team lost the bid to continue their research to a consulting firm that had used the Freedom of Information Act to obtain Houghton's original shoreline ecology proposal from NOAA. Marine Resource

Services underbid Houghton's team with a proposal for the same work. Houghton and his co-workers, however, continued to actively follow the research in the Sound. This turned out to be a good thing, because media events surrounding the ten-year spill commemorative in 1999 reopened the debate over ecological recovery of the Sound's beaches. This time, Houghton had to defend his team's work against both Exxon and the new NOAA contractors as discussed below.

Residual Oil and Lingering Harm

For ten years Rice and his team of federal scientists at the Auke Bay Lab had assumed, along with others, Exxon and public-trust scientists alike, that the distribution of residual oil would mirror where *it looked like* the oil had stranded initially (Neff et al. 1995; Short et al. 2002). Scientists thought most of the oil stranded in the upper intertidal zone, because it is drier there, so the oil would stick better. They also observed that the high tides floated the buoyant oil up the beaches, where it stayed, as evidenced by a so-called "bathtub ring" of oil. They had focused on this visual oil to assess the amount and condition of residual oil. Residual oil deposited high on the beaches was largely above the richly productive and sensitive life teeming in the mid- and lower-intertidal zones, and this surface oil weathered relatively quickly into tarry asphalt mats. During the cleanup from 1990 to 1992, scientists had assumed that this *visual surface* oil would pose little threat to wildlife—and that this surface oil was the extent of residual oil on the beaches.

After receiving insistent reports of buried oil from Natives and fishermen for several years after the cleanup, Rice and his team led a project to evaluate a beach cleaning attempt in Sleepy Bay, Latouche Island, in 1997. They unexpectedly encountered substantial deposits of subsurface oil (Broderson et al. 1999; Carls et al. 2001a). This raised concerns about persistence of subsurface oil on other beaches. Their concerns were heightened in 1999 when scientists with the NVP project concluded that recovery of sea ducks and sea otters was delayed due to continued oil exposure.

In 2001, Rice and his team set out to determine how much oil was still on the Sound's beaches and whether it still posed a threat to sealife. Assisted by Natives from Chenega and Tatitlek in the Sound, Rice's team dug some 9,000 pits at ninety-one beaches selected at random, which were representative of the most heavily oiled beaches in the Sound (Short et al. 2002). They focused this intense effort in the intertidal zone, where the *Exxon Valdez* oil had initially stranded.

What they found surprised them. Poking around with their shovels in the quadrants and transects dictated by their study design, they discovered liquid oil at fifty-three of the ninety-one beaches. The oil was buried just below the surface and it welled into the pits, leaving a rainbow sheen on the water surface. Most of the subsurface oil was in the mid-intertidal zone—well below the bathtub ring, the visual stain in the upper intertidal area, and directly within the richly productive biological region.

Using forensic chemistry, Short analyzed dozens of typical sediment samples and determined that 90 percent of the surface oil and 100 percent of the subsurface oil was from the *Exxon Valdez* (Short et al. 2004). The remaining surface oil was from the Monterey (California) Formation—heating oil spilled during the 1964 earthquake when storage tanks ruptured in Valdez. Rice and his team estimated the total beach area contaminated by residual *Exxon Valdez* oil, counting both surface and subsurface deposits, was twenty-eight acres (ibid.). They reasoned this was a low-end estimate; it did not include the lower intertidal zone where they had not sampled very extensively because they had not expected to find oil there. Instead they found this was where more of the buried oil was located. They conservatively estimated the weight of the intertidal residual oil was over 56 tons (122,320 pounds), but felt a more realistic number was probably twice that (ibid.).

While this may not seem like much in the greater scheme of things, anyone who has struggled with cancer knows it doesn't take much to threaten life. Further, the subsurface oil is harbored in the biological equivalent of a critical organ in the Sound—marshes and gravel beaches. As Phil Mundy, the science director of the EVOS Trustee Council told Meg McKinney in an April 2004 interview on KCHU public radio, these areas are relatively rare in the steep, rocky-walled fjord system and they are critical habitat for wildlife.

The stunned team mulled over their discovery. How could hundreds of scientists completely have missed the mother lode of oil for twelve years? This was not oil that had migrated down slope from the upper to the lower intertidal over time—there was too much of it and it was consistently there, buried in the middle intertidal zone, regardless of whether beaches were wide or narrow, steep or shallow. They finally decided that the subsurface oil in the lower beaches had simply been overlooked, mainly because of where it was. The lower intertidal zones were covered by seawater more often than the upper-intertidal zone so there was less time to sample these elevations. The rockweed and kelp that blanketed the lower beaches made walking treacherous and obscured surface oil from view. Rice and his team realized that scientists simply had gravitated to study the oil they could see in the areas that were easier to access.

The team also wondered why the buried oil wasn't mostly beneath the surface oil in the upper intertidal zone as everyone had assumed. They knew much of the oil had initially stranded during the ferocious three-day storm that occurred three days after the oil spilled in March 1989. They also knew from all their work in the Sound during 1989 that the stranding was not a one-time event, but rather a series of repeated coatings over many tidal cycles, for up to two months in the most heavily oiled bays. Rice's team realized now that gravity, capillary action, and months of daily tides had pulled the oil into the beach substrate during low tides, where it got stuck by capillary action, literally held in place by surface tension between the liquid oil and solid substrate. The buried oil was only partially removed during high tides. There, in the middle intertidal zone, significant quantities of oil had accumulated largely out of sight.

Rice and his team realized, over a decade after the spill, that *everyone had assumed wrong*: the long-term presence of surface oil was in fact a poor indicator of buried oil! Given the liquid nature of this buried oil, they realized that everyone had probably also incorrectly assumed that oil remaining on beaches would not cause biological harm.

As part of their study in 2001, the team sampled a variety of intertidal sealife to test if the buried oil was bioavailable; that is, whether prey species were picking up the residual oil. The team found *Exxon Valdez* oil in everything they sampled—clams, mussels, dog and spindle whelks, ribbon worms, hermit crabs, and even gunnels, a small fish often consumed by pigeon guillemots (M. Lindeberg, researcher, NOAA/NMFS ABL, Juneau, AK, pers. comm., July 2003).

Surprised, Rice and his team realized they had found the key to account for the lingering harm observed in wildlife that lived, spawned, or foraged lower in the intertidal zone. Oil can be both long-lasting *and* bioavailable. This understanding completely shattered old scientific paradigms built on beliefs of short-term effects from oil.

Rice and his team extended their study in 2002 and 2003 to evaluate the biological availability of the buried oil. Their findings

validated results from their first field season. Buried pockets of
residual oil are still contaminating Prince William Sound's wildlife.
These findings support the evidence of persistent harm documented in such wildlife as harlequin ducks and sea otters. Scientific
papers from the most recent seasons were in preparation at the
time this book went to press.

Background Oil—Much Ado about Nothing

While Rice and his team were busy documenting residual *Exxon
Valdez* oil on the beaches in Prince William Sound, Exxon scientists
were looking very hard to find something other than Exxon's oil to
explain the mounting evidence of lingering oil effects on the Sound's
wildlife. Exxon scientists stayed the course that they had started
when the oil first spilled: that is, they claimed there was other oil in
the Sound back then, lots of other oil, and they published a series of
papers to this effect as discussed below.

In response to these papers, Short from the Auke Bay Lab teamed
up with scientists from the USGS in California to identify the source
of the Hinchinbrook hydrocarbons that they had first observed in
their baseline surveys over twenty years earlier (Chapter 11). The
bone of contention was whether these background hydrocarbons
were bioavailable. Hydrocarbons in oils are bioavailable and a source
of contamination for sealife; hydrocarbons in coals and hydrocarbon-
rich shales or "*source rocks,*" the geological precursors of oil
deposits, are not.

Over a period of five years (1996 to 2000), Short and his colleagues collected a series of sediment samples from Prince William
Sound east to Yakutat Bay. The samples included deep-water marine
sediments from the Sound and Gulf of Alaska (Short and Heintz
2003), terrestrial sediments from rivers and lake beds near known
oil seeps, outcrops of source rocks, and coal fields on the North
Gulf coast, and archived samples from oil-seep source rocks and
coal fields in the region (Van Kooten, Short, and Kolak 2002). They
also collected samples of juvenile coho salmon and mussels from
the Katalla oil-seep area to determine if this seep oil was bioavailable (Short et al. 1999). While collecting these samples, Short discovered that oil seeps at Katalla and later near Cape Yakataga were
actually nothing more than small slow trickles of oil. Exxon scientists had claimed, in papers and public speeches, that these seeps

were substantial sources of bioavailable hydrocarbons, blanketing the seafloor of the northern Gulf of Alaska and Sound!

Exxon scientists initially emphasized the Katalla oil seeps as the primary source of the seafloor hydrocarbons (Page et al. 1995a, 1996, 1997). They backed off this claim after Short and his colleagues demonstrated that the Katalla oil didn't even make it to the mouth of the Katalla River, less than six miles distant from the source, let alone to the bottom of Prince William Sound (Short et al. 1999). Then Exxon scientists emphasized the more distant oil seeps at Cape Yakataga (Boehm et al. 2000; Page et al. 1998). Short and his colleagues reported oil flow from these seeps was extremely low, at well under a barrel a day (Becker and Manen 1989); they found it was not nearly enough to be detectable in the sediments throughout the northern Gulf of Alaska (Short et al. 2000).

The crucial point, consistently avoided by Exxon scientists, but emphasized by Short and his colleagues, was how any bioavailable source of hydrocarbons could get into the marine sediments of the northern Gulf of Alaska without being chemically altered. Hydrocarbons either are bioavailable or they are not: ones that are, are bioavailable *immediately* from the point and time of release into the environment. Oil in the environment is attacked by bacteria and broken down in ways that produce characteristic chemical profiles. This proven concept was the basis for Alyeska's biological treatment system at the tanker terminal, discussed in Chapter 1, and for Exxon's bioremediation program during the oil spill cleanup, discussed elsewhere in this book.

Short and his colleagues showed that the hydrocarbons in the seafloor sediments of the Gulf of Alaska and Sound are strictly preserved—locked into a rock matrix—even the highly soluble ones (Short et al. 2004). This proved that these hydrocarbons are not bioavailable and that the source of these hydrocarbons is not oil seeps. These findings also confirmed Short's theory from twenty years earlier that the source of the low-grade contamination found during the baseline surveys along beaches in Hinchinbrook Entrance was coal, not oil (Chapter 11). This is also true for the hydrocarbons littering the Sound's seafloor. This means that, despite Exxon's filibustering, there is one dominant source of bioavailable oil on the beaches of Prince William Sound—*Exxon Valdez* oil—and this oil is the culprit for the persistent biological harm observed in the ecosystem studies. (This is patently clear from all of the studies examined and cited in this book.)

Exxon's Response
Ecological Recovery and Semantical Nonsense

In 1998 Exxon scientists re-sampled some of the sites from their initial assessment program and found, essentially, fewer animals and species at the heavily oiled sites (Page et al. 1999b). So they presented a new definition of recovery—"parallelism"—under which they proclaimed that sites had recovered as long as *the trends, not the absolute values,* in community structure and species abundance and diversity paralleled similar trends at the reference sites. Exxon scientists argued that "lack of convergence [at the oil sites] is not a lingering effect of the oil spill" (ibid., 124) They pointed out the PAHs had decreased over time. They reasoned that if there had been an oil spill effect, differences in the number and types of species between oiled and unoiled beaches "would have converged as the TPAH [total PAHs] disappeared" (ibid., 124). Under this new definition, Exxon scientists boldly claimed that the oiled beaches had indeed recovered.

Fishermen in Cordova saw it another way. "If the price of a barrel of oil dropped by half, then stabilized as a parallel trend with the prices before the drop, I bet the oil industry wouldn't consider the lower price to have 'recovered'," remarked fisherman, Danny Carpenter, dryly.

The NOAA database—the work of Jon Houghton and his colleagues—was particularly troubling to Exxon. For Exxon's new definition of recovery to work, Exxon scientists had to ignore the waves of ecological succession, documented by Houghton and his team members. Under Exxon's personally minted notion of parallelism, black could become white—at least on paper. The new group of NOAA contractors that replaced Houghton's team adopted Exxon's new definition of recovery (Hoff and Shigenka 1999, 112). They completely reanalyzed Houghton's eight-year database, added two more years of sampling, and concluded under parallelism that the beaches actually had "recovered" within one to two years after the spill.

Pete Peterson was among the many coastal ecologists who strongly criticized the theory of "parallelism." Peterson (2000) argued, "You could have only 10 percent of the animals and plants on the shoreline that you would have expected in the absence of the spill, and you would call that recovered." The scientific community eventually rejected Exxon scientists' theory of parallelism, but well after it had served its real purpose of creating controversy and muddying the clear picture of lingering oil spill effects in the Sound.

Looking for Oil in All the Wrong Places

Prince William Sound offered the world an opportunity to witness ecological impacts from Exxon's spill, because Exxon's oil was the overwhelmingly dominant source of pollution along the Sound's shorelines. This was a unique opportunity, in the chronicles of the planet, because the Sound was essentially devoid of other pollutants, yet it was not so remote as to preclude extensive monitoring of impacts and recovery—and, significantly, there was funding for long-term studies.

The key to realizing impacts from Exxon's spill was the *absence* of other petroleum hydrocarbons in the intertidal zone, specifically hydrocarbons that could be picked up by marine life. To confuse the stark findings of persistent harm from the spill, Exxon scientists repeatedly challenged the notion that Exxon's oil was the only source of pollution in the Sound (Bence and Burns 1995; Page et al. 1995a, 1999a). "Remnants of asphalt spilled during the 1964 earthquake ... [Kvenvolden et al. 1993], organic particles from shales, seep oil residues, coal, products of incomplete fuel combustion, soot, forest fire fallout, atmospheric fallout, and boat fuel/oil residues are potential sources of the PAH studied here" (Boehm et al. 2001, 473).

Exxon scientists looked high and low for other sources of oil. They found what they were looking for "low" in the marine sediments on the seafloor of the Sound—inert coal. Then they proceeded to make "much ado about practically nothing," as Huff (1954, 53), author of *How to Lie with Statistics*, aptly described such situations. It was ludicrous from the outset to try to shift the source of the Sound's woes from millions of gallons of fresh toxic crude oil that stranded overnight in a highly productive biological zone to miniscule quantities of hydrocarbons that had accumulated in deep-water sediments over thousands of years. But try Exxon did.

First, Exxon scientists made it seem that there were a lot of Gulf of Alaska hydrocarbons compared to very little spill hydrocarbons. This was relatively easy to do simply by looking in the wrong place for *Exxon Valdez* oil. Exxon's oil largely stayed in intertidal areas, while the Gulf of Alaska hydrocarbons largely collected in deep-water benthic sediments. So Exxon scientists looked for *Exxon Valdez* hydrocarbons on the Sound's seafloor where, not surprisingly, they found very little spill oil (Page et al. 1995a). They did, however, find some Gulf of Alaska hydrocarbons (coal dust), but only by counting everything that had drifted down and accumulated over the last 160 years or so (Page et al. 1996). This led them to state, "In those areas where they were detected, spill hydrocarbons were generally a small increment to the natural petroleum hydrocarbon background" (Page et al. 1995a, 42; see similar statements in Page et al. 1996, 1266, 1277).

Next, Exxon scientists set about "proving" that the PAHs in Gulf of Alaska hydrocarbons were bioavailable and, therefore, a potential threat to sealife. This meant that the Gulf of Alaska hydrocarbons had to originate from *oil* not coal or source rocks, because the PAHs in coal or source rocks are not bioavailable—they were locked inside the rock matrix. Initially, Exxon scientists looked no further east than Katalla seep-oil, which, they claimed, was the source of the Gulf of Alaska hydrocarbons in the Sound. When Short and others from the Auke Bay Lab pointed out there is a "substantial coal field" to the east of Katalla (Short and Heintz 1998, 1651), Exxon scientists claimed that those Alaska coal maps were "previously unavailable" to them (Page et al. 1998, 1651). When Short and others pointed out the chemical profiles that Exxon scientists used to identify seep-oil were "inadequate to distinguish seep-oil from coal" (Short et al. 1999, 41), Exxon scientists claimed, essentially, that the Katalla-area coal fields were too old and weathered to yield chemical signatures that fit their profile of PAHs in Gulf of Alaska sediments (Boehm et al. 2000, 2064).

The problem, of course, lay in the details of Exxon's chemical portfolio or library, which was used to identify hydrocarbon signatures. Their portfolio was incomplete. It did not include sediment samples from dominant sources of PAHs in Gulf of Alaska sediments such as the coal fields near Yakutat Bay or source rocks beneath the Malaspina Glacier and other glaciers (Boehm et al. 2001). Exxon scientists also *excluded* from their portfolio aromatic hydrocarbons with a chemical profile that would have eliminated oil seeps as significant contributors of Gulf of Alaska hydrocarbons. Huff (1954, 126) warns, "Watch out for evidence of a biased sample, one that has been selected improperly . . . ." Short and his colleagues (2004) pointed out that Exxon's models could not accurately identify key sources of PAHs in Gulf of Alaska sediments when the key sources and key chemical indicators were left out of the portfolio!

Huff encourages people to ask, "What's missing?" He notes that the absence of critical information "is enough to throw suspicion on the whole thing" (127). In this case, the younger and larger coal fields near Cape Yakataga and the source rocks beneath glaciers were missing from Exxon's oil portfolios. Short and his co-workers found that these omitted sources had hydrocarbon profiles that closely matched that of the hydrocarbons in the Sound's seafloor.

Readers may well wonder why such a seemingly minor issue assumed such epic proportions. The liability (and credibility) stakes in the oil versus coal controversy were extraordinarily high. Exxon was liable for up to $100 million for "unanticipated long-term harm" to wildlife and habitat in a clause tucked into the 1991 civil settlement. This liability wrinkle made it well worth Exxon's while to try to pin the blame for some of the lingering harm from the oil spill on other culprits. The Katalla seep-oil became the unlikely, but handy, candidate. The heated controversy attracted media attention, distracting the public from the residual *Exxon Valdez* oil in the beaches. The public relations battle over the coal-versus-source-rocks-versus-oil controversy was in full swing as this book went to press; the political ramifications are discussed in Chapter 24.

Sound Truth

In 2003, the last year covered by this book, *Exxon Valdez* oil was still present in Prince William Sound, mostly buried in the lower intertidal reaches of dozens of bays and beaches that had been heavily oiled in 1989. This residual oil still poses a potential threat to wildlife that spawns, forages, or lives on or near the contaminated beaches.

The Sound has spoken its truth. Will the world have the wisdom and courage to listen?

2004 Status of the Sound

Chapter 22.

Sound Truth

This part of the collective story, shared by dozens of individuals, ends with recovery—or at least with the status of injured species that are the subject of this book. Before a tidy summary of this status can be made from literally over 500 scientific papers spanning fourteen years of research, it is necessary first to define recovery and second to get an 'Eagle View,' as the Alaska Natives say, in order to keep things in perspective.

By "recovery," I mean when *all parts* of the ecosystem are functioning as they would have been absent the spill *and* when species have replaced individuals lost to the spill *in areas where the individuals were lost.* The EVOS Trustee Council and public-trust scientists used the federal NRDA regulations' definition for recovery under Superfund: "The length of time required to return the services of the injured resource to their baseline conditions" (U. S. Code for Federal Regulations, 1987, 11.60–11.73). Exxon scientists adopted various definitions depending on their needs (Page et al. 1995b, 1999b). I believe a new definition is in order in light of the new findings of long-term harm from oil.

The first part of the new definition assumes that it is possible to separate out ecosystem response to different disturbances; for example, the oil spill versus the climate regime shift. The second part of this definition makes it impossible to declare, for example, sea otters as "recovered" when they are still showing lingering harm in the most heavily oiled areas. Ignoring the lack of recovery in oiled areas is the equivalent of declaring the world's human population "recovered" from AIDS, while ignoring the few heavily affected areas. In either case, heavily affected areas should not be treated as anomalies—they are the barometers of true and full recovery.

Regarding the 'Eagle View,' the Natives reference to the big picture perspective, it is important to keep in mind that beaches in the Sound were contaminated unevenly. During the initial oiling, beaches, bays, and fjords that faced northwest to northeast acted like baseball mitts, catching the oil as it swept down from Bligh Reef. Beaches and bays that faced southwest to southeast were lightly oiled or even unoiled. Recovery of habitat and wildlife was also uneven—it mirrored degree of oiling. Habitat and wildlife in lightly oiled areas recovered faster than habitat and wildlife in heavily oiled areas, where wildlife are still experiencing persistent oil effects after fifteen years.

With these two items in mind, let us now zoom in on the 'Mouse View,' the Natives' way of referring to details (Table 3, p. 376). The details are the status of recovery of habitat and individual species of wildlife in the Sound, and the nature of the lingering harm, as presented by the EVOS Trustee Council in its 2003 Status Report.

1–4 Years Post-Spill

No species discussed in this book recovered within the first four years after the *Exxon Valdez* oil spill. (According to the EVOS Trustee Council, only the bald eagle recovered within this time period.)

5–9 Years Post-Spill

Several species recovered within five to nine years of the spill, including pink salmon, river otters, and black oystercatchers. Reasons for the delayed recovery included persistent oil effects from **oiled habitat** in all three species; and **oiled food** for black oystercatchers, which eat mostly invertebrates, and river otters, which supplement their fish diet with some invertebrates.

For **pink salmon**, oiled habitat alone was the source of delayed recovery. Buried residual oil, mobilized by tides and storms, drained downstream through salmon redds (nests) and was picked up by eggs and developing embryos, which bioconcentrated PAHs to many times over the level in the water incubating the redds. Exposure to even infinitesimal levels of PAHs at this critical life stage in pink salmon killed eggs, deformed embryos, stunted growth and development of surviving juveniles, and ultimately led to fewer surviving adults. It wasn't until 1994, five years after the spill, that the PAHs associated with the residual oil dissipated to the point where scientists could no longer measure lingering harm to salmon eggs. And the truly frightening thing was: this all happened at PAH levels many

Table 3
Status of Recovery as of 2003
(for habitat and wildlife discussed in this book)

Time	Species	Status	Reason for Delayed Recovery
1–4 years	None		
5–9 years	Pink salmon	Recovered	Oiled habitat (buried oil)
	Black oystercatchers	Recovered	Oiled habitat; oiled food
	River otters	Recovered	Oiled habitat; (some) oiled food
14+ Years	Mussel beds	Recovering	Oiled habitat
	Beach communities	Recovering	Oiled habitat; cleanup effects
	Pacific herring	Not recovered	Oiled habitat
	Sea otters	Recovering	Oiled habitat; oiled food
	Murres	Recovered	Loss of forage fish
	Harlequin ducks	Not recovered	Oiled habitat; oiled food
	Pigeon guillemots	Not recovered	Oiled habitat; loss of forage fish
	Harbor seals	Not recovered	Loss of forage fish
	Orcas, fish-eating	Recovering	Disrupted social behavior; low reproductive rates
	Orcas, mammal-eating	Not recovered	Contaminants (PCBs); zero reproductive rates; loss of harbor seals

Source: EVOS Trustee Council 2003 Status Report. For full list of species studied by public-trust scientists see www.oilspill.state.ak.us

Economic Impacts of the Disrupted Salmon Fisheries

The Prince William Sound salmon fisheries fuel Cordova's economy—nearly half of the community's residents (population 2,500) are directly employed in the commercial fisheries and the revenue from fish harvests supports the local businesses (Grabacki 1998). The *Exxon Valdez* oil spill occurred just as the Prince William Sound salmon ranching program was reaching peak production (Sidebar 13, p. 290); fishermen never realized the program's promise of stable income and shared wealth.

Instead salmon fisheries were thoroughly disrupted for five years:

1) in 1989 due to spill-related closures and reduced market value from public perception of oiled fish (affected all salmon species) (The *Exxon Valdez* Case);

2) in 1990 due to 25 percent reduced return of pink salmon from oil effects (Geiger et al. 1996; Willette 1996);

3) in 1991 due to disrupted (unprecedented) run timing of pink salmon (Chapter 16);

4) in 1992 and 1993 due to stock collapses of pink salmon (Chapters 16 and 20).

These most likely spill-related disruptions, primarily to pink salmon, translated to loss of millions of dollars for Prince William Sound fishermen and the Cordova community. Using pink salmon as an example, the Sound's largest salmon fishery and the bread and butter of Cordova—the seine fishery—had a ten-year average pre-spill harvest value of $21 million; the five-year average post-spill value (1989–1993) was $9 million (ADFG 2002). This amounts to an annual loss of $12 million for five years—just from reduced pink salmon seine harvests. Once highly-prized Prince William Sound limited entry permits for the salmon seine fishery dropped from a high market value of $310,000 in 1989 to less than $20,000 in 2003 (Alaska Commercial Fisheries Entry Commission 2004). Many fishermen who bought their permits before or just after the spill have faced huge debt for permit loans that the fishery no longer supports.

Cordova depends on its commercial fisheries to generate "fish bucks," which bounce their way through the local economy as they are spent on groceries, fuel, clothing, services, labor, and local and state taxes, among other things (Impact Assessment, Inc. 1990). Studies by the Pacific Sea Grant College Program and the Oregon State University calculated "output (sales) multipliers" to estimate the economic activity generated from fish bucks (and "tree bucks," "tourism bucks," etc.) in small rural communities (Radtke, Dewees, and Smith 1981). The output multiplier value for commercial fishing ranged from 1.5 to 2.5. Applying the median value (2.0) to the economic model from the studies, and accounting for the processors' margin (1.5), I estimated the lost income from the pink salmon fisheries alone translated to a loss of economic activity to Cordova of nearly $36 million per year for these five years.

In a small town with one primary business—commercial fishing, this loss is keenly felt by everyone. For example, the city and state share a 3 percent "raw fish tax," which generated about 15 percent of the city's general operating revenue before the oil spill. Disrupted fisheries (including herring, see Sidebar 15, p. 380) meant less revenue to the city, which made up the shortfall by increasing the cost of services (garbage collection, dock fees, etc.), property taxes, and sales taxes. This created, essentially, a secondary spill disaster, because many of the businesses in town were owed money from fishermen who had no money to give. Disrupted fisheries thus contributed to what sociologists call, "a downward economic spiral," from which the town has yet to recover. Although the pink salmon fisheries recovered in volume after 1993, the value of the harvest remains depressed due to global market changes from introduction of farmed fish.

times lower than the levels currently thought to be "safe" for fish! (Sidebar 14, p. 376)

Pink salmon populations crashed in the Sound in 1992 and 1993—and nowhere else in the state. The Auke Bay Lab tests suggest that the pink salmon run collapses in the Sound may have been related to oil exposure of early life stages of the parent stocks.

For **black oystercatchers**, oiled habitat and oiled food delayed recovery until about eight years after the spill. Oystercatchers live, breed, and forage in the intertidal zone and so were vulnerable to persistent effects from residual oil along with other wildlife with similar habits. Oystercatchers were physically oiled when they uncovered residual pockets of oil by digging, poking, or just moving in the intertidal zone. They also ate oiled food: invertebrates from heavily oiled areas soaked up and concentrated subsurface oil and passed the poison on to the predators that ate them. Chicks fed oiled mussels from oiled beaches grew more slowly and fledged later (decreasing their chances of survival) than chicks from unoiled beaches. Wildlife that consumed oiled invertebrates, such as mussels, clams, snails, and crabs, generally took longer to recover than wildlife that preyed upon fish.

For **river otters**, oiled habitat and, to a lesser extent, oiled food also delayed recovery until about eight years after the spill. Many river otters in the Sound live and breed in old-coastal forest immediately adjacent to beaches. They encountered oil as they traversed the intertidal zone during their commutes to the sea, and also when they foraged for invertebrates, which they eat to supplement their fish diet. Fish, unlike invertebrates, break down and metabolize toxic PAHs. So wildlife such as river otters that ate predominately fish—even in heavily oiled areas—were, in general, consuming prey that was relatively free of oil, especially compared to invertebrates in the same oiled areas.

10–14 Years Post-Spill

The EVOS Trustee Council listed **murre colonies** in the spill path as "recovered" in 2002. According to John Piatt, the *census counts* and *reproductive success (productivity)* of the large colonies at the Barren Islands and Semidis Islands are, indeed, similar to pre-spill numbers, but not so at several smaller colonies such as in Puale Bay (Dragoo, Byrd, and Irons 2003). Piatt, however, prefers to use a different measure of recovery: that of demographics, not just numbers. He points out that a stable age distribution with a spread of different age classes may be important for a species like murres where successful *survival of young* (not just productivity) depends on experienced breeders (pers. comm., 26 February 2004). By his definition, the murres are still "recovering," and this process could take twenty to seventy years, exactly as he originally predicted. The climate in the North Pacific has shifted, once again, back to a cool regime that supports shrimp, capelin—and seabirds. Piatt hopes this will hasten recovery.

15+ Years, Recovering

Habitat and wildlife showing positive trends towards recovery include mussel beds, beach animals and sea plants, sea otters, murres, and fish-eating orcas. Reasons for the delayed recovery included persistent oil effects from **oiled habitat** and **oiled food; cleanup effects;** and **loss of forage fish**.

Mussel beds and **intertidal beach communities** are recovering as the buried oil slowly redistributes and diminishes with time. The sea plants and animals on beaches that were treated by the hydraulic water washes are also slowly recovering as former communities are re-established through recruitment and cycles of succession.

Sea otters show signs of recovery that mirrors the degree and recovery of oiled habitat: these animals have recovered fully in lightly oiled areas, but not at all in some of the most heavily oiled areas of the Sound. Sea otters are vulnerable to persistent oil effects through encounters with buried pockets of residual oil while foraging and consuming oiled prey.

The **resident AB pod of fish-eating orcas** lost thirteen of its thirty-six members in the two years following the spill and was observed to have extremely disrupted social behavior for several years thereafter. The pod remains split into two subgroups, one of which is seen only infrequently. Naturally low reproductive rates require years to replace large losses; only half of the whales lost since 1989 have been replaced. In contrast, the Gulf of Alaska population of resident whales has increased since the spill.

(Author's note: Many people wrote comments or testified against *the EVOS Trustee Council's decision to list the AB pod of*

orcas as "recovering" when, in fact, only half the whales lost have been replaced. At least this trend is hopeful and only time, not politics, will dictate the true status of recovery.)

15+ Years, Not Recovering

Wildlife not recovering from the spill include harlequin ducks, pigeon guillemots, harbor seals, and mammal-eating orcas. Reasons for the delayed recovery are varied and directly or indirectly related to oil. Spill-related reasons include **oiled habitat, oiled food,** and **loss of prey species**.

Over-wintering adult female **harlequin ducks** continue to have lower survival rates in oiled versus unoiled beaches. Harlequins forage for invertebrates in the intertidal zone and shallow nearshore seas. Harlequin ducks are vulnerable to persistent oil effects through encounters with residual pockets of oil while foraging and eating oiled food.

The Sound's **herring** population has had problems since the spill. At a minimum, oil exposure in 1989 killed lipid-rich eggs, incubating along oiled beaches; maimed and killed embryos adrift in surface waters; and reduced fertility in survivors of the 1989-year class. PAH exposure also may have wreaked havoc with the immune system of surviving 1989 year-class and adults, making them more susceptible to diseases (Marty et al. 1999). The herring stocks collapsed in Prince William Sound and nowhere else in the state in 1993, the year that survivors of the ill-fated 1989-year class matured and joined the adult stocks. Viral outbreaks decimated the Sound's remaining herring stocks again in 1998 and 2001 (Gary Marty, pers. comm., 1 March 2004).

However, herring *may* be starting to come back; in fall 2003 nearly 80 percent of the stocks were composed of three-year-old fish. *If* these juveniles successfully recruit into the adult stocks in 2004, without triggering another disease outbreak, they could finally start to rebuild the Sound's herring population, currently at 30,000 tons—less than one-quarter of the pre-spill population (ADFG biologist Steve Moffitt, Cordova, AK, pers. comm., 2 March 2004). Only time will tell whether this is wishful thinking or the Sound's newest truth (Sidebar 15, p. 380).

Pigeon guillemots have not been able to replace the individuals lost during the spill. Adults forage for fish in shallow nearshore seas and they supplement their diet with invertebrates. Evidence suggests that adults are exposed to persistent oil effects through oiled habitat and oiled food (invertebrates). Young are fed only fish and

Economic Impacts of the Lost Herring Fisheries

Loss of the Sound's lucrative herring fisheries has affected fishermen, families, and the Cordova community. Herring fisheries included sac roe and spawn-on-kelp fisheries in spring and food-and-bait fisheries in fall. These fisheries provided hundreds of jobs in harvesting, processing, and exporting. For example, there were roughly 100 limited-entry permits for purse seine sac roe fishery (ADFG 1998, Appendix H.4). Each of these permit holders had usually four crew members. Harvests were processed at local canneries and exported. The timing of these jobs was also critical: the spring fisheries generated the first big revenue after a long winter and the fall fisheries provided income primarily to local residents.

The herring fisheries were first closed in 1989 because of the oil spill; they were closed again from 1993 to 1996 after a virus outbreak decimated the stocks (Marty et al. 1998, 2003). There were very small harvests in 1997 and 1998. The fisheries have been closed since 1999 to allow the stocks to recover.

The ten-year average annual harvest value for all fisheries combined from 1982 to 1992 was $16.3 million (ADFG 2002, Appendix H.13; Cohen 1997, Table 9.3), but the loss to the community was much more. Applying the economic model explained elsewhere (Sidebar 14, p. 376), I estimated the lost income from the herring fisheries translates to a loss of economic activity to Cordova of nearly $50 million per year.

Of course the fishermen and their families were the hardest hit by the loss of the herring fisheries. Financially, many herring fishermen were devastated by the loss of income and rapid devaluation of limited-entry permit values. Limited-entry permits are like stock: the value reflects the strength of the fishery. According to Alaska's Commercial Fisheries Entry Commission (2004), the estimated permit value of Prince William Sound sac roe purse seine fisheries plunged from a high of $245,000 in 1989 to a low of $21,160 in 2003 and the price is still dropping. Fishermen who bought in when the permits were high have been trapped into escalating debt on permit loans they cannot pay because of the closed herring fisheries and general low prices in other fisheries. A growing number of these fishermen are giving their anticipated share of the punitive damage award from the consolidated lawsuit against Exxon to the State of Alaska as a lien in order to obtain debt relief and restart their lives (Platt 2002). This financial havoc has also taken an emotional toll, which has separate economic costs (Picou and Gill 1997; Rodin et al. 1997).

Prince William Sound herring fishermen are asking the parties to the 1991 civil settlement to reopen it and use part of the $100 million available for unanticipated long-term damages to buy-back herring permits to give the fishermen—and the fish—a chance to recover. Fifteen years ago, Exxon spokesperson Don Cornett told the Cordova community, "You have had some good luck and you don't realize it. You have Exxon and we do business straight. We will make you whole!" (Mullins 1994). Fulfilling this promise is long overdue show no signs of oil exposure, however, they are showing signs of food stress throughout the Sound. Pigeon guillemot populations were in decline before the spill, possibly because of loss of high-lipid forage fish due to the climate regime shift. It is possible that additional losses of herring (and possibly sand lance) after the spill exacerbated this problem.

Harbor seals are listed as not recovering, primarily because they have not been able to replace the individuals lost during the spill. Like many species of piscivorous (fish-eating) marine birds, seal populations were declining before the spill due to loss of energy-rich forage fish from the climate regime shift. Although seal pups in the Sound appear healthy, it is possible that loss of high-fat forage fish (such as herring) after the spill may be affecting survival of other ages of seals.

The **transient AT1 group of mammal-eating orcas** is not recovering, primarily because they have not been able to replace individuals lost from the spill. This group lost 11 of its 21 members in the two years following the spill. In contrast to the fish-eating orcas, this group of transient whales has not produced any new calves since 1984. This reproductive failure is most likely caused by high levels of PCBs stored in the whales' blubber. PCBs are known to disrupt reproductive hormones in mammals. Reproductive failure may also be related to loss of their preferred prey, harbor seals; the harbor seal population in the Gulf of Alaska has declined by 90 percent during the past twenty years. According to whale researcher Craig Matkin (pers. comm., 26 February 2004), these whales have little chance of recovery, largely because of the contaminant problem. NMFS is considering a proposal to list this group as depleted under the Marine Mammal Protection Act.

Indirect Effects

Besides the direct effects of oiled habitat and oiled food on injured species, there were **indirect effects** of the oil spill that also caused injury and delayed recovery. Two examples stand out most clearly in the dynamic Prince William Sound ecosystem. First, as a direct effect of the spill, beaches scoured clean of sea plants and animals by Exxon's pressurized hot water wash took far longer to recover than beaches left untouched. Treated beaches endured waves of succession as freshly recruited single-age stands of sea plants died en masse

every three to four years, literally uprooting the shelter and food for
beach-dwelling animal life. These violent die-offs also created indirect effects on wildlife like young salmon and herring that used
these areas for protected nurseries and on predators that foraged
along these beaches. The indirect effects diminished as the treated
beaches slowly regained a stable and diverse assemblage of sea
plants and animals.

Second, the spill-related loss of herring wreaked havoc in the
Sound's ecosystem. Herring are a keystone species in the food web
where over forty species of fish, birds, and mammals rely on these
lipid-rich, surface-schooling fish to feed their young and to build
blubber or fat reserves for critical times of the year. Estimated loss of
half of the young-of-the-year herring in 1989, few surviving one-year
old herring in 1990, and generally smaller size year classes since the
spill created hardships for a variety of species.

For example, **black-legged kittiwakes** in the Sound were not
injured initially by the spill and their population was not in decline
before the spill, unlike many other piscivorous birds. The distribution
of kittiwakes, however, had shifted before the spill to allow colonies
better access to herring nurseries in response to loss of other lipid-
rich forage fish. With the decline in herring abundance since the spill
(first young herring, then adults), kittiwake colonies that did not have
access to alternative prey started to decline. Other colonies switched
to pink salmon fry and other less preferred forage fish as alternative
prey for their young. Thus, loss of herring also caused indirect effects
to **pink salmon**.

For well over a decade after the spill, herring failed to reproduce
successfully and the population was still plagued by devastating disease outbreaks. Until the herring population has fully recovered,
there will continue to be indirect effects on many other species.

Recovery Unknown

These neat and tidy classifications of the status of recovery belie the
fact that the overwhelming majority of creatures in the Sound fall
into the final category of "recovery unknown." For most wildlife,
there simply was not enough known about their habits, population,
health, or spill injury, if any, to even hazard the faintest guess at recovery status. For this reason, **subtidal seafloor communities** of
plants and animals were finally moved into this category—there was

not sufficient baseline or pre-spill information to determine spill effects or recovery. Other examples include **shellfish**; the initial spill may very well have wiped out the entire year-class of larval shellfish—shrimp, crabs, scallops, and their kin—as well as noncommercial but important **forage species** adrift in the water column with the larval herring. The spill most likely devastated **sand lance**, intertidal fish that are known to be sensitive to oil; residual oil likely delayed recovery of this important forage fish in heavily oiled areas. Spill effects on **capelin**, another important forage fish, are also largely unknown.

Understanding the natural history of the species in an ecosystem and the complex relationships among the species is of utmost importance to being able to monitor ecosystem health and detect impacts from oil spills, climate change, contaminants, and other threats. The only way to overcome our gaping knowledge is to embrace a more comprehensive and holistic approach to monitoring and management. Instead of just doing "ecosystem studies," as the fishermen of Prince William Sound demanded in 1993, we need to do "long-term ecosystem monitoring." This approach and the new understanding of oil's harmful effects to life are legacies of the *Exxon Valdez* oil spill and are discussed in Chapter 23.

Part 3
The Legacy and Beyond

"We have been slow to realize we are all in harm's way."

Daniel Yergin
The Prize: The Epic Quest for Oil, Money & Power

Chapter 23.

The Legacy of the Exxon Valdez Oil Spill: Emerging Science and Policy

One of the scientists whom I interviewed for this book warned, "Our job is not done." Many people who shared their personal stories with me expressed this sense of unfinished business. They feel it was not enough to simply understand or experience how the oil spill caused lingering harm to ecosystems and individuals for fifteen years. They believe either there are fundamental wrongs that need to be fixed to prevent history from repeating itself, or there are fundamental truths that need to be told to improve our quality of life on this planet.

The last two chapters of this book are about the unfinished business of conveying these lessons and recommendations to the public—the stewards of wildlife and wild lands. In this final section, the responsibility shifts from us, the messengers, to you, the reader, because although this book ends, the story of oil will continue for a while longer. We all, especially people in developed nations, will have a role in crafting the final chapter on oil history. As you read 'The Legacy' of the *Exxon Valdez* oil spill, please tuck this question in the back of your mind: what would have happened if no one had listened to Paul Revere?

Shifting Paradigms: Oil Causes Persistent Harmful Effects

A scientist's view of the natural world is held as a paradigm—an understanding, based on theories, studies, models, and other generalizations, which seeks to accurately reflect and explain observations from the natural world. Paradigms are dynamic, not static; that is, they shift to accommodate new observations, and when paradigms shift, science advances. For example, a scientific paradigm once held that the world was flat, but we no longer believe this.

It should not be surprising in this day of rapid technological advancements that, in the time span of one generation or thirty years, since most of our protective standards were first established, our understanding of the effects of oil on humans and wildlife have radically evolved.

In 1999, the U.S. EPA identified twenty-two PAHs as *"persistent, bioaccumulative, and toxic (PBT) pollutants"* (U.S. EPA 2001). PBT pollutants are 'the worst of the worst' known human health hazards—the list includes mercury, dioxin, PCBs, DDT, and now PAHs. The EPA (2000) states that PBT pollutants are "highly toxic, long-lasting substances that can build up in the food chain to levels that are harmful to humans and ecosystems." In other words, they are persistent *and* they are bioavailable; that is they are able to spread throughout the ecosystem. Further, "PBTs are associated with a range of adverse human health effects, including the nervous system, reproductive and developmental problems, cancer, and genetic impacts. Reducing risks from PBTs presents a challenge ... because of the pollutants' ability to travel long distances, move easily from air to water or land and linger for generations in people and the environment."

This listing of PAHs on the 'worst of the worst' chemicals' inventory reflects a shift in scientific understanding about the toxic nature of PAHs (ATSDR 2002; Colborn, Dumanoski, and Myers 1996; Steingraber 2001). The *Exxon Valdez* oil spill played a central role in this drama. In the fifteen years since the *Exxon Valdez* oil spill, researchers have advanced two new and mutually supporting paradigms in oil toxicity—one in humans and one in wildlife. These advanced understandings of oil toxicity show that acute and chronic symptoms from oil exposure in vertebrates—humans, fish, birds, and mammals—are quite similar, often disabling, and occur at much lower levels of oil than previously thought harmful to life. With wildlife, scientists have completed the circle of understanding from individual to population-level effects. With humans, this full circle has yet to be drawn, but the implications are clear—hence, the new listing for PAHs as persistent bioaccumulative toxins.

Human Harm

The journey to discovery of the effects of oil on human health started nearly a century ago. For example, the aromatic hydrocarbon benzene, a solvent, was considered dangerous as early as the 1920s. As a potent carcinogen, "The only absolutely safe level for benzene is zero" states a 1948 health review prepared for American Petroleum Institute (Rampton and Stauber 2001, 85). Over fifty years later, the risk has not changed (ATSDR 1997). NIOSH, the research arm for OSHA, lists no recommended exposure limit for benzene is its Health Hazard Evaluation Report for the *Exxon Valdez* oil spill. Instead NIOSH investigators state, "The American Conference of Governmental Industrial Hygienists [ACGIH] considers benzene to be a suspected human carcinogen and recommends that exposures should be kept to a minimum" (NIOSH 1991, 41, Table 2, footnote e).

According to Exxon's own air quality monitoring data (Med-Tox 1989c), the highest levels of benzene to which cleanup workers were exposed occurred on beaches treated with pressurized hot water wash where benzene levels exceeded the legally enforceable OSHA permissible exposure limit (PEL) by almost eight times (Table A.1, p. 450).

With medical advancements in diagnostic tools and increased appreciation of the subtle functions of the human body, crude oil was found to be even more hazardous to humans than previously thought. Inhalation of oil mist and PAH aerosols was found to result in short- and long-term respiratory damage and central nervous system disorders as well as chronic blood (anemia, leukemia), liver, and kidney disorders; endocrine disruption; and immune suppression (Chapter 10). In part because of these concerns, oil spills were declared hazardous waste cleanups in March 1989 (OSHA, 1989).

According to Exxon's air quality monitoring data (Med-Tox 1989a–c), the highest levels of the carcinogen benzene and oil mist to which workers were exposed occurred on beaches treated with pressurized hot water wash and exceeded the legally enforceable OSHA PEL by eight and four times, respectively (Table A.1, p. 450).

Maximum exposure to PAH aerosols exceeded the legally enforceable OSHA PEL by two times. Further, Exxon scientists had published a study before the spill showing that standard PELs need to be reduced to adequately protect workers with extended work shifts (Exxon 1986). According to Exxon's study, the OSHA PELs should have been—but were not—reduced for cleanup workers by at least two to three times for their daily shifts of twelve to eighteen hours. Occupational health physician Dr. Daniel Teitelbaum, who served as expert witness in *Stubblefield v. Exxon* (1994), stated the OSHA PELs should have been reduced by 80 percent (Teitelbaum 1994). *This means the over-exposures stated above are conservative by a factor of two to five times.*

Solvents are also potent human health hazards with effects that largely overlap those from inhalation of oil mists and PAH aerosols. Solvents are key ingredients in industrial dispersants, used on oil spills, and in commercial degreasers. For example, the solvent 2-butoxyethanol is an ingredient in several products used during the *Exxon Valdez* cleanup such as Inipol, Corexit 9527, and Simple Green (ATSDR 1998). Known health hazards of products used during the cleanup include acute and chronic respiratory damage and central nervous system disorders, chronic liver, kidney, and blood (anemia) disorders; immune suppression; and acute skin disorders (dermatitis) (Chapter 10). Simple Green also damages developing fetuses, according to the EPA's website on janitorial products to avoid (U.S. EPA 2003).

According to Exxon's air quality monitoring data (Med-Tox 1989c), the highest levels of 2-butoxyethanol to which workers were exposed exceeded by two times the legally enforceable OSHA PEL (Table A.1, p. 450). *This over-exposure is conservative*—it does not consider the NIOSH recommendation or reductions for extended work hours. Exxon did not monitor exposure to commercial degreasers such as De-Solv-It, Citra-Solv, Simple Green, Limonene, and CitroKleen, which were used liberally to clean everything from workers' skin and clothes to skiffs, booms, and large vessels. Dr. Teitelbaum (1994) was appalled by the lack of monitoring of solvent exposures in Exxon's worker safety program.

The oil industry was well aware of the health hazards of inhalation of various fractions of crude oil and its refined products before the spill. Human health effects were documented in the *Amoco Cadiz* spill off the northern coast of France in 1978. At least one petrochemical company recommended a PEL for crude oil that was twenty-five times lower than the OSHA PEL (Lyondell Petrochemical Co. 1990). Exxon had developed an extensive library on health effects from inhalation of oil vapors, mists, and aerosols—and proper protection for company employees as evidenced in the toxic tort lawsuit *Stubblefield v. Exxon* (1994).

Exxon's pressurized hot water wash *created* oil mists and PAH aerosols. As shown in Chapter 2, Exxon's worker safety program failed to adequately protect cleanup crews from overexposure to dangerous chemicals. According to Exxon's clinical data (Exxon 1989b), 6,722 cleanup workers reported respiratory symptoms similar to cold and flu symptoms—or symptoms of chemical poisoning from inhalation of oil mists and aerosols. It seems that the Exxon and VECO medical staff on the cleanup and the cleanup workers themselves should have been trained to recognize—and treat in the case of the medical staff—symptoms associated with chemical exposures.

All 6,722 respiratory illnesses were reported by Exxon as "Upper Respiratory Infections" (URIs), rather than work-related illnesses, and dubbed by workers as "the Valdez Crud" (Stranahan 2003). Since OSHA *does not* require reporting of URIs or specifically, colds and flu (OSHA 2004a), Exxon dodged the long-term health monitoring requirements for hazardous waste cleanups. Workers did not receive early (in most cases, any) treatment for chemical poisoning. Many became sick with classic *chronic* symptoms of exposure to crude oil and cleaning products (solvents) used on the beaches. As one worker put it, "I just kept wondering, 'how come I got the Valdez Crud for the next ten years?'"

In 2003 Annie O'Neill, a master's student at Yale Medical School's Department of Epidemiology and Public Health, conducted a health survey of EVOS cleanup workers (O'Neill 2003). She found workers who had jobs with more exposure to oil mists and PAH aerosols have a greater prevalence of self-reported *chronic* symptoms of difficulty breathing (chronic airway disease), neurological impairment, and chemical sensitivity than workers with less exposure. Workers at risk of exposure to oil mists and PAH aerosols were classified as ones who sprayed beaches or boats and gear (DECON), operated skiffs, deployed booms, or collected animals or carcasses. One-third of the 169 participants in her survey reported symptoms similar to those listed on the MSDS for crude oil and the other chemicals of concern present during the cleanup. She submitted her study for publication in a peer-reviewed journal.

O'Neill's survey may be just the tip of the iceberg. It is likely that inhalation of oil and oil-solvent aerosols and mists has compromised the health of thousands of cleanup workers. This happened under a supposed best case scenario: state-of-the-art cleanup equipment (such as it is), pre-authorized and approved chemicals, a worker safety training program approved by the state and federal officials, an air quality sampling program, and a state-of-the-art worker safety program conducted by one of the most sophisticated oil companies in the world and monitored by the state and federal officials.

Cleanup workers' health certainly has been compromised in other spills. Cleanup workers and residents of towns misted by oily sea spray reported respiratory difficulties, sore throats, stinging eyes, nausea, and headaches after the 1992 *Braer* oil spill in Shetland and the 1999 *Sea Empress* oil spill in Wales (Campbell et al. 1993, 1994; Lyons et al. 1999). This caught the attention of researchers who found that oil aerosols are a cause for concern to spill responders in windy conditions and immediately following dispersant (solvent) application (Zhou and Liu 2001).

When politics, corporate public relations, big money, and oil spills intersect, governments look away while liability for damages is foisted onto the public, workers, and wildlife. Therefore, evidence from oil spill cleanups shows that acute and chronic health problems are likely *not unusual* for spill responders and others downwind of oily spray and mist. Instead, health problems *should be anticipated* by care providers for *any* oil spill cleanups that involve people working on beaches.

Wildlife Harm

The journey to discovery of oil effects on wildlife began during the 1970s, fifteen years or so before the *Exxon Valdez* oil spill. Scientists developed a paradigm that held that oil had mostly short-term toxic effects to sealife (Table 4, p. 394). Regarding wildlife, scientists thought oil effects were short-term and occurred *only* through hypothermia, drowning, or ingestion of toxic oil through preening or grooming. Regarding fish, scientists thought that the water-soluble fraction (WSF) caused short-term death from narcosis at concentrations of parts per million. Regarding habitat, scientists believed that most losses of plants and invertebrates from oiling would be due to smothering or short-term toxic exposure to the WSF; and that oil stranding on rocky intertidal beaches would rapidly "weather" or degrade and disperse by microbes and ultraviolet (UV) light. Visible weathered oil remaining on the beaches was thought to be environmentally benign.

In 1989, the *Exxon Valdez* spill started as scientists expected: with a staggering death toll of seabirds and marine mammals. Scientists poked about in some of the carcasses of harbor seals and sea otters to confirm death by inhalation of oil vapors, as evidenced by the hallmark brain lesions, and by ingestion of oil, as evidenced by the state of their internal organs. But then came the cleanup and scientists observed the unanticipated: the cleanup caused more damage to beach life than the original spill. Years passed with more surprises. The lingering pools of liquid oil, trapped under mussel beds and layers of gravel, were not anticipated, nor was the lingering harm to fish, seabirds, and mammals in areas with buried pools of oil. And finally, there was the unmistakable imprint of persistent oil effects in sickly rehabilitated sea otters and population crashes of pink salmon.

By 2004, fifteen years after the *Exxon Valdez* spill, scientists developed a new oil toxicity paradigm to explain the persistent and harmful effects to sealife, which could not be explained by the old paradigm (Peterson et al. 2003). The emerging paradigm holds that oil has both short-term direct effects on sealife, predicted by the old paradigm, *as well as long-term direct, indirect, and delayed effects* in the low *parts per billion*. The latter occur because oil is persistent in habitat types such as rocky beaches (the predominant shoreline in the Sound), peat marshes, and mussel beds; still bioavailable (readily picked up by sealife); and still toxic. This new paradigm shifts from a simplistic understanding based largely on short-term toxicity tests with single species to a more refined understanding based on a synthesis of ecosystem studies over time.

Scientists discovered oil can retain its toxicity for decades when buried in areas with little disturbance, oxygen required to support microbes, and UV light for photolysis (Table 4, p. 394). They found the persistent toxicity is caused by PAHs, the three- to five-ringed aromatic hydrocarbons that dissolve into water slowly over time and act as a slow poison to compromise health of individual organisms. Low levels of PAHs in buried pockets of oil were still delaying recovery of harlequin ducks, pigeon guillemots, and sea otters, among other wildlife, in some areas of the Sound in 2003, fourteen years after the spill and the last research season covered by this book.

The new paradigm holds that oil causes delayed and indirect

Table 4
Changing Paradigms in Understanding Oil Effects in Marine Ecosystems

Reprinted with permission from Peterson, Rice, Short, Esler, Bodkin, Ballachey, and Irons 2003. Long-term ecosystem response to the EVOS. *Science* 302:2082-2086. Copyright 2003 American Association for the Advancement of Science (AAAS).

Physical shoreline habitat

OLD PARADIGM: Oil that grounds on shorelines other than marshes dominated by fine sediments will be rapidly dispersed and degraded microbially and photolytically.

EMERGING APPRECIATION: Oil degrades at varying rates depending upon environment, with subsurface sediments physically protected from disturbance, oxygenation, and photolysis retaining contamination by only partially weathered oil for years.

Oil toxicity to fish

OLD PARADIGM: Oil effects occur solely through short-term (~4 day) exposure to water-soluble fraction (1–2 ringed aromatics dominate) through acute narcosis mortality at parts per million concentrations.

EMERGING APPRECIATION: Long-term exposure of fish embryos to weathered oil (3–5 ringed PAHs) at parts per billion concentrations has population consequences through indirect effects on growth, deformities, and behavior with long-term consequences on mortality and reproduction.

Oil toxicity to seabirds and marine mammals

OLD PARADIGM: Oil effects occur solely through short-term acute exposure of feathers or fur and resulting death from hypothermia, drowning, or ingestion of toxics during preening.

EMERGING APPRECIATION: Oil effects also are substantial (independent of means of insulation) over the long term through interactions between natural environmental stressors and compromised health of exposed animals, through chronic toxic exposure from ingesting contaminated prey or during foraging around persistent sedimentary pools of oil, and through disruption of vital social functions (care giving or reproduction) in socially organized species.

Oil impacts on coastal communities

OLD PARADIGM: Acute mortality through short-term toxic exposure to oil deposited on shore and the shallow seafloor or through smothering accounts for the only important losses of shoreline plants and invertebrates.

(continued)

EMERGING APPRECIATION: Clean-up attempts can be more damaging than the oil itself, with impacts recurring as long as clean-up (including both chemical and physical methods) continues. Because of the pervasiveness of strong biological interactions in rocky intertidal and kelp forest communities, cascades of delayed, indirect impacts (especially of trophic cascades and biogenic habitat loss) expand the scope of injury well beyond the initial direct losses and thereby also delay recoveries

(Author's Notes for non-scientists: "Photolysis" and "photolytically" refer to breakdown (weathering) of oil by ultraviolet light. "Ringed aromatics" and "ringed PAHs" refer to aromatic hydrocarbons; ppm and ppb are concentrations at parts per million or billion, respectively. Examples of "socially organized species" include common murres and orcas. "Trophic cascades" refer to food web interactions. "Biogenic habitat loss" refers to loss living habitat [i.e., coastal sea plants and invertebrates].)

effects by unraveling bits of the complex tapestry of life that we simply call "ecosystems." Oil disrupts food web interactions such as the prey-switching phenomena among herring, pink salmon, and apex predators; complex social behaviors in whales and seabirds; and critical dependencies on "biogenic" habitat—the sea plants and invertebrates that blanket rocky intertidal beaches and provide shelter and food for other fish and wildlife. There are "synergistic" effects in which oil acts in concert with other ecosystem disturbances—climate regime shifts, global warming, other PBT pollutants—to knock entire populations of species (such as transient orcas) down to tiny flames of their former blaze that could easily be put out by more hardships. It's as if, instead of greasing these intricate cogs of life, oil spills gum them up and cause the whole marine engine to malfunction for years.

The new paradigm shatters several tenets of the old paradigm. The new understanding holds that: (1) oil has short-term and **long-term toxic effects** to fish and wildlife; (2) subsurface buried oil *is not* environmentally benign; and (3) oil is both **persistent and bioavailable**. Oil was found to be harmful to fish and wildlife at 1,000 times lower levels than those thought to be "safe" for wildlife under the old paradigm. Further, extraordinarily low levels of oil—PAHs in the low parts per billion range—cause persistent and measurable **population-level harm** to sea life. On the basis of these scientific advancements, environmental policies are grossly under-protective of aquatic life.

Recent studies from this country and others validate the public-trust scientists' findings and confirm the emergence of the new oil toxicity paradigm (Couillard 2002). Some of these findings are summarized by National Research Council in its 2002 publication, *Oil *and the Sea III*, and in the technical literature. This book, *Sound Truth*, introduces the public to this scientific revolution in our understanding of oil toxicity.

Reassessing our Indicators of Individual Health

Under the old oil toxicity paradigm, scientists used indicators such as "death" and "cancer" in lab tests and risk assessments to determine levels of oil (and dispersant) that were supposedly safe for wildlife and humans. For example, to establish water quality standards for aquatic life, scientists in the 1970s divided the lowest concentrations of oil (WSF) that caused death to test organisms by 100—and used this as a safe standard, hoping it was sufficient to protect life. As scientists found out nearly thirty years later, it was not. Further, the endpoint of "cancer" in humans, it turns out, was also too blunt a tool to protect life from other more subtle diseases such as central nervous system malfunction, endocrine disruption, and chemical sensitivities that compromise health, disable, and kill.

Advances in medicine and science produced sensitive diagnostic tools to monitor individual health of humans, domestic animals, and wildlife. These tools allowed scientists to detect subtle effects of pollutants at the sub-cellular level. From the perspective of the new paradigm, the flaws in the bricks and mortar of the old paradigm are obvious.

Biomarkers and Human Health

One of the reasons EVOS cleanup workers became sick from inhalation of oil and solvent aerosols is because risk assessments, which health officials depend upon to establish "safe" exposure limits to chemicals, give only the illusion of safety. Risk assessments cannot deliver on their promise of public protection, because their ability to give honest answers depends on how closely the underlying studies reflect the reality of multiple chemical exposures in our home, workplace, and world. None do.

Rampton and Stauber (2001) in *Trust Us: We're the Experts* charge that "[d]etermining the cumulative effect of these insults is a scientific impossibility . . . resulting risk assessment is bogus" (109). It follows that the supposedly safe permissible exposure limits (PELs) for workers are also "bogus" because they are based on the faulty risk assessments.

Crude oil illustrates the problem. As occupational medicine physician Teitelbaum pointed out in his court deposition (1994), there are OSHA PELs for many individual hydrocarbons in crude oil, such as benzene and some of the PAHs. However, there are no PELs for the whole—for crude oil mists or aerosols, because these are complex mixtures of hundreds of hydrocarbons. Exxon cleanup workers were not just exposed to individual hydrocarbons—they were bombarded by hundreds of hydrocarbons as has been shown.

Our ability to predict "safe" levels of hazardous chemicals such as crude oil is further weakened by bad decisions. Examples abound from the EVOS cleanup (Teitelbaum 1994). The decision to use surrogates to determine safe levels of exposure of workers to crude oil mist and PAH aerosols instead of the real thing was bad enough. The choices of surrogates—the relatively benign mineral oil and dust particulates—made bad decisions worse, because the surrogates did not even come close to accurately representing the health hazards of the original chemicals. (For example, crude oil is a *"tumorigen"*—it produces carcinogenic tumors—when applied to skin; mineral oil is something mothers rub on babies' bottoms!) Exxon never voluntarily reduced the PELs to accommodate its extended work hours—a given in crisis situations; this was another bad decision that increased the risk to workers.

To make matters worse, Exxon's medical doctors, who should have recognized work-related exposure problems, sent ill workers to regional hospitals and traditional doctors who *are not* trained to link work-related exposures to health problems (U. S. OSHA 1994). The traditional doctors often fail to diagnose and treat chemical-induced symptoms and diseases. PAH aerosols and solvents such as benzene and 2-butoxyethanol are neurotoxins and endocrine disrupters that can cause systemic effects, which often are not connected by the patient or doctor back to the root cause, and therefore not treated as chemical exposures. The traditionally trained doctors sent the workers back out into the conditions that made them sick in the first place.

Chemical sensitivity and the emerging new disease paradigm these symptoms represent—toxic induced loss of tolerance or TILT—are slowly breaking through a wall of recognition-resistance formed by traditional doctors, allergists, and the petrochemical industry (Ashford and Miller 1998). Chemical-induced symptoms are also ill-defined in OSHA workers' compensation programs, which routinely dismiss such problems (at least in Alaska).

It's as if we don't want to admit that crude oil really is a health hazard, so we set lax worker exposure standards and then refuse to recognize health problems from overexposure. Despite the denial issues, the bottom line is the cleanup worker's human body and in it, at the cellular and sub-cellular level, lies the proof of past PAH and solvent exposures, clues of causation of mysterious chemical-induced symptoms, and predictors of future cancers and other health problems.

Biomarkers are the latest tool for ferreting out evidence of chemical exposures in the human body. Human blood, urine, and body tissues all lend evidence of exposure to PAHs, oil vapors and mists, and 2-butoxyethanol, among others (Accu-Chem Laboratories 1992b; Spence 1989a, 1989b). Sophisticated equipment, designed to sleuth out subtle clues in blood counts, hormone levels, DNA *"adducts"* (adhesions from PAH exposure) and developing fetuses, has accelerated identification of health problems caused by PAH exposures (Eubanks 1994; Perera 1992; Perera et al. 1999; Steingraber 1998, 2001).

Sandra Steingraber, author of *Living Downstream* (1998), refers to biomarkers as the crown jewel of molecular epidemiology (245). When these tools find their way into the courtroom, corporate lawyers will be hard pressed to sweep the stark evidence of chemical-induced illnesses under the carpet—and the policy tide will turn to favor increased protections for worker and public health.

The EVOS legacy is just starting to emerge in former cleanup workers who are experiencing mild to disabling symptoms of respiratory distress, neurological impairments, and chemical sensitivities. Long-term health monitoring of former cleanup workers and epidemiology studies have the potential to yield solid evidence of devastating chemical-induced illnesses from oil and dispersant (solvent) exposures. Whether any of our government officials, academic researchers, Exxon—or private lawyers step forward to answer the sick workers' clarion call of distress remains to be seen.

Biomarkers and Wildlife Health

The journey of discovery of the new oil toxicity paradigm was very similar for wildlife. Scientists now realize it was wrong to ignore the larger aromatic hydrocarbons, the PAHs, just because they do not dissolve or degrade as quickly as the water-soluble fraction (WSF). The very persistence of the PAHs is what proved to be toxic.

Once the focus shifted to the more toxic PAHs, scientists also realized that "death" is the wrong endpoint to gauge what is safe for fish and wildlife. The WSF acts on cell membranes to cause a fast-acting narcosis, as observed in the dizzy harbor seals after the spill. The narcosis can lead to death in high enough doses (Exxon 1988). In contrast, PAHs operate within the cell on proteins, jamming functions of essential enzymes, hormones, and immunoglobins, and even damaging DNA, the basic hereditary molecules of life. PAHs reduce individual fitness and sickly animals are less capable of dodging predators, capturing prey, and successfully reproducing. Instead of direct death from narcosis, PAHs result in indirect death such as getting eaten and populations that dwindle slowly over time.

The seminal ecosystem studies, conducted after the *Exxon Valdez* oil spill, prove that short-term bioassays are not the proper tools to assess ecological harm to wildlife. This is especially true when chemical exposures (oil spills or chronic pollution) are involved where there are complex long-term interactions among growth, body condition, maturation, diseases, reproduction, predation, and habitat condition including climate shifts—in other words, in a real life situation. The pioneering research on biomarkers, developed and refined these indicators into tools that are capable of measuring very subtle effects of chemical poisons at the cellular and sub-cellular level.

For example, blood chemistry tests were fine-tuned as metrics of health and hydrocarbon exposure assessment. Detection of the presence of the respiratory enzyme cytochrome P4501A was honed as a tracking tool to monitor exposure to low levels of PAHs. Measurement of concentrations of certain fecal porphyrins was developed as a tool to detect disruptions in heme synthesis from PAH exposure. Lipid compositions of food webs were catalogued into libraries to compare diet and nutrition in wildlife from different areas and times.

Ironically, Exxon scientists conducted standard 1970s-style laboratory toxicity tests with oiled sediments collected in 1990 to 1993 (Page et al. 1999) and, in so doing, demonstrated that these outdated tests no longer are relevant for predicting ecological harm. In Exxon's toxicity tests, increased mortality to the test organisms (adult amphipods) occurred at exposure levels above 2,600 parts per billion PAHs *in the sediments.* Public-trust scientists, using biomarkers and embryo-toxicity tests, detected mortality to sensitive life stages of fish and wildlife at PAH levels *in the water* of less than 1 to 20 parts per billion.

Public-trust scientists found that the use of biomarkers to assess well-being of wildlife conveys several advantages over the old methods. First, the new biomarker tests can detect biological harm at levels well below the limits of the old toxicity tests. Second, most of the measures are relatively non-intrusive so animals no longer needed to be sacrificed in the name of science. Third, biomarkers yield subtle information about population health that cannot be obtained from gross dissections. Fourth, biomarkers can be precisely measured so their use eliminates guesswork and opinion in assessing health of individuals.

Scientists who developed these sensitive and powerful investigative tools warn that biomarkers vary among species, sex, age class, reproductive status, physical condition, and other variables and that their very sensitivity complicates data interpretation. The blood chemistry of an adolescent river otter, for example, looks a lot different from that of a prime-age breeding adult and toxic exposure history further complicates the picture. Scientists learned that to most accurately assess the effects of environmental pollution, biomarkers are best used in concert with population ecology to integrate information on individual health with population-level studies. The powerful combination of cellular biology and population ecology sets a new bar for monitoring ecosystem health.

Assessing Population Health

Before the *Exxon Valdez* oil spill, the integration of disciplines to determine effects of environmental pollution at the population or ecosystem level was virtually nonexistent. Scientists struggled to understand complex dynamics without the tools to accurately identify, much less treat, mass chemical poisoning epidemics. Industrial accidents such as the *Exxon Valdez* oil spill create opportunities for scientists and physicians to understand chemical-induced illnesses and effects at the population-level. This knowledge can then be applied to find solutions to broader-based but subtler chemical epidemics such as those associated with the ubiquitous petrochemicals (Chapter 10) or other PBT pollutants.

Wildlife scientists fully utilized the opportunity presented by the *Exxon Valdez* oil spill to complete the circle from individual health effects to population level effects; medical researchers have not—yet. In the following section, wildlife research is presented first to serve as an example of what might be done to track oil-related health effects in the population of spill responders (cleanup workers, volunteers, scientists, Coast Guard personnel, state and federal monitors, etc.).

Marine Ecosystems and Environmental Pollutants

Population effects following a spill are rare, but in the case of the *Exxon Valdez* spill, several species of birds and mammals had population effects caused by a high number of deaths in the first months of the spill. Most surprising were the population effects that persisted from a few years in some species to over a decade in other species (Chapter 22; EVOS Trustee Council 2000). Elevated egg mortalities in pink salmon in oiled streams continued for four years after the spill. River otters and pigeon guillemots showed measurable harm until about eight years after the spill. Sea otters are still struggling to recover in northern Knight Island and harlequin ducks are still dying at higher rates in areas once heavily oiled fifteen years ago. But it was the back-to-back collapses of the pink salmon and herring populations in Prince William Sound made it glaringly obvious that the old way of doing business was inadequate to explain the reality of persistent oil effects and delayed recovery.

The three seminal ecosystem studies described in this book clearly demonstrate the importance of a holistic approach to enhance scientific certainty and understanding of complex relationships between the environment, the oil spill, and wildlife populations. This level of effort is necessary if we wish to have solid baseline data, against which impacts of oil spills and other human activities on fish and wildlife populations can be accurately assessed. The common elements in each of these studies include a comprehensive and inclusive planning process, selection of key species, and comparisons over geographic areas and time.

Ideally, a holistic approach involves a broad spectrum of scientific disciplines *and the public*. Public participation and collaboration with scientists strengthened each of the three ecosystem studies. The Sound Ecosystem Assessment Program had its genesis in a community-wide public planning process that encouraged open sharing of wisdom and knowledge and resulted in a comprehensive program that significantly improved fisheries management. The SEA Program used flights of opportunity by experienced local pilots to glean historical information on herring ranges and to monitor stocks (Brown et al. 2002). The APEX study used local fishing charters to gather information on groundfish forage habits at little additional cost (Roseneau and Byrd 1997). The Nearshore Vertebrate Predator Program incorporated observations and knowledge from villagers in the Sound to design its program and select key species (Holland-Bartels 1999). Collaborative efforts that engage the public are a lot of work, especially with regards to changing attitudes, but this process reaps huge rewards in terms of everyone's understanding of ecosystem function, oil spill effects, and better management decisions, supported by the public.

Scientists learned two types of species are indispensable for studying pervasive ecosystem injury and recovery: sentinel species and keystone species. Sentinel species are sensitive early indicators of systemic change like the canary in the coal mine. The "canaries" of Prince William Sound include mussels, sea otters, pigeon guillemots, and harlequin ducks. Keystone species fill critical niches in the ecosystem, usually as food for other species, and function to lock everything else into place. Loss of a keystone species such as herring creates ripple effects for other species as witnessed by the population crashes of fish-eating predators in the Gulf of Alaska as discovered from the studies on apex predators (Chapter 19). Mussels and clams are keystone species for underwater bottom-feeding birds and mammals and for wildlife that feeds in the intertidal zone; contamination of these keystone species with oil caused lingering harm to their predators.

Scientists found that much information about oil spill effects could be gained or lost through proper choices of geographic scale. For example, a decade after the spill, sea otters and harlequin ducks showed broad recovery at the regional scale of western Prince William Sound, but no recovery and lingering harm at very local scales of bays heavily oiled in 1989. For some species such as murres, the oil spill was a relatively small-scale event that played out on the larger stage of climatic regime shifts. Selection of appropriate geographic scales is like an eye exam: what lens one uses determines how clearly one sees oil spill effects.

Similarly, scientists found time-scale issues were also central to understanding population dynamics in response to the oil spill.

Species are more vulnerable to pollutants during certain times of their lives or during specific seasons. For example, adult female harlequin ducks in Prince William Sound are most vulnerable during the winter when ingestion of oiled food saps limited energy reserves needed to stave off cold temperatures and starvation. The time scale of the disturbance itself must also be taken into account. For example, to improve our understanding of ecosystem effects of the climatic regime shift, studies should be designed to span decades. Effects of an oil spill should be monitored as long as residual oil is available to wildlife and there is measurable biological harm.

The public-trust scientists found that ecosystem-based science can produce a sufficient weight of evidence to overcome data limitations from studies on individual species and inherent natural variability. This can lead to scientific certainty and understanding of complex environmental issues. The EVOS ecosystem studies contributed to emergence of the new oil toxicity paradigm and to understanding the biological effects of the North Pacific Decadal Oscillation (the climate regime shift).

The EVOS Trustee Council eventually consolidated most of its public-trust studies into a comprehensive monitoring program, dubbed the Gulf Ecosystem Monitoring (GEM) program. The GEM program began in 2000 and is supported through a long-term endowment program managed by the EVOS Trustee Council. The goal of the GEM program is "to promote the sustainability of a recovered and healthy ecosystem by understanding how natural and man-caused perturbations influence the production of resources of high value sport, commercial, and subsistence users in the region." Interested readers are directed to the EVOS Trustee Council's website at www.oilspill.state.ak.us/gem/.

The GEM Program will provide solid baseline data against which to monitor and understand future ecosystem perturbations in the North Pacific. Better understanding of population-level injuries to fish and wildlife will lead to better risk assessment and more effective prevention measures—at least for activities that humans can influence such as oil spills and climate change.

Ideally, ecosystem-based management will be adopted by state and federal entities that manage all of Alaska's fish and wildlife resources. Alaska Natives realize this is *the only* way to safeguard fish and wildlife for future generations. They are concerned that "the best available science" does not include Alaska Native ways of knowing and observation. In 2001, the Native-run Rural Alaska Community Action Program (RurAL CAP) hosted the first summit meeting of Native leaders, Elders, hunters, and gatherers to share observations of climate change, dwindling stocks of wildlife, and increasingly sickly fish, birds, and animals (RurAL CAP 2002). The testimonials at this Summit clearly demonstrate how Alaska Native peoples, particularly their elders, understand and see things in the environment in terms of interconnections: how the whole is greater than the sum of its parts, because one part cannot move without affecting another. They discussed strategies to deal with these trends and how to prepare their people for change. This effort was the vision of Alaska Native leader Ilarion (Larry) Merculieff and it is the ultimate example of Native-style science (*"traditional environmental wisdom"*) in service to people.

Human Populations and Environmental Pollutants

Understanding population-level effects of environmental contaminants these days is very similar to the state of affairs with the cholera epidemic of 1854. We know that the well of our environment is becoming increasingly contaminated with an alarming soup of dangerous chemicals that are responsible for a growing number of insidious diseases—but unfortunately there is no single pump handle to break. Instead of feigning ignorance of these chemical-induced illnesses, we need to start looking for the pump handles, one at a time.

The *Exxon Valdez* oil spill offers an invaluable opportunity to understand the spectrum of symptoms from inhalation of oil mists, oil aerosols, and the dispersant Inipol. (Medical records of individual exposures were specifically kept for the latter). Because this was a hazardous waste cleanup, Exxon is required to keep all medical records and related data for thirty years—and these records could be obtained by federal subpoena (something which should have been done in 1989 but wasn't). Half this time has elapsed; however, the timing is ripe for an independent epidemiology study, because many diseases associated with crude oil exposure take over a decade to manifest. This is one definitive study that should be done—it would directly benefit the cleanup workers and it would be a model for limiting exposures, if necessary, for all future spills. The data exist, but the study needs to be done to retrieve the data.

To structure an epidemiology study, researchers should look to the wildlife studies for guidance. Ideally, any studies will be a collaborative venture—right from the planning stages through to completion—between medical researchers and members of the community of exposed individuals. Such collaborative and concerted community efforts, dubbed *"popular epidemiology,"* are increasingly prevalent (and successful) in breaking through the wall of denial and/or feigned ignorance long held in place by vested interests, which all too often includes the federal government (Brown and Mikkelsen 1990). Pioneers of popular epidemiology are the communities of Love Canal in the 1970s and Woburn, Massachusetts, in the 1980s.

The team that conducts the popular epidemiology on chronic symptoms from the *Exxon Valdez* oil spill and cleanup should also collaborate with public-trust researchers who conducted veterinary medicine on wildlife. Many of the acute and chronic symptoms from oil exposure among wildlife and workers are similar. Acute exposures include respiratory difficulty, central nervous system damage, and damage to internal organs. Chronic symptoms in humans will most likely include the systemic effects observed in chronically-exposed wildlife such as sea otters and harlequin ducks and self-reported by cleanup workers who participated in the Yale survey (O'Neill 2003).

In Alaska, Natives concerned about chemical pollutants in their traditional subsistence foods and the effects on human health initiated two popular epidemiology studies. One is sponsored by Pam Miller's group, ACAT, and the other by the Alaska Native Science Commission (I. Merculieff, pers. comm., August 2003). Rural Natives are one of the "sentinel" groups of the Alaska population—to draw a parallel with the wildlife studies, because of their heavy reliance on wild foods. ACAT's study, for example, found people of St. Lawrence Island have eight times more PCBs in their bodies than the national average (Zamzow 2002).[1] Similarly, members of villages in the oil spill region and subgroups of EVOS cleanup workers, such as those identified by the Yale epidemiology study, would make good choices for "sentinel" groups for future studies on effects of oil and the cleanup on the human population.

Of course, it's one thing to conduct a popular epidemiology study and another to have the results recognized, accepted, and acted upon by our greater society. Currently our society treats the misfortunes of a few as "isolated" events—Love Canal, Woburn, Prince William Sound. They are not. Industrial accidents are the "canaries" for our global society. They are chemical douse wake-up calls that we are poisoning ourselves at much lower levels and robbing ourselves of a future free from cancers and other chemical-induced illnesses as the authors of *Our Stolen Future* so eloquently point out.

In 1998, the National Academy of Science Institute of Medicine identified a need for physicians to be better trained in environmental medicine to meet anticipated health care needs (National Academy of Sciences 1991). In August 2002 the Southwest College of Naturopathic Medicine opened the Environmental Medicine Center of Excellence, the first such college accredited program in the country. According to the Center's website (www.scnm.edu), environmental medicine is now defined as, "diagnosis and care of individuals with illnesses secondary to cumulative burden of chronic exposure to environmental toxins in homes, communities, and workplaces." This college program, and others that will surely follow, offers hope of global recognition of our current chemical poisoning epidemic, proactive problem-solving, and a healthy future with clean air, clean water, and toxic-free food.

Science, Politics, and the Public Interest

Thomas Kuhn, a professor at Massachusetts Institute of Technology until his death in 1962, wrote a book describing the contentious nature of scientific progress or *The Structure of Scientific Revolutions*, as he called the normal shifts of theories and paradigms that define scientific advancements. According to Kuhn, early stages of a paradigm shift include denial, blame on faulty science, and derision of the proponents of the new paradigm. Eventually, the old paradigm is discredited and rejected by the scientific community, the new paradigm is heartily embraced, and the whole process begins anew.

In the case of the new oil toxicity paradigm for wildlife and chemical sensitivity disease paradigm for humans, the scientific revolution has not progressed much past the denial and derision phase. The reason, as succinctly stated by authors Nicholas Ashford and Claudia Miller, is that "[b]lind adherence to old paradigms, couple with vested financial interests . . . are powerful incentives militating against change" (1998, 287). Scientific advancements inherent in paradigm shifts are made more controversial, bitter, and drawn out when vested interests buy science to "prove" their point. An educated public is often the only counterbalance to these vested interests.

Abuse of the scientific process by vested interests is widespread. Bad science is easily cloaked in fancy statistics to masquerade as respectable. This ball has been rolling a long time. Half a century ago, Darrell Huff wrote a little book to teach "honest men" some of the common tricks in self-defense. Huff was fed up, because he realized, "The secret language of statistics, so appealing in a fact-minded culture, is employed to sensationalize, inflate, confuse, and oversimplify" (1954, 8). In his book, *How to Lie with Statistics*, Huff charges, "A well-wrapped statistic is better than Hitler's 'big lie'; it misleads, yet it cannot be pinned on you" (ibid., 9). Bad science has since been dubbed, *"tobacco science,"* after this industry was publicly defrocked by insiders who revealed the truth about its biased studies.

Biological Sciences and the EVOS

Exxon's spill studies on wildlife can be considered as stellar examples of advocacy science. Exxon scientists slanted their own studies with under-powered designs so their studies would not detect oil spill effects. Choices in study design ensured pre-determined conclusions. Many of the common tricks used by Exxon scientists are described by Huff. For example, one trick was to improperly preserve samples to reduce concentrations of aromatic hydrocarbons (Chapter 12). Another was to average data to erase distinct differences between oiled and unoiled areas (Chapters 12, 13). Another was to selectively report data—or to simply *omit* the most damaging data—to hide oil spill effects (Chapters 13, 14). Another was to collect a small number of samples with a physically tiny sampling device to blur differences between oiled and unoiled areas (Chapter 13). Another was to mismatch control and oil sites—again to blur oil effects (Chapter 13). Examples abound.

Exxon counted on seemingly small nuances in study design to pull off its magician's act of making the Sound appear to have fully recovered within a couple years after the spill—*abracadabra!* Without samples that accurately represented the larger whole—whether that "whole" was beach ecology, seabird communities, sea otters populations, or something else—Exxon's studies reveal little about the true nature of the Sound.

If you were fooled by press reports that Prince William Sound had recovered back in 1990 and 1991, don't feel bad. So were millions of

Americans and others around the world. In the years immediately following the spill, there was intense public interest to learn if the Sound had recovered. However, with public-trust scientists gagged by government lawyers for two years, the press received only one side of the story. Huff (1954) warns, "Public pressure and hasty journalism often launch a treatment that is unproven, particularly when the demand is great and the statistical background hazy" (41). He further explains that "without writers who use the words with honesty and understanding and readers who know what they mean, the result can only be semantic nonsense" (ibid., 8).

Unfortunately, "semantic nonsense" in the hands of a clever wordsmith can still be skillfully tailored to convey whatever story suits the person who is paying for it. Exxon hired a large public relations firm to put a positive spin on the badly stained Prince William Sound and its corporate image. In the hands of these skilled craftsmen, the Sound was given a facelift, which created the illusion of recovery, although real problems remained. As Huff (1954) observes, "When all the mistakes are in the cashier's favor, you can't help wondering" (71). It is a testament to these craftsmen that most people didn't wonder—they believed this interpretation. Most people still do.

When scientific controversies spill into the public arena, it is time for the public to pay attention—not to the controversy, aptly covered by the press—but to the *reason for the controversy*; i.e., what is at stake for the public interest. The latter is usually *not* covered by the press. In this case, Exxon had to leap three liability hurdles in the aftermath of the spill to clear short-term damages to public resources, short-term damages to private entities, and potential long-term damages to public resources. A brief review of Exxon's science relative to its outstanding liabilities is quite revealing.

Exxon cleared its first hurdle in 1991 with the civil settlement and a payment of $900 million to public trustees of the state and federal governments for short-term damages to wildlife and public lands. Exxon wrapped up most of its oil spill studies after closure of this liability. When the public-trust science finally was relieved of the gag order and openly publicized in 1993, Exxon appeared to have manufactured a razzle-dazzle scientific controversy, promoting its studies over the public-trust science (Exxon 1993b). The controversy captured widespread media attention. Exxon used the instrument of the press to sow seeds of doubt about the extent of harm caused by the spill in order to minimize its liability for the second hurdle.

Exxon cleared the second hurdle in 1994. Exxon skillfully used its "science" in court to defend itself against potentially billions of dollars of liability for damages to people injured by the spill (Barker 1994). Subsequent interviews with jurors for the case indicated that "they were overwhelmed with conflicting testimony from scientific experts, so they largely ignored the scientific findings in reaching their verdict" (Wiens 1996, 595). Exxon's liability was contained by the jury at $5.3 billion—one year's net profit at the time—and the following day, Exxon's shares rose on the stock market (*The Dallas Morning News* 1994). (Exxon continues to challenge the $5 billion punitive damage award; none of it has been paid. Exxon's annual net profits rose over $20 billion in 2003 [Antosh 2004].)

The third liability hurdle is in play and actually has the most at stake for Exxon, the oil industry, and the public. The 1991 civil settlement contains a provision for reopening the settlement based on "unanticipated long-term harm," which is not defined, because, ironically, none was anticipated based on the understanding of oil toxicity at the time of the settlement. Although the price tag on the "reopener clause" is capped at $100 million, reopening the settlement would be invaluable from the public's perspective (Carleton 2003). It would secure the ability for the public and policy-makers to strengthen federal (and state) oil pollution laws, which are based on old science and outdated by the new oil toxicity paradigm. The reopener clause expires on 1 September 2006.

In preparation for the third and final liability hurdle, Exxon scientists returned to Prince William Sound in 1998, just prior to the ten-year anniversary media events, and again in 2002 and 2003 after the NOAA Auke Bay Lab scientists confirmed local reports of extensive toxic subsurface oil. Exxon scientists renewed their attacks on the public-trust studies and their personal attacks on federal scientists with barrages of critical press and Freedom of Information Act demands in attempts to discredit the federal scientists' work prior to publication.[2]

With the specter of unanticipated long-term damage looming (Rosen 2002), Exxon scientists changed the definition of "recovery" to suit its purposes. The scientific community was outraged. The press loved it—the controversy, that is—and once again the American public was presented with Grade A science and Grades "D" and "F" science as if it all carried the same measure of truth. Small wonder most people are confused. Exxon is losing the technical battle over persistent harm from oil, but it is winning, so far, the liability war for the reopener clause, because the state and federal political administrations currently in power have refused to request of Alaska federal district Judge Holland that the settlement be reopened.

In choosing to conduct its business in this manner, Exxon is following a worn trail that it helped blaze with other American corporate leaders such as the tobacco industry (Glantz et al. 1996). Like "tobacco science" (ibid., 3), Exxon's oil spill science has been suspect—the professed goal of pursuing the "truth" about oil spill effects was always subservient to litigation and federal policy concerns. Like the tobacco industry (ibid., 319, 321), Exxon used its spill "science" to muddy and confuse the clear evidence of very damaging oil spill effects. Like the tobacco industry (ibid., 4, 41, 169), Exxon created a false controversy to manipulate the press and public to forestall policy changes. Exxon perpetuates this fraud through the instrument of the press. And, best of all from the industry perspective, Exxon writes-off its shallow science, topped with legal glitz and public relations sprinkles, as a cost of doing business at taxpayers' expense.

Tobacco science with its attendant hype is also called "advocacy science." The authors of Trust Us, We're Experts! refer to advocacy science as "the best science money can buy" (2001, 195) to illustrate their point that these experts are actually hired guns: they are paid to present the best possible case for their client. They do not have to win an argument; they just simply have to confuse the argument, the media, and the public. The murk of confusion obfuscates clear understanding of the issue. In the doldrums of public confusion, the policy ship of state stalls, while the vested interests continue to steam full-speed ahead, profiting by polluting. There is a lot to be gained by vested interests such as Exxon in the confusion-based preservation of an existing system.

The lesson from this battlefield of conflicting science and paradigm shifts is that people—and other life on the planet—are the real losers unless and until public policies advance to match the current science. New science protects no one until outdated laws are updated. The good news is that the public can actually do something to correct this (Chapter 24).

Social Sciences and the EVOS

Exxon has yet to conduct any studies—"tobacco science" or otherwise—on the sick cleanup workers. Instead Exxon has focused its effort on legal manipulations to deny public access to medical records and to deny workers' health claims from the cleanup (Chapters 8, 9). Unfortunately, the U.S. Supreme Court's 1993 *Daubert* decision particularly disadvantages workers with chemical illnesses (Tellus report 2003). *Daubert* essentially appoints judges as amateur scientists and gatekeepers to screen admissibility of medical evidence and science to the courtroom before trial.

Exxon and many other corporations have wielded the *Daubert* sledgehammer to derail sick workers' toxic tort cases and prevent them from going to trial. Most of the sick workers from the *Exxon Valdez* oil spill who filed toxic tort cases, such as Phyllis La Joie and Ron Smith, lost their cases or were forced to settle because of legal technicalities. The 2003 Tellus investigation of *Daubert* abuses reports that, in the case of *Chambers v. Exxon*, Exxon successfully used *Daubert* to defend itself against an oil refinery worker who developed leukemia from benzene exposure.

The Tellus report summarizes the state of affairs in proving chemical illnesses in the court: "[T]he scientific community needs to become much more aware that an obscure procedural decision intended to provide clarity has instead given rise to a serious social imbalance" (2003, 17). Further, "[A]pplication of *Daubert* and *Daubert*-like challenges threaten to paralyze the systems we use to protect public health and the environment" (ibid., 17). Once again, public health and the environment are at risk from outdated legal decisions that cannot accommodate new science.

Designing Laws to Deter Spills

"Lies, damn lies, and the public relations industry" is the subtitle for John Stauber and Sheldon Rampton's book *Toxic Sludge is Good for You!* (1995), but it is also what the public gets when the spiller is left in charge of the cleanup.

As has been documented in the Introduction (Sidebar 1, p. 4, and Table 1, p. 7), Exxon officials and its contractor Caleb Brett appears to have substantially underestimated the amount of oil spilled. Underreporting of spill volumes is common, according to industry analysts (Schmidt Etkin 2001). In the United States, lying about self-reported spill volume is illegal and a breach of public trust, since spill volume is used to gauge the size of penalty for damages to public resources. Caleb Brett, an oil industry service provider, has been caught and fined $1 million (in 2001) for lying to investigators and falsifying documents for profit in other oil-related matters (Associated Press 2001; Margasak 2001). Underreporting saves the spiller penalties—if the spiller doesn't get caught.

In Exxon's case, an alert Homer resident, Findlay Abbott, filed a claim on behalf of the public interest under the False Claims Act for three times as much civil penalties since three times as much oil spilled (*U.S.A., ex rel., W. Findlay Abbott v. Exxon* [filed in 1996]). He has argued the case himself and it was pending in the Ninth Circuit Court of Appeals as of May 2004. At this point, it is unknown whether Exxon will succeed in avoiding addition penalties, potentially up to $2 billion, from its potential 20 million-gallon-plus lie.

Exxon and VECO personnel overstated the ability of their safety program to adequately protect workers' health from chemical exposures (Chapter 2), and they failed to adequately inform workers about the health hazards of aerosolized crude oil and many other chemicals present during the 1989 cleanup (Chapter 3). Exxon withheld its damaging air quality and clinical data from federal and state oversight agencies (and workers) and in so doing, evaded the long-term health monitoring requirements of hazardous waste cleanups (Chapter 3). This saved Exxon money and—so far— it also has shielded Exxon from potential liability problems from sick workers.

While failing to adequately protect worker health and to report work-related illnesses such as the 6,722 cases of respiratory illness (many likely from chemical exposure), Exxon aggressively sought to re-coup its cleanup expenses. According to one investigative reporter, Exxon recovered at least half of its cleanup expenses (Introduction; Curridan 1999).

Exxon officials and scientists invoked Huff's "well-wrapped statistics" to purposely mislead the public about the environmental damage from the spill and cleanup and about the extent and scope of recovery of sealife. This also saved Exxon money—to wit the amount of the compensatory damage award in the third party lawsuit. This strategy may also save Exxon from having to pay a penny of the outstanding $100 million from the 1991 civil settlement, but this remains to be seen. Exxon has used its spill studies to stir public controversy and forestall policy changes that would inevitably follow successful reopening of the 1991 civil settlement for unanticipated long-term harm.

The lesson from the Exxon Valdez spill is that federal and state laws designed to prevent and respond to spills and to protect workers' health during oil spill cleanups do not work to protect the public interest when rich corporations spill oil. Not only do the laws not work, they actually create situations in which the spiller can recover significant sums through accounting and tax strategies: this clearly undercuts the goal of deterrence. Obviously, unless polluters—not the public—are held accountable to pay for environmental mishaps, oil spills will continue to happen.

At the turn of the twenty-first century, scientists stepped on the firm shore of a new understanding of oil toxicity (Table 5, p. 414). Other scientists and doctors from across the United States and around the world who initiated similar journeys of exploration have verified that oil is indeed more deadly to wildlife and humans than previously thought. The final section of this book links the wildlife and human stories from the *Exxon Valdez* oil spill into a new framework of understanding our brave new world—and what can be done to reduce the toxic threat of oil pollution.

Table 5
The Legacy—Key Lessons Learned from *Exxon Valdez* Oil Spill

Shifting Paradigms: Oil Causes Persistent Harmful Effects

1. Oil acted like the persistent, bioavailable, toxic (PBT) pollutant that it is; it took over a decade for researchers to reach a new understanding of oil effects on humans and marine ecosystems.

2. Cleanup workers appear to have been overexposed to oil mist, PAH aerosols, and other chemicals relative to NIOSH *recommended* exposure limits and the OSHA permissible exposure limits.

3. There was a higher prevalence of acute respiratory symptoms, stomach distress (poisoning), and potential neurological symptoms among 1989 cleanup workers compared to those reported by the Alaska workforce in 1987; there is a higher prevalence of chronic airway distress, neurological symptoms, and chemical sensitivity among 1989 cleanup workers who were at a higher risk of oil and/or chemical exposure than less exposed workers.

4. Oil is harmful to fish and wildlife at 1,000 times lower levels than those thought to be the toxicity threshold in the 1970s; PAH concentrations in the low parts per billion cause persistent and measurable population-level harm to fish, seabirds, and marine mammals.

5. Public policies (laws) on oil pollution are grossly under-protective of life.

Reassessing our Indicators of Individual Health

1. Risk assessments are bogus: they cannot deliver on their promise of worker/public protection because they are not based on realistic parameters.

2. We do not know the full and true adverse health effects from crude oil exposures, one of the oldest known human health hazards; we do not know the full and true adverse health effects of most chemicals on our markets—and in our environment—today.

3. Traditional medical training does not prepare doctors to recognize, diagnose, and treat chemical-induced illnesses and diseases; because of this, our worker/public health policies and our legal system do not support victims of chemical exposures and illnesses; because of this, we feign ignorance of chemical illnesses at the expense of worker/public health and the environment.

4. Biomarkers are effective and accurate diagnostic tools for subtle chronic and systemic effects of crude oil, solvents, and many other environmental pollutants.

Assessing Population Health

1. Unrealistic risk assessments and short-term bioassays are *not* the proper tools to assess population-level and ecological harm to people and wildlife from crude oil, solvents, and other environmental pollutants.

2. To most accurately assess effects of environmental pollution, biomarkers should be used in concert with epidemiology or ecosystem-based (population) studies.

3. Effective population studies involve a holistic approach with five elements: collaborative efforts among multiple scientific disciplines and among scientists and the public; an inclusive and comprehensive planning process; identification and selection of key (sentinel and indicator) species (or subgroups of human population); and comparisons over time and geographic areas.

4. Wildlife scientists fully utilized the "opportunity" presented by the *Exxon Valdez* oil spill to complete the circle from individual to population-level health effects, which resulted in a new understanding of oil effects; medical researchers have not yet conducted parallel studies on human health effects from this spill. The opportunity is still there.

Science, Politics, and the Public Interest

1. Scientific advancements through paradigm shifts are a confusing messy business for all involved, including the public, as starkly demonstrated by the polarized interpretations and conclusions of Exxon scientists and public-trust scientists from the EVOS studies.

2. It is, unfortunately, quite an accepted practice for entities with interests vested in the old paradigm to resist change by conducting bad science or "tobacco science" to subvert the public process (which depends upon public understanding of the risk/issues) and stall undesirable policy changes; this drama is still in play in the case of the EVOS wildlife science. (Given Exxon's track record, it is anticipated that a similar drama would begin should an independent epidemiology study commence on EVOS cleanup workers.)

3. When scientists disagree and technical controversies spill into the public arena, the media is largely unable to sort out the truth and relevant details—the media focuses instead on the controversy itself; whenever scientists duel in public, it is time for the public to pay attention and try to understand what is at stake for the public interest, irrespective of media reports.

Designing Laws to Deter Spills

1. Spillers should not be left in charge of the cleanup; they have a basic conflict of interest between their economic self-interest and the public interests of environmental restoration and protection of worker/public health.

2. Human health problems are likely *not unusual* for spill responders and should be anticipated for *any* oil spill cleanup that involves people working on beaches or with dispersants (i.e., products with solvents or hazardous chemicals); this liability is largely unrecognized at this point because response efforts largely focus on the environment.

3. Laws designed to prevent and respond to spills and to protect workers' health during oil spill cleanups do not work to protect the public interest when rich corporations spill oil. Further, they actually create situations in which the spiller can recover significant sums through accounting and tax strategies: this clearly undercuts the goal of deterrence.

4. Industrial self-reporting of spill volumes should be verified by the U.S. Coast Guard with independent surveyors.

Chapter 24.

Beyond the Exxon Valdez Oil Spill: Recommendations for Strengthening Oil Pollution Prevention

In our journey that began on 24 March 1989, we have discovered that oil is much more toxic to life than previously thought. This chapter focuses on what we can do to reduce oil in our environment to protect life. Prevention measures are framed as a series of recommendations. Before we can turn to these steps, however, we need to understand where the oil that enters our sea comes from.

Inputs and Influences

According to the National Research Council's 2002 book, *Oil in the Sea III,* about 100,000 tons of *"anthropogenic oil"* (oil derived from human activities) entered the coastal seas of North America on average every year from 1990 to 1999. This is equivalent in volume to 29.4 million gallons (National Research Council 2004, 3, 13)—or about one *Exxon Valdez* spill (with the correct volume of 30 million gallons) per year for ten years. Just for the record, natural seeps contributed another estimated 160,000 tons (47 million gallons) per year during this decade, but well over 90 percent were from deep-sea, offshore seeps that did not endanger coastal sealife or human health. The flow rate from these natural seeps tends to be very slow.[1]

Of the 29 million gallons that are preventable, *"consumers account for nearly 90 percent of the discharge"* (emphasis in original), including "an estimated 92 percent of the PAH load"(50), according to the National Research Council. The Council notes, "PAH in urban runoff are automobile exhaust based" (I-13). The lion's share of this input is from land-based river and urban runoff; atmospheric deposition from internal combustion engines, power plants, and other facilities; and recreational marine vessels (two-stroke engines). The remaining anthropogenic input derives from petroleum extraction and transportation, including tanker spills. Here's the real kicker—the average annual input from our fleet of two-stroke recreational craft (including jet skis) equaled the input from tanker spills—about 1.5 million gallons!

I hasten to point out that the oil discharged from two-stroke engines occurs throughout North America, while oil spills occur in localized areas and so present a greater immediate risk to environment, but who can say which source presents a greater long-term threat to life? The National Research Council estimates that ten of the major life-giving, sweet water rivers in North America contain an average load of 820 parts per billion PAHs—well beyond what we now know kills aquatic life.[2]

"We have met the enemy and he is us!" I realized as Pogo, the cartoon character, so aptly stated. Dealing with the dregs of our oil consumption boils down two main approaches—through the consumers and through the suppliers, in this case the oil companies and the U.S. government.

U.S. oil shippers took measures to reduce oil spills when Congress forced their hand. The National Research Council (2002) notes that for tankers, "[O]il spillage dropped off significantly after 1991. This improvement followed the grounding of the *Exxon Valdez* in 1989, and the subsequent passage of the Oil Pollution Act of 1990 (OPA 90)" (E-11). Industry watchers attribute the reduced spillage to preventative measures and increased industry concerns over *escalating financial liability*, specifically, the potential for astronomical costs from unlimited liability under OPA 90 (de Bettencourt et al.

2001). As one senior USCG officer put it, the "requirement for some ships to assume a higher level of financial liability for spilling oil has likely had a greater impact on reducing the amount spilled than the plethora of 'command and control' regulations that (preceded or) followed OPA 90" (Elliott 2001, 31). OPA 90 prescribed the one medicine that worked best: economic sanctions instead of more paper regulations.

Reducing oil spills and oil pollution is a matter of making the polluter pay. When OPA 90 required shippers to assume more of the cost—the risk or financial liability—for spilling oil, shippers took note. OPA 90 did not attempt to identify and account for social and environmental costs, which were not fully recognized in 1990 (Elliott 2001). Since then, much has been learned; this book discusses just some of these costs as harm to human health, salmon, herring, sea otters, and other sealife. The principle of making the polluter pay can be applied to any liability—financial, social, or environmental.

From the perspective of financial liability, the "costs" of our oil dependency can be viewed as economic hot potatoes that are tossed among four players—government, industry, the public, and the environment. Under the existing system, those who profit handsomely from oil production and use, the suppliers, have foisted much of the risk of our oil dependency on an unwitting public and the environment. Tossing some of the social and environmental liability back to the industry and the government will provide the necessary incentives for these two suppliers to reduce the risk of oil pollution—and, more importantly, to begin to consider realistic alternatives to fossil fuels. Tossing some of the liability hot potatoes to the public should have the same effect, especially if people are given viable options.

Thus, the recommendations to strengthen oil pollution prevention are framed from two sides of the coin of oil dependency: supply-side economics and demand (consumer)-based economics.

Supply-Side Economics

Recommendation 1: The spiller should not be left in charge of the cleanup; all spills should be federalized.

Federalizing spills conveys many advantages to the public and environment. First, it strips away the oil industry's profit incentive. The *Exxon Valdez* cleanup was not unique in making a sizeable profit for the spiller and the oil industry support contractors. In fact, industry watchers have identified removal of *"pass-through profits"* (the "cost-plus" in the oil spill contractors' world) to reduce costs of oil spill cleanup (de Bettencourt et al. 2001). Mark-ups of 10 percent by the primary contractor for profit are not uncommon, according to industry consultants and the USCG. Federalizing spills would turn cleanups into not-for-profit operations.

Spillers would have to reimburse the government for cleanup expenses. Rather than a direct reimbursement, cleanup funds could be moved through the Oil Spill Liability Trust Fund created under OPA 90. This would help prevent the spiller from directly controlling expenditure of cleanup funds. The shuffling of financial liability provides another disincentive to spill, because the government is a notoriously expensive contractor—but at least the money does not further enrich the polluter. This incentive would work best if the costs of spill cleanup are *not* allowed to be deducted as tax write-offs, as under the current system.

Federalizing oil spill cleanups means the government, not the spiller, would determine which methods and products to use to minimize harm to the environment. Methods harmful to intertidal sealife, such as the pressurized hot water wash, could be stopped. There would be no bullying to use dispersants in sensitive habitat. The government would be free to choose which cleanup products worked best, rather than the spiller's products. (Exxon used primarily its own dispersants—various Corexits and Inipol— during the *Exxon Valdez* cleanup.)

Federalizing oil spill cleanups eliminates the embarrassing but paramount need for the government to secure the spiller's cooperation in order to clean up a spill. Currently the government has neither the dedicated resources nor the cash flow to undertake large cleanups. The NRDA process and OPA 90 do not have dedicated funds; our government's damage assessment is at the mercy of promises of reimbursement by the spiller! During the *Exxon Valdez* cleanup, Exxon withheld $20 million for funding government science during the first year, because the company was not allowed to participate in the development and design of the scope of the studies and the study results did not give the company "comfort," according to testimony of a senior Exxon official (Harrison 1990). Resource managers were forced to come begging to Congress for funds (Collinsworth 1990; Stewart 1990; Suuberg 1990).

Federalizing spills resolves this untenable issue; OPA 90 needs to be amended to ensure that the government also has access to NRDA funds through the Oil Spill Liability Trust Fund or some other dedicated funding account that is reimbursable through the spiller. The government could offer discounts off the full cost of the spill studies to encourage prompt payments and discourage lawsuits.

Federalizing an oil spill cleanup also ensures the same standard of cleanup whether the spiller is unknown, poor, or rich and whether the spill occurs in a little known area with little wildlife or a place that captures the public's heart, such as Prince William Sound. If the oil industry is put on notice that all spills *will* be cleaned up at the spiller's or industry's expense (the latter in the case of unknown spillers) to the same standard, then the industry itself is more likely to apply peer pressure to reduce the risks of spills. Further, as part of federalizing spills, the government should calculate the volume spilled, using independent contractors who don't have "a close working relationship with industry," (the excuse given by Caleb Brett for withholding its surveys of the *Exxon Valdez* from the State of Alaska [Hennelly 1990]).

Finally, federalizing spill cleanups addresses one of the biggest inequities of our oil dependency: that the benefits are spread among all oil users, while the risks are borne by the few communities, workers, or innocents (human and non-human) directly in the path of an accident. Simply put, oil disasters cost victims and their communities and they also cost all taxpayers in the form of tax write-offs for corporate cleanup expenses (at cost-plus). If the oil industry were required to reimburse the government for cleanup expenses and spill studies *without the benefit of tax deductions*, then the industry would take further action to reduce risk of spills.

Federalizing oil spill cleanups has been successfully done in other countries such as Norway that value their coasts, their seaports, and their sea-based economies. Some countries such as Shetland in Great Britain have stellar examples of shared oversight of the oil industry among the government, industry, and local citizens. This type of participatory democracy should be incorporated into reducing risk of oil spills at every phase of oil and gas development, including general oil industry oversight, monitoring programs, and oil spill cleanups. The *Exxon Valdez* experience provided a good model for participatory oversight: the Regional Citizens' Advisory Committees, established by OPA 90, in Prince William Sound (www.pwsrcac.org) and Cook Inlet. Other communities would benefit by establishing similar citizen oversight of oil and gas industries in their region.

Recommendation 2: Response preparation should include revised guidelines for dispersant use, stockpiling equipment, and treating injured wildlife.

Dispersant Use Guidelines

Dispersant use is an issue that comes loaded with liability—which both the federal government and the industry have managed to dodge, leaving people and the environment at risk—a very hot potato. During the past decade, the oil industry has stepped up its efforts to gain government pre-approval to use dispersants (Aurand, Coelho, and Steen 2001), presumably a cheap, alternative response method, instead of the expensive, labor-intensive mechanical pickup methods. Based on the history of the industry, one would assume that less surface oil equates to lower cleanup costs and less damning public relations exposure.

The National Academy of Sciences is currently gathering information to better understand and make recommendations for dispersant use in freshwater, estuarine, and marine environments (dels.nas.edu/dispersants/index.html). It is time to inject a dose of reality from the *Exxon Valdez* experience into this discussion. (For the purpose of this discussion, I define "dispersants" as commercial products with solvents as active ingredients; this includes Inipol.)

There are several new perspectives that must be considered in the nearly thirty-year old dispersant debate. First, there are worker health concerns, which have never been much of a factor in past debates, but now must be for reasons discussed below. Second, protecting intertidal habitat quality over the long-term has emerged as a new concern that must be balanced against short-term losses of fish and other wildlife from dispersant use.

The problem with this latter concern is that no one has ever monitored chemically-treated beaches to determine whether treated beaches recovered faster in terms of less oil and so less damage to wildlife over time than the untreated beaches. NOAA scientists, including those at the Auke Bay Lab, are just now considering such a study in Prince William Sound. Further, no one has determined whether dispersant use is a net gain or loss to wildlife populations already struggling to survive in more contaminated areas than the Sound. This concern echoes Jane Goodall's wisdom that in a declining population, every individual counts. Industry and agency scientists are debating the environmental costs of dispersants like blind men discussing an elephant, each without enough information to see the whole.

Given these new concerns of human health effects and intertidal habitat quality, the dispersant debate should be re-examined from an onshore versus offshore cost/benefit perspective. Dispersant composition has evolved during the last three decades, however, the arguments for and against dispersant use remain unchanged—and focus *almost exclusively* on potential harm to wildlife. The National Research Council stated in its 1989 book, "in shallow water with poor circulation, and in protected bays and inlets, the acute biological effects on some organisms and habitats from high concentrations of dispersed oil may be greater than the effects of untreated oil" (256). A decade of oil industry research found little difference: "Dispersant use involves trade-offs, some habitats may be at greater risk, while others may be protected" (Aurand, Coelho, and Steen 2001, 432). By 2001, it was widely accepted that dispersants harm shallow nearshore and intertidal habitats—that is, the costs outweigh the benefits of use in these areas (Boyd, Scholz, and Walker 2001).

Before considering the human health effects of dispersants, it is instructive to review how industry and government handle dispersant use from a liability perspective. The EPA maintains a schedule of chemical products for use in the National Oil and Hazardous Substances Pollution Contingency Plan. The schedule includes information on the manufacturer, special handling instructions, shelf life, and results of toxicity tests (on two standard laboratory animals) and effectiveness tests, which are usually accepted without independent verification by the EPA. The EPA specifically states, "The listing of a product on the Schedule does not constitute approval of the product" (Nichols 2001, 1483). Listed products must carry disclaimers stating, in part, that "listing does NOT mean that EPA approves, recommends, licenses, certifies, or authorizes use of (Product Name) on an oil discharge" (ibid., 1483, emphasis in original).[3]

The EPA requires dispersants to be at least 45 percent effective to be listed (ibid., 1483; 40 CFR 300.915 [7]); there are no thresholds for the toxicity tests. However, "EPA makes no claim that any of the listed products work exactly as they are supposed to"(Nichols 2001, 1483) and it turns out there are exemptions for even these "thresholds." Dispersants are only 10–15 percent effective with the viscous Alaska Prudhoe Bay crude spilled in cold water (Spiess 2001). Most dispersants do, however, meet the effectiveness threshold when applied to South Louisiana crude. So the EPA created a loophole for dispersant manufacturers (in many cases oil companies): it allows the results of the effectiveness tests for Prudhoe Bay crude and South Louisiana crude to be *combined and averaged* so that some dispersants can pass the effectiveness test! (Nichols 2001, 1480). Readers may well wonder what this means for spill response—do responders order up a tanker of South Louisiana crude to mix with any spilled Prudhoe Bay crude just so the dispersants will be effective?

A more basic problem with EPA's Schedule is the chemical products are designed for specific purposes. Some products are opposite in action and purpose to other products. For example, dispersants work to spread oil *into the water* column; surface-washing agents work to break up and lift the oil so it will float *on the water* to be collected and removed. Some products listed as bioremediation agents (such as Inipol) contain enough solvents to disperse oil. Interchanging products creates problems for wildlife *and workers*.

EPA knows its system is rife with abuse. According to a 2001 paper by William Nichols, who manages the Schedule for EPA's Oil Program Center, "EPA is concerned that these categories are often interchanged, which leads to misuse of the products. Surface-washing agents (SWAs) have been used on open-water spills, while dispersants have been used to wash oil from sandy beaches driving the oil deeper into the substrate. *Both misuses may cause further harm to the environment than the oil alone*" (Nichols 2001, 1481, emphasis added). Further, EPA does not enforce but rather "*encourages* the prudent and effective use of listed products" (1480, emphasis added). This is simply not adequate to protect wildlife—or workers as the *Exxon Valdez* cleanup demonstrated.

Even with all the exemptions and disclaimers, dispersants are still essentially industrial solvents and solvents are toxic to wildlife and people. This is why several of the dispersants used during the *Exxon Valdez* cleanup came with Material Safety Data Sheets (MSDSs) that warned to keep the product out of watercourses. (The warnings were obviously ignored, because Prince William Sound *is* a watercourse.) One dispersant, Corexit 9580M2, even came in barrels with labels that warned, "toxic to fish" (Wells and McCoy 1989). (This product also was used on beaches.)

When dispersants are applied to spills in open water—the proper use encouraged by the EPA, there is maximum mixing and dilution to protect wildlife and there is minimal contact with cleanup workers and other response personnel (because dispersants can be

sprayed from planes).When dispersants are applied to beaches, however, there is minimal mixing and little dilution, which threatens sealife, and the labor-intensive process places workers in harm's way as
demonstrated by the health consequences of the Disc Island dispersant test (Chapter 1) and use of Inipol (Chapters 6 and 7).

Regarding human exposure risks, the *Exxon Valdez* cleanup
demonstrates *unequivocally* that adverse health effects from exposure to dispersants of workers in realistic cleanup conditions, including extended work-shifts, shortages of respirators and goggles, and
lack of opportunities for proper personal hygiene, *must be included*
in the dispersant debates. From the data available after the *Exxon
Valdez* cleanup, it was impossible to differentiate between oil exposures and oil-dispersant exposures. (The medical records on Inipol
are an exception, but they were not yet available.) Since the physical
symptoms largely overlap, this may seem of academic interest only,
however it is not. Dispersants *do not have to be used* in the intertidal and nearshore areas.

Cleanup workers are not the only people at risk from use of solvents on the beaches. EPA does not currently authorize use of dispersants in freshwater, because of concerns about contamination of
groundwater and drinking water. During the *Exxon Valdez* cleanup,
at least one vessel created drinking water by desalinating seawater
immediately offshore of a beach being treated with Inipol! It is
unlikely that the desalination process removed this chemical.

Further, Alaska Natives and area residents subsist on seafood collected from beaches. Alaska Natives have a saying, "When the tide is
out, the table is set." Villagers understood that solvents are not good
for wildlife or people. There is a classic story of one tribal leader, Gail
Evanoff, who sat on a beach near her village, Chenega, to prevent
workers from spraying Inipol (Hyce 2003). No one studied the "shelf-
life" of these chemicals on the beaches and scientists operated under
the assumption that the most toxic effects would diminish rapidly
within the first 24 to 48 hours. However, there is anecdotal evidence
to suggest that the residual effects lasted much longer. In 1990 two
ADFG contract researchers developed headaches and nausea after
camping on one beach in the Sound. They discovered a withered
scary-faced balloon on the beach and later learned that the beach
had been treated with Inipol two weeks before they camped on it
(Susan Ogle, Cordova, AK, pers. comm. January 2004). (The balloons
were used to scare away wildlife and humans from treated beaches.)
Further, in 1990 two USFWS employees were not notified that the

beach where they were conducting a study was still being treated with Inipol (U.S. Coast Guard 1993, 364).

It is extremely irresponsible of the EPA to list potentially hazardous products for use in public multiple use areas where the chemicals can potentially threaten subsistence users, visitors, residents, and researchers who gather foods, picnic, camp, hike, or conduct studies on beaches. In fact, it is irresponsible for industry and the public agencies to even debate this issue without a broader focus of the full risks to people and the environment.

In addition, EPA should require independent and verifiable toxicity and effectiveness testing, without averaging the effectiveness on two vastly different crude oils before any product is listed as acceptable on the EPA product schedule. Toxicity testing should be required to be performed on species present in the areas where the dispersant may be applied.

De-listing of dispersants from Schedule C should be a formal process requiring a *written explanation* from the manufacturer of the reason for the de-listing. *Manufacturers should be responsible for the consequences of product use as long as the product is listed in the EPA's Schedule C* and *as long as the product is available for use (unless the product is formally recalled).* As it stands now, dispersants can be de-listed without industry explanation as to why the product has been taken off the list. For example, Inipol EAP 22 was discontinued as of January 1996 (U.S. EPA 2004a). One is left to wonder if the *Exxon Valdez* cleanup workers were unwitting guinea pigs in an experiment that failed to produce desired results.

The bottom line is that the federal government currently offers industry a virtual risk-free, rubber stamp process by which to use industry's preferred response option—chemical products. Neither government nor industry is willing to accept liability for the very real human and environmental costs of these products, as evidenced by EPA's disclaimer and Exxon's indemnity form, which it *paid* workers to sign after the disastrous Disc Island dispersant test in late July 1989 (Figure 1, p. 33). This liability-free ride creates a disincentive for industry to develop something that doesn't have the high health and environmental costs—or, if this can't be done, then to shoulder more responsibility to not spill oil in the first place.

Anything for which both industry *and* government are not willing to assume liability should not be used in environmentally sensitive habitat or handled by humans! *At a minimum, dispersants and other chemical products with high-test industrial solvents should be permanently BANNED in nearshore areas and on beaches.* In offshore situations, where dispersant use may lessen the risk of oil stranding on beaches, dispersant use could still be considered—but only with full knowledge of human risks, proven protection for workers in realistic situations, and realistic independently-verified toxicity and effectiveness tests.

In addition, any time dispersants are used, human health effects should be monitored *by government oversight agencies and non-industry funded researchers* over the long-term—and medical records, raw data, and results made public on a regular basis—to build our knowledge base. Funds could be appropriated from the Oil Spill Liability Trust Fund for this purpose, because the studies would benefit the public interest.

Stockpiling Equipment Guidelines

Changes in Alaska laws after the *Exxon Valdez* oil spill require the oil industry to stockpile oil spill response equipment in various key locations to be better prepared for spills of up to about 11 million gallons—supposedly the size of the *Exxon Valdez*. Equipment includes boom, skimmers, and even dispersants; *it does not include* personal protective gear for cleanup workers. Given the experience from the *Exxon Valdez* spill, industries that routinely handle materials that are governed by hazardous waste cleanup laws, such as crude oil, should be required to stockpile respirators, replacement filter cartridges for the respirators, goggles, gloves, and other protective clothing in strategically located warehouses in amounts necessary to respond to loss of a full tanker load. Equipment should be state-of-the-art, so inventory should be reviewed every three to five years and replaced, if necessary. Mobilization of thousands of workers to respond to accidental cleanups leaves zero lead-time for manufacturers to produce gear critical for protecting cleanup workers.

Further, Alaska now has some of the oldest stockpiles of dispersants in the country. Industry must be required to demonstrate effectiveness of this stockpiled product every three to five years, as well, and product that no longer meets the EPA effectiveness tests—or product that has been de-listed—must be discarded (at hazardous waste disposal sites) and replaced.

Wildlife Treatment Guidelines

Every public-trust scientist with whom I spoke about sea otters, *without exception*, was very critical of Exxon's wildlife "rehabilitation" centers (Bayha and Kormendy 1990). Treating and releasing sea otters was done to appease an irate public. Like the pressurized hot water wash, it took on a political life of its own. The evidence shows that most of the heavily oiled animals suffered horribly and died despite "treatment." Healthy unoiled animals were brought into the centers and suffered from the stress of capture. Surviving captive animals never fully recovered from the "treatment" and had shortened life spans upon release into the wild or captivity. Further, "treated" animals released into the wild triggered a disease outbreak that claimed the lives of dozens of healthy adult animals. The Marine Mammal Commission summed up the situation for sea otters in a book published three years after the spill: "the rescue and rehabilitation program was not very effective" (Hofman 1994, xiv).

In light of the *Exxon Valdez* experience and the collective experiences from other spills, wildlife treatment after an oil spill should be re-examined by the industry, government, and public. At a minimum, collection of live animals should be left to professionals. Human empathy for injured oiled wildlife is better directed into action to prevent spills from happening in the first place rather than misdirected (albeit well-intended) efforts to "save" the innocent victims of our oil mishaps after the fact.

Recommendation 3: Old laws should be revised based on new science.

The EPA listing (2001; 2004b) of twenty-two PAHs as persistent, bioaccumulative, toxic (PBT) pollutants is profound and affects every aspect of oil use from discovery (lease sales and exploratory drilling) to disposal of wastes (by industry and individual consumers). This type of use monitoring is called *"cradle-to-grave"* and it applies to hazardous substances, which we now understand crude oil to be, even though it does not yet carry this legal label and the liability stigma associated with it. Modern day researchers have expanded the concept of cradle-to-grave to *"cradle-to-cradle"* to accommodate the latest science, which shows that the PBT pollutants do not have a "grave," because they are persistently recycled through the biological web of life for many generations.

It is well beyond the scope of this book—and this author!—to consider the vast re-tooling of our nation's laws governing oil leasing, exploration, drilling, production, transportation (by tankers, pipelines, and trucks), distribution, consumption, and waste disposal in order to regulate oil as a PBT pollutant. Suffice to say that it must be done and that this information should become part of every discussion that concerns oil use—the energy debate, the car fuel efficiency debate, the tri-annual review of water quality standards, on and on. It already has been done in regions where people and political leaders are more aware of the true costs of our oil dependency. For example, the State of California adopted the "California-only hazardous waste law," which regulates crude oil vapors and liquid waste as hazardous. At a minimum, *crude oil should be reclassified as a hazardous substance and a hazardous material* under all pertinent federal laws and regulations.

A logical starting place for policy-makers is to weigh the elevated risk to public health and the environment from oil exposure against the perceived benefits and to then revamp laws and regulations from this new perspective. By way of example, I will focus on three laws that are discussed in this book: Superfund, OPA 90, and the Clean Water Act's water quality standards.

Superfund and OPA 90

Superfund and OPA 90 share a common "fix:" the need to institute a long-term ecosystem-based approach to assess damage and recovery from oil (or other hazardous waste) spills. Polluters should pay the cost of *all damages* to public lands and wildlife, including both short- and long-term injuries. Currently these laws only assess short-term damage and do not recognize long-term damage; the latter cost is borne by the public and the environment. Insisting on full payment for all damages provides another incentive for industry to reduce risk of spillage.

OPA 90 actually took us a step backwards, regarding assessment of long-term damages—it ensures the resource managers will never "see" the damage in the first place! To secure the spiller's cooperation and eliminate a problem created by Superfund (as discussed above), policy-makers removed most of the litigation risk—the roadblock to cooperation. They accomplished this by allowing the spiller to participate in the design and gathering of the NRDA data and to review, comment, and approve any restoration plans, all of which have significant financial benefits for the spiller. Disputes are "settled" with 1970s-style toxicity tests, which guarantees that NRDA studies will be capped at low levels and long-term damages will not be identified, which again benefits the spiller. When OPA 90 passed, long-term damages from oil spills were not anticipated, based on the science at the time, so this may have seemed like a good trade then—the spiller's cooperation in trade for something that was thought not to exist. We know better now and this policy needs to be fixed to reflect our new knowledge.

It may be possible to determine estimates for long-term damage based on new types of bioassays, sprouting from the main stem of studies conducted by the Auke Bay Lab. For example, one group in Quebec, Canada, has developed a sensitive and reliable eleven-day test for fish embryos with different types of oils (Couillard 2002). In the public interest of settling damages for public lands and wildlife quickly, it is necessary to find suitable indicators of long-term damages. Alternatively, in the interim, such damage could be calculated, based on a sliding scale of anticipated harm and using indicators such as percent of PAHs in the oil, volume spilled, volume stranded in sensitive habitat, and the time of year the spill occurred.

It is worthy of note that Superfund and OPA 90 deal with hazardous substances *after* they spill. The same type of ecosystem-based monitoring should be incorporated into the *pre-spill* permits of various phases of oil development and production, including wastewater discharge. As the public-trust scientists found out after the *Exxon Valdez* spill, the ability of post-spill studies to assess damage and recovery was much greater when pre-spill baseline studies were available for comparison. The federal government should conduct or contract the baseline monitoring studies with reimbursements from permitees as a tax-deductible cost of doing business. This way the cost of the baseline monitoring studies is shared among all the consumers of oil, while the cost of any *post-spill* studies are borne by the spiller.

Clean Water Act

Water quality criteria were established under the Clean Water Act to protect everyday use of our nation's waters. This use includes human needs, such as drinking water and recreation, and non-human needs, such as protecting aquatic life from pollutants. It is clear from the studies on early life stages of fish and from the ecosystem-based studies that a criterion for total aqueous PAHs is overdue. Further, the current water quality guideline for PAHs of 300 parts per billion is set way too high to protect sealife and presumably freshwater life. We now know that PAH levels of 1 to 20 parts per billion sicken individuals and reduce entire populations of fish, birds, and animals. These low levels of PAHs are washing off streets in urbanized areas and draining into our rivers and coastal seas on a daily basis (National Research Council 2002).

Applying the 1970s logic of dividing the lowest concentration at which an effect is observed—1 part per billion PAHs—by 100 yields an impossibly small standard of 10 parts per *trillion*. This is below the limits of most analytical equipment to even detect and it is beyond the ability of even the best wastewater technology in the United States to remove from highway runoff or industry effluent prior to discharge. Meeting this standard requires essentially zero discharge of PAHs. This creates an almost unimaginable situation in which we seem to lack the proper tools to adequately protect our nation's waters—and ultimately, life.

Where do we go from here? The answer, of course, is to start taking steps to tighten all the leaky taps that drip, drip, drip PAHs into our rivers and oceans. There are a lot of taps and each one is associated with someone or some industry that has learned to live with the annoying drip and so will be resistant to change. Several of the following recommendations suggest ways to approach our national plumbing crisis; that is, to institute fundamental change to reduce chronic oil pollution.

Recommendation 4: The 1991 civil settlement in the Exxon Valdez case should be re-opened and the entire $100 million available for unanticipated long-term harm should be claimed for educational purposes.

The 1991 civil settlement provides a key tool to initiate a fundamental overhaul of our national laws regulating oil use: a re-opener clause based on unanticipated long-term harm to wildlife from the *Exxon Valdez* oil spill. Stipulations for reopening the settlement are twofold: first there must be proof of unanticipated long-term harm, and second, there must be a realistic way to restore the injured resource. Up to $100 million is available for the latter purpose. The reopener clause sunsets on 1 September 2006.

If the settlement is successfully reopened before then, and funds are secured to address the harm, then this combined action would force policy-makers to acknowledge that oil causes persistent long-term harm. This action alone would upset the legal apple cart, because all of our national laws regulating oil use are predicated on short-term damages from oil—the scientific understanding in place when our body of laws were established.

The public-trust scientists are confident that there is long-term harm, evident fourteen years after the spill in intertidal habitat and species such as harlequin ducks and sea otters. They are confident this harm was unanticipated based on the science available at the time of the settlement. In 2003 they started to discuss restoration options to mitigate this harm in injured species.

However, at this point, the issues of oiled habitat and lingering harm pales in comparison to the more basic issue: that oil is much more harmful to life than previously thought and immediate steps need to be taken to curtail oil pollution. Action requires education; people and policy-makers cannot take any action until they are aware of the new risk.

It seems, then, that the highest and best use of the settlement money is to educate the public about the persistent harmful effects of oil in the same way that money from the tobacco case was used for public education about the harmful effects of smoking. The money could be used by the state and federal governments—or by a court-designated non-governmental organization—to establish an *Exxon Valdez* Oil Spill Legacy Educational Fund (EVOS ELF) to produce educational material on oil effects to wildlife and people for our nation's K-12 schools.

Educator and former Cordova fisherman Rick Steiner first proposed this idea in June 2001. None of the parties to the 1991 settlement and the ones responsible for reopening it—the United States Department of Justice, the State of Alaska, and Exxon—support it, or, in fact, have indicated any enthusiasm for reopening the settlement at all. The problem is politics. For example, 85 percent of the State of Alaska's general operating revenue is derived from North Slope oil activities. Given the current political climate, it is better to establish the EVOS Legacy Educational Fund through a non-governmental organization.

How will this happen when the governments are dragging their feet about reopening the settlement? When governments are controlled by special interest corporations, it is the people who must act to initiate change. When the public is organized into a clear majority about a certain position, politically entrenched mountains move, because politicians are sensitive to popular pressure. The fate of the reopener funds will be decided by Judge Russel Holland who will evenly consider all proposals, but first, at least one of the parties to the 1991 settlement needs to request that the settlement is reopened to consider long-term damages and appropriation of funds.

Ordinary people—YOU!—can influence this process by writing the President of the United States, your congressional delegation, the EVOS Trustee Council, and federal Judge Russel Holland to insist that the 1991 civil settlement is reopened and that the funds are directed to a non-governmental organization to establish an EVOS Legacy Educational Fund. (Contact information is provided in Appendix B, p. 455.)

Recommendation 5: We need to take the politics (corporations) out of science in order for science to best serve society.

In her Presidential Address at the annual meeting of the American Association for the Advancement of Science in 1997, Jane Lubchenco, PhD, challenged science to meet the needs of a rapidly changing society. She warned, "All too many of our current environmental policies and much of the street lore about the environment are based on the science of the 1950s, 1960s, and 1970s, not the science of the 1990s" (Lubchenco 1998, 495).

I believe that much of the environmental science (and social sciences as well; e.g., see Zarembo 2003) is being held hostage by powerful interests vested in oil, petrochemicals, the auto industry, and other corporations that grow rich on polluting our commons—the air, water, and soil we all share freely. It is not in corporations' best interest to have science educate society about the true risks—the threat to life on earth—posed by side effects of their businesses. So these corporations buy scientists, university professors and others, to spin counter stories, create public confusion, and stall unfavorable policy changes. "Whose truth are we talking about, your truth or my truth?" public relations specialist, John Scanlon, retorted to a reporter who had asked him whether he served his clients or the truth (in Rampton and Stauber 2001, 57). Until science deals with its Achilles' heel of advocacy science, *it cannot meet society's needs to protect life* and we will continue to pollute our soil, air, and water, ourselves, and other life on this planet. Like lemmings, we are all racing towards the cliff.

Based on the *Exxon Valdez* experience, four actions could be taken to control corporate abuse of science. First, as Jane Lubchenco pointed out, "Strong efforts should be launched to better communicate scientific information already in hand" (1998, 495). The targets of this effort should not be limited to policy-makers, but to ordinary people—laborers, consumers, students—who by their individual actions, multiplied by tens of millions, can reshape the world that we all share. This effort should also target the media, which in its ignorance of basic science and statistics, naively treats all science as equal, rarely examines motives (conflicts of interest), and focuses on superficial controversies rather than the real life-threatening issues at stake (Stauber and Rampton 1995). The public needs more science first-hand from credible groups of scientists. This effort is so critical that it deserves the full attention of a group such as the National Academy of Sciences to issue guidelines for public education as a mandatory component of publicly funded research.

Second, whenever a scientific controversy spills over into the media (which is the route used by corporations and advocacy scientists to confuse the public), *immediate steps should be taken* by a neutral party to referee the science and separate the grains of truth from the chaff of public relations. Unless challenged immediately, the press and public will develop their own understanding of the issue— the "street lore" referred to by Lubchenco—which tends to be based on the information which is the easiest to understand and which requires the least personal change. As evidenced by the *Exxon Valdez* controversy, corporations, with their army of public relations specialists, can make things that aren't true very appealing to believe.

Fake controversies stall important environmental policy changes and guarantee continued environmental degradation to support short-sighted economic interests. The lemmings race closer to the cliff. Again, this scientific refereeing is so critical that the National Academy of Sciences should designate a special group just to referee, for society's sake, environmental controversies that have global implications such as global warming, the ozone hole, and the *Exxon Valdez* spill effects (persistent oil effects).

Third, scientists need to become accountable to the public. The current system of "peer review," in which experts pass judgement on the work of other researchers supposedly to help protect the public from scientific errors and bias, doesn't work. In 1997, the researcher lamented in the *British Medical Journal* that peer review is "expensive, slow, prone to bias, open to abuse, possibly anti-innovatory, and unable to detect fraud" (Smith 1997, 759-760). The formal requirement for disclosure in many scientific journals is often ignored: one study that examined 62,000 articles published in 210 journals found only one-half of one percent included financial information (Krimsky 1999). Journals need to enforce their own policies and the media needs to take responsibility for providing the public with this information when quoting "experts."

In 1999, the White House Office of Science & Technology Policy adopted a more stringent definition of scientific misconduct for federal researchers—private research is not held to a similar standard (Hileman 1999). Since the voluntary system of peer review doesn't work, perhaps it is time for the National Science Foundation to adopt similar standards for private research. The standards should include a section on study design to eliminate the "common tricks," purposely used to deceive the public, as illustrated in *How to Lie with Statistics* and as demonstrated by Exxon's spill studies. "Peer-review" journals should only accept papers from authors whose work meets these standards.

Adopting standards for private research would also defuse *Daubert:* judges would no longer have to judge the adequacy of science. The scientists could take care of that themselves. For example, it wouldn't matter which journal published the papers (traditional or alternative medicine) as long as the research was held to the same scientific standards.

Finally, fourth, the Freedom of Information Act no longer protects the public interest. This act was originally designed so the public could access publicly funded research, but corporations are increasingly using the act to obtain information from the government in order to strengthen their advocacy science. For example, changes to this act allow "the public" to obtain government information before the original researchers even complete their study. This way, corporations ("the public") can publish advocacy science papers to interpret or discredit the original work before the original authors have time to do quality assurance and interpret the data themselves! Exxon has repeatedly demanded data from the NOAA Auke Bay Lab on studies still in progress, specifically to undermine studies that demonstrate persistent oil effects. Such abuse of the federal law is a tragedy for the public interest and a victory for industry.

In the new age of advocacy science, the Freedom of Information Act needs to become a two-way street. If industry demands studies from the government, then the public (including the government) should be allowed to use the Act to demand studies from the industry making the request. This would even the playing field for science, while still protecting the public interest because it would open industry's science for public scrutiny without having to file a lawsuit.

Recommendation 6: The legal/medical system designed to protect worker and public health needs to shift from a risk assessment approach to prevention and a precautionary approach. It also needs to be over-hauled to recognize and address chemical-induced illnesses.

Risk Assessment

We need a better system to protect *public* health. The basic tests, models, and formulas that form the foundation of risk assessment are outdated and fatally flawed, incapable of predicting "safe" levels of exposure of any single chemical in the chemical blizzard of our environment. Forget the standard 70-kilogram human (150-pound man) used in risk assessment; we need a system that protects the most vulnerable among us. We need more comprehensive tests, designed to assess harm, for example, to complex immune system functions and embryos. We need tests to match the very pathways that specific chemicals are designed to disrupt. For example, absent established testing protocol for endocrine disrupters such as PAHs, these endocrine disrupters should not be allowed on the market until such tests are designed and conducted, and the chemical proven safe *for non-target species including humans,* by the agencies responsible for protecting public and environmental health.

Also, we need realistic exposure limits. We need to determine the most common chemicals in the "normal population" and in the environment and we should test exposures of new chemicals against this mix of common chemicals already in our bodies and environment rather than in a vacuum. The National Centers for Disease Control and Prevention (CDC) under the U.S. Department of Health and Social Services recognized this problem and has established a baseline level of chemical residues in humans (NIOSH 2003). The CDC hopes to update its information every couple of years. We need tests that accurately predict effects of *chronic* low-level exposure, not just acute exposure.

Every new chemical should be thoroughly tested, as described, by public agencies responsible for protecting public health *before* the chemical is released. Until public health agencies are adequately funded for these tasks, new chemicals with unknown health risks should not be approved for market and workers should not be put in harm's way. Given the uncertainties inherent in the risk assessment process, OSHA should adopt the precautionary principle ("better safe than sorry") with respect to approval of new chemicals into our environment.

This challenge is so great and so life-threatening that the National Academy of Sciences should convene an interdisciplinary committee of scientists and public representatives, all without affiliation to the petrochemical industry. The committee should make recommendations to Congress on how to protect public health from chronic exposure to chemical poisons, including PAHs.

Similarly for workers, risk assessment needs to be revamped so OSHA can establish realistic personal exposure limits (PELs) for compounds of concern—not surrogates!—based on multiple simultaneous chemical exposures and more subtle measures of harm such as endocrine disruption or immune function suppression. Further, OSHA should *require* that PELs are adjusted to fit the actual hours worked, especially during disaster response, and OSHA should *require* health monitoring on the job and for the long-term if workers are exposed to hazardous materials.

Hazardous Waste Cleanups

We currently do not do enough to protect workers' health before, during, and after hazardous waste cleanups. Federal oversight should be *mandated*. The exemption for recording "colds and flu" as work-related illness *should not apply* during hazardous waste cleanups. This exemption creates a loophole that allows many symptoms of chemical exposure to slip through unrecognized and unreported such as most likely occurred during the *Exxon Valdez* cleanup. If there are illness outbreaks of *any* type, including "colds and flu," during a hazardous waste cleanup, then long-term monitoring should be *required* until we better understand chronic effects of acute exposures.

The spiller should not be the primary responsible party for worker health and safety (Sidebar 16, below); instead, during hazardous waste cleanups, the federal government should hire an independent occupational medicine physician to lead the worker safety program. The party in charge of the worker safety program should be *required* to provide a copy of all exposure assessment data and medical records to OSHA and the spiller.

OSHA should establish a central repository for holding records from hazardous waste cleanups. The records should be maintained for thirty years, as currently required, but the data should be available to the public, such as researchers who wish to track the course of

---

Picking Up the Pieces of a Failed Worker Safety Program

The well-being of workers is addressed in the National Contingency Plan, subpart C, 40 CFR 300.38 (as well as federal and state occupational safety and health laws). The National Contingency Plan directs the spiller to be in charge of worker safety and requires the Federal On-Scene Coordinator (a high-ranking U.S. Coast Guard official) to be "alert" for problems (U.S. Coast Guard 1993, 397).

During the Exxon Valdez cleanup, the U.S. Coast Guard appeared to have focused on physical injuries rather than illnesses (ibid., 399, Table 17.1), because of a belief that the "levels of toxics . . . had diminished significantly since the spill; thus, the greatest remaining danger from contact with petroleum residues was skin irritation" (ibid., 400). The Coast Guard reported that a "monitoring program was maintained throughout the response, with exposure tests (lead, hazardous compounds, etc.), screening tests (for biochemical changes associated with tissue damage), and other analyses . . . " (ibid., 402), but failed to mention that public health officials never saw or obtained the records, as per the NIOSH investigators' original plan (NIOSH 1991).

The Coast Guard noted that state OSHA authorities did not recognize Inipol EAP 22 and Corexit 9580 as hazards "likely to cause death or serious harm if appropriate monitoring and measuring is not performed" until 1990 (U.S. Coast Guard 1993, 403). The Coast Guard reported "a number of [legal] cases of serious illness among former cleanup workers" in 1992. The agency concluded that the matter of whether there were long-term or delayed ill effects from the cleanup on worker health "is likely to remain unresolved for some time, and worker health issues may ultimately be litigated, perhaps in significant numbers" (ibid., 404).

Such after-the-fact reporting does not seem to be in the best interests of protecting worker well-being. Instead of leaving the matter up to individual workers to litigate—or to the public health officials (OSHA) that fumbled the ball of worker health protection during the 1989 cleanup, Congress should subpoena Exxon for all the medical records from the cleanup and authorize a long-term health monitoring program (Recommendation #6, p. 436).

exposures and illnesses over time. This way, we might actually learn something from these unfortunate toxic releases.

Workers' Compensation Laws

Our national workers' compensation system is not currently designed to recognize or treat chemical illnesses or to adequately compensate chemically injured workers. Some of the symptoms take years and generations to manifest.

To facilitate better understanding of chemical illnesses, workers' claims from hazardous waste cleanups should be processed different from the other workers' claims. They should be coded as hazardous waste cleanup claims. Only personnel with training in environmental medicine or occupational medicine should diagnose and code these claims. The coding system should be expanded to include the symptoms of chemical injuries. For example, headaches and dizziness should be coded as central nervous system symptoms (if workers were exposed to chemicals with these effects), rather than simply coded as "other" or "ill-defined." When workers' compensation boards are forced to recognize chemical illnesses, then the legal system will no longer provide a refuge for corporate polluters.

Exxon Valdez cleanup workers filed injury and illness claims through several different state and federal agencies. This makes it difficult for public health agencies and researchers to piece together these valuable data to determine if there are any long-term effects from these hazardous waste cleanups. Therefore, *all* workers' claims and records from hazardous waste cleanups should be sent to OSHA's central repository, recommended above, for use by public health agencies and researchers.

AkPIRG, the Alaska Injured Workers' Alliance, among others, contend that the Alaska Workers' Compensation Program is currently biased against workers: the program favors insurance companies and industry and dismisses ("controverts") the vast bulk of claims. It is outside the scope of this book to cover recommendations to fix this program in detail. However, at a minimum, the Alaska Workers' Compensation Board should be reconfigured so *more than half* of its members are occupational medicine physicians or industrial hygienists, and delegates of groups that represent injured workers. A board composed solely of insurance companies and industry representatives is a deck stacked against injured workers. The reconfigured board could then begin to address the myriad of ills that rots the workers' lifeline to compensation and robs society of a healthy workforce (Appendix C, p. 459).

Remedies for Exxon Valdez Oil Spill Cleanup Workers

Congress should hold an oversight hearing to evaluate OSHA's handling of the EVOS cleanup and other mass disasters, such as the 9/11 cleanup, to determine how to improve agency oversight and worker protection. You can request such a hearing by contacting your congressional delegation as well as key members of committees that oversee labor and public health (Appendix B, p. 455).

As part of its investigation, Congress should subpoena the payroll records of Exxon and its primary contractors, all the clinical and medical records, and Exxon's air quality monitoring data. Congress should lift the confidentiality order in the personal injury lawsuit *Stubblefield v. Exxon* (1994), filed in Superior Court, Third Judicial District, State of Alaska as a matter of urgent public interest.

If the investigation finds that a long-term health monitoring program of EVOS cleanup workers is warranted, Congress should require Exxon to fund such a study (as the company would have been required to do in 1989). However, Congress should request OSHA to contract a team of independent epidemiologists to conduct the long-term monitoring program and Exxon should not be allowed to participate in any phase of the study, including, notably, the study's scope or design.

Demand (Consumer)-Side Economics

I have no delusions that this book will convince every reader that there is a problem. But if you heed the startling truth about crude oil's persistent harmful effects to all life, you will be wondering what you as an individual can do at this point to make a difference. The truths revealed in this book call for fundamental sweeping change in our society. The great irony of such societal-level change is that it starts with individuals who commit to making personal changes, one by one. The change gains momentum as hundreds, thousands, and millions of people make individual commitments, and finally the change happens when the sheer mass of people sweep out the old and bring in a new way of living.

We don't have to wait for scientists to agree. We don't have to wait for Congress to pass significant legislation. We don't have to wait for nations to sign global treaties to reduce or ban dangerous chemicals. We don't have wait for non-governmental organizations to do the hard work of social change. We can make up our own minds in the privacy of our own homes to make a difference today!

Recommendation 7: Be socially responsible—get informed and get involved.

Helen Keller once said, "This is the time for a loud voice, open speech, and fearless thinking. I rejoice that I live in such a splendidly disturbing time." Let us rejoice with Helen Keller that we, too, live in such times when we can exercise our democratic rights and make a difference.

Vote.

Inquire. Find out what public interest groups are in your area and see if one suits your tastes and talents.

Volunteer. Share your time and talents with others whether it's mentoring students or working with a public interest group. That's how I got started!

Read. Go to your local library or spend some time on the web and search for topics that concern you. Find out who's doing what. Strive to keep abreast of the news on that topic and see if you might be able to weigh in with a letter or phone call to your local state leader or congressional delegates to request some of the law changes discussed earlier (Appendix C, p. 459). Targeted letter-writing leads to law changes—and letters are especially powerful when they come from individuals.

Question authority. Rampton and Stauber, authors of *Trust Us, We're Experts!*, warn, "If something is too complicated to explain, maybe it's also too complicated to be safe" (2001, 300). Don't rely on newspapers and television for news! The authors of *Trust Us* found that much of the "so-called news is actually canned material supplied by PR [public relations] firms" (308). Further, with competition down because of all the media mergers, a lot of important information fails to get reported in the "news."

Exercise your right of choice. A good place to get and stay

informed is through journals and communications of public interest groups that focus on specific topics. For example, the Union of Concerned Scientists (www.ucs.org) is one of my favorites to learn about the latest developments on alternative energy vehicles. The authors of *Trust Us* state, "The people who are most easily manipulated are those who have not studied a subject much and are therefore susceptible to any argument that sounds plausible" (Rampton and Stauber 2001, 311).

"Activate yourself!" cry the authors of *Trust Us*. They charge, "Activism, in our opinion, is not just a civic duty. It is a path to enlightenment" (ibid., 311). I found this to be so: seventeen years ago I volunteered to serve on the board of Cordova District Fishermen United and learn about oil issues that concerned the fishermen of Prince William Sound . . .

Recommendation 8: Be personally accountable—take action to reduce oil pollution.

Walk. Bike. Take public transit. Car-pool or car-share. This may seem like an inconvenience at first, but once a new routine is established, you may find that you actually miss the walk to work on days when you just can't do it.

Take care with what you do with your personal oil wastes. Does your car leak oil or burn oil? Fix it. Dispose of your waste oil in waste oil containers, not down sewers. Will this make a difference? You bet. Remember that it takes less than one one-hundredth of a teaspoon of crude oil in a pond the size of an Olympic swimming pool to reach PAH levels that kill aquatic life.

For those folks who can afford to do more, there are other options. I was once asked, after I gave a talk on my preliminary research findings for this book, what a person could do to help. When I replied, "Purchase an alternative energy car," the audience sat in stunned silence. Finally someone protested, "But you've got us all charged up to *do* something! I don't just want to hear that same old message again!" The sad truth is that if we all had just done this ten years ago, the petrochemical problem would look a lot differently today. I have made a personal pledge to buy an alternative energy or hybrid SUV before the end of 2005.

We can "vote with our pocketbook," as the public interest groups like to say, to reduce oil pollution. Besides the car example, you could trade in older equipment (such as jet skis, snow-machines, and outboard motors) with conventional two-stroke engines (the only type available before 1998) for that with new, direct injected two-stroke engines, which have 80 percent less hydrocarbon emissions than their predecessors, or even more efficient four-stroke engines. We can also vote with our pocketbooks to support companies that match our values whether we are purchasing gas for our cars or company stock.

It is true that the federal government could "help" us vote by offering to partially subsidize products that pollute less or, alternatively, taxing products that pollute more. For example, fossil fuel-burning vehicles could carry a carbon tax, based on the life expectancy, type of the vehicle, and fuel efficiency. But our government is currently betting on the wrong horse—it is heavily subsidizing fossil fuel-burning vehicles—and this bet is tightly bound by special interests, heavy-duty politics, and election campaigns. We are free to place our own bets.

Recommendation 9: Be empowered—focus on the positive.

I think one of the greatest challenges facing people today is to overcome our own sense of insignificance. Many people seem discouraged by an overwhelming number of life-threatening issues and feel there is little one can do to make a difference. We are fed "news" that numbs the senses with fear and pain. My mother's mother called the newspaper, "the Daily Saddener," and that was fifty years ago!

You can choose to focus on the positive. I gave up on television and mostly on newspapers as well—instead I read about what people around the world, individually and in groups, are doing to solve problems in their own backyards. Sometimes people's sense of "backyard" extends beyond their immediate neighborhood and across regions and borders of neighboring countries. All the better. This way I learn about the challenges and about what people are doing to meet these challenges at the same time. It leaves one with a powerful sense that people can and are making a difference, every day, everywhere. One of my favorite upbeat journals in this vein is *Yes!* magazine, sponsored by the Positive Futures Network (www.yesmagazine.org).

Finally, you already *have done* something positive if you purchased this book. A portion of the proceeds are dedicated to

Epilogue

Reflections from the Sound

"I never had an animus against their size and wealth, never objected to their corporate form. I was willing that they should combine and grow as big and rich as they could, but only by legitimate means. But they had never played fair, and that ruined their greatness for me."

Ida Tarbell
All in a Day's Work

My father introduced me to the writings of Ida Minerva Tarbell after I became involved with the politics of Big Oil. I was fascinated by this woman whose work captured the minds of a nation and a president and eventually led to the dissolution of Standard Oil Company in 1909. Oil historian, Daniel Yergin, called her 1904 book, *The History of Standard Oil Company*, "arguably . . . the single most influential book on business ever published in the United States" (Yergin 1991, 105).

It has been one hundred years since Ida Tarbell published her trust-busting book. Oddly, I feel as if the players in this great drama have been reassembled. The Exxon/Mobil merger in 1999 reunited the two largest chunks of the dissolved Standard Oil Company

445

(Standard Oil of New Jersey [Exxon] and Standard Oil of New York [Mobil]). David Rockefeller, Jr., the great grandson of John D. Rockefeller who created Standard Oil Trust, uses the fortunes of his ancestor to bring public awareness to the plight of our great oceans to help safeguard our seas for future generations.

And now this book—raking through the muck of Exxon's indiscretions after the *Exxon Valdez* oil spill, I found the amoral, predatory behavior described by Ida Tarbell still applies to Exxon's behavior today. I can only write of it; whether the public and politicians judge this behavior to be as unacceptable today as it was so judged one hundred years earlier will be up to the collective mind and heart of our nation. What we decide to do about this behavior will be a reflection of the morals of our times. But I believe that if we chose to do nothing, then Ida Tarbell's writings of the Standard Oil Company were prescient: "I am convinced that their brilliant example has contributed not only to a weakening of the country's moral standards but to its economic unsoundness" (Tarbell 1939, 230).

Bill Watterson (1992) summarized the challenge we face in his Calvin & Hobbes cartoon dialogue:

Calvin and Hobbes are careening in their red wagon through the woods. Calvin: "It's true, Hobbes, ignorance is bliss! Once you know things, you start seeing problems everywhere. And once you see problems, you feel like you ought to try to fix them. And fixing problems always seems to require personal change. And change means doing things that aren't fun! I say phooey to that!"

They pick up speed, charging downhill. Calvin looks back at Hobbes and says, "But if you're willfully stupid, you don't know any better, so you can keep doing whatever you like! The secret to happiness is short-term, stupid self-interest!"

Hobbes, looking concerned, points out, "We're heading for that cliff!" Calvin puts his hands over his eyes and says, "I don't want to know about it." They fly over the cliff. "Waaaugghhh!"

After the crash landing, Hobbes says, "I'm not sure I can stand so much bliss." Calvin responds, "Careful! We don't want to learn anything from this."

I believe that the curtain is starting to fall on the Age of Oil. Slow learners though we may be, as Daniel Yergin pointed out, more and more people realize that we are all in harm's way. Our "Hydrocarbon Society" (as Yergin refers to it) has become sickly with chemical-induced illnesses from the "petrochemical problem" (low levels of aromatic hydrocarbons present in our everyday environment). Yergin's "Hydrocarbon Man," Woman, and Child pay for oil spills and pipeline explosions with loss of income, loss of quality of life, and loss of life itself, while the corporations profit obscenely. Our system fails to mete out justice to corporate polluters. Our coastal sealife slowly sickens and disappears from low levels of contaminants, oil among others, which dribble into our ocean. Our planet coughs and sputters from the fossil fuel-driven effects of global warming, spewing forth intense storms in strange places and swelling her seas with fresh water from melting polar icecaps.

With so much "bliss," how much longer can we afford our oil dependency and what author David Korten calls our "suicide economy?" (Korten 2002)

Shifting energy sources is not as hard as it seems from the perspective of standing at the base of the mountain of fossil fuel power and looking up to the summit of alternative energy. After all, we shifted energy sources once before: in the 1920s, oil was introduced as the "clean" alternative to dirty lung-clogging coal fires and streets littered with horse droppings. People made the shift with the help of the government and business. Now we know there are clean alternatives to oil. The mountain looks a little different this time, because the people must initiate this shift without the help of government and the big energy businesses. But once we start, they will have no choice but to join us.

Climbing a mountain begins with taking the first steps. I invite you to join the pilgrims who have already started to climb this mountain of corporate greed and fear. Step one: do not purchase any ExxonMobil gas or other products, including stock—and do not take this action out of anger or revenge, but because you wish to make other choices that match your values. Step two: write a letter to the President and your congressional delegation in support of opening the 1991 civil settlement to claim $100 million for education on long-term effects of oil. Write another letter in support of a congressional investigation of the human health effects of Exxon's cleanup on workers (contact information in Appendix B, p. 455). Your letters *will* make a difference. Step three: walk, bike, car-pool, or pledge that your next car will be a hybrid or some alternative that is completely free of fossil fuels—and purchase it soon!

If the collective will of the people makes these conscious choices, I have no doubt that together—the people, nations, and corporate businesses—we will scale the mountain before us. It is time to turn

Appendix A

Cleanup Workers' Exposure and Illness Data

Table A.1

Exposure Levels of Some Hazardous Compounds Present during 1989 EVOS Cleanup Compared to OSHA PEL (Permissible Exposure Limits) and NIOSH REL (Recommended Exposure Limit)

(Numbers in **bold** indicate a violation of OSHA PEL)

Airborne Chemical	Geometric mean $(\bar{x}) \pm 95\%$ CI	Range (Maximum Exposure)	Sample Size	OSHA PEL[a]	NIOSH REL[b]	Overexposure[c] Max./OSHA	Overexposure[c] Max./NIOSH
Crude Oil							
Benzene	0.069 ± 0.596 ppm	0 - **7.8** ppm	1,611	1 ppm	0.1 ppm	8	78
Oil mist, mineral[d]	0.615 ± 4.0 mg/m^3	0 - **20** mg/m^3	114	5 mg/m^3	5 mg/m^3	4	4
PAH aerosol[e,f]	2.297 ± 1.15 mg/m^3	0 - **8.6** mg/m^3	29	5 mg/m^3	—	2	—
Other							
2-butoxyethanol	1.66 ± 19.2 ppm	0 - **99** ppm	112	50 ppm	5 ppm	2	20
Carbon monoxide	1.19 ± 16.6 ppm	0 - **100** ppm	711	50 ppm	35 ppm	2	3
Hydrogen sulfide	2.11 ± 30.6 ppm	0 - **199** ppm	471	20 ppm	10 ppm	10	20

Sources: Med-Tox 1989b, 1989c (in *Stubblefield v. Exxon* [1994]); NIOSH 1991, Table 3. Also, see below.

[a]OSHA PELs for 2-butoxyethanol; carbon monoxide; oil mist (mineral); and PAH aerosols (nuisance dust) are from 29 CFR 1910.1000 Subpart Z, Table Z-1 Limits for Air Contaminants (OSHA 2004b)
www.osha.gov/pls/oshaweb/owadisp.show_document?p_table=STANDARDS&p_id=9992
OSHA PELs for benzene and hydrogen sulfide are from Table Z-2 (OSHA 2004c)
www.osha.gov/pls/oshaweb/owadisp.show_document?p_table=STANDARDS&p_id=9993
PELs are for a standard 8-hour workday TWA (time-weighted average) except for hydrogen sulfide; concentrations must not be exceeded during any 8-hour shift of a 40-hour week. There is no established PEL for hydrogen sulfide based on TWA. Instead the PEL is for the acceptable ceiling concentration for a 10-minute exposure.

[b]NIOSH RELs are from U.S. Dept. of Health and Social Services, NIOSH (2004) www.cdc.gov/niosh/npg/nengapdx.html#d

[c]Over-exposures ratios were calculated by dividing the maximum range exposure ("Max.") by either the OSHA PEL or the NIOSH REL, as indicated.

[d]Exxon used mineral oil as a surrogate for crude oil. (OSHA has not established PELs for crude oil.)

[e]Exxon used nuisance dust ("particulates not otherwise regulated" [respirable value] in Table Z-1) as a surrogate for PAH aerosols. (OSHA has not established PELs for PAH aerosols.)

[f]Med-Tox air samples for polycyclic aromatic hydrocarbons (PAHs) were analyzed with EPA Method 5515, which measures only 17 priority pollutants. This method does *not* measure alkylated PAHs. This constitutes a serious oversight that significantly under-represents actual PAH exposures to workers. Further, geometric mean and 95 percent confidence intervals (CI) were *not* provided in Med-Tox statistical summary for PAHs; I calculated them from the raw data.

Table A.2
Exxon's Clinical Data—Upper Respiratory Infections (URIs) by Personnel Type[a]

Week	Incidence of URIs			Number of Personnel			
	Offshore	Onshore	Total	PWS	Onshore	Other[b]	Total
5/7	5	8	13	430	5,306	180	5,916
5/14	39	16	55	590	5,994	200	6,784
5/21	95	6	101	848	7,162	224	8,234
5/28	54	219	273	939	7,442	280	8,661
6/4	46	205	251	1,064	7,786	371	9,221
6/11	132	279	411	1,144	7,700	467	9,311
6/18	182	315	497	1,467	7,978	541	9,986
6/25	265	283	548	1,636	7,661	661	9,958
7/2	217	231	448	1,724	7,623	765	10,112
7/9	233	206	439	1,923	7,541	809	10,273
7/16	243	181	424	2,112	7,743	727	10,582
7/23	219	193	412	2,205	7,792	855	10,852
7/30	198	165	363	2,441	7,599	886	10,926
8/6	205	173	378	2,177	7,568	884	10,629
8/13	214	167	381	2,213	7,476	863	10,552
8/20	256	111	367	2,152	7,037	749	9,938
8/27	209	139	348	1,868	7,417	609	9,894
9/3	175	187	362	1,556	7,223	491	9,270
9/10	178	197	375	765	8,090	140	8,995
Totals	3,237	3,485	6,722				
Mean[c]	182	203	385				

Source: Exxon Company USA 1989b. In *Stubblefield v. Exxon* (1994).

[a] Personnel type not defined in available Exxon records.

[b] Other =Unaccounted workers in Exxon data presentation.

[c] Average reported URIs per week = 385 ± 39 @ 95% CI for every week pressurized water washes were primary beach treatment methods (excluding first three weeks of data).

Riki Ott

Table A.3
Reported Injuries & Illnesses by Part of Body Affected

Comparison of Time Loss Claims for 1989 EVOS Cleanup and 1987

Part of Body Affected	1989 Cleanup Time Loss Cases	1989 Cleanup Time Loss Percent	1987 Time Loss Cases	1987 Time Loss Percent	Ratio[a] 1989:1987
Total	518	100.0	9,661	100.0	
Head	36	6.9	757	7.8	0.9
Neck	5	1.0	288	3.0	0.3
Upper Extremity	62	12.0	2,141	22.2	0.5
Trunk	124	23.9	3,637	37.6	0.6
Lower Extremity	130	25.1	2,041	21.1	1.2
Multiple Parts	31	6.0	547	5.7	1.1
Body System	113	21.8	225	2.3	9.5
Body System Breakout					
Unspecified	27	5.2	89	0.9	5.8
Digestive system	7	1.4	14	0.1	14.0
Excretory system	7	1.4	0	0	*
Musculo-skeletal system	0	0	1	0.0	*
Nervous system	4	0.8	35	0.4	2.0
Respiratory system	65	12.5	55	0.6	20.8
Circulatory system	1	0.2	28	0.3	0.7
Other body system	2	0.4	3	0.0	*
Nonclassifiable	6	1.2	25	0.3	11.0[b]
Not Coded	11	2.1	0	0	*

Source: ADOL 1990a, 27, Table 3.2, (includes ratio column).
[a]Ratios compare the percent of cases per body part affected. For example, for head injuries, 6.9 percent (1989) divided by 7.8 percent (1987) yields a ratio of 0.9 or almost one, indicating no difference in the occurrence of this claim in this body part between years. A ratio greater than one indicates that occurrences were higher in 1989; a ratio less than one indicates that occurrences were higher in 1987.
[b]This ratio represents combined categories of "Nonclassifiable" and "Not Coded."
*Ratio could not be calculated because of zero in either denominator or numerator.

Table A.4

Nature of Reported Injuries and Illnesses During 1989 EVOS Cleanup

(including time loss and non-time loss claims)

Nature of Illness or Injury	ADOL Claims Subtotal	Total	Percent
TOTAL		1,771	
Injuries [a]		1,024	n/a
Illnesses [a]		747	100%
Respiratory symptoms		317	42%
Respiratory system	264		
Exposure to low temperatures	6		
Infection, parasite	46		
Pneumoconiosis	1		
Chemical symptoms		106	14%
Poisoning, systemic	34		
Eye, other diseases	15		
Burn, chemical	13		
Dermatitis	44		
CNS (Central nervous system)		24	3%
Nervous system	19		
Cerebrovascular	5		
Other, defined		37	5%
Inflammation of joints	35		
Mental disorders	2		
Other, undefined		260	35%
Symptoms, ill-defined	129		
Other disease, NEC[b]	108		
"No illness"	20		
Occupational disease, NEC[b]	3		

Source:ADOL 1990a, 30,Table 3.5, (includes percent column).
[a]Where possible, author separated out injury (fractures, sprains, etc.) and illness (respiratory symptoms, etc.) data subjectively based on nature of claim.
[b]NEC "not elsewhere classified."

Table A.5
Source of Reported Injuries and Illnesses During the 1989 EVOS Cleanup

Source of Illness or Injury[a]	Subtotal No. of Claims	Total Claims	Percent
TOTAL		1,771	
Injury		1,147	n/a
Illnesses		624	100%
Chemical symptoms		127	20%
Chemical	55		
Petroleum	52		
Liquid	8		
Clothing	6		
Radiation	6		
Respiratory symptoms		255	41%
Cold environment	206		
Infection	49		
Other		242	39%
Non classified	234		
Other sources	8		

Source: ADOL 1990a, 31, Table 3.6, (includes percent column).

[a]Where possible, author separated out injury and illness data subjectively based on source of claim. Injuries were selected from codes of amputation, heat (burn), bruise, cut, fracture, abrasion, and strain. Illnesses were selected from codes of chemical, occupational disease, respiratory system, other classifiable, non-classifiable.

Appendix B

Contact Information

To Pursue the $100 Million Reopener

President of the United States of America
Postal Mail: The White House
                1600 Pennsylvania Avenue NW
                Washington, DC 20500
Telephone: 202-456-1111; TTY/TDD: 202-456-6213
Fax: 202-456-2461
Email: president@whitehouse.gov

Your Congressional Delegates
Postal Mail: Senator (Name)   Representative (Name)
                US Senate   US House of Representatives
                Washington, D.C. 20515   Washington, D.C. 20515
Telephone Switchboard: 202-224-3121; TTY/TDD: 202-225-1904
Other: To locate address, telephone, fax, email, and web pages for specific persons:

US Senator by State:
http://www.senate.gov/contacting/index_by_state.cfm

US Representative by State: http://www.house.gov/writerep/

Exxon Valdez Oil Spill Trustee Council

Postal Mail:	EVOSTC
	441 West Fifth Avenue, Suite 500
	Anchorage, AK 99501
Telephone:	800-283-7745 (outside AK); 800-478-7745 (in AK);
	907-278-8012 (local)
Fax:	907-276-7178
Email:	restoration@evostc.state.ak.us

Federal District Judge Russel H. Holland

Postal Mail:	Honorable Judge Russel H. Holland
	U.S. District Court, Federal Building
	U.S. Courthouse
	222 W. 7th Ave.
	Anchorage, AK 99513
Telephone:	866-243-3814 (toll-free); 907-677-6144 (local)

To Request a Congressional Oversight Hearing on Human Health Effects of the EVOS Cleanup

Senator Edward Kennedy (D-MA)
Ranking Democrat on Senate Health, Education, Labor, and Pensions Committee

Postal Mail:	315 Russell Senate Office Building
	Washington, DC 20510
Telephone:	224-4543
Fax:	202-224-2417
Email:	www.kennedy.senate.gov

Representative John Dingell (D-MI)
Ranking Democrat on House Energy and Commerce Committee

Postal Mail:	2328 Rayburn House Office Building
	Washington, DC 20515
Telephone:	202-225-4071
Email:	www.house.gov/dingell

Your Congressional Delegate
(contact information provided above)

Please mail copies of your letter to the following people:

Alaska Community Action on Toxics
Postal Mail: 505 W. Northern Lights Blvd., Suite 205
                     Anchorage, AK 99503
Telephone: 907-222-7714
Fax: 907-222-7715
Email: info@akaction.org

Brian O'Neill, attorney
Postal Mail: Faegre & Benson, LLP
                     2200 Norwest Center, 90 South Seventh St.
                     Minneapolis, MN 55402-3901
Telephone: 1-800-328-4393 (toll free)

Appendix C

Recommendations for the Alaska Workers' Compensation Program

Provided by Barbara Williams, Executive Director, Alaska Injured Workers' Alliance
(Phone: 907-278-3661; Mailing address: POB 101093, Anchorage, AK 99510)

1. Provide employees with access to the laws and regulations for safety and work-related injuries before and during employment and after injury.

2. Educate employees about the laws, regulations, and compensation programs for safety and work-related injuries as part of safety training.

3. Enforce stiff fines and penalties for employers who do not report injuries or OSHA information. (Set up a cross-reference system to ensure reporting.)

4. Require insurance companies to report legal and medical costs related to workers' claims with penalties for non or fraudulent reporting.

5. Provide government oversight on reporting of health and safety issues, including claims. (Set up a cross-reference

system to ensure reporting.)

6.   Provide government oversight of insurance companies and employers who provide coverage privately or through self-insurance.

7.   Make the compensation process less adversarial for injured workers.

8.   Establish meaningful public participation, including involvement by community advocates of injured workers, in compensation board operation and administration. (This is a system supposedly set up to assist and compensate injured workers, yet workers are not allowed to make suggestions in the way the system works. Both union and non-union workers should be included in legislative panels and decisions that affect workers' rights and benefits.)

9.   Establish a system to evaluate workers' compensation process in terms of providing relief to workers and mandate annual evaluations by the Ombudsman's Office. Create an Ombudsman's position just to deal workers' compensation laws, regulations, and issues.

10. Provide consumers with information regarding the claim and compensation processes to educate the general public about the needs of our workforce.

11. Have meaningful participation from the Attorney General's Office to ensure fair settlement of claims as per Alaska's fair trade practices.

12. Add medical diagnostic codes to allow better access to diagnostic treatment and to better track long-term symptoms.

Appendix D

The People—Where are They Now?

Paul Anderson still works for National Marine Fisheries Service in Kodiak, Alaska. His small-mesh trawl surveys have been approved for long-term funding from the *Exxon Valdez* Oil Spill Trustee Council as part of the GEM project.

Jim Bodkin lives with his wife in Anchorage where he still participates in aerial surveys of sea otters for the Alaska Biological Science Center, USGS, in Anchorage. He spends more time, though, leading coastal ecosystem research for USGS from California to Russia, with a focus in Monterey, California, and Glacier Bay, Prince William Sound, and Kodiak, Alaska. He enjoys working with wood, playing ice hockey, and surfing.

Terry Bowyer is a Professor of Wildlife Ecology who holds joint appointments as Associate Director of the Institute of Arctic Biology and Head of the Department of Biology and Wildlife at the University of Alaska Fairbanks. There, he has studied the ecology and behavior of large mammals for the past seventeen years, and he has received numerous awards and recognition for his work. He and his wife, Karolyn, live in Fairbanks. He enjoys fishing, and waterfowl and upland game bird hunting with Pepper, his black Labrador Retriever.

461

Evelyn Brown (formerly **Biggs**) and her family live in Fairbanks. After a decade as a part-time student and full-time mother, Brown completed her PhD dissertation on herring ecology in Prince William Sound. She is now a full-time research associate at the University of Alaska Fairbanks, where she studies marine ecology through remote sensing, incorporating airborne and satellite measurements to better understand food web relationships among forage fish and a wide range of predators. In her spare time, she coaches Nordic skiing for Special Olympics (her oldest son is autistic) and is an avid gardener during the hot interior Alaskan summers.

Mark Carls works for the NOAA Auke Bay Lab near Juneau where he is developing a new passive sampler tool for cost-effective hydro-carbon monitoring in aquatic environments. He paddles his kayak to and from work and is an avid outdoorsman and photographer who enjoys mountain climbing, cross country skiing, and backpacking. When the weather keeps him inside, he builds furniture. During the long winters, he sings in choirs and plays a violin with the local symphony.

"Sara Clarke" (alias) lives in Canada. She collects 100 percent disability and is re-organizing her life based on her physical limitations. She is also exploring alternative spiritual pathways to healing.

Ted Cooney retired from the University of Alaska Fairbanks in 1999 and now lives in Choteau, Montana. He assists stakeholders and managers apply the results of SEA program to activities designed to sustain pink salmon stocks in Prince William Sound in the face of climate change and human influences. In his spare time, he coaxes flowers out of the prairie desert, captures the mood of the country with digital photography, and sings in a local choir and for residents of a retirement home.

Bill Driskell volunteers half-time at his son's elementary school in Seattle to take care of the computers and network. For his other day job, he works as a "white-hat" consultant, primarily on the oil-monitoring programs of the Regional Citizens' Advisory Councils in Prince William Sound and Cook Inlet, Alaska. He still deals with marine biology and oil chemistry issues related to the *Exxon Valdez* oil spill.

David Duffy is the director of the Pacific Cooperative Studies Unit with the University of Hawaii in Honolulu, Hawaii, where he works hard to control invasive species to Hawaii's endangered ecosystems. He also studies marine birds, sometimes in Prince William Sound.

Dan Esler is a University Research Associate with the Center for Wildlife Ecology at Simon Fraser University in British Columbia. He and his graduate students study an array of issues in water bird conservation along the Pacific Coast. When he can, Dan and his family enjoy boating among the coves and islands of the Strait of Georgia.

Kathy Frost and her husband Lloyd Lowry retired from ADFG in 2000 and from dog mushing shortly thereafter. They returned to their first passion, doing occasional contract work on marine mammals in the Arctic Ocean, and they spend time at their new home in Hawaii, where they have found "unlimited issues" with native plants, birds, and marine mammals to work on together.

Shelton Gay Jr., III, remains with the Prince William Sound Science Center, where he works on physical oceanography of the Sound. He hopes to start a doctoral program in physical oceanography at the University of Alaska, Institute of Marine Sciences, Fairbanks in fall 2004. He shares his love of outdoor activities and Folk and Celtic music wherever he travels.

Greg Golet now works as Senior Ecologist for The Nature Conservancy in northern California on the Sacramento River Project. Since leaving Alaska, his focus has shifted from "assessing damage" to "promoting recovery" in the largest and most important river in California. Golet was forced to shift his research priorities ("somewhat begrudgingly") from seabirds to neotropical migrants. He found "tremendous value" in his Alaska lessons and he uses a multi-species approach in his new studies.

Ron Heintz works at the Auke Bay Lab where he leads a variety of projects ranging from cataloging the occurrence of persistent organic compounds in fish collected from Alaska's coastline to understanding the *"bioenergetics"* (caloric budget) of juvenile forage fish. In his spare time he volunteers for the Southeast Alaska Regional Science Fair and coaches soccer. He enjoys hiking, hunting, running, and all forms of skiing when he is not modifying or maintaining his house.

Leslie Holland-Bartels and her family left Alaska in 1998 when she was promoted to be director at the USGS' Upper Midwest Environmental Sciences Center, one of eighteen national biological research centers for USGS in the country and a sister center to the Alaska Biological Science Center. She is in charge of interdisciplinary work on the Mississippi River and invasive species in the Great Lakes. She and her family live in LaCrosse, Wisconsin, where she started as a scientist over twenty-five years ago. She hopes to return to Alaska when the opportunity arises.

Jon Houghton has not been back to Prince William Sound since 1997. He primarily keeps busy in Puget Sound, working on marine shoreline inventories, salmon habitat restoration planning, and design and implementation of projects to restore habitats disturbed by decades of industrial discharges or by agricultural dikes. He tries to stay young hiking, skiing, fishing, and bird watching.

Dave Irons continues to work as a biologist for the USFWS in Anchorage, Alaska, but one of his current projects takes him around the world to collect data from eight Arctic countries to examine the effects of climate change on murres at a circumpolar scale. He is also studying auklets on St. Lawrence Island in the Bering Sea and he is still studying black-legged kittiwakes and other seabirds in Prince William Sound.

Sara Iverson is a professor with the biology department at Dalhousie University in Halifax, Nova Scotia (Canada) where she continues to develop the potential of fatty acid signatures to determine pinniped diets, changes in diet and local prey availability over time, and effects of diet diversity on pinniped health.

Phyllis "Dolly" La Joie lives in Honolulu, Hawaii. She has "just enough energy to get through the laundry, cooking, and other chores each day." She budgets carefully so that she can take one "big trip" each year to visit her ninety-five–year-old mother, four children, four grandchildren, and five great grandchildren. The trips exhaust her, but she says she's sick wherever she is and she would rather be with her family.

"Evan Lange" (alias) lives with his family in Fairbanks, Alaska. He is a mechanical engineer who enjoys building houses, flying, and hunting. His health is stable, however he is still slightly anemic with low hemoglobin and red and white blood cell levels. He intends to someday revisit many of the cleanup sites in Prince William Sound with his family, and he believes he will find evidence of the spill to show them.

Dennis Lees is working as a private consultant from his home base in San Diego, California. He continues to be involved in monitoring, baseline, and ground-truth surveys throughout southcentral Alaska and is examining the effects of the EVOS shoreline clean-up program on intertidal bivalves in western Prince William Sound. When the waves are good in southern California, he can often be found in the line-up trying to pick off the set waves on his trusty "Paipo" belly board.

Gary Marty is board certified in veterinary pathology (1993) and completed his PhD in Comparative Pathology from the University of California, Davis (1996). He stayed at UC Davis as a faculty researcher (non-tenure track), where he is completing studies on the effect of disease on Pacific herring populations that began in Prince William Sound in 1994. He also teaches gastrointestinal anatomy to first year veterinary students.

Craig Matkin lives in Homer, Alaska, and is studying orcas (killer whales) and humpback whales from Prince William Sound through the Aleutians. He is director of the North Gulf Oceanic Society, a charitable nonprofit organization (www.whalesalaska.org).

Ed Meggert is currently the ADEC State On-Scene Coordinator in charge of emergency response for oil and hazardous substances for the northern two-thirds of the state, including Prudhoe Bay. He lives with his son in Fairbanks where they are logging and clearing land on a homestead about ten miles out of town. They enjoy hunting, fishing, and playing music together.

Dennis Mestas lives and practices law in Anchorage. He continues to handle serious personal injury, wrongful death, and insurance cases. These have included aviation crashes, product defects cases, maritime injuries, oil field injuries, including a $2,500,000 jury verdict in Barrow, Alaska against a major oil company, and suits against

insurers for refusing to fully pay injured insured persons. He is a pilot
and a serious sports fisherman. He is a member of Trial Lawyers for
Public Justice.

Pamela K. Miller, MEn (Master's in Environmental Science/
Aquatic Biology), lives in Anchorage and is the executive director of
Alaska Community Action on Toxics. She is actively engaged in com-
munity-based research, advocacy, and other ACAT activities, including
taking care of an organic garden. Every summer she spends time
kayaking and exploring Alaska; in winter she skis on local trails with
her dog Darby.

Don Moeller lives in Valdez, Alaska, and is a primary care provider
for handicapped folks through the group home, Connecting Ties.
Sometimes, he takes a summer off to work odd jobs in remote wilderness lodges. He spends "as much time as possible" hiking with his
dogs or kayaking. He hopes to finish his cabin "on the Copper River
bluffs" soon. He manages his activities and life around his lingering
health issues from the oil spill cleanup.

Richard Nagel lives in Florida and is currently collecting 100 percent disability. His thoughtful, sincere, and persistent letters about
the treatment of sick *Exxon Valdez* oil spill cleanup workers by the
worker compensation system have been well received by several
congressional committees, which may trigger an investigation. His
health continues to decline and his traditionally-trained medical doctors have been unable to diagnose or treat his condition.

Riki Ott lives in Cordova, Alaska, with her cat. She (and her cat)
retired from commercial fishing in 1994 to organize community
approaches to deal with the lingering social, economic, and environmental problems from the *Exxon Valdez* oil spill. She co-founded
three nonprofit organizations: the Alaska Forum for Environmental
Responsibility (TAPS oversight); the Copper River Watershed Project
(sustainable development); and the Oiled Regions of Alaska
Foundation (social and economic concerns). She enjoys camping,
rafting, and afternoon treks on foot, skates, or snowshoes with her
friends—dogs and humans. This is her second book.

Vince Patrick is now at the University of Maryland, Institute for
Systems Research, although he feels like he never really left Prince

William Sound. He continues to work on "the progeny of the SEA Program," seeking to refine and apply his models to improve salmon and herring forecasting in collaboration with Cordova fisherman and others. He commits much of his time to grassroots organizing around issues of affordable housing, inclusive zoning, living wages, co-housing, and, in particular, cooperatives. He set up a small press, CFIMS Press, to publish the theoretical results from the SEA program (www.cfims.org:8080/nfrhforum/).

Sam Patten retired from ADFG in 2001 when he turned fifty-five and qualified for full retirement benefits. He now works full time as a fire management officer for the USFWS in the Yukon Flats, Kanuti, and Arctic National Wildlife Refuges. His sons are both college graduates. He has been happily married for over thirty years.

Charles "Pete" Peterson lives on the Southern Outer Banks of North Carolina with his wife, two teenage boys, and two Labradors. In warm months he swims 3K daily in the sound or ocean with the dogs in the lead. His work involves marine restoration and conservation ecology, as well as serving as Water Quality Chair of the North Carolina Environmental Management Commission.

John Piatt lives with his large family of adopted children and a menagerie of pets in Anchorage, AK. He still works as a seabird research biologist for USGS, where he studies complex dynamics and relationships among climate, forage fish, seabirds and marine mammals in Cook Inlet, Glacier Bay, and the Bering Sea.

William Rea, MD, is still actively treating patients with chemical sensitivity as president of the Environmental Health Center-Dallas (www.ehcd.com). In his spare time, he writes medical and technical books to help doctors provide better care and treatment for their patients. His latest book is *Optimum Environments for Optimum Health and Creativity* (Crown Press: Dallas, Texas, 2002).

Stanley "Jeep" Rice continues as the Habitat Division Program Director at the Auke Bay Lab near Juneau, Alaska. Jeep and his research team work on a variety of contaminants issues in Alaska, including oil spills and the long-term persistence of oil from the *Exxon Valdez* spill. To relax, he coaches high school football—something his son urged him to do more than twenty years ago.

Daniel Roby is a professor with Oregon State University, Department of Fisheries and Wildlife, in Corvallis, Oregon. He works on the breeding biology and nesting ecology of colonial seabirds, focusing on the diet composition and reproductive energetics and how these factors influence nesting success.

Sam Sharr now lives "in the land of potatoes" (Nampa, Idaho) where he does fish research as the Principal Research Biologist for Salmon and Steelhead with the Idaho Department of Fish and Game. He sees fewer salmon in one year in the entire state of Idaho than he used to see in Sheep Bay during one round of his escapement surveys. He claims to "know practically all of the fish on a first name basis."

Jeff Short completed his doctorate from the University of Alaska based on ecosystem research he did in Prince William Sound following the *Exxon Valdez* oil spill. He continues to lead research at the Auke Bay Laboratory on the long-term effects of oil pollution and other organic contaminants. He rebuts annual challenges to his oil spill research from Exxon scientists. To relax, he enjoys the Alaskan outdoors with his wife and son.

Ron Smith lives in Soldotna, Alaska, where he and his wife Shirley run a scrap metal recycle business and a septic disposal business. He describes this work as "doing all the things nobody else wants to do." Overall, his health has improved and he is "not as sick as he was ten years ago," which he attributes to his experience at the Environmental Health Center–Dallas with Dr. Rea. However, his chemical sensitivities, especially to diesel exhaust and oil fumes have worsened—just a whiff can trigger horrible bouts of nausea. His life is full with his children and grandchildren who all live nearby. He is looking forward to retirement and "getting away from chemicals."

Rick Steiner is based in Anchorage, Alaska, where he works as a professor and conservation specialist for the University of Alaska Marine Advisory Program. He works on international conservation issues, roaming the world from Russia, to South America, to Asia. He produces and hosts the Alaska Resource Issues Forum public television debates. He continues to press the state and federal governments to reopen the 1991 settlement and claim the $100 million for an Education Endowment, among other things.

Daniel Teitelbaum, MD, continues to practice as a medical toxicologist in Denver, Colorado. He has participated in numerous medico-legal activities on behalf of persons who were injured by occupational and environmental toxic exposures. Among the problems he has worked on were the community exposures to hexavalent chromium, fictionalized in the movie *Erin Brockovich*, and the groundwater pollution problems in Redlands, California. He has also worked on TCE pollution in several communities in Illinois and elsewhere. He was the deponent on behalf of the plaintiffs in a groundwater pollution case, in the longest deposition ever taken in California, which concluded very recently after 56 days. Still he endures. Tilting at windmills aside, he feels that once in a while justice is achieved and it pleases him to be part of it.

Lynn (Thorne) Weidman lived and worked in Cordova and Prince William Sound for thirty-six years before moving, gradually, to increasingly warmer latitudes to accommodate her health needs. She battles with bouts of depression, which, she claims, started with the spill. Because of the cleanup work, she was not able to care for her father, who suffered a heart attack three days after the spill (he had fished for fifty years, twenty of those years out of Cordova). He passed away a year later. She feels that her "mental health was more scarred than anything" from the spill and everything that went with it. For her (and many others), the oil spill is "unfinished business." She is rebuilding her life—again—in Santa Clarita, California.

Mark Willette lives in Soldotna and is the ADFG research project leader in upper Cook Inlet. He focuses on ways to improve forecast techniques for adult sockeye salmon returns to major river systems, drawing from lessons learned in during the SEA program. He tries to spend "as much time as possible" outdoors, skiing, fishing, and camping.

Bruce Wright lives in Santa Cruz, California, where he directs the nonprofit corporation Conservation Science Institute. He has written three books, including one on the oil spill. His latest book covers the topic of predators in Alaska and highlights the role of sharks as predators in Prince William Sound and the North Pacific Ocean.

Appendix E
Chronology

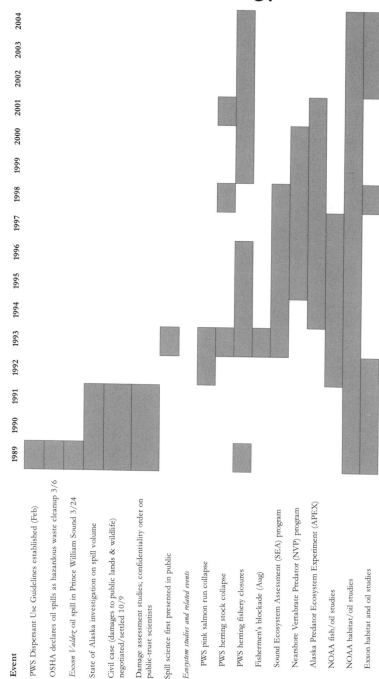

Event

PWS Dispersant Use Guidelines established (Feb)

OSHA declares oil spills as hazardous waste cleanup 3/6

Exxon Valdez oil spill in Prince William Sound 3/24

State of Alaska investigation on spill volume

Civil case (damages to public lands & wildlife) negotiated/settled 10/9

Damage assessment studies; confidentiality order on public-trust scientists

Spill science first presented in public

Ecosystem studies and related events

PWS pink salmon run collapse

PWS herring stock collapse

PWS herring fishery closures

Fishermen's blockade (Aug)

Sound Ecosystem Assessment (SEA) program

Nearshore Vertebrate Predator (NVP) program

Alaska Predator Ecosystem Experiment (APEX)

NOAA fish/oil studies

NOAA habitat/oil studies

Exxon habitat and oil studies

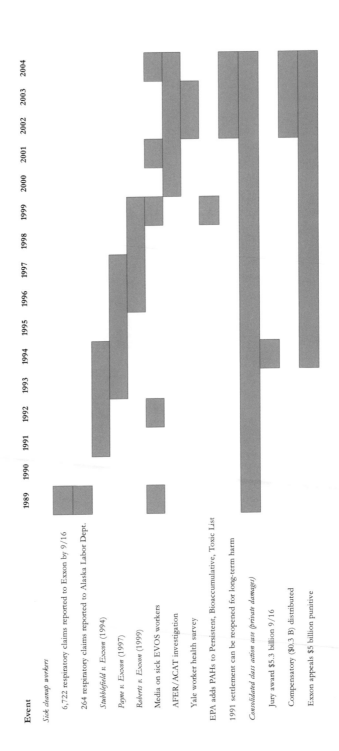

Notes

Introduction

1. John Keeble was the first author to report the more accurate spill estimate in the tenth anniversary edition of his book, *Out of the Channel*
   (1999, 59, footnote 24).

Chapter 1. Overview: The Cleanup

1. Among other things, these tapes recorded conversations among Alyeska
   scientists and others that questioned whether the dispersant Corexit
   9527 would be toxic to wildlife, whether it is water soluble, and how to
   clean contaminated equipment (APSC 1989, KWY001042402–409).
2. Alaska Oil Spill Commission 1990, IV (Day 25), 10; IV (Day 26), 11; Sims
   1989, 95. In addition to causing severe headaches and vomiting, studies of
   an experimental burn with Prudhoe Bay crude found both the oil and the
   residual material were mutagenic (Georghiou 1989).
3. Originally Exxon sued for $850 million in cleanup expenses, plus interest
   (Freemantle 1995). However, the company asked for triple damages or
   $2.5 billion, because the insurers acted in bad faith by denying the claim.
   The lawsuit was filed in state court in Dallas, TX. Lloyd's of London and
   100 other underwriters filed a counter suit in New York, claiming among
   other things that Exxon had not purchased enough coverage under its
   policy ($400 million worth) and that the spill resulted from Exxon's
   intentional misconduct, which was not covered in the insurance agreement. (See also ADN staff 1996; Dragoo 1996; Mibank 1994.)

Chapter 3. Dangerous Chemicals at Dangerous Levels

1. Mike Barsan, an industrial hygenist at the Center for Disease Control,
   NIOSH, provided the background on the OSHA and NIOSH standards
   (pers. comm, 3 March 2004). Evidence that the more stringent OSHA standards went into effect on 1 March 1989 can be found at:
   www.osha.gov/pls/oshaweb/owadisp.show_document?p_table=FEDER-
   AL_REGISTER&p_id=12908.

Chapter 6. A Collection of Stories

1. One of the EPA's studies found that the fertilizer was much more toxic
   than the oil (EVS Consultants 1990, Table 1); however this study was

plagued with problems such as only one or two replicates, poor control survival, sloppy chemical analyses conducted several months after the experiment with no indication of storage conditions, etc.

Chapter 8. Vanishing Claims

1. The Alaska Workers' Comp Board reported 1,797 cases from the 1989 cleanup; the NIOSH Health Hazard Evaluation (1991) reported 1,811, but the NIOSH numbers add up to 1,814 (30-31). I could find no explanation for these discrepancies.
2. The Alaska Department Of Labor analyzed 1,771 claims, not the full 1,797. No explanation is given for the discrepancy.

Chapter 9. Toxic Tort and Justice Denied

1. In September 1994, an Alaskan jury awarded plaintiffs $286 million in compensatory claims and $5 billion in punitive damages (Mulligan and Parrish 1994). Exxon still holds the punitive award hostage with numerous appeals and challenges; in the latest round appeals (January 2004), the jury-award was reduced by Judge Russel Holland to $4.5 billion and returned to the Ninth Circuit Court of Appeals (Reuters Network 2004; D'Oro 2004). (Exxon had argued for a punitive award of $25 *million*.) The compensatory award was finally fully paid by Exxon to the Qualified Settlement Fund Administrators (plaintiffs' attorneys) in 2002; distribution to plaintiffs is expected to be largely completed in 2004.
2. Judge Holland went on to say, "More generally, rule 26(a)(2)(b) requires that witnesses who are 'retained . . . to provide expert testimony in the case' disclose 'a written report prepared and signed by the witness.' Dr. Rea in substance disavowed the report which he may have signed, and . . . Dr. Ewing appears to have adopted the report written by counsel without signing it. Neither doctor 'prepared' a report of his opinion" (90, 4).

Chapter 10. An Occupational Health Disaster

1. U.S. EPA Janitorial Products Pollution Prevention Program web site: www.westp2net.org/janitorial/tools/haz2.htm.
2. The estimate of former EVOS cleanup workers is calculated as follows: 11,000 workers x 73 percent (workers in high-risk jobs [Carpenter, Dragnich, and Smith 1991]) x 0.3 (Yale worker health survey [O'Neill 2003]) = 2,409. *This is a low-end estimate as it is based on the peak, not the total, number of cleanup workers.*

Chapter 11. 1970s Science

1. An Olympic swimming pool holds 800,000 gallons (3 million liters) of water. Ten parts per billion is 0.00008 gallons or 0.0017 teaspoons (0.00008 gal x 4 quarts/gallon x 2 pints/quart x 4 gills/pint x 4 fluid ounces/gill x 1/6 teaspoons/fluid ounce). Prudhoe Bay crude contains 18.9 percent total aromatic hydrocarbons in its WSF, so it would take 0.0017 teaspoons/0.189 or 0.009 teaspoons of crude oil to yield 0.0017 teaspoons WSF. One teaspoon equals 1/6 fluid ounce (5 milliliters), so 0.034 teaspoons equals 0.05 ml. Thus, it takes 0.009 teaspoons (0.05 ml) of crude oil in an Olympic swimming pool to achieve 10 parts per billion (ppb) WSF (J. Short, NOAA/NMFS Auke Bay Lab, Juneau, AK, pers. comm., 18 December 2001).

2. During the TAPS authorization process, representatives of Alyeska and its owner companies made numerous promises to Congress and in various written documents to allay fishermen's fears about potential oil pollution (Townsend Environmental 1994). For example, TAPS Project Director George Hughes told a U.S. Senate committee, "Ballast water storage and treating facilities will be provided to treat all dirty ballast which is brought in by the tankers so that the effluent will meet all water quality control requirements of the State and Federal government" (U.S. Congress, Senate 1969, 17). Some of these promises wound up in the official agreement and grant of right-of-way: "Water discharged from the waste-water treatment facility shall not contain more than 10 parts of oil per million parts of water, on a weekly (seven day) average" (U.S.A. 1974, 23.B).

3. Two of the original researchers, Richard Myren and George Perkins, stated they could not attribute the cause of the *Macoma* decline to any one event (R. Myren, pers. comm., 3 March 2004; G. Perkins, pers. comm., 2 March 2004). High levels of particulate-bound PAHs, often in excess of permitted values (ADEC 1988), entered the receiving waters of Port Valdez for well over a decade—at least until a more stringent wastewater discharge permit was authorized by the EPA in 1989. However, for a number of years, the study site also received sediment from an eroding intertidal stream.

Chapter 13. Coastal Ecology

1. The species most responsible for this general pattern were the rockweed *Fucus*, the barnacle *Balanus glandula*, and the limpet *Tectura persona* in the upper middle intertidal zone, and the mussel *Mytilus trossulus* in low intertidal zone (Highsmith et al. 1996; Houghton and Elbert 1991; Houghton et al. 1996).

2. The four major studies analyzed by Peterson, Erickson, and Strickland (2001) include two public trust studies—the Coastal Habitat Injury Assessment (McDonald, Erickson, and Strickland 1995; Highsmith et al. 1995; Stekoll et al. 1996; and Sundberg et al. 1996) and the NOAA Hazardous Materials program (Driskell et al. 1996; Houghton et al. 1996; and Lees, Houghton, and Driskell 1996)—and two Exxon studies—the

Shoreline Ecology Program (Page et al. 1995b, Gilfillan et al. 1995a) and the Gulf of Alaska study (Gilfillan et al. 1995b).

3. Peterson and his team (2002) point out that the total area of shoreline habitat sampled per year in any given stratum of habitat type and shoreline elevation varied dramatically between the studies conducted by Exxon scientists and those conducted by public-trust scientists. For example, in their Shoreline Ecology Program, Exxon scientists sampled a total of about 1.36 square meters or an area equivalent to the top of a grand piano. In the comparable Coastal Habitat Injury Assessment, public-trust scientists sampled a total of about 60 square meters or the equivalent of nearly one-quarter of a football field. Public-trust scientists also took up to 6 times as many samples as Exxon scientists at each beach, and the samples were up to 8 times larger in size. These differences implied, as Peterson and others stated, not only a varying level of sampling effort, but also "differing ability to characterize the biological communities" (Peterson et al. 2002, 310).

4. Exxon scientists' paper-fix of ecological recovery is demonstrated in a series of figures in Gilfillan et al. (1995a, 413–417, Figures 2–4 for fixed design study and Figures 5–7 for stratified random sample design study).

5. Exxon scientists challeneged the critical review by Peterson and his colleagues (Gilfillan, Harner, and Page 2002). Peterson and his team replied (Peterson et al. 2002): their comments and conclusions were essentially unchanged.

Chapter 14. Marine Mammals

1. See reduction in scope of work and data analysis in Monnett and Rotterman 1995c, final report, compared to that of original work plan in Monnett and Rotterman 1990, draft paper.

2. Accounting for sea otters: 343 captured and taken to treatment centers + 18 pups born in captivity = 361 total animals in centers. 123 died in centers + 196 released into wild + 37 sent to zoos and aquariums + 5 escaped from centers = 361 total animals (Hofman 1994, xiv).

3. "The sample size for this study differs markedly from that outlined in the study plan ... Too few radios were ordered to permit implantation of 100 pups" (Rotterman and Monnett 1995, 4).

4. "Differences were observed in the overwinter survival rate of juvenile sea otters in eastern and western PWS in 1990–1991 (Rotterman and Monnett 1991) and again in 1992–1993 (USFWS, unpublished data). Both studies demonstrated lower survival for pups inhabiting western areas of PWS, relatiave to the eastern areas. However, the survival rates for both eastern and western areas were lower in 1990–1991 relative to 1992–1993" (Ballachey, Bodkin, and DeGange 1994, 52).

Chapter 15. Marine Birds

1. ADEC issued a citation for excessive use of Inipol on island beaches and stream mouths in Herring Bay on Knight Island (KN115) on 31 August 1989 (Patten et al. 2000b,Supplement 3, 3).Inipol was *not* intended for use in salmon streams, but in 1991 and 1992, Patten reported, "beach treatment activities subsided markedly and were more localized in the areas of salmon streams. Inipol was still being applied to harlequin habitat sites" (Patten et al. 2000a, 36.)

Chapter 16. Fish

1. Eventually coded-wire tags were replaced by simply temperature shocking hatchery fish at a certain stage of their development, which scarred the *odoliths* (tiny ear bones) with an extra heavy calcium deposit.ADFG biologist Mark Willette developed this technique to eliminate the labor-intensive tagging procedure, while improving data collection as the new technique marked 100 percent of the hatchery fish.Temperature shocking is now a standard management tool used by aquaculture specialists throughout the world. Lab technicians still collect fish heads for scientific analyses.

2. Foot surveys were expensive and intense. Sharr designed and installed eight intertidal weirs in 1994 as a pilot project to improve the accuracy of foot surveys. Using the weir data, Sharr discovered that wild salmon stocks in the southwest Sound had been grossly *under-estimated*—by at least half—in the past, a finding that led to more fishing time and improved management. Weirs became a standard management tool to augment and improve the accuracy of foot surveys (Bue et al. 1998).

3. Comparison of sample size and replicates in the fall egg surveys between public-trust scientists (Bue et al. 1996) and Exxon scientists (Brannon et al. 1995) demonstrates small sample bias in Exxon's study design.

	Public-trust	Exxon
Sample size (number of streams)	31	9
Number of replicates per stream	10-14	3
Number of years of study	4	1

4. Exxon scientists (Brannon et al. 1995) report that in 1991 the proportion of live alevins was significantly lower in oiled versus reference streams at the lowest tidal height, but not at the middle or highest tidal height.They correctly state, "Because of these mixed results, a test on the proportion of live alevins for all three tide levels combined was not valid." They then go on to average and present the data anyway "for visual comparison only" (Table 4, 567). However, in the discussion, they report that "no statistically significant effects of oil were found in those streams identified to have the heaviest level of oil contamination" (ibid., 575).

Chapter 19. Apex Predators

1. Much of the information on sand lance was wholly new for Alaska and a significant contribution towards ecosystem-based management (Armstrong et al. 1999; Litzow et al. 2000; Robards et al. 1999a–c; Robards and Piatt 1999; Robards, Piatt, and Rose 1999; Robards, Rose, and Piatt 2002; Wilson et al. 1999).

2. The fatty acid signature work is used to understand food web relationships in a variety of species. For example, the NOAA Auke Bay Lab scientists have applied it to fisheries research (Heintz and Larsen 2003).

3. The acrimonious controversy is covered in the following papers: Boersma and Clark 2001; Irons et al. 2001; Parrish and Boersma 1995a, 1995b; Piatt 1995, 1997; Weins et al. 1996, 2001; Wiens and Parker 1995.

4. Exxon scientists wrote, "A high degree of philopatry would cause birds to return to contaminated areas, regardless of the condition of those areas. Under such circumstances, we would fail to detect negative impacts on habitat use . . ." (Day et al. 1997, 609).

5. Exxon scientists' manipulated data had the effect of obscuring oil spill effects. "To keep sites with the highest counts from overly influencing our examination of population trends, we standardized on 1984 counts . . . ." (Hoover-Miller et al. 2001, 115).

Chapter 20. Fish and Oil Toxicity

1. While embryonic exposure to low levels of oil clearly caused reproductive damage, the question of whether this was the result of genetic damage remained elusive. One study, independent from the federal research, found evidence of genetic damage in second-generation offspring of pink salmon exposed to oil during embryonic development. While this study used offspring of fish from the Little Port Walter experiments in 1993, the federal researchers have yet to replicate the findings, but they are working towards this end (Smoker, Crandell, and Malecha 2000).

2. The initial 1993 crash of the Prince William Sound herring stocks, and the subsequent crashes and generally unstable population during the next decade, led to a significant number of studies on herring, disease, and oil. Many were funded by the EVOS Trustee Council (Carls et al. 1998, 2000; Davis et al. 1999; Kocan et al. 1997; Kocan and Hershberger 2001; Kocan, Herschberger, and Elder 2001; Marty et al. 1998; Thomas et al. 1997).

Chapter 23. The Legacy

1. ACAT's study also found that people who conducted subsistence harvests of wild foods near the military site of NE Cape, St. Lawrence Island, had higher PCB levels than other island residents. The four-year study, funded by NIEHS, is in its final year; scientists are seeking to identify the exposure pathways (P. Miller, ACAT, pers. comm., February 2003).

2. The Auke Bay Lab team first presented their findings of lingering toxic oil on the beaches at the annual EVOS Trustee Council symposium in January 2002 (Short et al. 2002). Their study findings were reported by the press (O'Harra 2002) and immediately attacked by Exxon scientists. The controversy was covered by the Alaska press (ADN editorial 2002a, 2002b; Page 2002; Short 2002a, 2002b; Spies, McCammon, and Balsinger 2002). Exxon's attacks in technical journals is covered in Chapter 22.

Chapter 24. Beyond the *Exxon Valdez* Oil Spill

1. Estimated annual inputs are the "best estimates" cited by the National Research Council. The Council also gives a range as minimum and maximum: total oil, 260,000 to 2300,000 tons; natural seeps, 160,000 to 240,000 tons; tank vessel spills, 5,300 to 6,400 tons; land-based river and runoff, 2,600 to 1,900 tons; recreational marine vessels, 2,000 to 9,000 tons (National Research Council 2002, 28, Table 2-2).
    "The U.S. Coast Guard data base, which contains generally conservative estimates of outflow from all reported incidents, is assumed to establish the minimum estimate" (National Research Council 2002, 213).
2. The ten rivers included in the National Research Council study are the Columbia, Delaware, Hudson, James, Mississippi, Neuse, Sabine, Sacramento, Susquehanna, and Trinity (National Research Council 2002, 2 51, Table I-21).
3. These disclaimers are from the National Oil and Hazardous Substances Pollution Contingency Plan, Subpart J (Use of Dispersants and Other Chemicals) Product Schedule (40 CFR 300.900); specifically §300.920(e) for disclaimers.

References

Abbreviations Used in References

In referencing works, short titles and other abbreviations have generally been used. Authors, works, and entities frequently cited have been identified by the abbreviations below.

Legal citations for briefs, records, motions, and memoranda presented a challenge for the two cases referenced in depth: *Roberts v. Exxon* (1999), a federal case, and *Stubblefield v. Exxon* (1994), a state case. Alaska's federal and state cases are not available in electronic form on legal search engines; Alaska State Courts did not even docket their cases until recently. Therefore, for citation purposes, I used the Clerk's Docket Number for the federal case and the "unofficial" docket number for the state case. The latter is based on year-month-day; e.g., 910919 is 1991, September 19. Otherwise, I included standard information in the citation. Full dates follow each citation.

ABL	Auke Bay Lab
ADEC	Alaska Department of Environmental Conservation
ADNR	Alaska Department of Natural Resources
ADFG	Alaska Department of Fish and Game
ADN	Anchorage Daily News
ADOL	Alaska Department of Labor

481

AFSS	Rice, S. D., B. Spies, D. A. Wolfe, and B. A. Wright, eds. 1996. *Proceedings of the EVOS Symposium.* American Fisheries Society Symposium 18, Bethesda, MD.
AkPIRG	Alaska Public Interest Research Group
AMOP	Arctic and Marine Oilspill Program, Technical Seminar, Environment Canada, Ottawa, ON.
ARLIS	Alaska Resource Library and Information Services, Anchorage, AK www.arlis.org
ASTM	American Society for Testing and Materials
ASTM STP 1219	Wells, P. G., J. N. Butler, and J. S. Hughes, eds. 1995. *EVOS: Fate and Effects in Alaskan Waters.* 1995. Eds P. G. American Society for Testing and Materials STP 1219, Philadelphia, PA.
ASTM Symposium	Third ASTM symposium on Environmental Toxicology and Risk Assessment: Aquatic, Plant, and Terrestrial, 26–28 April 1993, Atlanta, GA.
ATSDR	Agency for Toxic Substances and Disease Registry
AWCB	Alaska Workers' Compensation Board
BMJ	used to be known as: *British Medical Journal*
CFR	Code of Federal Registrations
CFS	Cordova Fact Sheet
DHHS	U.S. Department of Health and Human Services
DOI	U.S. Department of the Interior
EVOS	*Exxon Valdez* Oil Spill
EVOS Symposium Abstracts	EVOS Trustee Council, University of Alaska Sea Grant College Program, and American Fisheries Society, Alaska Chapter. 1993. EVOS Symposium Program and Abstracts, Anchorage, 2–5 February.
Forage Fish Proceedings	*Proceedings Forage Fish in Marine Ecosystems.* Alaska Sea Grant College

	Program. AK–SG–97-01, 1997.
GAO	U.S. General Accounting Office
GPO	U.S. Government Printing Office
HEW	U.S. Department of Health, Education, and Welfare
IOSC	International Oil Spill Conference Proceedings. American Petroleum Institute: Washington, DC.
MSDS	Material Safety Data Sheet
NMFS	National Marine Fisheries Service
NOAA	National Oceanic and Atmospheric Administration
NRDA	Natural Resource Damage Assessment
NIOSH	National Institute of Occupational Safety and Health
OSHA	U.S. Occupational Safety and Health Act
PAHS	polycyclic aromatic hydrocarbons
PWS	Prince William Sound
Roberts v. Exxon (1999)	*Scott Roberts, Richard Merrill, Phyllis La Joie, and Ron Smith v. Exxon Corp., SeaRiver Maritime, and VECO.* A96-040-CV, U.S. District Court for the District of Alaska (1999).
Sea Otter Symposium	Bayha, K. and J. Kormendy, eds. 1990. Proceedings of a symposium to evaluate the response effort on behalf of sea otters after the T/V EVOS in PWS, Anchorage, AK. April 1990. Biological Report 90(12). Washington, DC: U.S. DOI, USFWS, and National Fish and Wildlife Foundation.
Stubblefield v. Exxon (1994)	*Garry Stubblefield and Melissa Stubblefield v. Exxon Shipping Company, Exxon Corp., VECO, Inc., and Norcon, Inc.* 3AN-91-6261 CV, Alaska Superior Court, Third Judicial District at Anchorage (1994).
USFWS	U.S. Fish and Wildlife Service
USGS	U.S. Geological Survey
WSJ	*Wall Street Journal*

References

Abookire, A. A., J. F. Piatt, and M. D. Robards. 2000. Nearshore fish distribution in an Alaskan estuary in relation to stratification, temperature and salinity. *Estuarine, Coastal, and Shelf Science* 51:45-59.

Accu-Chem Laboratories. 1992a. Blood evaluation for La Joie, 12 June. In *Roberts v. Exxon* (1999), No. 26, Exhibit E.

Accu-Chem Laboratories. 1992b. Laboratory report for Smith, 23-24 July. In *Roberts v. Exxon* (1999), No. 101, attachment.

Adams, D. 1991a. Co-op bankrupt. *Cordova Times,* 7 March.

Adams, D. 1991b. Chugach bankruptcy shocks Cordovans. *Cordova Times,* 14 March.

Adams, D. 1991c. Pinks may jeopardize hatcheries, canneries. *Cordova Times,* 15 August.

Adams, D. 1991d. Governor offers surplus pinks to Soviets. *Cordova Times,* 22 August.

Adams, D. 1991e. No buyers for 3M [million] pink salmon, corporation says. *Cordova Times*, 22 August.

Adams, D. 1991f. Fishermen lament 1991 season, year of the fish spill. *Cordova Times,* 29 August.

Adams, D. 1991g. Coop shuts down: Hard times hit another cannery. *Cordova Times,* 10 October.

ADEC. 1992. Shoreline treatment/cleanup monitoring: Review of field activities during the EVOS treatment operations. Unpublished ADEC review, spring 1992. In ADEC 1993, 151-153, footnotes 61, 64, 66, and 87.

ADEC. 1993. The EVOS. *Final Report, State of Alaska Response.* Prepared by E. Piper. Anchorage, AK. June.

ADFG. 1993. The EVOS: What have we learned? *Alaska's Wildlife,* Special Issue 25(1): January–February.

ADFG. 1998. PWS Management Area: 1997 Annual Finfish Management Report. Commercial Fisheries Management and Development Division, Central Region, Anchorage, AK. Regional Information Report No. 2A98-05.

ADFG. 2002. PWS Management Area: 2001 Annual Finfish Management Report. Commercial Fisheries Management and Development Division, Central Region, Anchorage, AK. Regional Information Report No. 2A02-20.

ADN editorial. 2002a. Exxon's science attack on government research was unfounded. *ADN,* 8 October.

ADN editorial. 2002b. Sound warning. *ADN,* 24 January.

ADN staff. 1996. Exxon recoups $300 million. Insurers ordered to pay in partial settlement of oil-spill suit. *ADN,* 18 January.

ADN. 1990. Cleanup makes VECO biggest Alaska firm. 27 September.

ADOL. 1990a. PWS oil spill. In *Occupational Injury and Illness Information-AK, 1989,* 25-34. Juneau, AK: ADOL.

ADOL. 1990b. Controversion [denial] notice from Surety of Alaska, April 24. In ADOL, AWCB 1992a.

ADOL. 1990c. Letter to E. Meggert. 11 May. In ADOL, AWCB 1992a.

ADOL, AWCB. 1992a. Case No. 9009878 [Ed Meggert].

ADOL, AWCB. 1992b. Case No. 9104004 [Ron S. Smith].

ADOL, AWCB. 1992c. Partial compromise and release regarding recovery of overpayment of benefits. Filed 2 October. In ADOL, AWCB 1992b.

ADOL, AWCB. 1995. Case No. 9034054 [Scott R. Roberts].

Agler, B.A , S. J. Kendall, D. B. Irons, and S. P. Klosiewski 1999. Declines in marine bird populations in PWS, AK, coincident with a climatic regime shift. *Waterbirds* 22(1):98-103.

Ainley, D. G., R. G. Ford, E. D. Brown, R. M. Suryan, and D. B. Irons. 2003. Prey resources, competition, and geographic structure of kittiwake colonies in PWS. *Ecology* 84(3):709-723.

Alaska. 2000. Workers' Compensation Laws and Regulations, Annotated. 2000-2001 ed. Charlottesville, VA: Lexis Publishing.

Alaska Commercial Fisheries Limited Entry Commission. 2004. Estimated permit value report for G01E, herring roe, purse seine, PWS, AK www.cfec.state.ak.us/pmtvalue/X_G01E.htm and S01E, salmon, purse seine, PWS, AK www.cfec.state.ak.us/prtvalue/X_S01E.htm (accessed 1 March).

Alaska Department of Health and Social Services. 1989a. Public health advisory on crude oil. 28 April. In U.S. Congress House 1989a, 1063-65.

Alaska Department of Health and Social Services. 1989b. Public health advisory on crude oil. Juneau, AK. 5 May.

Alaska Department of Law. 1991. Files on "ACE" investigation 1989-1991. ARLIS, Anchorage, AK.

Alaska Health Project. 1989. Glycol ethers (Cellosolves). Anchorage, AK. 2 June.

Alaska Oil Spill Commission. 1990. *Spill: The wreck of the* Exxon Valdez, *Appendix N.* State of Alaska. February. ARLIS, Anchorage, AK.

Alaska Public Interest Research Group. 1995. Working without a net: The failure of workers' compensation in Alaska. Anchorage, Alaska.

Alaska Sea Grant College Program. 1993. Is it food? Addressing marine mammal and seabird declines. Workshop Summary. University of Alaska Fairbanks, Fairbanks, AK. AK-SG-93-01.

Alaska State Legislature, Legislative Budget and Audit Committee. 1999. ADOL and Workforce Development, Division of Workers' Compensation. Division of Legislative Audit, Audit Control Number 07-4601-00.

Alyeska Pipeline Service Company. 1989. Alyeska Emergency Center telephone transcripts, 24–27 March. ARLIS, Anchorage, AK.

Ames, J. 1990. Impetus for capturing, cleaning, and rehabilitating oiled or potentially oiled sea otters after the T/V EVOS. *Sea Otter Symposium* 90(12):137–140.

Anderson, P. J., J. E. Blackburn, and B. A. Johnson. 1997. Declines of forage species in the Gulf of Alaska, 1972–1995, as an indicator of regime shift. Abstract. *Forage Fish Proceedings,* 531–543.

Anderson, P. J. and J. F. Piatt. 1999. Community reorganization in the Gulf of Alaska following ocean climate regime shift. *Marine Ecology Progressive Series* 189:117–123.

Andres, B. 1997. The EVOS disrupted the breeding of black oyster-catchers. *Journal of Wildlife Management* 61(4):1322–1328.

Andres, B. 1998. Shoreline habitat use of black oystercatchers breeding in PWS, AK. *Journal of Field Ornithology* 69(4):626–634.

Andres, B. 1999. Effects of persistent shoreline oil on breeding success and chick growth in black oystercatchers. *The Auk* 116(3):640–650.

Anthony, J. A. and D. D. Roby. 1997. Variation in lipid content of forage fishes and its effect on energy provisioning rates to seabird nestlings. Extended abstract. *Forage Fish Proceedings,* 725–729.

Anthony, J. A., D. D. Roby, and K. R. Turco. 2000. Lipid content and energy density of forage fishes from the northern Gulf of Alaska. *Journal of Experimental Marine Biology and Ecology* 248:53–78.

Antosh, N. 2004. Exploration and production. In times of plenty, they don't flash cash. *Houston Chronicle,* 5 February.

Armstrong, R. H., M. F. Willson, M. D. Robards, and J. F. Piatt. 1999. Sand lance: Annotated Bibliography. In Robards et al. 1999.

Ashford, N. and C. Miller. 1998. *Chemical Exposures: Low Levels and High Stakes.* 2d. ed. New York: John Wiley & Sons.

Associated Press. 1991. Report: Spill cleanup method was harmful.

Anchorage Times, 10 April.

Associated Press. 2001. Houston firm fined $1 million in cover-up attempt. *Houston Chronicle,* 13 April.

Associated Press. 2002. Appeals court upholds ban on *Exxon Valdez.* Panel keeps ship out of PWS. *ADN,* 31 October.

ATSDR. 1997. Toxicological profile for benzene. Atlanta, GA: U.S. DHHS, Public Health Services.

ATSDR. 1998. Toxicological profile for 2-butoxyethanol and 2-butoxyethanol acetate. Atlanta, GA: U.S. DHHS, Public Health Services.

ATSDR. 2002. Toxicological profile for polycyclic aromatic hydrocarbons. Atlanta, GA: U.S. DHHS, Public Health Services.

Atlas, R. M., P. D. Boehm, and J. A. Calder. 1981. Chemical and biological weathering of oil from the *Amoco Cadiz* spillage within the littoral zone. *Estuarine Coastal and Shelf Science* 12:589-608. In Braddock et al. 1996.

Aurand, D. V., G. M. Coelho, and A. Steen. 2001. Ten years of research by the U.S. oil industry to evaluate the ecological issues of dispersant use: An overview of the past decade. IOSC 2001, 429-434.

Babcock, M. M., G. V. Irvine, P. M. Harris, J. A. Cusick, and S. D. Rice. 1996. Persistence of oiling in mussel beds three and four years after the EVOS. AFSS 18:286-297.

Babcock, M. M., G. V. Irvine, S. D. Rice, P. Rounds, J. A. Cusick, and C. Brodersen. 1993. Oiled mussel beds two and three years after the EVOS. Abstract. *EVOS Symposium Abstracts,* 184-85.

Baker, J. M., R. B. Clark, and R. F. Kingston. 1990. Environmental recovery in PWS and the Gulf of Alaska. Institute of Offshore Engineering, Heriot-Watt University, Edinburgh EH14 4AS Scotland. June.

Baker, J. M., R. B. Clark, and R. F. Kingston. 1991. Two years after the spill: Environmental recovery in PWS and the Gulf of Alaska. Institute of Offshore Engineering, Heriot-Watt University, Edinburgh EH14 4AS Scotland. November suppl.

Ballachey, B. E., J. L. Bodkin, and A. R. DeGange. 1994. An Overview of sea otter studies. In Loughlin 1994, 47-59.

Ballachey, B. E., J. L. Bodkin, S. Howlin, A. M. Doroff, and A. H. Rebar. 2003. Correlates to survival of juvenile sea otters in Prince William Sound, AK, 1992-1993. *Canadian Journal of Zoology* 81:1494-1510.

Ballachey, B. E., J. J. Stegeman, P. W. Snyder, G. M. Blundell, and 10 others. 2000. Oil exposure and health of nearshore vertebrate predators in Prince William Sound following the 1989 *Exxon Valdez oil spill*. In Holland-Bartels 2000.

Barker, E. 1994. The Exxon Trial: A Do-It-Yourself Jury. *The American Lawyer*, November: 68–77.

Barron, M. G. and L. Ka'aihue. 2001. Potential for photoenhanced toxicity of spilled oil in PWS and Gulf of Alaska waters. *Marine Pollution Bulletin* 43:86–92.

Batten, B. T. 1990. Press interest in sea otters affected by the T/V EVOS: A star is born. *Sea Otter Symposium* 90(12):32–40.

Bayha, K. 1990. Role of the USFWS in the sea otter rescue. *Sea Otter Symposium* 90(12):26–28.

Bayha, K. and J. Kormendy, eds. 1990. Proceedings of a symposium to evaluate the response effort on behalf of sea otters after the T/V EVOS in PWS, Anchorage, AK. U.S. DOI, USFWS, and National Fish and Wildlife Foundation. Springfield, VA: National Technical Information Service (distributor).

Becker, P. R. and C.-A. Manen. 1989. Natural oil seeps in the Alaskan marine environment. U.S. Department of Commerce, NOAA Outer Continental Shelf Environmental Assesment Program Final Report 62:1–126.

Beder, S. 1998. *Global Spin: The Corporate Assault on Environmentalism*. White River Junction, VT: Chelsea Green Publishing Company.

Begley, S. and T. Waldrop. 1989. Microbes to the rescue! Bioremediation, bacteria and fungi are used to clean up aquifers, toxic dumps and oil spills. *Newsweek*, 19 June.

Bence, A. E. and W. A. Burns. 1995. Fingerprinting hydrocarbons in the biological resources of the EVOS area. ASTM STP 1219:84–140.

Ben-David, M., G. M. Blundell, and J. E. Blake. 2002. Post-release survival of river otters: Effects of exposure to crude oil and captivity. *Journal of Wildlife Management* 66:1208–1223.

Ben-David, M., R. T. Bowyer, and L. K. Duffy. 2001. Biomarker responses in river otters experimentally exposed to oil contaminants. *Journal of Wildlife Diseases* 37:489–508.

Ben-David, M., T. M. Williams, and O. A. Ormseth. 2000. Effects of oiling on exercise physiology and diving and behavior of river otters: A captive study. *Canadian Journal of Zoology* 78:1380–1390.

Benji, C. 1990. Workshop V. Release. [Transcription of panel discussion.] *Sea Otter Symposium* 90(12):457-465.

Bickert,T. 1993a. Protest shuts port. *Cordova Times*, 26 August.

Bickert,T. 1993b. Coast Guard: No fines for fishermen protesters. *Cordova Times*, 14 October.

Bigg, M.A., P. F. Olesiuk, G. M. Ellis, J. K. B. Ford, and K. C. Balcomb III. 1990. Social organization and genealogy of resident orcas (*Orcinus orca*) in the coastal waters of British Columbia and Washington State. Report of the International Whaling Commission, special issue 12:386-406.

Biggs, E.,T. Baker, L. Brannian, and S. Fried. 1993. PWS herring—what does their future hold? *Alaska's Wildlife* 25(1):37-39.

Biggs, E.—see Brown, E..

Bishop, M.A. and S. P. Green. 2001. Predation on Pacific herring (*Clupea pallasi*) spawn by birds in PWS,AK. *Fisheries Oceanography* 10 (Suppl. 1):149-158.

Bodkin, J. 2000. Research Wildlife Biologist,Alaska Biological Science Center, USGS,Anchorage,AK. Interview by Riki Ott, 12 July.

Bodkin, J., B. E. Ballachey,T.A. Dean,A. K. Fukuyama, S. C. Jewett, L. McDonald, D. H. Monson, C. E. O'Clair, and G. R.VanBlaricom. 2002. Sea otter population status and the process of recovery from the 1989 EVOS. *Marine Ecology Progress Series* 241:237-253.

Bodkin, J. L. and M. S. Udevitz. 1994. An intersection model for estimating sea otter mortality along the Kenai Peninsula. In Loughlin 1994, 81-95.

Boehm, P. D.,W.A. Burns, D. S. Page,A. E. Bence, P. J. Mankiewicz, J. S. Brown, and G. S. Douglas. 2002.Total organic carbon, an important tool in an holistic approach to hydrocarbon source fingerprinting. *Environmental Forensics* 3:243-250.

Boehm, P. D., G. S. Douglas, J. S. Brown, D. S. Page,A. E. Bence,W.A. Burns, and P. J. Mankiewicz. 2000. Comment on natural hydrocarbon background in benthic sediments of PWS,AK: Oil vs. coal. *Environmental Science and Technology* 34:2064-2065.

Boehm, P. D., D. S. Page,W.A. Burns,A. E. Bence, P. J. Mankiewicz, and J. S. Brown. 2001. Resolving the origin of the petrogenic hydrocarbon background in PWS,AK. *Environmental Science and Technology* 35:471-479.

Boehm, P. D., D. S. Page, E. S. Gilfillan,W.A. Stubblefield, and E. J. Harner. 1995. Shoreline ecology program for PWS,AK, following the EVOS: Part 2—Chemistry and toxicology.ASTM STP 1219:347-397.

Boehm, P. D., P. J. Mankiewicz, R. Hartung, J. M. Neff, D. S. Page, E. S. Gilfillan, J. E. O'Reilly, and K. R. Parker. 1996. Characterization of mussel beds with residual oil and the risk to foraging wildlife four years after the EVOS. *Environmental Toxicology and Chemistry* 15(8):1289-1303.

Boersma, P. D. and J. A. Clark. 2001. Seabird recovery following the EVOS: Why was murre recovery controversial? IOSC 2001, 1521-1526.

Boersma, P. D., J. K. Parrish, and A. B. Kettle. 1995. Common murre abundance, phenology, and productivity of the Barren Islands, AK: The EVOS and long-term environmental change. ASTM STP 1219:820-853.

Boesch, D. F. and N. N. Rabalais, eds. 1987. *Long-term Environmental Effects of Offshore Oil and Gas Development.* New York: Elsevier Science Publishing Company.

Boffetta, P., N. Jourenkova, P. Gustavsson. 1997. Cancer risk from occupational and environmental exposure to PAHs. *Cancer Causes and Control* 8:444-472.

Bowyer, T., PhD. 2003. Professor of Wildlife Ecology and Associate Director, University of Alaska, Institute of Arctic Biology, Fairbanks, AK. Interview by Riki Ott, 12 May.

Bowyer, R. T., G. M. Blundell, M. Ben-David, S. C. Jewett, T. A. Dean, and L. K. Duffy. 2003. Effects of the EVOS on river otters: Injury and recovery of a sentinel species. *Wildlife Monographs* 153:1-53.

Bowyer, R. T., J. W. Testa, and J. B. Faro. 1995. Habitat selection and home ranges of river otters in a marine environment: Effects of the EVOS. *Journal of Mammalogy* 76(1):1-11.

Bowyer, R. T., J. W. Testa, J. B. Faro, C. C. Schwartz, and J. B. Browning. 1994. Changes in diets of river otters in PWS, AK: Effects of the EVOS. *Canadian Journal of Zoology* 72:970-975.

Bowyer, R. T., J. Ward, J. B. Faro, and L. K. Duffy. 1993. Effects of the EVOS on river otters in PWS. Abstract. *EVOS Symposium Abstracts,* 297-99.

Boyd, C. J., ed. 1999. American Society for the Advancement of Science (AAAS) Annual Meeting and Science Innovation Exposition: Challenges for a New Century. NY: AAAS.

Boyd, J. N., D. Scholz, and A. H. Walker. 2001. Effects of oil and chemically dispersed oil in the environment. IOSC 2001, 1213-1216.

Braddock, J. F., J. E. Lindstrom, T. R. Yeager, B. T. Rasley, and E. J. Brown. 1996. Patterns of microbial activity in oiled and unoiled sediments in PWS. AFSS 18:94-108.

Bragg, J. R., R. C. Prince, J. B. Wilkinson, and R. M. Atlas. 1992. Bioremediation for shoreline cleanup following the 1989 Alaska oil spill. Exxon Research and Engineering Company, Florham Park, NJ.

Bragg, J. R. and S. H. Yang. 1995. Clay-oil flocculation and its role in natural cleansing in PWS following the EVOS. ASTM 1219:178-214.

Brannon, E., K. Collins, L. Moulton, and K. R. Parker. 2001. Review of studies on oil damage to PWS pink salmon. IOSC 2001, 569-575.

Brannon, E. J., L. L. Moulton, L. G. Gilbertson, A. W. Maki, and J. R. Skalski. 1995. An assessment of oil spill effects on pink salmon populations following the EVOS: Part 1—Early life history. ASTM STP 1219:548-584.

Brazil, E. 1996. Belli's law practice sold to a S. F. firm. *San Francisco Gate News,* 16 August.

Broderson, C., J. Short, L. Holland, M. Carls, J. Pella, M. Larsen, and S. Rice. 1999. Evaluation of oil removal from beaches eight years after the EVOS. Proceedings of the 25th AMOP, 325-336.

Brown, E. 1999. Doctoral graduate student, University of Alaska, Institute of Marine Sciences, Fairbanks, AK. Interview by Riki Ott, 22 April.

Brown, E., PhD. 2001. Research associate, University of Alaska, Institute of Marine Sciences, Fairbanks, AK. Interview by Riki Ott, 27 March.

Brown, E., D. T. Baker, J. E. Hose, R. M. Kocan, G. D. Marty, M. D. McGurk, B. L. Norcross, and J. Short. 1996. Injury to the early life history stages of Pacific herring in PWS after the EVOS. AFSS 18:448-462.

Brown, E. D., J. H. Churnside, R. L. Collins, T. Veenstra, J. J. Wilson, and K. Abnett. 2002. Remote sensing of capelin and other biological features in the North Pacific using lidar and video technology. *ICES Journal of Marine Science* 59:1120-1130.

Brown, P. and E. J. Mikkelsen. 1990. *No Safe Place. Toxic Waste, Leukemia, and Community Action.* Berkley: University of California Press.

Brown, E. D., S. M. Moreland, G. A. Borstad, and B. L. Norcross. 1999. Estimating forage fish and seabird abundance using aerial surveys: Survey design and uncertainty. Appendix VI in Cooney 1999.

Brown, E., D. Norcross, and J. Short. 1996. An introduction to studies on the effects of the EVOS on early life history stages of Pacific herring, *Clupea pallasi,* in PWS, AK. *Canadian Journal of Fisheries and Aquatic Sciences* 53:2337-2342.

Brown, E. D., J. Seitz, B. L. Norcross, and H. P. Huntington. 2002. Ecology of herring and other forage fish as recorded by resource users of PWS and the outer Kenai Peninsula, AK. *Alaska Fishery Research Bulletin* 9(2):75-101.

Bue, B. G., S. M. Fried, S. Sharr, D. G. Sharp, J. A. Wilcock, and H. J. Geiger. 1998. Estimating salmon escapement using area-under-the-curve, aerial observer efficiency, and stream-life estimates: The PWS pink salmon example. *North Pacific Anadromous Fish Commission Bulletin* No. 1:240-250.

Bue, B., S. Fried, S. Sharr, and M. Willette. 1993a. Pinks in peril: Declining wild stocks in PWS. *Alaska's Wildlife* 25(1):34-36.

Bue, B. G., S. Sharr, S. D. Moffitt, and A. Craig. 1993b. Assessment of injury to pink salmon eggs and fry. Abstract. *EVOS Symposium Abstracts,* 101-103.

Bue, B. G., S. Sharr, S. D. Moffitt, and A. K. Craig. 1996. Effects of the EVOS on pink salmon embryos and pre-emergent fry. AFSS 18:619-627.

Bue, B. G., S. Sharr and J. E. Seeb. 1998. Evidence of damage to pink salmon populations inhabiting PWS, AK, two generations after the EVOS. *Transactions of the American Fisheries Society* 127:35-43.

Burn, D. M. 1994. Boat-based population surveys of sea otters in PWS. In Loughlin 1994, 61-80.

Burnham, K. and M. Bey. 1991. Effects of crude oil and ultraviolet radiation on immunity within mouse skin. *Journal of Toxicolology and Environmental Health* 34:83-93.

Burnham, K. and M. Rahman. 1992. Effects of petrochemicals and ultraviolet radiation on epidermal 1A expression *in vitro. Journal of Toxicolology and Environmental Health* 35:175-185.

Burns, J. J. 1993. Data report: Harbor seal surveys in northern and western PWS, August 26 to September 2, 1992. Report to Exxon Company, Houston, TX, May 1993. Living Resources, Inc., POB 83570, Fairbanks, AK 99708.

Caleb Brett U.S.A., Inc. 1989. Letter from Paul Kellett, Caleb Brett, to John Tompkins, Exxon Shipping Company; re: breakdown of cargo lightered and spill volume. 18 May. In Alaska Dept. of Law 1991, ACE 577444. ARLIS, Anchorage, AK.

California Bar Journal. 1996. 'King of Torts' Melvin Belli leaves controversial legacy. August.

Campbell, D. M. 1993. Shetland oil spill—letter. *BMJ* 306:519.

Campbell, D. M., D. Cox, J. Crum, K. Foster, P. Christie, D. Brewster. 1993. Initial effects of the grounding of the tanker *Braer* on health in Shetland. *BMJ* 307:1251-1255.

Campbell, D. M., D. Cox, J. Crum, K. Foster, A. Riley. 1994. Later effects of the grounding of tanker *Braer* on health in Shetland. *BMJ* 309:773-774.

Capuzzo, J. 1987. Biological effects of petroleum hydrocarbons: Assessments from experimental results. In Boesch and Rabalais 1987, 343-410.

Capuzzo, J., J. Farrington, and S. Kellogg. 1990a. Summary of reviewers' comments. In Prince, Clark, and Lindstrom 1990.

Capuzzo, J., J. Farrington, and S. Kellogg. 1990b. Reviewers' executive summary. In Prince, Clark, and Lindstrom 1990.

Carleton, J. 2003. Exxon may face more payments from Alaska spill. *WSJ*, 3 October.

Carls, M. G. 1987. Effects of dietary and water-borne oil exposure on larval Pacific herring (*Clupea harengus pallasi*). *Marine Environmental Research* 22:253-270.

Carls, M. G. 2001. Fisheries Biologist, NOAA/NMFS ABL, Juneau, AK. Interview by Riki Ott, 5 April.

Carls, M. G., Babcock, M. M. P. M. Harris, G. V. Irvine, J. A. Cusick, and S. D. Rice. 2001a. Persistence of oiling in mussel beds after the EVOS. *Marine Environmental Research* 51:167-190.

Carls, M. G., R. A. Heintz, A. Moles, S. D. Rice, and J. W. Short. 2001b. Long-term biological damage: What is known, and how should that influence decisions on response, assessment, and restoration? IOSC 2001, 399-403.

Carls, M. G., L. Holland, M. Larsen, J. L. Lum, D. G. Mortensen, S. Y. Wang, and A. C. Wertheimer. 1996b. Growth, feeding, and survival of pink salmon fry exposed to food contaminated with crude oil. AFSS 18:608-618.

Carls, M. G., G. D. Marty, and J. E. Hose. 2002. Synthesis of the toxicological and epidemiological impacts of the EVOS on Pacific herring in PWS, AK. *Canadian Journal of Fisheries and Aquatic Sciences* 59:153-172.

Carls, M. G., G. D. Marty, T. R. Meyers, R. E. Thomas, and S. D. Rice. 1998. Expression of viral hemorrhagic septicemia virus in prespawning Pacific herring (*Clupea pallasi*) exposed to weathered crude oil. *Canadian Journal of Fisheries and Aquatic Sciences*

55:2300-2309.

Carls, M., S. D. Rice, and J. E. Hose. 1999. Sensitivity of fish embryos
to weathered crude oil: Part I. Low-level exposure during incubation causes malformations, genetic damage, and mortality in larval Pacific herring *(Clupea Pallasi). Environmental Toxicology
and Chemistry* 18(3):481-493, 1999.

Carls, M. G., A. C. Wertheimer, J. W. Short, R. M. Smolowitz, and J. J.
Stegeman. 1996a. Contamination of juvenile pink and chum
salmon by hydrocarbons in PWS after the EVOS. AFSS
18:593-607.

Carlsen, W. 1995. Belli, 88, files for bankruptcy. Fabled career nearing
end, some fear. *San Francisco Chronicle,* 9 December.

Carpenter, A. D., R. G. Dragnich, and M. T. Smith. 1991. Marine operations and logistics during the EVOS cleanup. IOSC 1991,
205-211.

Carson, R. 1962. *Silent Spring.* Boston: Houghton Mifflin Company.

Castellini, M. A., J. M. Castellini, and S. J. Trumble. 2000. Recovery of
harbor seals. Phase II: Controlled studies of health and diet. EVOS
Trustee Council Restoration Project Annual Report (Restoration
Project 99341). ADFG, Habitat and Restoration Division,
Anchorage, AK.

Castellini, J. M., H. J. Meiselman, and M. A. Castellini. 1996.
Understanding and interpreting hematocrit measurements in pinnipeds. *Marine Mammal Science* 12(2):251-264.

Catania, P., ed. 2000. *Energy 2000: The Beginning of a New
Millennium.* Lancaster, England: Technomic Publishing Co.

Celewycz, A. G. and A. C. Wertheimer. 1996. Prey availability to juvenile salmon after the EVOS. AFSS 18:564-577.

CFS. 1989 1(26)a. Health investigation, workers' compensation and
taxes. 12 May.

CFS. 1989 1(26)b. Time to organize. 12 May.

CFS. 1989 1(29). Health concerns. 16 May.

CFS. 1989 1(35). PWS shoreline survey. 24 May.

CFS. 1989 1(58). [A]DEC report. 23 June.

CFS. 1989 2(14). ADEC report, field test of Corexit 9580M2. 29 July.

CFS. 1989 2(15). Exxon promotes bioremediation. 1 August.

CFS. 1989 2(20). ADEC report, Corexit 9580M2 test resumes. 10
August.

CFS. 1989 2(21). House subcommittee holds hearing in Cordova. 12
August.

CFS. 1989 2(24)a. ADEC report, bioremediation will not be used on

Seward shorelines. 17 August.

CFS. 1989 2(24)b. Corexit 9580[M2] use denied. 17 August.

CFS. 1989 2(34). ADEC report: Notice of violation issued to Exxon for improper application of Inipol. 6 September.

CFS. 1989 2(39). Exxon chairman visits shorelines. 15 September.

Citra-Solv, LLC. 2001. MSDS: Citra-Solv. Citra-Solv, LLC., P.O. Box 2597, Danbury, CT.

"Clarke, S." *(alias)*. 2003. Former cleanup worker. Vancouver, Canada. Interview by Riki Ott, 19 July.

Clipp, A. 1993. Job-damaged people: How to survive and change the workers compensation system. Environmental Health Network and Louisiana Injured Workers' Union.

Coates, P. A. 1993. *The Trans-Alaska Pipeline Controversy. Technology, Conservation, and the Frontier.* First paperback ed. Anchorage: University of Alaska Press. First ed. London: Associated University Presses, Inc., 1991.

Cohen, M. J. 1997. Economic impacts of the EVOS. In Picou, Gill, and Cohen 1997, 133–160.

Cohen, Y. 1992. Multimedia fate and effects of airborne petroleum hydrocarbons in the Port Valdez region. Report by Multimedia Envirosoft Corp. for PWS Regional Citizens' Advisory Committee, Anchorage, AK. 14 March.

Colborn, T., D. Dumanoski, and J. P. Myers. 1996. *Our Stolen Future.* New York: Dutton.

Collinsworth, D. W. 1990. Testimony of ADFG Commissioner. In U.S. Congress House 1990, 180–184.

Cooney, T., PhD. 1999. Professor of Marine Science, Emeritus, University of Alaska, Institute of Marine Sciences, Fairbanks, AK. Interview by Riki Ott, 22 April.

Cooney, R. T. (compiler). 1999. Sound Ecosystem Assessment (SEA)–An integrated science plan for the restoration of injured species in PWS. EVOS Trustee Council Final Report (Restoration Project 00320). Anchorage, AK.

Cooney, R. T., J. R. Allen, M. A. Bishop, D. L. Eslinger, T. Kline, Jr., B. L. Norcross, C. P. McRoy, J. Milton, J. Olsen, E. V. Patrick, A. J. Paul, D. Salmon, D. Scheel, G. L. Thomas, S. L. Vaughan, and T. M. Willette. 2001a. Ecosystem control of pink salmon *(Onchorhynchus gorbuscha)* and Pacific herring *(Culpea pallasi)* populations in PWS, AK. *Fisheries Oceanography* 10 (Suppl. 1):1–13.

Cooney, R. T., K. O. Coyle, E. Stockmar, and C. Stark. 2001b. Seasonality in surface-layer net zooplankton communities in

PWS,AK. *Fisheries Oceanography* 10 (Suppl. 1):97–109.

Coughlin,W. P. 1992a. Illness tied to Exxon cleanup is cited in spate of lawsuits. *Boston Sunday Globe*, 12 April.

Coughlin,W. P. 1992b. Doctor says oil, cleanup toxins fatal. Mixture of chemicals responsible for illness and death from Alaska spill, suits claim. *Boston Globe*, 10 May.

Couillard, C. M. 2002.A micro-scale test to measure petroleum oil toxicity to mummichog embryos. *Environmental Toxicology* 17:195–202.

Cowan, R. M. 1993. Letter to Edward Stewart, U.S.Aviation Underwriters; re: settlement proposal. File No. GL-1512, Law Offices of Cowan & Gerry, Kenai,AK. 30 June. In ADOL,AWCB 1992b.

Crowley, D.W. and S. M. Patten,Jr. 1996. Breeding ecology of harlequin ducks in PWS,AK. EVOS State/Federal NRDA Final Report, Bird Study No. 71.ADFG,Anchorage,AK.

Crum,J. E. 1993. Peak respiratory flow rate in schoolchildren living close to *Braer* oil spill. *BMJ* 307:23–24.

Cullen, M. R. 1987. *Workers with Multiple Chemical Sensitivities, Occupational Medicine: State of the Art Reviews*. Philadelphia, PA: Hanley & Belfus.

Cummings,A. D. 1992.The EVOS and the confidentiality of NRDA data. *Ecology Law Quarterly* 19(2):363–412.

Curriden, M. 1999. Exxon finds slow pace of Valdez (sic) case profitable. Company says fairness, not money, is the issue. *Dallas (TX) Morning News*, 14 March.

Dahlheim, M. and C. Matkin. 1994.Assessment of injuries to PWS orcas. In Loughlin 1994, 163–171.

Daubert v. Merrell Dow Pharmaceuticals, Inc., 509 U.S. 579 L. Ed. 2d 469, 113 S. Ct. 2786 (1993).

Davidson,A. 1990. *In the Wake of the* Exxon Valdez:*The Devastating Impact of the Alaska Oil Spill*. San Francisco: Sierra Club Books.

Davis, C. R., G. D. Marty, M.A.Adkinson, E. F. Freiberg, and R. P. Hedrick. 1999.Association of plasma IgM with body size, histopathologic changes, and plasma chemistries in adult Pacific herring *Clupea pallasi. Diseases of Aquatic Organisms* 38:125–133.

Davis, R.W. 2000. Harbor seal recovery. Phase III: Effects of diet on lipid metabolism and health. EVOS Trustee Council Restoration Project Annual Report (Restoration Project 99441). ADFG,

Habitat and Restoration Division, Anchorage, AK.

Day, R. H., S. M. Murphy, J. A. Wiens, G. D. Hayward, E. J. Harner, and L. N. Smith. 1995. Use of oil-affected habitats by birds after the EVOS. ASTM STP 1219:726-759.

Day, R. H., S. M. Murphy, J. A. Wiens, G. D. Hayward, E. J. Harner, and L. N. Smith. 1997. Effects of the EVOS on habitat use by birds in PWS, AK. *Ecology Applied* 7:593-613.

de Bettencourt, M., G. Merrick, T. Deal, and B. Travis. 2001. Safeguarding the public interest: A look at government policies that affect the OPA 90 oil spill liability trust fund and oil spill costs. IOSC 2001, 725-729.

Dean, T. A., J. L. Bodkin, A. K. Fukuyama, S. C. Jewett, D. H. Monson, C. E. O'Clair, G. R. VanBlaricom. 2002. Food limitation and the recovery of sea otters following the EVOS. *Marine Ecology Progress Series* 241:255-270.

Dean, T. A., J. L. Bodkin, S. C. Jewett, D. H. Monson, and D. Jung. 2000. Changes in sea urchins and kelp following a reduction in sea otter density as a result of the EVOS. *Marine Ecology Progress Series* 241:281-291.

Dees, C. W. 1991. Deposition. In *The Exxon Valdez* Case (filed in 1989).

DeGange, A. R. and C. J. Lensink. 1990. Distribution, age, and sex composition of sea otter carcasses recovered during the response to the T/V EVOS. *Sea Otter Symposium* 90(12):124-129.

DeGange, A. R. and T. D. Williams. 1990. Procedures and rationale for marking sea otters captured and treated during the T/V EVOS. *Sea Otter Symposium* 90(12):394-399.

DeGange, A. R., D. H. Monson, D. B. Irons, C. M. Robbins, and D. C. Douglas. 1990. Distribution and relative abundance of sea otters in Southcentral and Southwestern Alaska before or at the time of the T/V EVOS. *Sea Otter Symposium* 90(12):18-25.

Didriksen, N. A., M.D. 1993. Psychological consultation report. 30 January. In *Roberts v. Exxon* (1999), No. 101, attachment.

Didriksen, N. A. and J. R. Butler. 1997. Psychological concomitants of chemical sensitivity: Evaluation and treatment. In Rea 1997a, 2771-2901.

D'Oro, R. 2004. Judge orders Exxon Mobil to pay nearly $7 billion in spill damages. JUDGMENT: Money to go to people, cities affected by 1989 disaster. *The Associated Press*, 28 January.

Doroff, A. M. and J. L Bodkin. 1994. Sea otter foraging behavior and hydrocarbon levels in prey. In Loughlin 1994, 193–208.

Dowler, L. 1992. Work Therapy Enterprises, PPI rating measurement on Smith for Dr. Davidhizar, 17 April. In *Roberts v. Exxon* (1999), No. 62, Exhibit C.

Drago, M. 1996. Exxon wins insurance case. Company due $250 million for spill cleanup expenses. *ADN,* 11 June.

Dragoo, D. E., G. V. Byrd, and D. B. Irons. 2003. Breeding status, population trends, and diets of seabirds in Alaska, 2001. USFWS, Alaska Maritime National Wildlife Refuge, Homer, AK. Report AMNWR 03/05.

Drew, G. 2002. Primary and secondary production in lower Cook Inlet. In Piatt 2002, 26–31.

Drew, G. and J. Piatt. 2002. Oceanography of lower Cook Inlet. In Piatt 2002, 17–25.

Driskell, B., MS. 2000. Marine Biology Consultant, Seattle, WA. Interview by Riki Ott, 15 January.

Driskell, W. B., A. K. Fukuyama, J. P. Houghton, D. C. Lees, A. J. Mearns, and G. Shigenaka. 1996. Recovery of PWS intertidal infauna from *Exxon Valdez* oiling and shoreline treatments, 1989–1992. AFSS 18:362–378.

Driskell, W.B., J. L. Ruesink, D. C. Lees, J. P. Houghton, and S.C. Lindstrom. 2001. Long-term signal of disturbance: *Fucus gardneri* after the EVOS. *Ecological Applications* 11(3):815–827.

Duffy, L. K., R. T. Bowyer, J. W. Testa, and J. B. Faro. 1993. Differences in blood haptoglobin and length-mass relationships in river otters *(Lutra canadensis)* from oiled and nonoiled areas of PWS, AK. *Journal of Wildlife Diseases* 29(2):353–359.

Duffy, L. K., R. T. Bowyer, J. W. Testa, and J. B. Faro. 1994a. Chronic effects of the EVOS on blood and enzyme chemistry of river otters. *Environmental Toxicology and Chemistry* 4:643–647.

Duffy, L. K., R. T. Bowyer, J. W. Testa, and J. B. Faro. 1994b. Evidence for recovery of body mass and haptoglobin values of river otters following the EVOS. *Journal of Wildlife Diseases* 30(3):421–425.

Duffy, L. K., R. T. Bowyer, J. W. Testa, and J. B. Faro. 1996. Acute phase proteins and cytokines in Alaskan mammals as markers of chronic exposure to environmental pollutants. *AFFS* 18:806–813.

Duffy, L. K., M. K. Hecker, G. M. Blundell, and R. T. Bowyer. 1999. An analysis of the fur of river otters in PWS, AK: Oil related hydrocarbons 8 years after the EVOS. *Polar Biology* 21:56–58.

Ecological Consulting, Inc. 1991. Assessment of direct seabird mortality in PWS and the western Gulf of Alaska resulting from the

EVOS. Final report to the U.S. Department of Justice, Ecological Consulting Inc., Portland, OR.

Elder, D., G. Killam, and P. Koberstein. 1999. The Clean Water Act. An owner's manual. River Network, POB 8787, Portland, OR. www.rivernetwork.org

Elliott, J. E. 2001. Preventing oil spills in the twenty-first century: An ecological economics perspective. IOSC 2001, 27–33.

Environmental Health Center–Dallas. 2004. Services and treatment. www.ehcd.com/services/treatments.htm

Epler, P. 1985a. Report: Not all pollutants limited in Alyeska permit. Discharge flows into Port Valdez, but effect on environment unknown. *ADN*, 3 October.

Epler, P. 1985b. Slipping through the cracks. Wastewater from Alaska oil terminal eludes agency control. *ADN*, 27 October.

Epler, P. 1988a. Port Valdez study stirs controversy. Report finds no pollution; one scientist doesn't buy it. *ADN*, 29 July.

Epler, P. 1988b. Scientists say Port Valdez fish contaminated. Federal researchers find bottom-dwelling fish tainted with hydrocarbons from oil pollution. *ADN*, 2 October.

Erikson, D. E. 1995. Surveys of murre colony attendance in the northern Gulf of Alaska following the EVOS. ASTM STP 1219:780–819.

Esler, D. 1999. Research Wildlife Biologist, USGS, Alaska Biological Science Center, Anchorage, AK. Interview by Riki Ott, 3 December.

Esler, D., T. D. Bowman, T. A. Dean, C. E. O'Clair, S. C. Jewett, and L. L. McDonald. 2000a. Correlates of harlequin duck densities during winter in PWS, AK. *Condor* 102:920–926.

Esler, D., T. D. Bowman, K. A. Trust, B. E. Ballachey, T. A. Dean, S. C. Jewett, and C. E. O'Clair. 2002. Harlequin duck population recovery following the EVOS: Progress, process and constraints. *Marine Ecology Progressive Series* 241:271–286.

Esler, D., J. A. Schmutz, R. L. Jarvis, and D. M. Mulcahy. 2000b. Winter survival of adult female harlequin ducks in relation to history of contamination by the EVOS. *Journal of Wildlife Management* 64(3):839–847.

Eslinger, D. L., R. T. Cooney, C. P. McRoy, A. Ward, T. C. Kline, Jr., E. P. Simpson, J. Wang, and J. R. Allen. 2001. Plankton dynamics: Observed and modelled responses to physical conditions in PWS, AK. *Fisheries Oceanography* 10 (Suppl. 1):81–96.

Eubanks, M. 1994. Biological markers: The clues to genetic susceptibility. *Environmental Health Perspectives* 102:50-56.

EVOS Trustee Council. 1994. EVOS restoration plan, update on injured resources and services. Anchorage, AK. November.

EVOS Trustee Council. 1999. 1999 Status report: Legacy of an oil spill 10 years after *Exxon Valdez*. Anchorage, AK.

EVOS Trustee Council. 2000. 2000 Status report: Ecosystem research. Anchorage, AK.

EVOS Trustee Council. 2002. 2002 Status report. Anchorage, AK.

EVOS Trustee Council. 2003. 2003 Status report. Anchorage, AK.

EVOS Trustee Council, University of Alaska Sea Grant College Program, and American Fisheries Society, AK Chapter. 1993. EVOS symposium program and abstracts. Anchorage, AK, 2-5 February.

EVS Consultants. 1990. Measurement of acute toxicity of the oleophilic fertilizer Inipol EAP22. Prepared for EPA, Environmental Research Laboratory, Gulf Breeze, FL. Project No. 2/294-05.

Ewing, G., MD. 1992. Allergy evaluation for La Joie. 20 October. In *Roberts v. Exxon* (1999), No. 26, Exhibit E.

Ewing, G., MD. 1994. Letter to Stenson; re: La Joie. 18 April. In *Roberts v. Exxon* (1999), No. 101, attachment.

Ewing, G., MD. 1997a. Deposition. 9 December. In *Roberts v. Exxon* (1999).

Ewing, G., MD. 1997b. Expert witness report. 24 October. In *Roberts v. Exxon* (1999), No. 63, Exhibit L. (This report was not written by Ewing: see *Roberts vs. Exxon*, No. 90.)

Exxon Company, USA. 1986. Occupational health aspects of unusual work schedules: A review of Exxon's experiences. *American Industrial Hygiene Assoc. Journal*, 47(4):199-202.

Exxon Company 1988. MSDS for crude oil. 15 May. Houston, TX, T1252-2180

Exxon Company, USA. 1989a. MSDS for Inipol EAP22. 28 July. Houston, TX

Exxon Company, USA. 1989b. Clinical data on upper respiratory infections: URIs-Breakdowns. In *Stubblefield v. Exxon* (1994).

Exxon Company, USA. 1989c. Partial release. In ADOL, AWCB 1992a.

Exxon Company, USA. 1991a. Sea otters thrive in PWS, AK. Brochure. Houston, TX, 77252-2180. February.

Exxon Company, USA. 1991b. Two years after: Conditions in PWS and the Gulf of Alaska. Houston, TX. October.

Exxon Company, USA. 1992. MSDS for Corexit 9527. Houston, TX, 14 June.

Exxon Company, USA. 1993a. Long awaited science studies on Valdez (sic) oil spill effects to be presented at Atlanta symposium. Media advisory. Anchorage, AK, 7 April.

Exxon Company, USA. 1993b. Ten posters depicting the results of scientific studies of PWS and the Gulf of Alaska following the EVOS. Anchorage, AK.

Fadely, B. S., J. M. Castellini, and M. A. Castellini. 1998. Recovery of harbor seals from EVOS: Condition and health status. EVOS Restoration Project Final Report (Restoration Project 97001). ADFG, Habitat and Restoration Division, Anchorage, AK.

Falk-Filipsson, A., A. Lof, M. Hagberg, E. W. Hjelm, and Z. Wang. 1993. d-limonene exposure to humans by inhalation: Uptake, distribution, elimination, and effects on pulmonary function. *Journal of Toxicology and Environmental Health* 38(1):77–88.

Feuston, M. H., L. K. Low, C. E. Hamilton, and C. R. Mackerer. 1994. Correlation of systemic and developmental toxicities with chemical component classes of refinery streams. *Fundamentals of Applied Toxicology* 22:622–630.

Fingas, M. F., M. A. Bobra, and R. K. Velicogna. 1987. Laboratory studies on the chemical and natural dispersibility of oil. IOSC 1987, 241–246.

Ford, R. G., M. L. Bonnell, D. H. Varoujean, G. W. Page, H. R. Carter, B. E. Sharp, D. Heinemann, and J. L. Casey. 1996. Total direct mortality of seabirds from the EVOS. AFSS 18:684–711.

Ford, R. G., G. W. Page, and H. R. Carter. 1987. Estimating mortality of seabirds from oil spills. IOSC 1987, 747–757.

Foy, R. J. and B. L. Norcross. 1999. Feeding behavior of herring *(Clupea pallasi)* associated with zooplankton availability in PWS, AK. *Ecosystem Approaches for Fisheries Management,* Alaska Sea Grant College Program, AK-SG-99-01:129–135.

Foy, R. J. and A. J. Paul. 1999. Winter feeding and changes in somatic energy content of age-0 Pacific herring in PWS, AK. *Transactions of the American Fisheries Society* 128:1193–1200.

Francis, R. C. and S. R. Hare. 1994. Decadal-scale regime shifts in the large marine ecosystems of the northeast Pacific: A case for historical science. *Fisheries Oceanography* 3:279–291.

Francis, R. C., S. R. Hare, A. B. Hollowed, and W. S. Wooster. 1998. Effects of interdecadal climate variability on the oceanic ecosystems of the NE Pacific. *Fisheries Oceanography* 7:1–21.

Freemantle, T. 1995. Billion-dollar battle looms over spill costs. Exxon Corp. trying to collect from its insurance companies. *ADN* from *Houston Chronicle*, 5 September.

Frost, G. 1989. VECO purchases *The Anchorage Times*. *ADN,* 21 September.

Frost, K. J. 1997. Harbor seal (*Phoca vitulina richardsi*). Restoration notebook. EVOS Trustee Council, Anchorage, AK.

Frost, K. J. 2003. Affiliate Associate Professor of Marine Science, University of Alaska, Institute of Marine Science, Fairbanks, AK; Retired Marine Mammal Biologist, ADFG, Fairbanks, AK. Interview by Riki Ott, 19 July.

Frost, K. J. and L. F. Lowry. 1993. Assessment of damages to harbor seals caused by the EVOS. Abstract. *EVOS Symposium Abstracts,* 300-302.

Frost, K. J., L. F. Lowry, E. H. Sinclair, J. Ver Hoef, and D. C. McAllister. 1994. Impacts on distribution, abundance, and productivity of harbor seals. In Loughlin 1994, 97-118.

Frost, K. J., L. F. Lowry, and J. M. Ver Hoef. 1999. Monitoring the trend of harbor seals in PWS, AK, after the EVOS. *Marine Mammal Science* 15(2):494-506.

Frost, K. J., L. F. Lowry, J. M. Ver Hoef, S. J. Iverson, and M. A. Simpkins, eds. 1999. Monitoring, habitat use, and trophic interactions of harbor seals in PWS, AK. EVOS Restoration Project (Restoration Project 98064), ADFG, Division of Wildlife Conservation, Fairbanks, AK.

Frost, K. A., L. L. Lowry, J. M. Ver Hoef, S. J. Iverson, and M. A. Simpkins, eds. 1998. EVOS Trustee Council Restoration Project Annual Report (Restoration Study 97064). ADFG, Division of Wildlife Conservation, Fairbanks, AK.

Frost, K. J., C.-A. Manen, and T. Wade. 1994. Petroleum hydrocarbons in tissues of harbor seals from PWS and the Gulf of Alaska. In Loughlin 1994, 331-358.

Frost, K. J., M. A. Simpkins, and L. F. Lowry. 2001. Diving behavior of subadult and adult harbor seals in PWS, AK. *Marine Mammal Science* 17(4):813-834.

Funk, F. C., D. W. Carlile, and T. T. Baker. 1993. The PWS herring recruitment failure of 1989: Oil spill or natural causes? Abstract. *EVOS Symposium Abstracts,* 258-61.

Garloch, K. 1995. The Gulf War and birth defects. *ADN,* 22 January.

Garrott, R. A., L. L. Eberhardt, and D. M. Burn. 1993. Mortality of sea otters in PWS following the EVOS. *Marine Mammal Science* 9(4):343-359.

Gay, S. M. III and S. L. Vaughan. 2001. Seasonal hydrography and tidal currents of bays and fjords in PWS, AK. *Fisheries Oceanography* 10 (Suppl. 1):159-193.

Geiger, H. J., B. G. Bue, S. Sharr, A. C. Wertheimer, and T. M. Willette. 1996. A life history approach to estimating damage to PWS pink salmon caused by the EVOS. AFSS 18:487-498.

Geiser, K. 2001. *Materials Matter: Toward a Sustainable Materials Policy.* Cambridge, MA: MIT Press.

General Electric Co. v. Joiner, 522 U.S. 136 (1997).

George, S. E., G. M. Nelson, M. J. Kohan, S. H. Warren, B. T. Eischen, and L. R. Brooks. 2001. Oral treatment of Fischer 344 rats with weathered crude oil and a dispersant influences intestinal metabolism and microbiota. *Journal of Toxicolology and Environmental Health,* 63(4):297-316.

Georghiou, P.E. 1989. Mutagenicity of Prudhoe Bay crude and weathered products. Paper presented at the Alaskan Oil Spill and Human Health conference, sponsored by National Institute of Environmental Health Sciences, NIOSH, University of Washington School of Public Health, US EPA, and ATSDR, Seattle, WA , 28-30 July.

Gerde, P. and P. Scholander. 1987. An experimental study of the penetration of PAHs through the bronchial lining layer. *Environmental Research* 44:321-344.

Gerde, P. and P. Scholander. 1989. An experimental study of the penetration of PAHs through a model of the bronchial layer. *Environmental Research* 48:287-295.

Gibeaut, J. C. and E. Piper. 1993. Shoreline oiling assessment of the EVOS. EVOS Restoration Project Final Report (Restoration Project 93038). ADEC, Anchorage, AK.

Gilfillan, E. S., E. J. Harner, and D. S. Page. 2002. Comment on Peterson et al. (2001) 'Sampling design begets conclusion.' *Marine Ecology Progress Series* 231:303-308.

Gilfillan, E. S., D. S. Page, E. J. Harner, and P. D. Boehm. 1995a. Shoreline ecology program for PWS, AK, following the EVOS: Part 3—Biology. ASTM STP 1219:398-443.

Gilfillan, E. S., T. H. Suchanek, P. D. Boehm, E. J. Harner, D. S. Page, and N. A. Sloan. 1995b. Shoreline impacts in the Gulf of Alaska region following the EVOS. ASTM STP 1219:444-481.

Gilfillan, E. S., D. S. Page, K. R. Parker, J. M. Neff, and P. D. Boehm. 2001. A 10-year study of shoreline conditions in the EVOS zone, PWS, AK. IOSC 2001, 559-567.

Glantz, S. A., J. Slade, L. A. Bero, P. Hanauer, and D. E. Barnes. 1996. *The*

Cigarette Papers. Berkeley, CA: University of California Press.

Golet, G. H., PhD. 1999. Senior Ecologist, The Nature Conservancy, Northern Central Valley Office, Chico, CA. Interview by Riki Ott, 3 December.

Golet, G. H. and D. B. Irons. 1999. Raising young reduces body condition and fat stores in black-legged kittiwakes. *Oecologia* 120:530-538.

Golet, G. H., D. B. Irons, and D. P. Costa. 2000. Energy costs of chick rearing in black-legged kittiwakes *(Rissa tridactyla)*. *Canadian Journal of Zoology* 78:982-991.

Golet, G. H., D. B. Irons, and J. A. Estes. 1998. Survival costs of chick rearing in black-legged kittiwakes. *Journal of Animal Ecology* 67:827-841.

Golet, G. H., K. J. Kuletz, D. D. Roby, and D. B. Irons. 2000. Adult prey choice affects chick growth and reproductive success in pigeon guillemots. *The Auk* 117(1):82-91.

Golet, G. H., P. E. Seiser, A. D. McGuire, D. D. Roby, J. B. Fischer, K. J. Kuletz, D. B. Irons, T. A. Dean, S. C. Jewett, and S. H. Newman. 2002. Long-term direct and indirect effects of the EVOS on pigeon guillemots in PWS, AK. *Marine Ecology Progress Series* 241:287-304.

Gomer, C. J. and D. M. Smith. 1980. Acute skin phototoxicity in hairless mice following exposure to crude shale oil or natural petroleum oil. *Toxicology* 18:75-85.

Goodwin, T. 2001. Owner/Manager of Seagoville Ecological Housing, Seagoville, TX. Interview by Riki Ott, 22 May.

Gordon, R. I., 1998. No balm in Gilead: Why workers' compensation fails workers in a toxic age. In Matthews 1998, 61-100.

Grabacki, S. 1998. Making the most of the Copper River resources: Fisheries resource options. Prepared by Graystar Pacific Seafood, Ltd. for the Copper River Watershed Project, Cordova, AK.

Gruber, J. A. and M. E. Hogan. 1990. Transfer and placement of non-releasable sea otters in aquariums outside AK. *Sea Otter Symposium* 90(12):428-431.

Haebler, R. J., R. K. Wilson, and C. R. McCormick. 1990. Determining health of rehabilitated sea otters before release. *Sea Otter Symposium* 90(12): 390-393.

Hansen, D. J. 1997. Shrimp fishery and capelin decline may influence decline of harbor seal *(Phoca vitulina)* and northern sea lion *(Eumetopias jubatus)* in western Gulf of Alaska. Abstract. *Forage Fish Proceedings,* 197-207.

Harding, A. 2002. Horned puffin biology on Duck (Chisik) Island. In Piatt 2002, 110–121.

Harding, A., J. F. Piatt, and K. C. Hamer. 2003. Breeding ecology of horned puffins *(Fratercula corniculata)* in Alaska: Annual variation and effects of El Niño. *Canadian Journal of Zoology* 81:1004–1013.

Harris, R. K., R. B. Moeller, T. P. Lipscomb, J. M. Pletcher, R. J. Haebler, P. A. Tuomi, C. R. McCormick, A. R. DeGange, D. Mulcahy, and T. D. Williams. 1990. Identification of a herpes-like virus in sea otters during rehabilitatiion after the T/V EVOS. *Sea Otter Symposium* 90(12):366–368.

Harrison, C. M. 1990. Statement of Executive Vice President, Exxon Company, USA. In U.S. Congress House 1990, 185–188.

Harrison, O. R. 1991. Overview of the EVOS. IOSC 1991, 313–319.

Harte, J., C. Holdren, R. Schneider, and C. Shirley. 1991. *Toxics A to Z: A Guide to Everyday Pollution Hazards.* Berkeley, CA: University of California Press.

Hayes, D. L. and K. J. Kuletz. 1997. Decline of pigeon guillemot populations in PWS, AK, and apparent changes in distribution and abundance of their prey. Abstract. *Forage Fish Proceedings,* 669–702.

Hayes, M. O. and J. Michel. 1997. Evaluation of the condition of PWS shorelines following the EVOS and subsequent shoreline treatment, 1997 geomorphology monitoring survey. NOAA Technical Memorandum, NOS ORCA 126, Hazardous Materials Response and Assessment Division, NOAA, Seattle, WA.

Heinemann, D. 1993. How long to recovery for murre populations, and will some colonies fail to make the comeback? Abstract. *EVOS Symposium Abstracts,* 139–41.

Heinrich, P. C., J. V. Castell, and T. Anders. 1990. Interleukin-6 and the acute phase response. *Biochemical Journal* 265:621–636.

Heintz, R. 1998. Fishery Research Biologist, NOAA/NMFS ABL, Juneau, AK. Interview by Dorothy Shepard, 28 June.

Heintz, R. A. and M. Larsen. 2003. Classification of fish diet composition and quality with lipid class and fatty acid analysis. Poster. www.afsc.noaa.gov/posters.htm

Heintz, R. A., S. D. Rice, A. C. Wertheimer, R. F. Bradshaw, F. P. Thrower, J. E. Joyce, and J. W. Short. 2000. Delayed effects on growth and marine survival of pink salmon *Oncorhynchus gorbuscha* after exposure to crude oil during embryonic development. *Marine Ecology Progressive Series* 208:205–216.

Heintz, R. A., J. W. Short, and S. D. Rice. 1999. Sensitivity of fish embryos to weathered crude oil: Part II. Incubating downstream from weathered *Exxon Valdez* crude oil caused increased mortality of pink salmon (*Oncorhynchus gorbuscha*) embryos. *Environmental Toxicology and Chemistry* 18:494-503.

Helle, E., M. Olsson, and S. Jensen. 1976. PCB levels correlated with pathological changes in seal uteri. *Ambio* 5:261-263.

Hennelly, R. 1990. Split wide open: Did the *Valdez* (sic) spill 11 million gallons-or 27 million? *Village Voice,* 2 January.

Hickey, D. 1992. Letter from Advanced Metabolic Imaging, North Dallas, Inc., to Dr. Rea; re: brain scan and SPECT test on Smith. 4 December. In *Roberts v. Exxon* (1999), No. 101, attachment.

Highsmith, R., PhD. 1999. Professor and Director of West Coast and Polar Regions Undersea Research Center, University of Alaska, School of Fisheries and Ocean Sciences, Fairbanks, AK. Interview by Riki Ott, 22 April.

Highsmith, R. C., T. L. Rucker, M. S. Stekoll, S. M. Saupe, M. R. Lindeberg, R. N. Jenne, and W. P. Erickson. 1996. Impact of the EVOS on intertidal biota. AFSS 18:212-237.

Hileman, B. 1991. Multiple chemical sensitivity: A special report. *Chemical and Engineering News* 69(29):26-42.

Hileman, B. 1999. White House defines scientific misconduct. *Chemical and Engineering News* 77(43):12.

Hill, E. 1989. Photo. ADN, 27 March.

Hill, P. S., D. P. DeMaster, and R. J. Small. 1996. Draft Alaska marine mammal stock assessments. NMML/NMFS/NOAA, Seattle, WA.

Hobbie III, R. and A. J. Garger. 2001. Oil spills and criminal sanctions: An insurer's perspective. IOSC 2001, 825-828.

Hoff, R. Z. and G. Shigenaka. 1999. Lessons from 10 years of post-*Exxon Valdez* monitoring on intertidal shorelines. IOSC 1999, 111-117.

Hofman, R. J. 1994. Foreword. In Loughlin 1994, xiii-xvi.

Holland, J. M., M. J. Whitaker, and L. C. Gipson. 1980. Chemical and biological factors influencing the skin carcinogenicity of fossil liquids. In Rom and Archer 1980, 463-480.

Holland-Bartels, L. E. 1998. Mechanisms of impact and potential recovery of nearshore vertebrate predators. EVOS Trustee Council Restoration Project Annual Report (Restoration Project 97025). USGS, Biological Resources Division, Anchorage, AK.

Holland-Bartels, L., PhD. 1999. Director, USGS' Upper Midwest Environmental Sciences Center, LaCrosse, WI. Interview by Riki

Ott, 9 December.

Holland-Bartels, L. E. 2000. Mechanisms of impact and potential recovery of nearshore vertebrate predators. EVOS Trustee Council Restoration Project Final Report (Restoration Project 98025). USGS, Biological Resources Division, Anchorage, AK.

Holland-Bartels, L. E, ed. 2002. Mechanisms of impact and potential recovery of nearshore vertebrate predators following the EVOS, volume 1. EVOS Trustee Council Restoration Project Final Report (Restoration Project 99025). USGS, Alaska Biological Science Center, Anchorage, AK.

Homer News staff. 1989. Fumes sicken security men. *Homer News,* 17 August.

Hong, S., MD. 1997. Deposition. 10 December. In *Roberts v. Exxon* (1999).

Hong, S., MD. 1997. Letter to Daniel Stenson; re: La Joie symptoms. 5 November. In *Roberts v. Exxon* (1999), No. 101, attachment.

Hoover-Miller, A., K. R. Parker, and J. J. Burns. 2001. A reassessment of the impact of the EVOS on harbor seals *(Phoca vitulina richardsi)* in PWS, AK. *Marine Mammal Science* 17(1):111–135.

Hopkins, M. 1991. Work Therapy Enterprises' report on Smith, Work Hardening Program for Eagle Pacific Insurance Company. 25 July. In *Roberts v. Exxon* (1999), No. 101, Exhibit Q.

Hose, J. E., E. Biggs and T. T. Baker. 1993. Effects of the EVOS on herring embryos and larvae: Sublethal assessments, 1989–1991. Abstract. *EVOS Symposium Abstracts,* 247–249.

Hose, J. E., M. D. McGurk, G. D. Marty, D. E. Hinton, E. D. Brown, and T. T. Baker. 1996. Sublethal effects of the EVOS on herring embryos and larvae: Morphological, cytogenetic, and histopathological assessments, 1989–1991. *Canadian Journal of Fisheries and Aquatic Sciences* 53:2355–2365.

Houghton, J., PhD. 2000. Senior Marine Biologist, Pentec Environmental, a Division of Hart Crowser, Inc., Edmonds, WA. Interview by Riki Ott, 24 March.

Houghton, J. D. and T. Elbert. 1991. Evaluation of the condition of intertidal and shallow subtidal biota in PWS following the EVOS and subsequent shoreline treatment. NOAA Report No. HMRB91-1, March.

Houghton, J. P., R. H. Gilmour, D. C. Lees, W. B. Driskell, S. C. Lindstrom, and A. Mearns. 1997. Intertidal biota seven years later—Has it recovered? IOSC 1997, 679–686.

Houghton, J. P., D. C. Lees, W. B. Driskell, S. C. Lindstrom, and A. J.

Mearns. 1996. Recovery of PWS intertidal epibiota from *Exxon Valdez* oiling and shoreline treatments, 1989 through 1992. AFSS 18:379-411.

Howe, G. R., D. Fraser, J. Lindsay, B. Presnal, and S. Z. Yu. 1983. Cancer mortality in relation to diesel fume and coal exposure in a cohort of retired railway workers. *Journal of the National Cancer Institute* 70(6):1015-1019.

Huff, D. 1954. *How to Lie with Statistics.* New York: W. W. Norton & Company, Inc., paperback reissue 1993.

Hyce, L. 2003. Director of Training, PWS Community College, Valdez, AK. Interview by Riki Ott, 23 September.

Impact Assessment, Inc. 1990. Economic, social, and psychological impact assessment of the EVOS. Final report. Prepared for Oiled Mayors' Subcommittee, Alaska Conference of Mayors. ARLIS, Anchorage, AK.

Irons, D. B. 1996. Size and productivity of black-legged kittiwake colonies in PWS before and after the EVOS. AFSS 18:738-747.

Irons, D. B., S. J. Kendall, W. P. Erickson, L. L. McDonald, and B. K. Lance. 2000. Nine years after the EVOS: Effects on marine bird populations in PWS, AK. *Condor* 102:723-737.

Irons, D. B., S. J. Kendall, W. P. Erickson, L. L. McDonald, and B. K. Lance. 2001. A brief response to Wiens et al. twelve years after the EVOS. *Condor* 103:892-894.

Isleib, M. E. and B. Kessel. 1973. *Birds of the North Gulf Coast-PWS, AK.* Biological papers of the University of Alaska, No. 14. Reprint, Fairbanks, AK: University of AK Press, 1989, 1992.

Iverson, S. J., C. Field, W. D. Bowen, W. and Blanchard. 2004. Quantitative fatty acid signature analysis: A new method of estimating predator diets. *Ecological Monographs* 74(2):211-235.

Iverson, S. J, K. J. Frost, and S. L. C. Lang. 1999. The use of fatty acid signatures to investigate foraging ecology and food webs in PWS, AK: Harbor seals and their prey. In Frost et al. 1999, 40-127.

Iverson, S. J., K. J. Frost, and S. L. C. Lang. 2003. High body energy stores and condition are linked with diet diversity in juvenile harbor seals in PWS, AK: New insights from quantitative fatty acid signature analysis (QFASA). Abstract. 15th Biennial Conference on the Biology of Marine Mammals, Greensboro, NC.

Iverson, S. J., K. J. Frost, S. L. C. Lang, C. Field, and W. Blanchard. 1998. The use of fatty acid signatures to investigate foraging ecology and food webs in PWS, AK: Harbor seals and their prey. In Frost et al. 1998, 38-117.

Iverson, S. J., K. J. Frost, and L. F. Lowry. 1997. Fatty acid signatures reveal fine scale structure of foraging distribution of harbor seals and their prey in PWS, AK. *Marine Ecology Progressive Series* 151:255-271.

Jewett, S. C., T. A. Dean, B. R. Woodin, M. K. Hoberg, and J. J. Stegeman. 2002. Exposure to hydrocarbons ten years after the EVOS: Evidence from cytochrome P450-1A expression and biliary FACs in nearshore demersal fishes. *Marine Environmental Research* 54:21-48.

Johanson, G., A. Boman, B. Dynesius. 1988. Percutaneous absorption of 2-butoxyethanol in man. *Scandanavian Journal of Work, Environment, and Health* 14(2):101-109.

Johanson, G. and P. Fernstrom. 1988. Influence of water on the percutaneous absorption of 2-butoxyethanol in guinea pigs. *Scandanavian Journal of Work, Environment and Health* 14(2):95-100.

Johnson, A. 1993. Letter from Environmental Health Center-Dallas to Robert Cowan; re: Smith's physical exam and chiropractic adjustment. 6 January. In *Roberts v. Exxon* (1999), No. 101, attachment.

Johnson, C. B. and D. L. Garshelis. 1995. Sea otter abundance, distribution, and pup production in PWS following the EVOS. ASTM STP 1219:894-929.

Johnson, S. W., M. G. Carls, R. P. Stone, C. C. Brodersen, and S. D. Rice. 1997. Reproductive success of Pacific herring, *Clupea pallasi*, in PWS, AK, six years after the EVOS. *Fisheries Bulletin* 95:368-379.

Jones, L. 1989. Former cleanup worker. Interview with Susan Ogle, 2 May. In U.S. Congress House 1989a, 1141-1143.

Juday, G. and N. Foster. 1990. A preliminary look at effects of the EVOS on Green Island research natural area. *Agroborealis* 22:10-17.

Kanner, A. 1999. Toxic tort causation–A new frontier. In King 1999, 124-128.

Karinen, J. 1988. Sublethal effects of petroleum on biota. In Shaw and Hameedi 1988, 293-328.

Karinen, J. 1998. Oceanographer, NOAA/NMFS ABL, Juneau, AK. Interview by Dorothy Shepard. 29 June.

Karinen, J. F. and M. M. Babcock. 1991. Pre-spill and post-spill concentrations of hydrocarbons in sediments and mussels at intertidal sites within PWS and the Gulf of Alaska. NRDA Draft Status Report, Coastal Habitat Study No. 1B. NOAA/NMFS, ABL, Juneau,

AK.

Karinen, J. F., M. M. Babcock, D. W. Brown, W. D. MacLeod, Jr., L. S. Ramos, and J. W. Short. 1993. Hydrocarbons in intertidal sediments and mussels from PWS, AK, 1977–1980: Characterization and probable sources. NOAA Technical Memorandum NMFS-AFSC-9.

Keeble, J. 1999. *Out of the Channel: The EVOS in PWS.* Tenth anniversary ed. Spokane, WA: Eastern Washington University Press.

Kelder, B. 1988. Mapco, VECO fines urged for waste violations. *Fairbanks Daily News-Miner,* 17 February.

Kilburn, K. H. 1998. *Chemical Brain Injury.* New York: Van Nostrand Reinhold, a division of International Thomson Publishing, Inc.

King, L. P. 1999. *Chemical Injury and the Courts, A Litigation Guide for Clients and Their Attorneys.* Jefferson, NC: McFarland & Company.

Kitaysky, A. S., J. F. Piatt, and J. C. Wingfield. 1999. The adrenocortical stress-response of black-legged kittiwake chicks in relation to dietary restrictions. *Journal of Comparative Physiology* B 169:303–310.

Kitaysky, A. S., J. C. Wingfield, and J. F. Piatt. 1999. Dynamics of food availability, body condition, and physiological stress response in breeding black-legged kittiwakes. *Functional Ecology* 13:577–584.

Klosiewski, S. P. and K. K. Laing. 1994. Marine bird populations of PWS, AK, before and after the EVOS. EVOS State/Federal NRDA Final Report, Bird Study Number 2. USFWS, Anchorage, AK. June.

Kocan, R. M., T. T. Baker, and E. Biggs. 1993. Adult herring reproductive impairment following the EVOS. Abstract. *EVOS Symposium Abstracts,* 262–63.

Kocan, R., M. Bradley, N. Elder, T. Meyers, W. Batts, and J. Winton. 1997. North American strain of viral hemorrhagic septicemia virus is highly pathogenic for laboratory-reared Pacific herring. *Journal of Aquatic Animal Health* 9(4):279–290.

Kocan, R. M. and P. Hershberger. 2001. Epidemiology of viral hemorrhagic septicemia (VHS) among juvenile Pacific herring and Pacific sand lances in Puget Sound, WA. *Journal of Aquatic Animal Health* 13(2):77–85.

Kocan, R. M., P. Hershberger, and N. Elder. 2001. Survival of the North American strain of viral hemorrhagic septicemia (VHS) virus in filtered seawater and seawater containing ovarian fluid, crude oil and serum-enriched culture medium. *Diseases of Aquatic Organisms* 44:75–78.

Kocan, R. M., J. E. Hose, E. D. Brown, and T. T. Baker. 1996b. Pacific herring (*Clupea pallasi*) embryo sensitivity to Prudhoe Bay petroleum hydrocarbons: Laboratory evaluation and in situ exposure at oiled and unoiled sites in PWS. *Canadian Journal of Fisheries and Aquatic Sciences* 53:2366–2375.

Kocan, R. M., G. D. Marty, M. S. Okihiro, E. D. Biggs, and T. T. Baker. 1996a. Reproductive success and histopathology of individual PWS Pacific herring three years after the EVOS. *Canadian Journal of Fisheries and Aquatic Science* 53:2388–2393.

Korten, D. 2002. Economies for life. *Yes!* A Journal of Positive Futures, fall 2002 (issue #23), 12–17.

Krimsky, S. 1999. Will disclosure of financial interests brighten the image of entrepreneurial science? Abstract A-29. In Boyd 1999. Quoted in Rampton and Stauber 2001, 204.

Kubaiewicz, M., A. Starzynski, and W. Symczak. 1991. Case-referent study on skin cancer and its relation to occupational exposure to PAHs. II. Study Results. *Polish Journal of Occupational Medicine and Environmental Health* 4:141–147.

Kuhn, T. S. 1996. *The Structure of Scientific Revolutions.* 3rd ed. Chicago, IL: University of Chicago Press.

Kuletz, K. 1997. Restoration notebook: Marbled murrelet (*Brachyramphus marmoratus marmoratus*). EVOS Trustee Council, Anchorage, AK.

Kuletz, K. J., D. B. Irons, B. A. Agler, and J. F. Piatt. 1997. Long-term changes in diets and populations of piscivorous birds and mammals in PWS, AK. Extended abstract. *Forage Fish Proceedings,* 703–706.

Kumho Tire Co. v. Carmichael, 526 U.S. 137 (1999).

Kvenvolden, K. A., F. D. Hostettler, J. B. Rapp, and P. R. Carlson. 1993. Hydrocarbons in oil residues on beaches of islands of PWS, AK. *Marine Pollution Bulletin* 26:24–29.

Laborers' International Union of North America. 1989. Letter from Angelo Fosco, general president, to Congressman G. Miller (D-CA), enclosing prepared statement of Mano W. Frey, executive president, AFL-CIO; re: recommendations for the *EVOS* cleanup and other attachments. In U.S. Congress House 1989a, 1029–1069.

Laborers' National Health and Safety Fund. 1989a. Letter to Jim

Sampson, ADOL; re: VECO worker training program deficiencies.
28 April. In U.S. Congress House 1989a, 1067–1068.

Laborers' National Health and Safety Fund. 1989b. Report of the
public health team assessing the EVOS cleanup. Washington, DC,
24 April. In U.S. Congress House 1989a, 1036–1062.

La Joie, P. 1996. Deposition. 2 October. In *Roberts v. Exxon* (1999),
No. 64, Exhibit A.

Lamming, J. 1989a. Spill stench permeates Aleut village. *Anchorage
Times*, 28 March.

Lamming, J. 1989b. Oil spill crews apprised of peril, but crews 'don't
comprehend'. *Anchorage Times,* 16 April.

Lancaster, J. 1991. Exxon plea bargain thrown out by judge, $100
million oil spill fine called too lenient. *Washington Post,* 25 April.

Lanctot, R., B. Goatcher, K. Scribner, S. Talbot, B. Pierson, D. Esler, and
D. Zwiefelhofer. 1999. Harlequin duck recovery from the EVOS: A
population genetics perspective. *The Auk* 116(3):781–791.

"Lange, E." (*alias*). 2001. Former cleanup worker, Fairbanks, AK.
Interview by Riki Ott, 23 November.

Langford, T. 1996. Exxon, Lloyd's settle. Insurers to pay $480 million.
ADN, 1 November.

Larsen, K.S., M. Hougaard, M. Hammer, Y. Alarie, P. Wolkoff, P.A.
Clausen, C. K. Wilkins, and G. D. Nielson. 2000. Effects of r-(+) and
s-(–) limonene on the respiratory tract in mice. *Human and
Experimental Toxicology* 19:457–466.

Lamoreaux, B. and B. Baker. 1989. ADEC and ADFG memorandum to
Commander McCall, U.S. Coast Guard. ADEC, Anchorage, AK. In
Lethcoe and Nurnberger 1989, 45.

Leahy, J. G. and R.R. Colwell. 1990. Microbial degradation of hydrocarbons in the environment. *Microbiological Reviews*
54:305–315. In Braddock et al. 1996.

Lees, D. C., W. B. Driskell, and J. P. Houghton. 1999. Response of infau-
nal bivalves to EVOS and related shoreline treatment. IOSC 1999,
999–1002.

Lees, D. C., J. P. Houghton, and W. B. Driskell. 1996. Short-term effects
of several types of shoreline treatment on rocky intertidal biota
in PWS. AFSS 18:329–348.

Lethcoe, J. 1990. *An Observer's Guide to The Geology of PWS.*
Valdez, AK: PWS Books.

Lethcoe, N. 1987. *An Observer's Guide to The Glaciers of PWS.*
Valdez, AK: PWS Books.

Lethcoe, N. and L. Nurnberger, eds. 1989. *PWS Environmental*

Reader, 1989—T/V EVOS. Valdez, AK: PWS Conservation Alliance.

Liberman, M. S., B. J., DiMuro, and J. B. Boyd. 1999. Multiple chemical sensitivity: An emerging area of law. In King 1999, 129-139.

Lipscomb, T. P., R. K. Harris, R. B. Moeller, J. M Pletcher, R. J. Haebler, and B. E. Ballachey. 1993. Histopathologic lesions in sea otters exposed to crude oil. *Veterinary Pathology* 30:1-11.

Lipscomb, T. P., R. K. Harris, A. H. Rebar, B. E. Ballachey, and R. J. Haebler. 1994. Pathology of sea otters. In Loughlin 1994, 265-279.

Little, R. 2002. Infamous oil tanker hung out to dry. It may be end of line for old *Exxon Valdez. San Francisco Chronicle,* 17 October.

Litzow, M. A. 2002. Pigeon guillemot biology in Kachemak Bay. In Piatt 2002, 100-109.

Litzow, M.A., J. F. Piatt, A. A. Abookire, A. K. Prichard, and M.D. Robards. 2000. Monitoring temporal and spatial variability in sandeel (*Ammodytes hexapterus*) abundance with pigeon guillemot (*Cepphus columba*) diets. *ICES Journal of Marine Science* 57:976-986.

Lobdell, H. 1989. Testimony of Laborers' International Union of North America. U.S. House of Representatives Committee on Merchant Marine and Fisheries, Subcommittee on Coast Guard and Navigation, Seattle, WA. 16 June.

Loughlin, T. R., ed. 1994. *Marine Mammals and the Exxon Valdez.* San Diego: Academic Press.

Lowry, L. F. and K. J. Frost. 1993. Harbor seals: Were they injured and will they recover? *AK's Wildlife* 25(1):20-21.

Lowry, L. F., K. J. Frost, and K. W. Pitcher. 1994. Observations of oiling of harbor seals in PWS. In Loughlin 1994, 209-226.

Lowry, L. F., K. J. Frost, J. M. Ver Hoef, and R. A. DeLong. 2001. Movements of satellite-tagged subadult and adult harbor seals in PWS, AK. *Marine Mammal Science* 17(4):835-861.

Lubchenco, J. 1998. Entering the century of the environment: A new social contract for science. *Science* 279:491-497.

Lyondell Petrochemical Company. 1990. MSDS for crude oil. MSDS No. HCR00001.

Lyons, R.A., J. M. F. Temple, D. Evans, D. L. Fone, and S. R. Palmer. 1999. Acute health effects of the *Sea Empress* oil spill. *Journal of Epidemiology and Community Health* 53:306-310.

MacFarland, H. N., et al., eds. 1984. *Applied Toxicology of Petroleum Hydrocarbons.* Vol. 6 of *Advances in Modern Environmental Toxicology.* Series Editor M. A. Mehlman. Princeton, NJ: Princeton

Scientific Publishers.

Maki, A. W. 1991. The EVOS: Initial environmental impact assessment. *Environmental Science and Technology* 25(1):24-29.

Maki, A. W., E. J. Brannon, L. G. Bilbertson, L. L. Moulton, and J. R. Skalski. 1995. An assessment of oil spill effects on pink salmon populations following the EVOS: Part 2—Adults and escapement. ASTM STP 1219:585-625.

Margasak, L. 2001. Environmental labs caught faking data. *Seattle Times*, 22 January.

Marty, G. D., E. F. Freiberg, T. R. Meyers, J. Wilcock, T. B. Farver, and D. E. Hinton. 1998. Viral hemorrhagic septicemia virus, *Ichthyophonus hoferi*, and other causes of morbidity in Pacific herring *(Clupea pallasi)* spawning in PWS, AK, USA. *Diseases of Aquatic Organisms* 32:15-40.

Marty, G. D., R. A. Heintz, and D. E. Hinton. 1997. Histology and teratology of pink salmon larvae near the time of emergence from gravel substrate in the laboratory. *Canadian Journal of Zoology* 75:978-988.

Marty, G. D., J. E. Hose, M. D. McGurk, E. D. Brown, and D. E. Hinton. 1997a. Histopathology and cytogenetic evaluation of Pacific herring larvae exposed to petroleum hydrocarbons in the laboratory or in PWS, AK, after the EVOS. *Canadian Journal of Fisheries and Aquatic Sciences* 54:1846-1857.

Marty, G. D., M. A. Okihiro, E. D. Brown, D. Hanes, and D. E. Hinton. 1999. Histopathology of adult pacific herring in PWS, AK, after the EVOS. *Canadian Journal of Fisheries and Aquatic Sciences* 56:419-426.

Marty, G. D., T. J. Quinn, II, G. Carpenter, T. R. Meyers, and N. H. Willits. 2003. Role of disease in abundance of a Pacific herring *(Clupea pallasi)* population. *Canadian Journal of Fisheries and Aquatic Sciences* 60(10):1258-1265.

Marty, G. D., J. W. Short, D. M. Dambach, H. H. Willits, R. A. Heintz, S. D. Rice, J. J. Stegeman, and D. E. Hinton. 1997b. Ascites, premature emergence, increased gonadal cell apoptosis, and cytochrome P4501A induction in pink salmon larvae continuously exposed to oil-contaminated gravel during development. *Canada Journal of Zoology* 75:989-1007.

Masry, E., E. Brockovitch, M. Schneider, and R. Ott. 2001. Letter to U.S. Senator H. Clinton (D-NY), U.S. Senator H. Reid (D-NV), U.S. Congressman G. Miller (D-CA), and U.S. Congressman M. Owens (D-NY); re: Protecting cleanup workers and lessons learned from

Exxon Valdez. 21 November.

Matkin, C. 1998. Executive director and marine mammal biologist, North Gulf Oceanic Society, Homer, AK. Interview by Riki Ott, 5 June.

Matkin, C. 1994. *An Observer's Guide to the Orcas of PWS.* Valdez, AK: PWS Books.

Matkin, C. 1997. Comprehensive killer whale investigation. EVOS Restoration Project Annual Report (Restoration Project 96012). North Gulf Oceanic Society, Homer, AK.

Matkin, C., G. Ellis, M. Dahlheim, and J. Zeh. 1994. Status of killer whale pods in PWS 1984-1992. In Loughlin 1994, 141-162.

Matkin, C. O., G. Ellis, P. Olesiuk, and E. Saulitis. 1999. Association patterns and inferred genealogies of resident killer whales, *Orcinus orca,* in PWS, AK. *Fisheries Bulletin* 97:900-919.

Matkin, C., D. R. Matkin, G. M. Ellis, E. Saulitis, and D. McSweeney. 1997. Movements of resident killer whales in southeastern Alaska and PWS, AK. *Marine Mammal Science* 13(3):469-475.

Matkin, C. and E. Saulitis. 1997. Killer whale *Orcinus orca.* Restoration notebook series, EVOS Trustee Council, Anchorage, AK.

Matkin, C., R. Steiner, and G. Ellis. 1986. Photo-identification and deterrent experiments applied to orcas in PWS, AK. Unpublished report to the University of Alaska, Sea Grant, Anchorage, AK.

Matthews, B. L., ed. 1998. *Defining Multiple Chemical Sensitivity.* Jefferson, NC and London: McFarland and Company, Inc.

Matthews, D. 1993. The remarkable recovery of PWS. *The Lamp* 75(3):4-13. Exxon Corporation Public Affairs Department, P. O. Box 160369, Irving, TX 75016-0369.

Mauer, R. 1989. Allen begins a new chapter in powerful Alaska career. *ADN,* 21 November.

McCoy, C. 1989. Broken promises–Alyeska record shows how big oil neglected Alaska environment. Pipeline firm cut corners and scraped safeguards, raising risk of disaster. Allegations of fabricated data. *WSJ,* 6 July.

McDonald, L. L., W. P. Erickson, and M. D. Strickland. 1995. Survey design, statistical analysis, and basis for statistical inferences in coastal habitat injury assessment: EVOS. ASTM STP 1219:296-311.

McDowell, T. 1989. Seldovia town meeting with spill agency representatives. 24 August 1989. VHS. ARLIS, Anchorage, AK. 7 videos.

McGonigle, S. and E. Timms. 1992. Doctor detects oil poisoning in 2

ailing gulf war veterans. Army says it's studying possible exposure in Desert Storm. *The Dallas Morning News,* 2 August.

McGurk, M. and E. Brown. 1996. Egg-larval mortality of Pacific herring in PWS, AK, after the EVOS. *Canadian Journal of Fisheries
and Aquatic Sciences* 53:2343-2354.

Mearns, A. J. 1996. *Exxon Valdez* shoreline treatment and operations: Implications for response, assessment, monitoring, and
research. AFSS 18:309-328.

Medred, C. 1989. Biologists track a slaughter. Oil killed thousands of
Barren Islands murres, researchers say. *ADN,* 13 July.

Med-Tox. 1989a. Air monitoring results for oil mist: VOCs master by
task and VOCs master by date. In *Stubblefield v. Exxon* (1994).

Med-Tox. 1989b. Results of air sampling for PAHs. In *Stubblefield v.
Exxon* (1994).

Med-Tox. 1989c. Statistical summary of industrial hygiene monitoring. In *Stubblefield v. Exxon* (1994).

Meggert, E. 1991. Letter to Surety of Alaska Claim Manager. 9
January. In ADOL, AWCB 1992a.

Meggert, E. 2001. State On-Scene Coordinator, ADEC, Fairbanks, AK.
Interview by Riki Ott, 24 November.

Menez, J. F., F. Berthou, D. Picaut, C. Riche. 1978. Impacts of the oil
spill *Amoco Cadiz* on human biology. *Penn. Ar. Bed.* (French)
94:367-378.

Merrick, R. L., T. R. Loughlin, and D. G. Calkins. 1987. Decline in
abundance of the northern sea lion, *Eumetopias jubatus,* in AK,
1956-1986. *U.S. National Marine Fisheries Service Fishery
Bulletin* 85:351-365.

Michel, J. M. and M. O. Hayes. 1993. Evaluation of the condition of
PWS shorelines following the EVOS and subsequent shoreline
treatment, Vol. 1. NOAA Technical Memorandum NOS ORCA 67,
Seattle, WA.

Michel, J. M. and M. O. Hayes. 1999. Weathering patterns of oil
residues eight years after the EVOS. *Marine Pollution Bulletin*
38(10):855-863.

Milbank, D. 1994. Lloyd's investors win key ruling in British court.
WSJ, 5 October.

Moeller, D. S. 1989. Journal entries, 1-4 August. Papers. Private collection of D. S. Moeller, Valdez, AK.

Moeller, D. S. 2001. Former cleanup worker. Interview by Riki Ott,
13 January 2001.

Moles, A., M. M. Babcock, and S. D. Rice. 1987. Effects of oil exposure on pink salmon *Oncorhynchus gorbuscha* alevins in a simulated intertidal environment. *Maine Environmental Research* 21:49-58.

Moles, A. and S. D. Rice. 1983. Effects of crude oil and naphthalene on growth, caloric content, and fat content of pink salmon juveniles in seawater. *Transactions of the American Fisheries Society* 112(2A):205-211.

Moles, A., S. D. Rice, and M. S. Okihiro. 1993. Herring parasite and tissue alterations following the EVOS. IOSC 1993, 325-328.

Monahan, T. P. and A. W. Maki. 1991. The *Exxon Valdez* 1989 wildlife rescue and rehabilitation program. IOSC 1991, 131-139.

Monnett, C. 1990. Letter to Larry Pank, Chief, Mammals Section, Alaska Fish and Wildlife Research Center, USFWS, Anchorage, AK. 13 June.

Monnett, C. and L. M. Rotterman. 1990. Assessment of the fate of sea otters oiled and treated as a result of the EVOS. Draft Marine Mammal Study No. 7, USFWS, Anchorage, AK.

Monnett, C. and L. M. Rotterman. 1993. The efficacy of the T/V EVOS sea otter rehabilitation program and the possibility of disease introduction into recipient sea otter populations. Abstract. *EVOS Symposium Abstracts,* 273.

Monnett, C. and L. M. Rotterman. 1995a. Movements of weanling and adult female sea otters in PWS, AK, after the EVOS. EVOS State/Federal NRDA Final Report, Marine Mammal Study No. 6-12. USFWS, Anchorage, AK.

Monnett, C. and L. M. Rotterman. 1995b. Mortality and reproduction of female sea otters in PWS, AK. EVOS State/Federal NRDA Final Report, Marine Mammal Study No. 6-13. USFWS, Anchorage, AK.

Monnett, C. and L. M. Rotterman. 1995c. Mortality and reproduction of sea otters oiled and treated as a result of the EVOS. EVOS State/Federal NRDA Final Report, Marine Mammal Study No. 6-14, USFWS, Anchorage, AK.

Monnett, C. and L. M. Rotterman. 2000. Survival rates of sea otter pups in Alaska and California. *Marine Mammal Science* 16(4):794-810.

Monnett, C., L. M. Rotterman, C. Stack, and D. Monson. 1990. Postrelease monitoring of radio-instrumented sea otters in PWS. In Bayha and Kormendy 1990, 400-420.

Monson, D. H., D. F. Doak, B. E. Ballachey, A. Johnson, and J. Bodkin. 2000a. Long-term impacts of the EVOS on sea otters, assessed through age-dependent mortality patterns. *Proceedings of the*

National Academy of Science 97(12):6562-6567.

Monson, D. H., J. A. Estes, J. L. Bodkin, and D. B. Siniff. 2000b. Life history plasticity and population regulation in sea otters. *OIKOS* 90:457-468.

Morita, A., Y. Kusaka, Y. Deguchi, A. Moriuchi, Y. Nakanaga, M. Iki, S. Miyazaki, K. Kawahara. 1999. Acute health problems among people engaged in the cleanup of the *Nakhodka* oil spill. *Environmental Research* (Section A) 81:185-194.

Morris, B. F. and T. R. Loughlin. 1994. Overview of the EVOS, 1989-1992. In Loughlin 1994, 1-22.

Morton, A. B. 1990. A quantitative comparison of the behavior of resident and transient forms of killer whale off the central British Columbia coast. Report of the International Whaling Commission Special Issue 12:245-248.

Mulcahy, D. M. and B. E. Ballachey. 1994. Hydrocarbon residues in sea otter tissue. In Loughlin 1994, 313-330.

Mulligan, T. and M. Parrish. 1994. Exxon ordered to pay $5 billion for oil spill. *Los Angles Times*, 17 September.

Mullins, R. 1994. We will make you whole. VHS. Cordova town meeting, 28 March 1989. Ross Mullins, POB 436, Cordova, AK, 99574.

Murchison, J. 1991. Letter to Assistant Attorney General Mike Mitchell; re: preliminary review EVOS volume. In ADOL 1991, ACE 9486067-9486071. ARLIS, Anchorage, AK.

Murphy, K. 2001. Exxon spill's cleanup workers share years of crippling illness. *Los Angeles Times,* 5 November.

Murphy, S. M., R. H. Day, J. A. Wiens, and K. R. Parker. 1997. Effects of the EVOS on birds: Comparison of pre- and post-spill surveys in PWS, AK. *Condor* 99:299-313.

Murphy, M., R. A. Heintz, J. W. Short, M. L. Larsen, and S. D. Rice. 1999. Recovery of pink salmon spawning areas after the EVOS. *Transactions of the American Fisheries Society* 128:909-918. (See also poster: www.afsc.noaa.gov/abl/evos/pssynthe.htm.)

Myren, R. T. and J. J. Pella. 1977. Natural variability in distribution of an intertidal population of *Macoma balthica* subject of potential oil pollution at Port Valdez, AK. *Marine Biology* 41(4):371-382.

Myren, R. T., G. Perkins, and T. R. Merrell. 1992 unpublished manuscript. Reduced abundance of *Macoma balthica* near an oil tanker terminal in Port Valdez, AK, 1971-1984. NOAA/NMFS ABL, Juneau, AK.

Nagel., R, Capt. 2003. Former cleanup worker, Florida. Interview by Riki Ott, 22 April.

National Academy of Sciences. 1991. *Addressing the Physician Shortage in Environmental and Occupational Medicine.* Washington, DC: National Academies Press.

National Research Council. 1980. *International Mussel Watch.* E. D. Goldberg, ed. Washington, DC: National Academies Press.

National Research Council. 1985. *Oil in the Sea: Inputs, Fates, and Effects.* Washington, DC: National Academies Press.

National Research Council. 1989. *Using Oil Spill Dispersants on the Sea.* Washington, DC: National Academies Press.

National Research Council. 2002. *Oil in the Sea III: Inputs, Fates, and Effects.* Washington, DC: National Academies Press.

National Science Board. 1999. Statement on sharing of research data. Document number (NBS 99-30) PS 99-2. 23 February.

National Wildlife Federation, National Research Defense Council, Wildlife Federation of Alaska, and the Windstar Foundation. 1990. The day the water died. A Compilation of the November 1989 Citizens Commission Hearings on the EVOS. Available through NWF, Washington, DC.

Nauman, S. A. 1991. Shoreline cleanup: Equipment and operations. IOSC 1991, 141-147.

Neff, J. M. 1990. Water quality in PWS. Summary of findings from the report by J. M. Neff. Battelle Ocean Sciences, Duxbury Operations, 397 Washington Street, P.O. Drawer AH, Duxbury, MA 02332-0601.

Neff, J. M. 1991. Water quality in PWS and the Gulf of Alaska. Summary of findings from the report by J. M. Neff. Arthur D. Little, 20 Acorn Park, Cambridge, MA 02140-2390. March.

Neff, J. M., E. H. Owens, S. W. Stoker, and D. M. McCormick. 1995. Shoreline oiling conditions in PWS following the EVOS. ASTM STP 1219:312-346.

Neff, J. M. and W. A. Stubblefield. 1995. Chemical and toxicological evaluation of water quality following the EVOS. ASTM STP 1219:141-177.

New York Times. 1993. New trouble in PWS. Editorial, 25 August.

Nichols, W. J. 2001. The U.S. EPA: National oil and hazardous substances pollution contingency plan, subpart J product schedule (40 CFR 300.900). IOSC 2001, 1479-1483.

NIOSH. 1988. Current Intelligence Bulletin 50: Carcinogenic effects of exposure to diesel exhaust. U.S. DHHS, Centers for Disease Control, Cincinnati, OH. Publication No. 88-116.

NIOSH. 1991. Health Hazard Evaluation Report. Prepared by R. W.

Gorman, S. P. Berardinelli, and T. R. Bender. U.S. DHHS, May. HETA 89-200 & 89-273-2111, Exxon/Valdez Alaska Oil Spill.

NIOSH. 2003. Major findings for NHANES studies: National report on human exposure to environmental chemicals and national health and nutrition examination survey. U.S. DHHS, National Center for Environmental Health and Centers for Disease Control and Prevention. Atlanta, GA. www.cdc.gov/nchs/about/major/nhanes/findings.htm.

NIOSH. 2004. NIOSH Pocket Guide to Chemical Hazards, Appendix D. U.S. DHHS, www.cdc.gov/niosh/npg/nengapdx.htm#d.

Norcross, B. L., E. D. Brown, R. J. Foy, M. Frandsen, S. M. Gay III, T. C. Kline, Jr., D. M. Mason, E. V. Patrick, A. J. Paul, and K. D. E. Stokesbury. 2001. A synthesis of the life history and ecology of juvenile Pacific herring in PWS, AK. *Fisheries Oceanography* 10 (Suppl. 1):42-57.

Norcross, B. and M. Frandsen. 1996. Distribution and abundance of larval fishes in PWS, AK, during 1989 after the EVOS. AFSS 18:463-486.

Norcross, B., M. Frandsen, J. Hose, and E. Biggs [Brown]. 1996. Distribution, abundance, morphological condition, and cytogenetic abnormalities of larval herring in PWS, AK, following the EVOS. *Canadian Journal of Fisheries and Aquatic Sciences* 53:2376-2387.

Nysewander, D. R., C. Dippel, G. V. Byrd, and E. P. Knudtson. 1993. Effects of the T/V EVOS on murres: A perspective from observations at breeding colonies. EVOS State/Federal NRDA Final Report, Bird Study No. 3. USFWS, Migratory Bird Management, Homer, AK.

Oakley, K. L. and K. J. Kuletz. 1993. Effects of the EVOS on pigeon guillemots (*Cepphus columba*) in PWS, AK. Abstract. *EVOS Symposium Abstracts,* 144-46.

Oakley, K. L. and K. J. Kuletz. 1996. Population, reproduction, and foraging of pigeon guillemots at Naked Island, AK, before and after the EVOS. AFSS 18:759-769.

O'Clair, C. E., J. W. Short, and S. D. Rice. 1996. Contamination of intertidal and subtidal sediments by oil from the *Exxon Valdez* in PWS. AFSS 18:61-93.

O'Harra, D. 2002. Experts amazed at oil left in sound. *ADN*, 23 January.

O'Harra, D. 2003. Agency seeks protection for orca group. Seal-eating pod seems to be dying out. *ADN*, 28 October.

Oil Pollution Act of 1990. Public Law 101–380. U.S. Statutes at Large 104 (1990).

Olson, A. 1989. Telephone record of 5 May. In U.S. Congress House 1989a, 1126.

O'Neill, A. 2003. Self-reported exposures and health status among workers from the EVOS cleanup. Master's thesis M. P. H. Yale University, Department of Epidemiology and Public Health.

Orange-Sol, Inc. 1987. MSDS for De-Solv-It cleaner, solvent. 22 April.

Orange-Sol, Inc. 1991. MSDS for De-Solv-It (MSDS Serial Number BSPYW). Chandler, AZ

Ornitz, B. E. 2001. Sustainable shipping: The benefits of the safety culture far outweigh the costs. IOSC 2001, 839–843.

Ortega, B. 1989. Day-by-day account of the spill shows evolving tragedy. *Homer News,* 29 December.

OSHA. 1989. Hazardous Waste Operations and Emergency Response Standard, 29 CFR Part 1910.120 (Federal Register 54 [42]: 9294-9336). U.S. Department of Labor, 6 March.

OSHA. 1994. Final Rule. Hazard Communication, Section 1-I. Background 59(27):6126–6184. U.S. Department of Labor, 9 February.

OSHA. 1998. Chemical Hazard Communication (OSHA 3084), revised. U.S. Department of Labor,

OSHA. 2004a. Determination of work-relatedness; Recording and reporting occupational injuries and illness. Subpart: Record-keeping forms and recording criteria. U.S. Department of Labor, 29 CFR 1904.5, Subpart C, Section 1904.5(b)(2)(viii). www.osha.gov/pls/oshaweb/owadisp.show_document?p_table=STANDARDS&p_id=9636

OSHA. 2004b. Occupational Safety and Health Standards: Subpart Z—Toxic and Hazardous Substances, Table Z-1. U.S. Department of Labor, 29 CFR 1910.1000. www.osha.gov/pls/oshaweb/owadisp.show-document?p_table=STANDARDS&p_id=9992

OSHA. 2004c. Occupational Safety and Health Standards: Subpart Z—Toxic and Hazardous Substances, Table Z-2. U.S. Department of Labor, 29 CFR 1910.1000. www.osha.gov/pls/oshaweb/owadisp.show-document?p_table=STANDARDS&p_id=9993

Ostrand, W. D., K. O Coyle, G. S. Drew, J. M. Maniscalco, and D. B. Irons. 1997. Selection of forage-fish schools by murrelets and tufted puffins in PWS, AK. Extended abstract. *Forage Fish*

Proceedings, 171-173.

Ott, F. S. 1986. Amphipod sediment bioassays: Effects on response of methodology, grain size, organic content, and cadmium. PhD. diss., University of Washington.

Ott, R. 1989a. Spilled oil and the Alaska fishing industry: Looking beyond fouled nets and lost fishing time. Presented at the IOSC Panel on the Open Ocean and Coastal Spills, 12-15 February 1989, San Antonio, TX. (Also in U.S. Congress House 1989a, 1104-1123.)

Ott, R. 1989b. Testimony. In U.S. Congress House 1989a, 69-146.

Ott, R. 1989c. Oil in the marine environment. In Lethcoe and Nurnberger 1989, 30-35.

Ott, R. 1999. *Exxon Valdez* aftermath. A decade after the huge Alaskan oil spill, hardly any wild species have fully recovered. *Defenders* 74(2):16-24.

Ott, R. 2000. Letter to George Frampton, White House Council on Environmental Quality; re: amended Freedom of Information Act policy. 22 November.

Ott, R., C. Peterson, and S. Rice. 2001. EVOS legacy: Shifting paradigms in oil ecotoxicology. Briefing paper prepared for the Alaska Forum for Environmental Responsibility, POB 188, Valdez, AK 99686. Available from: www.alaskaforum.org.

Owens, E. H. 1991a. Shoreline conditions following the EVOS as of fall 1990. Proceedings of the 14th AMOP, 579-606.

Owens, E. H. 1991b. Shoreline conditions following the EVOS as of fall 1990. Brochure. Presented at the 14th AMOP Technical Seminar. June. Woodward-Clyde Consultants, 3440 Bank of California Center, 900 Fourth Avenue, Seattle, WA 98164.

Owens, E. H. 1999. SCAT-A ten-year review. Proceedings of the 22nd AMOP, 337-360.

Pagano, R. 1992. Workers allege illnesses tied to *Exxon Valdez* cleanup. *Anchorage Times,* 16 April.

Pagano, R. 1993. Commercial fishers among those hardest hit by Exxon disaster. *ADN,* 5 February.

Page, D. S. 2002. Point counterpoint: Has PWS recovered? Short study exaggerates; Sound is as healthy as ever. *ADN,* 31 January.

Page, D. S., P. D. Boehm, G. S. Douglas, and A. E. Bence. 1995a. Identification of hydrocarbon sources in the benthic sediments of PWS and the Gulf of Alaska following the EVOS. ASTM STP 1219:41-83.

Page, D. S., P. D. Boehm, G. S. Douglas, A. E. Bence, W. A. Burns, and P. J.

Mankiewicz. 1996.The natural petroleum hydrocarbon background in subtidal sediments of PWS,AK, USA. *Environmental Toxicology and Chemistry* 15(8):1266-1281.

Page, D. S., P. D. Boehm, G. S. Douglas,A. E. Bence,W.A. Burns, and P. J. Mankiewicz. 1997. An estimate of the annual input of natural petroleum hydrocarbons to seafloor sediments in PWS,AK. *Marine Pollution Bulletin* 34(9):744-749.

Page, D. S., P. D. Boehm, G. S. Douglas,A. E. Bence,W.A. Burns, and P. J. Mankiewicz. 1998. Reply to letter to editor: Source of polynuclear aromatic hydrocarbons in PWS,AK, USA, subtidal sediments. *Environmental Toxicology and Chemistry* 17(9):1651-1652.

Page, D. S., P. D. Boehm, G. S. Douglas,A. E. Bence,W.A. Burns, and P. J. Mankiewicz. 1999a. Pyrogenic plycyclic aromatic hydrocarbons in sediments record past human activity:A case study in PWS, AK. *Marine Pollution Bulletin* 38(4):247-260.

Page, D. S., E. S. Gilfillan, P. D. Boehm, and E. J. Harner. 1995b. Shoreline ecology program for PWS, AK, following the EVOS: Part 1—Study design and methods. ASTM STP 1219:263-295.

Page, D. S., E. S. Gilfillan, J. M. Neff, and S. W. Stoker. 1999b. 1998 shoreline conditions in the EVOS zone in PWS. IOSC 1999, 119-126.

Park, J. M. and M. G. Holliday. 1999. Occupational-health aspects of marine oil spill response. *Pure Applied Chemistry* 71(1):113-133.

Parrish, J. K. and P. D. Boersma. 1995a. Muddy waters. *American Scientist* 83:112-115.

Parrish, J. K. and P. D. Boersma. 1995b. Letters to the editor. *American Scientist* 83:398-399.

Pasztor,A. and R. E.Taylor. 1986. Unsafe harbor. Alyeska pipeline firm is accused of polluting sea water since 1977. *WSJ*, 20 February.

Patten, S. M., Jr. 1993. Acute and sublethal effects of the EVOS on harlequins and other seaducks.Abstract. *EVOS Symposium Abstracts,* 151-154.

Patten, S. M., Jr. 1993. Reproductive failure of harlequin ducks. *Alaska's Wildlife* 25(1):14-15.

Patten, S. 2003. Fire Management Officer, USFWS, Fairbanks,AK. Interview by Riki Ott, 18 March.

Patten, S. M., Jr.,T. Crowe, R. Gustin, R. Hunter, P.Twait, and C. Hastings. 2000a.Assessment of injury to sea ducks from hydrocarbon uptake in PWS and the Kodiak archipelago,AK, following the EVOS. Volume I, plus five appendices. EVOS State/Federal NRDA Final Report, Bird Study No. 11. ADFG in cooperation with USFWS, Anchorage, AK.

Patten, S. M., Jr., T. Crowe, R. Gustin, R. Hunter, P. Twait, and C. Hastings. 2000b. Assessment of injury to sea ducks from hydrocarbon uptake in PWS and the Kodiak Archipelago, AK, following the EVOS. Volume II, plus three supplements. EVOS State/Federal NRDA Final Report, Bird Study No. 11. ADFG in cooperation with USFWS, Anchorage, AK.

Jacqueline Payne, Jacob Payne, Randy W. Lowe, Ferdinand Samuel, and Daniel Saliors v. Exxon Corporation, Exxon Company USA, Exxon Shipping Company, and VECO, Inc. (1997). U.S. 9th Circuit Court of Appeals, No. 96-35043. D.C. No. CV-93-00107-JWS. Filed 6 August 1997. (laws.findlaw.com/9th/9635043.html)

Pearcy, W. G., ed. 2001. Introduction [to Ecosystem-level studies of juvenile Pacific herring and juvenile pink salmon in PWS, AK]. *Fisheries Oceanography* 10 (Suppl. 1):v.

Pearson, W. H., D. L. Woodruff, and P. C. Sugarman. 1984. The burrowing behavior of sand lance, *Ammodytes hexapterus:* Effects of oil-contaminated sediment. *Marine Environmental Research* 11:17-32.

Pearson, W. H., R. W. Bienert, E. Moksness, and J. R. Skalski. 1995. Potential effects of the *Exxon Valdez* Oil Spill on Pacific herring in Prince William Sound. *Canadian Technical Report of Fisheries and Aquatic Sciences* 2060: 63-68.

Pearson, W. H., E. Moksness, and J. R. Skalski. 1995. A field and laboratory assessment of oil spill effects on survival and reproduction of Pacific herring following the EVOS. ASTM STP 1219:626-661.

Pearson, W. H., R. A. Elston, R. W. Bienert, A. S. Drum, and L. D. Antrim. 1999. Why did the PWS, AK, Pacific herring *(Clupea pallasi)* fisheries collapse in 1993 and 1994? Review of hypotheses. *Canadian Journal of Fisheries and Aquatic Sciences* 56:711-737.

Perera, F. P. 1992. DNA adducts and related biomarkers in populations exposed to environmental carcinogens. *Environmental Health Perspectives* 98:133-137.

Perera, F. P., W. Jedrychowski, V. Rauh, and R. M. Whyatt. 1999. Molecular epidemiological research on the effects of environmental pollutants on the fetus. *Environmental Health Perspectives* 107(1999 suppl. 3):451-460.

Peterson, C. H. 1993a. Overview of intertidal processes, damages, and recovery. Abstract. *EVOS Symposium Abstracts,* 19–22.

Peterson, C. H. 1993b. Improvement of environmental impact analysis by application of principles derived from manipulative ecology: Lessons from coastal marine case histories. *Australian Journal of Ecology* 18 (1993):32–33.

Peterson, C., PhD. 2000. Alumni Distinguished Professor of Marine Sciences at the University of North Carolina at Chapel Hill, Institute of Marine Sciences. Interview by Riki Ott, 18 January.

Peterson, C. H. 2001. The EVOS in Alaska: Acute, indirect and chronic effects on the ecosystem. *Advances in Marine Biology* 39:1–103.

Peterson, C. H. and L. Holland-Bartels. 2002. Chronic impacts of oil pollution in the sea: Risks to vertebrate predators. *Marine Ecology Progressive Series* 241:235–236.

Peterson, C. H., L. L. McDonald, R. H. Green, and W. P. Erickson. 2001. Sampling design begets conclusions: The statistical basis for detection of injury to and recovery of shoreline communities after the EVOS. *Marine Ecology Progressive Series* 210:255–283.

Peterson, C. H., L. L. McDonald, R. H. Green, and W. P. Erickson. 2002. Reply comment. The joint consequences of multiple components of statistical sampling designs. *Marine Ecology Progressive Series* 231:309–314.

Peterson, C. H., S. D. Rice, J. W. Short, D. Esler, J. L. Bodkin, B. E. Ballachey, and D. B. Irons. 2003. Long-term ecosystem response to the EVOS. *Science* 302:2082–2086.

Phillips, N. 1992. Spill left murres without a clue. *ADN,* 2 June.

Phillips, N. 1999. Still painful. 10 years later, front-line spill workers link physical ailments to cleanup work. *ADN,* 23 March.

Piatt, J. F. 1987. Behavioral ecology of common murre and Atlantic puffin predation on capelin: Implications for population biology. PhD Diss. Memorial University of Newfoundland, St. John's.

Piatt, J. F. 1993. The oil spill and seabirds: Three years later. *Alaska's Wildlife* 25(1):11–12.

Piatt, J. F. 1995. Water over the bridge. Letters to the editor. *American Scientist* 83:396–398.

Piatt, J. F. 1997. Alternative interpretations of oil spill data. *BioScience* 47(4):202–203.

Piatt, J., PhD. 2000. Research Wildlife Biologist, USGS, Alaska Biological Science Center, Anchorage, AK. Interview by Riki Ott, 6 July.

Piatt, J. F. 2002. Response of seabirds to fluctuations in forage fish density: Can seabirds recover from effects of the EVOS? In Piatt 2002, 132–171.

Piatt, J. F., ed. 2002. Response of seabirds to fluctuations in forage fish density. EVOS Trustee Council Final Report (Restoration Project 00163M). Minerals Management Service (Alaska Outer Continental Shelf Region) and Alaska Biological Science Center, USGS, Anchorage, AK.

Piatt, J. F. and P. Anderson. 1996. Response of common murres to the EVOS and long-term changes in the Gulf of Alaska marine ecosystem. AFSS 18:720–737.

Piatt, J. F., G. Drew, T. van Pelt, A. Abookire, A. Nielsen, M. Shultz, and A. Kitaysky. 1999. Biological effects of the 1997/1998 ENSO event in lower Cook Inlet, AK. Proceedings of the 1998 symposium on the impacts of the 1997/98 El Niño event on the North Pacific ocean and its marginal seas. North Pacific Marine Science Organization (PICES) Scientific Report No. 10:82–86.

Piatt, J. F. and R. G. Ford. 1996. How many seabirds were killed by the EVOS? AFSS 18:712–719.

Piatt, J. F. and C. J. Lensink. 1989. *Exxon Valdez* bird toll. *Nature* 342:865–866, 21/28 December.

Piatt, J. F., C. J. Lensink, W. Butler, M. Kendziorek, and D. Nysewander. 1990. Immediate impact of the EVOS on marine birds. *The Auk* 107:387–397.

Piatt, J. F. and D. Roseneau. 1999. Can murres recover from effects of the EVOS? *Sisyphus News* 1999 (1):1–5. www.absc.usgs.gov/research/seabird_foragefish/products/

Piatt, J. F. and T. I. van Pelt. 1997. Mass-mortality of guillemots *(Uria aalge)* in the Gulf of Alaska in 1993. *Marine Pollution Bulletin* 34(8):656–662.

Picou, S. J. and D. A. Gill. 1997. Commercial fishers and stress: Psychological impacts of the EVOS. In Picou, Gill, and Cohen 1997, 211–232.

Picou, J. S., D. A. Gill, and M. J. Cohen, eds. 1997. *The* Exxon Valdez *Disaster: Readings on a Modern Social Problem.* Dubuque, Iowa: Kendall/Hunt Publishing Company.

Pirtle, R. B. and M. L. McCurdy. 1977. PWS general districts 1976 pink and chum salmon aerial and ground escapement surveys and consequent brood year egg deposition and preemergent fry index programs. ADFG, Division of Commercial Fisheries, Technical Data Report 9, Juneau, AK. In Bue et al. 1996.

Platt, J. 2002. Commercial fisherman, Area E, PWS and the Copper

River Delta, AK. Interview by Riki Ott, 22 August.

Postman, D. 1989. Curtain coming down on Atwood era. *ADN*, 21 November.

Prichard, A. K., D. D. Roby, R. T. Bowyer, and L. K. Duffy. 1997. Pigeon guillemots as a sentinel species: A dose-response experiment with weathered oil in the field. *Chemosphere* 35:1531-1548.

Prince, R. C., J. R. Clark, and J. E. Lindstrom. 1990. 1990 Bioremediation Monitoring Program. Water Research Center, 1990. Fairbanks, AK: University of Alaska.

Project on Scientific Knowledge and Public Policy. 2003. *Daubert:* The most influential supreme court ruling you've never heard of. Available from: www.defendingscience.org/pdf/DaubertReport.pdf.

Prosser, W. L. and W. P. Keaton 1984. *Prosser and Keaton on Torts.* St. Paul, MN: West Publishing Company.

PWS Science Center, Conservation International, Copper River Delta Institute, Ecotrust. 1991. *PWS Copper River North Gulf of Alaska: Ecosystem.* PWSSC, POB 705, Cordova, AK 99574.

Rachel's Environment and Health Weekly. 1999. Corporate manipulation of scientific evidence: A tale of two industries, tobacco and pesticides. In King 1999, 207-211.

Radtke, H., C. M. Dewees, and F. J. Smith. 1987. The fishing industry and Pacific coastal communities: Understanding the assessment of economic impacts. Pacific Sea Grant College Program, Marine Advisory Program Publication UCSGMAP-87-1.

Rahimtula, A. D., P. J. O'Brien, and J. F. Payne. 1984. Induction of xenobiotic metabolism in rats on exposure to hydrocarbon-based oils. In MacFarland et al. 1984, 71-80.

Rahimtula, A. D., Y.-Z. Lee, and J. Silva. 1987. Induction of epidermal and hepatic ornithine decarboxylase by a Prudhoe Bay crude oil. *Fundamentals of Applied Toxicology* 8:408-414.

Rall, D. P. 1989. Oil spill health effects—letter. *Science* 246:564.

Rampton, S. and J. Stauber. 2001. *Trust Us, We're the Experts! How Industry Manipulates Science and Gambles with Your Future.* New York: Jeremy P. Tarcher.

Rappoport, A. G., M. E. Hogan, and K. Bayha. 1990. Development of the release strategy for rehabilitated sea otters. *Sea Otter Symposium* 90(12):375-383.

Rea, W., MD. 1992. *Principles and Mechanisms.* Vol. 1 of *Chemical Sensitivity.* Boca Raton, FL: Lewis Publishers.

Rea, W. J., MD. 1993. Letter to Daniel Stenson; re: Smith's diagnosis and list of chemical sensitivities. 6 January. In *Roberts v. Exxon*

528 Riki Ott

528 Riki Ott

528 Riki Ott

528 Riki Ott

528 Riki Ott

528 Riki Ott

528 Riki Ott

528 Riki Ott

528 Riki Ott

528 Riki Ott

528 Riki Ott

528 Riki Ott

528 Riki Ott

528 Riki Ott

528 Riki Ott

(1999), No. 47, attachment.

Rea, W., MD. 1994. *Sources of Total Body Load*. Vol. 2 of *Chemical Sensitivity*. Boca Raton, FL: Lewis Publishers.

Rea, W., MD. 1995. Clinical Manifestations of Pollutant Overload. Vol. 3 of Chemical Sensitivity. Boca Raton, FL: Lewis Publishers.

Rea, W., MD. 1997a. *Tools of Diagnosis and Methods of Treatment*. Vol. 4 of *Chemical Sensitivity*. Boca Raton, FL: Lewis Publishers.

Rea, W. J., MD. 1997b. Expert witness report, 24 October. In *Roberts v. Exxon* (1999), No. 62, Exhibit L. (This report was not written by Rea: see *Roberts v. Exxon*, No. 90.)

Rea, W. J., MD. 1998. Deposition. 30 January. In *Roberts v. Exxon* (1999), No. 62, Exhibit A.

Rea, W., MD, FACS, FAAEM. 2001. Founder, Environmental Health Center-Dallas. Dallas, TX. Interview by Riki Ott, 21 May.

Rebar, A. H., B. E. Ballachy, D. L. Bruden, and K. Koelcher. 1994. Hematological and clinical chemistry of sea otters captured in PWS, AK, following the EVOS. NRDA Report, Marine Mammal Study 6. USFWS, Anchorage, AK.

Rebar, A. H., T. P. Lipscomb, R. K. Harris, and B. E. Ballachey. 1995. Clinical and clinical laboratory correlates in sea otters dying unexpectedly in rehabilitation centers following the EVOS. *Veterinary Pathology* 32:346–350.

Redburn, D. 1988. Scientific, technical, and regulatory considerations in environmental management. In Shaw and Hameedi 1988, 375–402.

Reller, Carl. 1993. Occupational exposures from oil mist during the EVOS cleanup. Abstract. *EVOS Symposium Abstracts*, 313–315.

Reuters Network. 2004. Exxon to Pay $4.5 Bln [Billion] for Valdez (sic) spill. *Reuters,* 28 January.

Rice, S., PhD. 2001. Habitat Division Program Director, NOAA/NMFS ABL, Juneau, AK. Interview by Riki Ott, 19 January.

Rice, S. D., M. M. Babcock, C. C. Brodersen, M. G. Carls, J. A. Gharrett, S. Korn, A. Moles, and J. Short. 1987a. Lethal and sublethal effects of the water-soluble fraction of Cook Inlet crude oil on Pacific herring *(Clupea harengus pallasi)* reproduction. NOAA Tech. Memo. No. NMFS F/NWC-111, NOAA, Juneau, AK. 63 pp.

Rice, S. D., M. M. Babcock, C. C. Brodersen, J. A. Gharrett, and S. Korn. 1987b. Uptake and depuration of aromatic hydrocarbons by reproductively ripe Pacific herring and the subsequent effect of residues on egg hatching and survival. In Vernberg et al. 1987, 139–154.

Rice, S. D., C. C. Brodersen, P.A. Rounds, and M. M. Babcock. 1993. Oiled mussel beds:A lasting effect. *Alaska's Wildlife* 25(1):28-29.

Rice, S. D. and T. Collier. 2000. Letter to Bruce Wright; re: Comments on OPA 90 and the PEW Oceans Commission report on North Cape oil spill. NOAA/NMFS ABL, Juneau, AK.

Rice, S. D. and R. Heintz. 2000. A shifting paradigm for impacts of oil pollution with a bibliography of papers relevant to the paradigm shift. NOAA/NMFS ABL, Juneau, AK.

Rice, S. D., S. Korn, C. C. Brodersen, S.A. Lindsay, and S.A.Andrews. 1981.Toxicity of ballast-water treatment effluent to marine organisms at Port Valdez, AK. IOSC. 1981, 55-61.

Rice, S. D.,A. Moles, and J. F. Karinen. 1979. Sensitivity of 39 Alaska marine species to Cook Inlet crude oil and No. 2 fuel oil. IOSC 1979, 549-554.

Rice, S. D.,A. Moles, J. F. Karinen, S. Korn, M. G. Carls, C. C. Brodersen, J.A. Gharrett, and M. M. Babcock. 1984. Effects of petroleum hydrocarbons on Alaskan aquatic organisms: A comprehensive review of all oil-effects research on Alaskan fish and invertebrates conducted by the Auke Bay Laboratory, 1970-1981. U.S. Department of Commerce, NOAA Technical Memorandum NMFS F/NWC-67. 128 pp.

Rice, S. D.,A. Moles, and J.W. Short. 1975.The effect of Prudhoe Bay crude oil on survival and growth of eggs, alevins, and fry of pink salmon *Oncorhynchys gorbuscha.* IOSC, 667-670.

Rice, S. D., J.W. Short, and R. Heintz. 2001. Oil and gas issues in Alaska: Lessons learned about long-term toxicity following the EVOS. Conference Proceedings: *Exploring the Future of Offshore Oil and Gas Development in BC: Lessons from the Atlantic,* Continuing Studies in Science at Simon Fraser University, Burnaby, British Columbia, 91-97.

Rice, S. D., J.W. Short, R.A. Heintz, M. G. Carls, and A. Moles. 2000. Life history consequences of oil pollution in fish natal habitat. In Catania 2000, 1210-1215.

Rice, S. D., J.W. Short, and J. F. Karinen. 1977. Comparative oil toxicity and comparative animal sensitivity. In Wolfe 1977, 78-94.

Rice, S. D., R. B. Spies, D.A.Wolfe, B.A.Wright, eds. 1996. *Proceedings of the EVOS Symposium.* American Fisheries Society Symposium 18. Bethesda, MD:American Fisheries Society.

Rice, S. D., R. E.Thomas, M. G. Carls, R.A. Heintz,A. C.Wertheimer, M. L. Murphy, J.W. Short, and A. Moles. 2001. Impacts to pink salmon following the EVOS: Persistence, toxicity, sensitivity, and controversy. *Reviews in Fishery Science* 9(3):165–211.

Rigg, R. W., MD. 1989. Letter to Cordova District Fishermen United (CDFU). 13 May. CDFU, Cordova, AK. (Also in CFS 1989 1[29].)

Robards, M. D., J. A. Anthony, G. A. Rose, and J. F. Piatt. 1999a. Changes in proximate composition and somatic energy content for Pacific sand lance *(Ammodytes hexapterus)* from Kachemak Bay, AK, relative to maturity and season. *Journal of Experimental Marine Biology and Ecology* 242:245–258.

Robards, M. D., and J. F. Piatt. 1999. Biology of the Genus *Ammodytes*—The Sand Lances. In Robards et al. 1999c, 1–16.

Robards, M. D., J. F. Piatt, A. B. Kettle, and A. A. Abookire. 1999b. Temporal and geographic variation in fish communities of lower Cook Inlet, AK. *Fisheries Bulletin* 97(4):962–977.

Robards, M. D., J. F. Piatt, and G. A. Rose. 1999. Maturation, fecundity, and intertidal spawning of Pacific sand lance (*Ammodytes hexapterus*) in the northern Gulf of Alaska. *Journal of Fish Biology* 54:1050–1068.

Robards, M. D., G. A. Rose, and J. F. Piatt. 2002. Growth and abundance of Pacific sand lance, *Ammodytes hexapterus,* under differing oceanographic regimes. *Environmental Biology of Fishes* 64:429–441.

Robards, M. D., M. F. Wilson, R. H. Armstrong, and J. F. Piatt, eds. 1999c. Sand lance: A review of biology and predator relations and annotated bibliography. Research Paper PNW-RP-521. Portland, OR, U.S. Department of Agriculture, Forest Service, Pacific Northwest Research Station. www.fs.fed.us/pnw/pubs.htm

Robbins, C. 1989. Overview of cleanup methods and operations. Paper presented at the Alaskan Oil Spill and Human Health conference, sponsored by NIEHS, NIOSH, University of Washington School of Public Health, US EPA, and the Agency for Toxic Substances and Disease Registry, 28–30 July, Seattle, WA, 2–3.

Scott Roberts v. VECO, Inc. and Eagle Pacific Insurance Company. 1996. AWCB Case No. 9034054, AWCB Decision No. 96-0029, Decision and order. 19 January.

Scott Roberts, Richard Merrill, Phyllis La Joie, and Ron Smith v. Exxon Corp., SeaRiver Maritime, and VECO. A96-040-CV (HRH), US District Court for the District of Alaska (1999).

——, No. 1. Plaintiffs' complaint for damages and demand for jury trial, jurisdiction and venue. 2 February 1996.

——, No. 10. HRH Order granting stipulation; re: confidentiality of documents. 30 May 1996.

——, No. 18. HRH Order denying VECO's motion to dismiss for laches. 7 August 1996.

——, No. 26. Defendants Exxon's and VECO's motion for summary judgment dismissing Roberts' claims on statute of limitation grounds. 25 September 1996.

——, No. 37. HRH Judgment that plaintiff Roberts' claims and complaints are dismissed with prejudice as to all defendants. 16 April 1997.

——, No. 47. Plaintiffs' amended disclosure of expert witnesses. 11 August 1997.

——, No. 57. Stipulation for dismissal of plaintiff Merrill's claims with prejudice. 14 November 1997.

——, No. 61. Defendant VECO's motion for summary judgment against plaintiff Smith. 5 March 1998.

——, No. 62. Defendant Exxon's and SeaRiver's motion in limine to exclude expert testimony of Dr. Rea. 5 March 1998.

——, No. 63. Defendant Exxon's and SeaRiver's motion in limine to exclude expert testimony of Dr. Ewing. 5 March 1998.

——, No. 64. Defendant Exxon's and SeaRiver's motion for summary judgment against plaintiffs La Joie and Smith. 5 March 1998.

——, No. 65. Defendant VECO's motion in limine to exclude expert testimony of Dr. Rea. 5 March 1998.

——, No. 72. HRH Order granting motion for order extending time for filing opposition to motions of Exxon (No. 66). 14 April 1998.

——, No. 73, HRH Order granting motion for order extending time for filing opposition to motions of VECO (No. 69). 14 April 1998.

——, No. 74. Plaintiffs' opposition to defendants' motion for summary judgment against plaintiffs La Joie and Smith. 21 May 1998.

——, No. 76. Defendants Exxon's and SeaRiver's reply to opposition defendants' motions in limine and motions for summary judgment. 3 June 1998.

——, No. 78. HRH Judgment granting VECO's motion for entry of final judgment under rule 54(b) against Smith; all claims against VECO dismissed with prejudice and VECO awarded undetermined costs and attorney's fees against Smith. 15 July 1998.

——, No. 79. Plaintiff's appeal to 9CCA (Ninth Circuit Court of Appeals) to No. #78 fld 07/15/98 (98-35844). 18 September 1998.

——, No. 83. Certified copy 9CCA mandate; U.S. District Court affirmed (97-35518). 30 September 1998.

——, No. 87. Defendant VECO's reply to plaintiff Smith's opposition

to motion in limine of VECO, Inc. to exclude expert testimony of
Dr. Rea. 12 November 1998.
——, No. 90. HRH Order granting motions in limine of VECO (Docs.
62, 63, 65). 8 February 1999.
——, No. 91. HRH Order granting motions for summary judgment
against plaintiffs La Joie and Smith (No. 64) and granting VECO's
motion for summary judgment against Smith (No. 61); plaintiffs'
complaint is dismissed. 24 February 1999.
——, No. 92. HRH Judgment that plaintiffs' complaint is dismissed.
24 February 1999.
——, No. 94. Defendant VECO's memorandum (bill of costs) in support of motion for attorneys' fees [case background]. 10 March
1999.
——, No. 96. Defendant VECO's motion for attorney's fees. 10 March
1999.
——, No. 101. Defendant Exxon's application re: Writ of execution
as to Ron Smith in the amount of $1,585.18. 26 March 1999.
——, No. 103. Defendant Exxon's application re: Writ of execution
as to Phyllis La Joie in the amount of $2,926.44. 26 March 1999.
——, Note. Issued: Writ of execution as to Phyllis La Joie to the U.S.
Marshal for the District of Alaska. 29 March 1999.
——, Note. Issued: Writ of execution as to Ron Smith to the U.S.
Marshal for the District of Alaska. 1 April 1999.
——, No. 105. HRH Order granting motion for attorneys' fees in the
amount of $57,418.00. 7 April 1999.
——, No. 107. Defendant VECO's judgment satisfaction; re: Smith in
the amount of $951.36 for costs and $57,418.00 for attorney's
fees. 20 April 1999.
Robinson, A. R. and K. H. Brink, eds. 1998. *The Global Coastal
Ocean/Regional Studies and Syntheses.* New York: John Wiley &
Sons.
Roby, D. D. and A. K. Hovey. 2002. Pigeon guillemot restoration
research at the Alaska SeaLife Center. EVOS Restoration Project
Final Report (Restoration Project 01327). USGS-Oregon
Cooperative Fish and Wildlife Research Unit, Department of
Fisheries and Wildlife, Oregon State University, Corvallis, OR.
Rodin, M., M. Downs, J. Petterson, and J. Russell. 1997. Community
impacts of the EVOS. In Picou, Gill, and Cohen 1997, 193–205.
Rogers, M. L. 1990. *Acorn Days. The Environmental Defense Fund
and How it Grew.* New York: Environmental Defense Fund.
Rolseth, V., R. Djurhuus, A. M. Svardal. 2002. Additive toxicity of

limonene and 50% oxygen and the role of glutathione in detoxification in human lung cells. *Toxicology* 170:75-88.

Rom, W. N. and V. E. Archer, eds. 1980. *Health Implications of New Energy Technologies*. Ann Arbor, MI: Ann Arbor Science Publishers.

Rooper, C. N. and L. J. Haldorson. 2000. Consumption of Pacific herring *(Clupea pallasi)* eggs by greenling *(Hexagrammidea)* in PWS, AK. *Fisheries Bulletin* 98:655-659.

Rosen, Y. 2002. *Exxon Valdez* oil still harmful, US studies say. *Planet Ark,* 16 January.

Roseneau, D. G. and G. V. Byrd. 1997. Using Pacific halibut to sample the availability of forage fishes to seabirds. Abstract. *Forage Fish Proceedings,* 231-241.

Rosenberg, D. H. and M. J. Petrula. 1998. Status of harlequin ducks in PWS, AK, after the EVOS, 1995-1997. EVOS Restoration Project Final Report (Restoration Project 97427). ADFG, Division of Wildlife Conservation, Anchorage, AK.

Rotterman, L. M. 1992. Patterns of genetic variability in sea otters after severe population subdivision and reduction. PhD. diss., Univ. of Minnesota.

Rotterman, L. M. and C. Monnett. 1991. Mortality of sea otter weanlings in eastern and western PWS, AK, during the winter of 1990-91. NRDA Report, Marine Mammal Study No. 6. USFWS, Anchorage, AK.

Rotterman, L. M. and C. Monnett. 1993. Health, reproduction and survival of adult, dependent, and weanling sea otters in PWS from October 1989 to December 1991. Abstract. *EVOS Symposium Abstracts,* 296.

Rotterman, L. M. and C. Monnett. 1995. Mortality of sea otter weanlings in eastern and western PWS, AK, during the winter of 1990-91. EVOS State/Federal NRDA Final Report, Marine Mammal Study No. 6-18. USFWS, Anchorage, AK.

Royer, T. C. 1979. On the effects of precipitation and runoff on coastal circulation in the Gulf of Alaska. *Journal of Physical Oceanography* 9:555-563.

Royer, T. C. 1981a. Baroclinic transport in the Gulf of Alaska: Part I— Seasonal variations of the Alaska Current. *Journal of Marine Resources* 39:239-249.

Royer, T. C. 1981b. Baroclinic transport in the Gulf of Alaska. Part II—A freshwater driven coastal current. *Journal of Marine Research* 39:251-265.

Royer, T. C. 1982. Coastal freshwater discharge in the Northeast Pacific. *Journal of Geophysical Research* 87:2017-2021.

Royer, T. C. 1993. High-latitude oceanic variability associated with the 18.6-year nodal tide. *Journal of Geophysical Research* 98:4639-4644.

Royer, T. C. 1998. Coastal processes in the northern North Pacific. In Robinson and Brink 1998, 395-414.

Royer, T. C. and W. J. Emery. 1987. Circulation in the Gulf of Alaska, 1981. *Deep Sea Research* 34:1361-1377.

Royer, T. C., D. V. Hansen, and D. J. Pashinski. 1979. Coastal flow in the northern Gulf of Alaska as observed by dynamic topography and satellite-tracked, drogued drift-buoys. *Journal of Physical Oceanography* 9:785-801.

RurAL CAP. 2002. Alaska Native fish, wildlife, habitat and environment statewide summit report. Anchorage, AK.

Sanger, G. A. 1986. Diets and food-web relationships of seabirds in the Gulf of Alaska and adjacent marine regions. NOAA OCSEAP (Outer Continental Shelf Environmental Assessment Program) Final Report 45:631-771.

Saulitis, E. 1993. The behavior and vocalizations of the at group of transient orcas in PWS, AK. MS thesis, Institute of Marine Science, University of Alaska, Fairbanks.

Saulitis, E., C. Matkin, L. Barrett-Lennard, K. Heise, and G. Ellis. 2000. Foraging strategies of sympatric killer whale *(Orcinus orca)* populations in PWS, AK. *Marine Mammal Science* 16(1):94-109.

Scheel, D., C. O. Matkin, and E. Saulitis. 2001. Distribution of killer whale pods in PWS, AK 1984-1996. *Marine Mammal Science* 17(3):555-569.

Schindler, D. W. 1987. Detecting ecosystem response to anthropogenic stress. *Canadian Journal of Fisheries and Aquatic Sciences* 44 (Suppl.):6-25.

Schmidt Etkin, D. 2001. Analysis of oil spill trends in the United States and worldwide. IOSC 2001, 1291-1300.

Schneider, K. 1991. Judge rejects $100 million fine for Exxon in oil spill as too low. *The New York Times,* 25 April.

Seiser, P. E., L. K. Duffy, A. D. McGuire, and D. D. Roby. 2000. Comparison of pigeon guillemot, *Cepphus columba,* blood parameters from oiled and unoiled areas of Alaska eight years after the EVOS. *Marine Pollution Bulletin* 40(2):152-164.

Sharp, B. E. and M. Cody. 1993. Black oystercatchers in PWS: Oil spill effects on reproduction and behavior. Abstract. *EVOS*

Symposium Abstracts, 155–58.

Sharp, B., M. Cody, and R. Turner. 1996. Effects of the EVOS on the black oystercatcher. AFSS 18, 748–758.

Sharp, D., S. Sharr, and C. Peckham. 1994. Homing and straying patterns of coded-wire-tagged pink salmon in PWS. AK Sea Grant Rep. 94-02:77–82.

Sharr, S. 2001. Principal Research Biologist for Salmon and Steelhead, Idaho Department of Fish and Game, Boise, ID. Interview by Riki Ott, 24 January.

Sharr, S., B. Bue, and S. Moffitt. 1989 (sic) [1990a]. Injury to salmon eggs and pre-emergent fry in PWS. EVOS State/federal NRDA Draft Preliminary Status Report, Fish/Shellfish Study No. 2. ADFG, Commercial Fisheries Division, in cooperation with USFS and ADNR, Anchorage, AK.

Sharr, S., B. Bue, and S. Moffitt. 1990b. Injury to salmon eggs and pre-emergent fry in PWS. EVOS State/federal NRDA Draft Preliminary Status Report, Fish/Shellfish Study No. 2. ADFG, Commercial Fisheries Fisheries Division, in cooperation with USFS and ADNR, Anchorage, AK.

Sharr, S., B. Bue, and S. Moffitt. 1991. Injury to salmon eggs and pre-emergent fry in PWS. EVOS State/federal NRDA Draft Preliminary Status Report, Fish/Shellfish Study No. 2. ADFG, Commercial Fisheries Division, in cooperation with USFS and ADNR. Anchorage, AK.

Shaw, D. G., and M. J. Hameedi, eds. 1988. *Environmental Studies on Port Valdez, Alaska. A Basis for Management.* Lecture Notes on Coastal and Estuarine Studies 24. Berlin: Springer-Verlag.

Sheppard, C., ed. 2000. *Seas at the Millennium: An Environmental Evaluation.* Amsterdam: Pergamon Press.

Shigenaka, G. and C. B. Henry. 1995. Use of mussels and semipermeable membrane devices to assess bioavailability of residual polynuclear aromatic hydrocarbons three years after the EVOS. *ASTM* 1219:239–260.

Short, J. 1998. Chemist, NOAA/NMFS ABL, Juneau, AK. Interview by Riki Ott, 16 June.

Short, J. W. 2002a. Letter to editor. Critic of oil spill study attempts to discredit government science. *ADN,* 3 February.

Short, J. W. 2002b. Point counterpoint: Has PWS recovered? Oil remains; appears to be affecting wildlife rebound. *ADN,* 31 January.

Short, J. W. and M. Babcock. 1996. Prespill and postspill concentrations of hydrocarbons in mussels and sediments in PWS. AFSS 18:149-166.

Short, J. W. and P. M. Harris. 1996a. Chemical sampling and analysis of petroleum hydrocarbons in near-surface seawater of PWS after the EVOS. AFSS 18:29-39.

Short, J. W. and P. M. Harris. 1996b. Petroleum hydrocarbons in caged mussels deployed in PWS after the EVOS. AFSS 18:29-39.

Short, J. W. and R. A. Heintz. 1997. Identification of *Exxon Valdez* oil in sediments and tissues from PWS and the northwestern Gulf of Alaska based on PAH weathering. *Environmental Science and Technology* 31:2375-2384.

Short, J. W. and R. A. Heintz. 1998. Letter to the editor. Source of polynuclear aromatic hydrocarbons in PWS, AK, USA, subtidal sediments. *Environmental Toxicology and Chemistry* 17(9):1651-1652.

Short, J. W. and R. A. Heintz. 2003. Normal alkanes and the unresolved complex mixture as diagnostic indicators of hydrocarbon source contributions to marine sediments of the northern Gulf of Alaska. Proceedings of the 26th AMOP, 55-168.

Short, J. W. and R. A. Heintz. 1997. Identification of *Exxon Valdez* oil in sediments and tissues from PWS and the northwestern Gulf of Alaska based on a PAH weathering model. *Environmental Science and Technology* 31:2375-2384.

Short, J. W., K. A. Kvenvolden, P. R. Carlson, F. D. Hostettler, R. J. Rosenbauer, and B. A. Wright. 1999. Natural hydrocarbon background in benthic sediments in PWS, AK: Oil vs coal. *Environmental Science and Technology* 33:34-42.

Short, J. W., K. A. Kvenvolden, P. R. Carlson, F. D. Hostettler, R. J. Rosenbauer, and B. A. Wright. 2000. Response to comment; re: The natural hydrocarbon background in benthic sediments of PWS, AK: Oil vs. coal. *Environmental Science and Technology* 34:2066-2067.

Short, J. W., M. R. Lindeberg, P. M. Harris, J. M. Maselko, J. J. Pella, and S. D. Rice. 2002. Vertical oil distribution within the intertidal zone 12 years after the EVOS in PWS, AK. Proceedings of the 25th AMOP, 57-72.

Short, J. W., M. R. Lindeberg, P. M. Harris, J. M. Maselko, J. J. Pella, and S. D. Rice. 2004. Estimate of oil persisting on the beaches of PWS 12 years after the EVOS. *Environmental Science and Technology* 38(1):19-25.

Shultz, M. 2002. Black-legged kittiwake biology in lower Cook Inlet.

In Piatt 2002, 86–99.

Shultz, M. and A. Harding. 2002. Black-legged kittiwake biology in lower Cook Inlet. In Piatt 2002, 86–99.

Shultz, M. and T. I. van Pelt. 2002. Biology of other seabird species in lower Cook Inlet. In Piatt 2002, 122–131.

Sims, G. 1989. A clot in the heart of the earth. Fighting the lost war of the Valdez (sic) oil spill. *Outside* June, 39–105.

Smith, R. 1996. Deposition. 3 October. In *Roberts v. Exxon* (1999), No. 62, Exhibit A.

Smith, R. 1997. Peer review: Reform or revolution? *BMJ* 315:759–760. Quoted in Rampton and Stauber 2001, 199.

Smoker, B., P. Crandell, and P. Malecha. 2000. Genetic analysis of development mortality in oiled and unoiled lines of pink salmon. Final Report Project 40HCNF700235, revised 20 May to NOAA/NMFS ABL, Juneau, AK.

Solomon, C. 1993. Exxon attacks scientific views of Valdez (sic) spill. Exxon seeks to change beliefs about Valdez (sic) oil spill damage. *WSJ*, 12 April.

Speckman, S. and J. Piatt. 2002. Hydroacoustic forage fish biomass and distribution in Cook Inlet. In Piatt 2002, 55–63.

Spence, H. 1989a. Fertilizer blamed for illnesses. *Homer News,* 24 August.

Spence, H. 1989b. Seldovians charge workers' health neglected. *Homer News,* 31 August.

Spence, H. 1989c. Feds ask state inspectors to get tough. *Homer News,* 31 August.

Spence, H. 1990. Was the spill 38 million gallons? *Homer News,* 12 April.

Spies, R., M. McCammon, and J. Balsinger. 2002. Letter to editor. PWS oil study critic's fraud charge is unfounded. *ADN*, 3 February.

Spies, R. B., S. D. Rice, D. A. Wolfe, and B. A. Wright. 1996. The effect of the EVOS on the Alaskan coastal environment. AFSS 18:1–16.

Spiess, B. 2001. Dispersants study casts doubt on effectiveness. OIL: Weapon against spills doesn't handle Slope oil or cold water well, tests say. *ADN*, 23 February.

Spraker, T. R. 1990. Hazards of releasing rehabilitated animals with emphasis on sea otters and the T/V EVOS. *Sea Otter Symposium* 90(12):385–389.

Spraker, T. R., L. F. Lowry, and K. J. Frost. 1994. Gross necropsy and histopathological lesions found in harbor seals. In Loughlin 1994, 281–311.

Springborn Institute for Bioresearch. 1985. 28-day subchronic dermal toxicity study in rats. American Petroleum Institute (API), Medical Research Publication 32-32652. Washington, DC.

Springer, A. M. and S. G. Speckman. 1997. A forage fish is what? Summary of the symposium. *Forage Fish Proceedings,* 773–801.

State of Alaska v. Exxon Corporation and Exxon Shipping. A91-083-CV (HRH), US District Court for the District of Alaska (1991).

Stauber, J. C. and S. Rampton. 1995. *Toxic Sludge is Good for You. Lies, Damn Lies, and the Public Relations Industry.* Monroe, ME: Common Courage Press.

Steiner, R. and D. Grimes. 1999. Professor and Conservation Specialist for the University of Alaska Marine Advisory Program (Steiner); Cordova fisherman and musician (Grimes). Interview by Riki Ott, March.

Steingraber, S. 1998. *Living Downstream: A Scientist's Personal Investigation of Cancer and the Environment.* New York: Vintage Books.

Steingraber, S. 2001. *Having Faith: An Ecologist's Journey to Motherhood.* Cambridge, MA: Perseus Publishing.

Stekoll, M. S. and L. Deysher. 1996. Re-colonization and restoration of upper intertidal *Fucus gardneri (Fucales, Phaeophyta)* following the EVOS. *Hydrobiologia* 326/327:311–316.

Stekoll, M. S., L. Deysher, R. C. Highsmith, S. M. Saupe, Z. Guo, W. P. Erickson, L. McDonald, and D. Strickland. 1996. Coastal habitat injury assessment: Intertidal communities and the EVOS. AFSS 18:177–192.

Stenson, D. 1998. Letter to Exxon. 6 February. In *Roberts v. Exxon* (1999), No. 63, Exhibit L.

Stewart, R. B. 1990. Testimony of Assistant Attorney General, Land and Natural Resources Division, Department of Justice, Washington, DC. In U.S. Congress House 1990, 202–206.

Stokesbury, K. D., E., R. J. Foy, and B. L. Norcross. 1999. Spatial and temporal variability in juvenile Pacific herring, *Clupea pallasi,* growth in PWS, AK. *Environmental Biology of Fishes* 56:409–418.

Stokesbury, K. D., E., J. Kirsch, E. D. Brown, G. L. Thomas, and B. L. Norcross. 2000. Spatial distributions of Pacific herring, *Clupea pallasi,* and walleye pollock, *Theragra chalcogramma,* in PWS, AK. *Fisheries Bulletin* 98:400–409.

Stokesbury, K. D., E., J. Kirsch, E. V. Patrick, and B. L. Norcross. 2002. Natural mortality estimates of juvenile Pacific herring *(Clupea pallasi)* in PWS, AK. *Canadian Journal of Fisheries and Aquatic Sciences* 59:416-423.

Stranahan, S. 2003. The Valdez Crud. Are crude oil and chemicals to blame for the health problems of workers who cleaned up Exxon's mess? *Mother Jones*, 3 March.

Stringer, W. J., G. Dean, R. M. Guritz, H. M. Garbeil, J. E. Groves, and K. Ahlnaes. 1992. Detection of petroleum spilled from the T/V *Exxon Valdez*. *International Journal of Remote Sensing* 13:799-824.

Stuart, T. 1989. ADOL letter to Dr. Knut Ringen, Director, Laborers' National Health and Safety Fund, 21 April. In U.S. Congress House 1989a, 1061-1062.

Garry Stubblefield and Melissa Stubblefield v. Exxon Shipping Company, Exxon Corporation, VECO, Inc., and Norcon, Inc. 3AN-91-6261 CV (HBS), AK Superior Court, Third Judicial District at Anchorage (1994).

——, No. 910919. Discovery Master's order relating to production of confidential materials by all parties. 19 September 1991.

——, No. 911030. Defendant Exxon Shipping's answer to plaintiff's requests for production. 30 October 1991.

——, No. 920810. Summons to Exxon to file written answer to complaint. 10 August 1992.

——, No. 920817. Plaintiffs' motion in support of second motion to compel. 17 August 1992.

——, No. 920903. Plaintiffs' reply to Exxon's opposition to request for production. 3 September 1992.

——, No. 920906. Plaintiffs' motion to remove from rule 16.1 designation. 6 September 1992.

——, No. 921104. HBS Order compelling production from Exxon. 6 September 1992.

——, No. 921231. Plaintiffs' motion for order requiring production of toxicological information in accordance with court order of 4 November 1992. 31 December 1992.

——, No. 930324. Stipulation for protective order protecting confidentiality of medical information. 24 March 1993.

——, No. 930526a. Defendant Exxon's motion for protective order for medical information. 26 May 1993.

——, No. 930526b. Defendant VECO's motion for protective order for medical information. 26 May 1993.

——, No. 930604. Plaintiffs' opposition to VECO's motion for protective order and motion to rescind VECO stipulation. 4 June 1993.

——, No. 930915. Master's report on Exxon's motion for protective order. 15 September 1993.

——, No. 930928. Order dismissing plaintiffs' punitive damage claims and Melissa Stubblefield's loss of consortium claim. 28 September 1993.

——, No. 931005. Plaintiffs' preliminary witness list. 5 October 1993.

——, No. 931015. Order granting protection of medical information. 15 October 1993.

——, No. 931018. Master's report; re: defendants' motions for protective order. 18 October 1993.

——, No. 941017. Plaintiffs' opposition to Exxon's motion for protective order. 17 October 1994.

——, No. 941219. Order dismissing case with prejudice. 19 December 1994.

Stubblefield, W. A., G. A. Hancock, W. H. Ford, H. H. Prince, and R. K. Ringer. 1995. Evaluation of the toxic properties of naturally weathered *Exxon Valdez* crude oil to wildlife species. ASTM STP 1219:665–692.

Stubblefield, W. A., R. H. McKee, R. W. Kapp, and J. P. Hinz. 1989. An evaluation of the acute toxic properties of liquids derived from oil sands. *Journal of Applied Toxicology* 9:59–65.

Sturdevant, M. V., A. L. J. Brase, and L. B. Hulbert. 2001. Feeding habits, prey fields, and potential competition of young-of-the-year walleye pollock *(Theragra chalcogramma)* and Pacific herring *(Clupea Pallasi)* in PWS, AK, 1994–1995. *Fisheries Bulletin* 99:482–501.

Sturdevant, M. V., A. C. Wertheimer and J. L. Lum. 1996. Diets of juvenile pink and chum salmon in oiled and non-oiled nearshore habitats in PWS, 1989 and 1990. AFSS 18:578–592.

Sundberg, K, L. Deysher, and L. McDonald. 1996. Intertidal and supratidal site selection using a geographical information system. AFSS 18:167–176.

Sunshine Makers, Inc. 2002. MSDS: Simple Green®. Version No. 1007. Sunshine Makers, Inc., Huntington Harbour, CA.

Surety of Alaska. 1990. Letter to Ed Meggert; re: controversion notice. 19 September. In ADOL, AWCB 1992a.

Suryan, R. M. and D. B. Irons. 2001. Colony and population dynamics of black-legged kittiwakes in a heterogeneous environment. *The Auk* 118(3): 636–649.

Suryan, R. M., D. B. Irons, and J. Benson. 2000. Prey switching and variable foraging strategies of black-legged kittiwakes and the effect on reproductive success. *Condor* 102:374-384.

Suryan, R. M. and J. T. Harvey. 1998. Tracking harbor seals *(Phoca vitulina richardsi)* to determine dive behavior, foraging activity, and haul-out site use. *Marine Mammal Science* 14:361-372.

Suryan, R. M., D. B. Irons, M. Kaufman, J. Benson, P. G. R. Jodice, D. D. Roby, and E. D. Brown. 2002. Short-term fluctuations in forage fish availability and the effect on prey selection and brood-rearing in the black-legged kittiwakes, *Rissa tridactyla. Marine Ecology Progressive Series* 236:273-287.

Suuberg, M. J. 1990. Testimony of Associate Solicitor for Conservation and Wildlife, U.S. DOI. In U.S. Congress House 1990, 214-221.

Taneda, S., H. Hayashi, A. Sukushima, K. Seki, A. Suzuki, K. Kamata, M. Sakata, S. Yoshino, M. Sagai, Y. Mori. 2002. Estrogenic and anti-estrogenic activities of two types of diesel exhaust particles. *Toxicology* 180:153-161.

Tarbell, I. M. 1904. *The History of Standard Oil Company.* 2 vols. New York: Macmillan.

Tarbell, I. M. 1939. *All in the Day's Work: An Autobiography.* New York: Macmillan Company.

Teal, A. R. 1991. Shoreline cleanup–reconnaissance, evaluation, and planning following the EVOS. IOSC 1991, 149-160.

Teitelbaum, D. T., MD. 1994. Deposition. October 12. In *Stubblefield v. Exxon* (1994).

Tellus report—see Project on Scientific Knowledge and Public Policy. 2003.

Testa, J. W., D. F. Holleman, R. T. Bowyer, and J. B. Faro. 1994. Estimating populations of marine river otters in Prince William Sound, AK, using radio-tracer implants. *Journal of Mammalogy* 75:1021-1032.

The Dallas Morning News staff. 1992. Summary: Two Desert Storm veterans with mysterious ailments are suffering from petroleum poisoning, a nationally recognized medical expert said. 2 August.

The Dallas Morning News staff. 1994. For investors, a sigh of relief. Exxon stock, prospects rise as uncertainty ends. 19 September.

The Exxon Valdez Case. A89-095-CV (HRH) (Consolidated). (Filed in 1989.)

Thomas, R. E., M. G. Carls, S. D. Rice, and L. Shagrun. 1997. Mixed fuction oxygenase induction in pre- and post-spawn herirng *(Clupea pallasi)* by petroleum hydrocarbons. *Comparative*

Biochemistry and Physiology 116C(2):141–147.

Thompson, M. 1993. Government to study Gulf vet illnesses. *ADN*, 2 November.

Thorne, L.—see Weidman, L.

Townsend Environmental. 1994. The promises issue. Commitments and representations by Alyeska and its owner companies regarding the Trans-Alaska Pipeline System. Prepared for private citizens Chuck Hamel and Dan Lawn. Promises I: www.Alaskaforum.org/rowhist/rt/101rt.pdf and Promises II: www.Alaskaforum.org/rowhist/rt/102rt.pdf.

Trust, K., D. Esler, B. R. Woodin, and J. J. Stegeman. 2000. Cytochrome P450-1A induction in sea ducks inhabiting nearshore areas of PWS, AK. *Marine Pollution Bulletin* 40(5):397–403.

Tyson, R. 1990. VECO International: Revenues skyrocket from oil spill cleanup. *Alaska Business Monthly* October, 53–58.

U.S.A. 1974. Agreement and Grant of Right-of-Way for Trans-Alaska Pipeline between the United States of America and Amerada Hess Corporation, ARCO Pipe Line Company, Exxon Pipeline Company, Mobil Alaska Pipeline Company, Phillips Petroleum Company, Sohio Pipe Line Company, and Union Alaska Pipeline Company. Washington, DC.

U.S.A. v. Exxon Corporation, Exxon Shipping Company, and Exxon Pipeline Company, and the T/V Exxon Valdez. A91-082 (HRH), US District Court for the District of Alaska (1991a).

U.S.A. v. Exxon Shipping Company and Exxon Corporation. A90-015-CR (HRH), US District Court for the District of Alaska (1991b).

U.S.A. v. State of Alaska. A91-081-CV (HRH), US District Court for the District of Alaska (1991).

U.S.A., ex rel., W. Findlay Abbott v. Exxon Corporation and Exxon Shipping Company. A96-0041-CV (HRH). US District Court for the District of Alaska (filed in 1996).

U.S. Coast Guard. 1993. T/V EVOS: Federal On-Scene Coordinator Report. Washington, DC: USCG, U.S. Department of Transportation.

U.S. Congress. House. 1973. Committee on Interior and Insular Affairs. Subcommittee on Public Lands. *Oil and Natural Gas Pipeline Rights-of-Way Hearings.* 93rd Cong., 1st sess., Serial No. 93-12. Washington, DC: GPO.

U.S. Congress. House. 1989a. Committee on Interior and Insular Affairs. *Investigation of the EVOS, PWS, AK, Part I.* 101st Cong.,

1st sess. 5 May, Cordova, AK, and 7–8 May, Valdez, AK. Serial No. 101-5. Washington, DC: GPO.

U.S. Congress. House. 1989b. Committee on Interior and Insular Affairs. *Investigation of the EVOS, PWS, AK, Part II. Oil Spill Cleanup Research and Technology.* 101st Cong., 1st sess. 18 July, Washington, DC. Serial No. 101-5. Washington, DC: GPO.

U.S. Congress. House. 1989c. Committee on Interior and Insular Affairs. *Investigation of the EVOS, PWS, AK, Part III. Status of Exxon AK Oil Spill Cleanup.* 101st Cong., 1st sess. 28 July, Washington, DC. Serial No. 101-5. Washington, DC: GPO.

U.S. Congress. House. 1990. Committee on Interior and Insular Affairs. *Investigation of the EVOS, PWS, AK, Part IV. Books 1 and 2. Oil Spill Cleanup Operation and Damage.* 101st Cong., 2nd sess. 22 March and 24 April. Serial No. 101-5. Washington, DC: GPO.

U.S. Congress. House. 1991a. Committee on the Budget. *Hearing before the Task Force on Urgent Fiscal Issues.* 102nd Cong., 1st sess. 31 October, Washington, DC. Serial No. 4-3. Washington, DC: GPO.

U.S. Congress. House. 1991b. Committee on Interior and Insular Affairs. *Oversight Hearings on Alyeska Pipeline Service Company Covert Operation.* 102nd Cong., 1st sess. November 4, 5 and 6. Serial No. 102-13. Washington, DC: GPO.

U.S. Congress. House. 1992. Committee on Interior and Insular Affairs. *Oversight Hearings on Alyeska Pipeline Service Company Covert Operation.* Part I, Session Report and Table of Contents; Part II, Appendix–Exhibits 1–83; Part III, Exhibits 84–200; Part IV, Appendix–Minority Exhibits 1–44. 102nd Cong., 1st sess., hearings held in Washington, DC, 4-6 November 1991. Serial No. 102-13. Washington, DC: GPO.

U.S. Congress. Senate. 1969. Committee on Interior and Insular Affairs. *Trans-Alaska Pipeline Hearings on the Status of the Proposed Trans-Alaska Pipeline.* 91st Cong., 1st sess. on 9 September (Part 1) and 16 October (Part 2). Washington, DC: GPO.

U.S. Department of Commerce, NOAA. 1989. National Status and Trends Program for Marine Environmental Quality, Progress Report. NOAA Technical Memorandum NOS OMA 49, NOAA, Rockville, MD.

U.S. Department of Health, Education, and Welfare. 1977. *Occupational Diseases. A Guide to their Recognition.*

Washington, DC: GPO Office. Stock number 017-033-00266-5.
U.S. EPA. 1991. Technical support document for water quality-based
toxics control. Office of Water Enforcement and Permits, Office
of Water Regulations and Standards, Washington, DC. EPA/505/2-
90-001, PB91-127415, March.
U.S. EPA. 1992. Oil tanker waste disposal practices: A review. Water
Division, Water Permits and Compliance Branch, Region 10,
Seattle, WA.
U.S. EPA. 2000. 1999 Persistent, Bioaccumulative and Toxic
Pollutants Initiative (PBTI) Report. Pollution Prevention
Information Clearinghouse, Washington, DC. www.epa.
gov/pbt/accomp99.htm.
U.S. EPA. 2001. Emergency Planning and Community Right-to-Know
Act, Section 313. Guidance for reporting toxic chemicals:
Polycylic aromatic compounds category. Final. Washington, DC:
U.S. EPA. EPA 260-B-01-03. August.
www.epa.gov/tri/guide_docs/2001/pacs2001.pdf.
U.S. EPA. 2003. Janitorial Products Pollution Prevention Program.
www.westp2net.org/janitorial/tools/haz2.htm.
U.S. EPA. 2004a. Oil Program, National Contingency Plan Product
Schedule and Notebook, bioremediation agents, Inipol EAP 22.
www.epa.gov/ceppo/ncp/inipoleap.htm.
U.S. EPA. 2004b. PBT Chemical List [PAHs listed as "PACs" or poly-
cyclic aromatic compounds]. TRI [Toxic Release Inventory] cate-
gory number N590.
U.S. GAO. 1987. Water pollution. EPA controls over ballast water at
Trans-Alaska Pipeline marine terminal. Report to U.S. House of
Representatives, Committee on Energy and Commerce,
Subcommittee on Oversight and Investigations. GAO/RCED-87-
118. Washington, DC. June.
U.S. GAO. 1993. Natural resources restoration: Use of EVOS settle-
ment funds. GAO/RCED-93-206BR.
U.S. Office of Management and Budget. 1999. Circular A-110.
VanBlaricom, G. R. 1990. Capture of lightly oiled sea otters for reha-
bilitation: A review of decisions and issues. *Sea Otter Symposium*
90(12):130–136.
Van Kooten, G. W., J. W. Short, and J. J. Kolak. 2002. Low maturity
Kulthieth formation coal: A possible source of polycyclic aromatic hydrocarbons in benthic sediment of the northern Gulf of
Alaska. *Environmental Forensics* 3:227–242.
van Pelt, T. and M. Shultz. 2002. Common murre biology in lower

Cook Inlet. In Piatt 2002, 71-85.

Van Tamelen, P. G. and M. S. Stekoll. 1996a. Population response of the brown alga *Fucus gardneri* and other algae in Herring Bay, PWS, to the *EVOS*. AFSS 18:210.

Van Tamelen, P. G. and M. S. Stekoll. 1996b. The role of barnacles in recruitment and subsequent survival of the brown alga, *Fucus gardneri (Silva). Journal of Experimental Marine Biology and Ecology* 208:227-238.

Van Tamelen, P. G., M. S. Stekoll, and L. Deysher. 1997. Recovery processes of the brown alga *Fucus gardneri* following the EVOS: Settlement and recruitment. *Marine Ecology Progress Series* 160:265-277.

Varanasi, U. 1990. Survey of subsistence finfish and shellfish for exposure to oil spilled from the *Exxon Valdez:* The first year. NOAA, NMFS. Northwest Fisheries Center, Environmental Conservation Division, Seattle. Memo F/NWC 191.

Varanasi, U., T. Hom, D. G. Burrows, C. A. Sloan, L. J. Field, J. E. Stein, K. L. Tilbury, B. B. McCain, and S.-L. Chan. 1993. Survey of Alaskan subsistence fish, marine mammal and invertebrate samples collected 1989-1991 for exposure to oil spilled from the *Exxon Valdez*. NOAA Technical Memorandum NMFS-NWFSC. National Technical Info. Services. U.S. Dept. Commerce. 5285 Port Royal Rd., Springfield VA 22161.

Vaughan, S. L., C. N. K. Mooers, and S. M. Gay III. 2001. Physical variability in PWS during the SEA study (1994-1998). *Fisheries Oceanography* 10 (Suppl. 1):58-80.

VECO, Inc. 1989. VECO EVOS hazardous waste cleanup training video. VHS. ARLIS, Anchorage, AK.

Vernberg, W. B., A. Calabrese, F. P. Thurberg, and F. J. Vernberg, eds. 1987. *Pollution Physiology of Estuarine Organisms*. Columbia, SC: Univ. SC Press.

Waldron, J. K. 2001. Stratety to cooperate and minimize financial liability. IOSC 2001, 835-838.

Wang, J., M. Jin, E. V. Patrick, J. R. Allen, D. L. Eslinger, C. N. K. Mooers, and R. T. Cooney. 2001. Numerical simulations of the seasonal circulation patterns and thermohaline structures of PWS, AK. *Fisheries Oceanography* 10 (Suppl. 1):131-148.

Ward, G. 1989. Memorandum to Oil Spill Response Steering Committee. In U.S. Congress House 1989a, 1135-40.

Warheit, K. I., C. S. Harrison, and G. J. Divoky. 1997. EVOS seabird restoration workshop. EVOS Restoration Project Final Report

(Restoration Project 95038). Pacific Seabird Group, Seattle, WA.

Watterson, B. 1992. *Calvin and Hobbes.* Distributed by Universal Press Syndicate. 17 May.

Weidman (formerly Thorne), L. 2001. Former cleanup worker, Santa Clarita, CA. Interview by Riki Ott, 12 February.

Weiser, B. 1989. Toxic waste, court secrets. How the American legal system covers up environmental hazards. *The Washington Post,* 2 April.

Wells, P. G., J. N. Butler, and J. Staveley Hughes, eds. 1995. *EVOS: Fate and Effects in Alaskan Waters.* Philadelphia, PA: ASTM.

Wells, K. and C. McCoy. 1989. Exxon confronted by mutiny in ranks of cleanup workers. Shore crews refuse to spray new chemical on shore. Safety may be concern. *WSJ,* 9 August.

Wertheimer, A. 1998. Fisheries Research Biologist, NOAA/NMFS ABL, Juneau, AK. Interview by Dorothy Shepard, 28 June.

Wertheimer, A. C. and A. Celewycz. 1996. Abundance and growth of juvenile pink salmon in oiled and non-oiled locations of western PWS after the EVOS. AFSS 18:518-532.

Wertheimer, A., R. A. Heintz, J. F. Thedinga, J. M. Maselko, and S. D. Rice. 2000. Straying behavior of adult pink salmon *(Oncorhynchus gorbuscha)* exposed as embryos to weathered *Exxon Valdez* crude oil. *Transactions of the American Fisheries Society* 129:989-1004.

Whitney, D. 1991. Hot washing oily beaches was mistake. NOAA says marine life "cooked." ADN, 10 April.

Wiedmer, M., M. J. Fink, J. J. Stegeman, R. Smolowitz, G. D. Marty, and D. E. Hinton. 1996. Cytochrome P-450 induction and histopathology in pre-emergent pink salmon from oiled spawning sites in PWS. AFSS 18:509-517.

Wiens, J. A. 1995. Recovery of seabirds following the EVOS: An overview. ASTM STP 1219:854-893.

Wiens, J. A. 1996. Oil, seabirds, and science. The effects of the EVOS. *Bioscience* 46(8): 587-597.

Wiens, J. A., R. H. Day, S. M. Murphy, and K. R. Parker. 2001. On drawing conclusions nine years after the EVOS. *Condor* 103:886-892.

Wiens, J. A., T. O. Crist, R. H. Day, S. M. Murphy, and G. D. Hayward. 1996. Effects of the EVOS on marine bird communities in PWS, AK. *Ecology Applied* 6:828-841.

Wiens, J. A. and K. R. Parker. 1995. Analyzing the effects of accidental environmental impacts: approaches and assumptions. *Ecological Applications* 5(4):1069-1083.

Wilkinson, S. L. 1998. Breather beware? Chemical sensitivity may
result from stress, learned behavior, or a new disease process.
Chemical and Engineering News 76(38):57-67.

Willette, M. 1996. Impacts of the EVOS on the migration, growth,
and survival of juvenile pink salmon in PWS. AFSS 18:533-550.

Willette, T. M. 2001. Foraging behavior of juvenile pink salmon
(Oncorhynchus gorbuscha) and size-dependent predation risk.
Fisheries Oceanography 10 (Suppl. 1):110-131.

Willette, T. M., R. T. Cooney, E. V. Patrick, D. M. Mason, G. L. Thomas,
and D. Scheel. 2001. Ecological processes influencing mortality of
juvenile pink salmon *(Oncorhynchus gorbuscha)* in PWS, AK.
Fisheries Oceanography 10 (Suppl. 1):14-41.

Wilson, J. 1991. Accident rate for oil spill cleanup not unusual.
Alaska Economic Trends, September.

Willson, M. F., R. H. Armstrong, M. D. Robards, and J. F. Piatt. 1999.
Sand lance as cornerstone prey for predator populations. In
Robards et al. 1999c, 17-44.

Wohlforth, C. 1989. Admiral wants to water-blast oiled beaches.
Coast Guard commandant's call to action gets a mixed response.
ADN, 15 April.

Wohlforth, C. 1990a. Hot-water spill cleanup kills shore life. Tests
show high death toll for organisms on beach and those under
water near shore. *ADN,* 17 February.

Wohlforth, C. 1990b. Spill scientists frustrated. Researchers say spill
science impedes knowledge. *ADN,* 4 March.

Wolfe, D. A., ed. 1977. *Fate and Effects of Petroleum Hydrocarbons
in Marine Organisms and Ecosystems.* New York: Pergamon
Press.

Wolfe, D. A., M. M. Krahn, E. Casillas, and S. Sol. 1996. Toxicity of
intertidal and subtidal sediments contaminated by the EVOS.
AFSS 18:121-139.

Wood, M. A. and N. Heaphy. 1991. Rehabilitation of oiled seabirds
and bald eagles following the EVOS. IOSC 1991, 235-239.

Wright, B., PhD. 1999. Executive Director, Conservation Science
Institute, Santa Cruz, CA. Interview by Riki Ott, 29 December.

Wright, B., PhD. 2003. Executive Director, Conservation Science
Institute, Santa Cruz, CA. Interview by Riki Ott, 9 July.

Wright, B. A., J. W. Short, T. J. Weingartner, and P. J. Anderson. 2000.
The Gulf of Alaska. In Sheppard 2000, 373-384.

Wuerth, S. 1993a. Fishermen blame spill for disaster, plan protest.
Cordova Times, 19 August.

Wuerth, S. 1993b. Officials seeking antidote for area's ailing econo-
my. *Cordova Times*, 26 August.
Yergin, D. 1991. *The Prize. The Epic Quest for Oil, Money, and
Power*. New York: Simon & Schuster.
Ylitalo, G. M., C. O. Matkin, J. Buzitis, M. M. Krahn, L. L. Jones, T.
Rowles, and J. E. Stein. 2001. Influence of life-history parameters
on organochlorine concentrations in free-ranging killer whales
(Orcinus orca) from PWS, AK. *The Science of the Total
Environment* 281:183–203.
Zador, S. G. and J. F. Piatt. 1999. Time-budgets of common murres at a
declining and increasing colony in Alaska. *Condor* 101:149–152.
Zamzow, K. 2002. Contaminants in wildlife and people of Saint
Lawrence Island, Alaska. A report of the Alaska Community Action
on Toxics, Anchorage, AK. www.akaction.org.
Zarembo, A. 2003. Funding studies to suit need. *Los Angles Times,* 3
December.
Zeh, J. E., J. P. Houghton, and D. C. Lees. 1981. Evaluation of existing
marine intertidal and shallow subtidal biological data. Prepared
by Mathematical Sciences Northwest, Inc., and Dames & Moore
for MESA Puget Sound Project, Office of Environmental
Engineering and Technology, Office of Research and
Development, U.S. EPA. EPA Interagency Agreement No. D6-E693-
EN, Seattle, WA.
Zenteno-Savin, T. and M. A. Castellini. 1998. Plasma angiotensin II,
arginine vasopressin and atrial natriuretic peptide in free ranging
and captive seals and sea lions. *Comparative Biochemistry and
Physiology* 119C(1):1–6.
Zenteno-Savin, T., M. A. Castellini, L. D. Rea, and B. S. Fadely. 1997.
Plasma haptoglobin levels in threatened Alaskan pinniped popu-
lations. *Journal of Wildlife Diseases* 33(1):64–71.
Zhou, T. and H. Liu. 2001. Modeling research of exposure to oil
aerosols during oil spills. IOSC 2001, 413–416.
Zimmerman, S. T., C. S. Gorbics, and L. F. Lowry. 1994. Response activ-
ities. In Loughlin 1994, 23–45.

Index

A

Accu-Chem Laboratory, 74, 81, 92–93, 136, 146–47, 150, 153, 398
Advocacy science, 407–13, 415, 433–36. *See also* Exxon studies; *How to Lie with Statistics;* Statistical tricks
Alaska Coastal Current, 185, 278–80, 283, 323, 328
Alaska Community Action on Toxics (ACAT), 38, 118, 164, 466
Alaska Department of Environmental Conservation (ADEC), 4–5, 11, 29, 30, 32, 34–35, 89, 97–98, 102, 175, 243, 477n1 (chap. 15)
Alaska Department of Fish and Game (ADFG), 245–46, 249, 264, 290, 300, 425.
Alaska Department of Health and Social Services, 27, 136
Alaska Department of Labor (ADOL), 27–28, 35, 128–30, 162, 165, 452–54, 474n2 (chap. 8). *See also* Alaska Workers' Compensation Board
Alaska Department of Law, 4, 7
Alaska Forum for Environmental Responsibility (AFER), 15–16, 38, 164
Alaska, Gulf of, 35, 336
oil spill studies and, 231–39, 246–48, 317–27, 337–341
physical oceanography of, 274, 277–78, 280–82, 288–91, 320–323, 326–29, 340
Alaska Injured Workers' Alliance, 131, 439, 459–60
Alaska Predator Ecosystem Experiment (APEX). *See* Ecosystem studies, on apex predators.
Alaska Public Interest Research Group (AkPIRG), 131–33, 439
Alaska Workers' Compensation Board. *See also* Alaska Department of Labor; OSHA injury/illness codes
coding of injury/illness claims, 128–30
deficiencies in, 131–33, 436–37, 439–40
injury/illness claims filed with, 35, 125–35, 142–43, 148, 439–40,
451–54, 474n1 (chap. 8)
recommendations to improve process, 459–60
Alyeska, 3, 14, 21–24, 186, 196, 278, 365
Amount of oil spilled, 4–7, 418, 473n1 (introduction)
recommendation to require independent estimates of, 411–12
Anderson, Paul, 238–39, 320–23, 461
Anemia. *See* Biomarkers; Illness symptoms
Apex predators, **238**–39, 277, 317–18, 323, 328, 334, 339–41, 395, 402. *See also* Ecosystem studies, on apex predators
Aromatic hydrocarbons, 21, 395, 475n1 (chap. 11). *See also* PAHs (Polycyclic aromatic hydrocarbons)
biomarkers and, 296, 309
ecological health concerns with, 182–84, 192, 256, 352–53, 393, 399, 407, 447
public health concerns with, 62, 296, 309, 389, 447
signatures (oil-weathering model), 346–49, 353, 355, 368
Arrhythmia. *See* Illness symptoms
Ashford, Nicholas, 168, 174–75, 406
Asthma. *See* Illness symptoms
Auke Bay Lab (NOAA), 181–85, 191–95, 198, 256, 259, 343–55, 361–65, 409, 422, 430, 435, 479n2 (chap. 23).

B

Baseline studies
Gulf Ecosystem Monitoring (GEM) Program, 287, 341, 403, 461
importance of, 201–02, 333, 383, 401, 403, 430
in Prince William Sound, (as pre-spill surveys), 122–23, 184–86, 226, 237–38, 251–53, 305, 329, 337, 364–65
Beach (intertidal zone). *See also* Sediments
amount of oiled, 6, 8, 202, 362
ecology, 10, **201**, 202, 208, 360
residual oil found on, 195, 207, 214,

303–4, 313, 315, 343, 348–50, 356, 358, 361–64, 369, 375–379, 383, 403
status of oiled: in 1993, 205–07, 214; in 2003, 375, 378, 381–82
studies, 185, 191, 195, 201–14, 244, 258–59, 311, 348, 350, 359–69, 475n2 (chap. 13)
wildlife uses of, 23, 226, 240, 242, 250–51, 309, 344, 377, 379, 383, 395, 402
Belli, Melvin, 75, 139–42, 152, 154
Benzene, 27, 45, 61, 65, 117, 182, 256, 389, 397, 411.
See also Toxic chemicals
Biggs, Evelyn, 259–66, 287–89, 462.
See also Brown, Evelyn
Bingham, Eula, 28, 34, 171
Bioaccumulate, 185, 194, 295, 354, 377.
See also Bioavailable; PBT pollutants
Bioassays, 183.
See also Oil toxicology
use of, in oil spill studies, 195, 197, 314
vs. ecosystem science, 399, 430
water quality standards and, 183–84, 192, 430
Bioavailable, 256, 315, 346, 354, 363–65, 368, 388, 393, 395, 414.
See also PAHs; PBT pollutants
Biomarkers, 136
analysis of, 136–37, 296, 309–13
cytochrome P450-1A, 256, 296–97, 303, 307, 312, 343, 399
emerging science and, 396–98
EROD (liver enzyme), 297
fatty acid signatures (lipid composition), 332–33, 335, 339, 399, 478n2 (chap. 19)
genetic makeup, 216, 222, 227, 259, 287, 336, 339, 341, 349
haptoglobin, 309–11, 313
hemoglobin, 111, 310–11, 313, 465
interleukin-6, 309, 311, 313
porphyrins, 311, 313, 399
use of, to monitor health, 396, 414: in humans, 136–37, 396–98; in wildlife, 137, 296, 308–9, 313, 398–400
Bioremediation (for toxic waste cleanups), 36, 97, 114, 196, 365.
See also Inipol
Bioremediation Application Team (BAT), 99, 103–06, 108–110, 165
Birds. See Marine birds
Black-legged kittiwakes.
See also Ecosystem studies, on apex

predators; Marine birds
ecology of, 231, 239, 247, 304, 318, 324–31
studies of spill effects on, 340–41, 382
Black oystercatchers, 244, 328, 331, 374–75, 377
Blood disorders. See Illness symptoms
Blundell, Gail, 312–13
Bodkin, Jim, 300–4, 461
Bowyer, Terry, 308–12, 461
Brady, James, 249, 252, 259
Brain fog. See Illness symptoms
Brown, Evelyn, 287–89, 320, 350, 462.
See also Biggs, Evelyn
Bue, Brian, 254, 257–59, 343
Bue effect, the, 343, 346, 350, 352, 356
Buried oil. See Residual oil

C

Caleb Brett (contractor), 4, 5, 411–12
Cancer. See Illness symptoms
Capelin, 305–6, 322–24, 328–30, 332, 340, 378, 383.
See also Forage fish
Carls, Mark, 351–54, 357–58, 462
Centers for Disease Control and Prevention (CDC), 63, 436–37
Central nervous system (CNS) problems.
See Illness symptoms
Chemical-induced illnesses.
See also Chemical sensitivity; Toxic chemicals
dismissed claims for, 131–38
legal/medical system and, 140–54, 170–75
military veterans and, 75, 93, 100, 133, 146, 165, 168, 165, 175, 196
misdiagnosis of, xvii, 93, 141
OSHA injury/illness codes and, 128–30
recommendations to help prevent, 436–40
from solvents, 149
Toxicant-Induced Loss of Tolerance (TILT), 159, 168–70, 397.
Chemical sensitivity.
See also Chemical-induced illnesses; Illness symptoms; Limbic kindling
about, 74–80
gaining recognition for, 167–70, 397–98
misdiagnosis of, as psychiatric problem, xvii, 81, 94, 141, 148–51
possible mechanisms of, 168–70
treatment for, 74–75, 80–83, 94, 117

Citra-Solv, 114–15, 118, 143, 166, 390
CitroKleen, 46, 72, 143, 166, 390
Clams
    ecology of, 202, 208, 219, 240, 302, 309,
        377, 402, 475n3
    oil spill and cleanup effects on, 187, 212,
        219, 244, 360, 363, 377
Clarke, Sara (pseudonym), 113–18, 462
Cleanup.
    See also Bioremediation; Dispersants;
        Ecological damage; Exxon Worker
        Safety Program; Illness claims;
        Illness symptoms; Pressurized hot
        water wash; Recommendations
    controversy over: bioremediation effec-
        tiveness, 97–101, 424; dispersant
        use, 15, 22–23, 32, 98, 100, 204;
        miles of beaches 'treated,' 29–31,
        35; oil removal during, 36; pres-
        surized hot water wash, 6, 31–32,
        34, 36, 203–7, 359–60, 381
    expenses, 36–37, 412, 419–20
    health risk concerns, during, 27–29,
        44–51, 62–63, 99, 135–36, 389–90
    leaving the spiller in charge of, 27–29,
        34, 206, 419–21
    logistics, 11–15, 21–24, 36
    toxic chemicals present during, 12–13,
        159–64 (*see also individual chemi-
        cals and products*)
Cleanup workers.
    See also Exxon Worker Safety Program;
        Illness claims; Illness symptoms
    need for continued monitoring of, 53,
        125, 175–76, 437–39
    perspectives of, on health risks during
        cleanup, 28, 35, 45–46, 48–52, 87,
        100, 104, 115, 118–19
    recommendations for injured, 440 (*see
        also* Recommendations)
Cleanup workers' health survey, 159,
        164–67, 391–92, 405, 474n2 (chap.
        10).
    See also Epidemiology studies
Clean Water Act (CWA), 182–3, 186, 206,
        429–31.
    See also Bioassays; Water quality stan-
        dards
Climate regime shift. *See* Pacific Decadal
        Oscillation
Coal
    fields, 185, 197, 364, 368
    in marine sediments, 185–86, 365,
        367–68
    vs. oil controversy, 185–86, 364–365,
        367–369 (*see also* Aromatic hydro-

carbon signatures)
Coastal ecology. *See* Beach ecology
Comprehensive Environmental Response,
        Compensation, and Liability Act
        (CERCLA) or " Superfund," 206–7,
        373, 429–30
Confidential/confidentiality.
    See also Lawsuits; Toxic torts
    cleanup workers and, 56, 58, 67, 71, 156,
        174
    public interest and, 56, 126, 160, 174,
        206, 408, 440
    spill science and, 8–9, 206, 220, 236, 256,
        258, 267, 348, 408
Controversy.
    See also Amount of oil spilled; Cleanup;
        Dispersants; Exxon studies
    background for, of spill science, 198–99,
        408–9
    over ecological damage/recovery, 187,
        206, 246, 286, 328, 351, 355, 366,
        476n5, 478n3 (chap. 19), 479n2
        (chap. 23)
    over PAH source (coal vs. oil), 185–85,
        198, 364–365, 367–369
    over use of water quality standards: as
        safe for wildlife, 197, 289, 313–14,
        354–56; as unsafe for wildlife, 345,
        348–49, 352, 377, 400, 431
    recommendations to help forestall,
        433–36
    use of, to manipulate public policy, 353,
        408–10, 412, 415
Cook Inlet
    oil spill studies in, 319–20, 323–26
    physical oceanography of, 321
Cooney, Ted, 273–76, 280–87, 291, 462
Cordova District Fishermen United
        (CDFU), 1–3, 10, 15, 24, 193
Corexit, 12, 15, 22–23, 32–34, 61, 98, 100,
        105, 134–35, 159–160, 163, 170,
        204, 390, 420, 424, 473n1 (chap.
        1).
    See also Dispersants; Solvents; 2-
        butoxyethanol
Crude oil.
    aerosols (PAHs), 17, 27, 44, 47, 53, 58,
        60, 62, 64–65, 88–89, 93, 130, 134,
        160–61, 165, 172, 389–92, 396–97,
        404, 412
    health hazards from, 12, 27, 44, 51, 53,
        57, 87, 136, 159–67, 217, 256,
        390–92, 397–98
    mists, 26, 42, 44, 49–50, 55, 58, 60–65,
        87–89, 103, 115, 130, 159–67,
        389–92, 397–98, 404

vapors and VOCs (volatile organic carbons), 12, 21, 44–45, 57–58, 64–65, 72, 87–88, 130, 160–61, 166, 224–25, 391, 393, 398, 429
water soluble fraction, 182–84, 186, 346–47, 352, 392, 396, 399, 475n1 (chap. 11)
weathered, 44, 51, 60, 120, 136, 160–61, 195, 243, 256, 314, 344–48, 352–54
Cullen, Mark, 164
Cytochrome P450-1A. *See* Biomarkers

D

Damage assessments. *See* Natural Research Damage Assessments (NRDA)
Daubert v. Merrel Dow Pharmaceuticals (1993), 153, 155, 171–73, 411, 435. *See also* Toxic torts
DDT, 2–3, 336, 341, 388.
Demand-side economics, 440–44
Depression. *See* Illness symptoms
Dermatitis. *See* Illness symptoms
De-Solv-It, 46, 60–1, 64, 143, 146, 160, 166, 390
Diesel exhaust/fumes, xvi, 42, 47, 56–7, 64–65, 59–60, 74, 77, 80, 82, 107, 144, 159, 168
Dispersants.
See also Corexit; Inipol
authorization to use, 15, 206, 423–24
controversy over use of, 10, 15, 22–23, 32, 34, 98, 100, 204
disclaimers, 33, 423, 479n3
effectiveness of, 14, 422–25
guidelines for use of, 10–11, 22–23, 32, 422–27
health risk concerns of: public 32–33, 51, 100, 159, 162, 164–65, 404, 422–23, 425–27; wildlife, 99, 204, 243, 473n1 (chap. 1), 477n1 (chap. 15)
possible chemical poisoning incidents, 32–34, 99–100
Dizziness. *See* Illness symptoms
Driskell, Bill, 201, 203–4, 462
Ducks. *See* Harlequin ducks.
Duffy, David, 319, 463
Duffy, Larry, 309, 312–13
Dyssynchronicity, 62–63, 390

E

Ecological damage from oil spill.
See also Ecosystem studies; Natural Resource Damage Assessment
to beach communities, 195, 201–14, 303–4, 313, 315, 343, 348–50, 356, 358, 361–64, 369, 375–379, 383, 403: clams, 187, 212, 219, 360, 363, 377; mussels, 219, 240–44, 363, 377, 393
to fish (*see* Pacific herring; Pink salmon)
to marine birds, 231–48 (*see also* Black-legged kittiwakes; Black oyster-catchers; Harlequin ducks; Murres; Pigeon guillemots
to marine mammals (*see* Harbor seals; Orcas; River otters; Sea otters)
unanticipated effects (delayed, indirect, lingering), 343–50, 373–83, 392–96, 432
Ecological recovery from oil spill, 227, 237, 299–300, 304, 306, 308, 315–16, 349, 356, 373–82
residual oil and, 361–64
unknown status of, 382–83
Economic impacts of disrupted fisheries, 275, 376, 380
Ecosystem-based management, 291, 341, 403, 429–31
Ecosystem studies, 6–8, 271, 291, 365, 383, 393, 399, 401, 403
on apex predators, 317–41, 402
fishermen's blockade and, 275
NVP Project, 293–316, 402
SEA Program, 273–91, 402
Endocrine disruption. *See* Illness symptoms
Environmental medicine, xv, 75–76, 113, 153, **406**
Environmental pollutants.
See also PAHs; PBT pollutants; Toxic chemicals
human health and, xvii, 404–6
indoor air pollution, 144, 147, 167–68, 396, 406
marine ecosystems and, 401–4
monitoring effects of, 396–400, 404–6, 414, 478n1 (chap. 23)
Epidemiology studies, 404–6, 478n1 (chap. 23).
Esler, Dan, 297–300, 315, 463
Ewing, George, 92, 95, 151–52, 154–55, 474n2 (chap. 9)
Exxon.
See also Controversy; Exxon Worker

Safety Program
contractors (*see* Caleb Brett; Med-Tox;
VECO)
cost of spill to, 36–37, 412, 419–20,
473n3
disclaimer, 32–33
liability of, from oil spill damages to:
cleanup workers' health, 33, 67,
143, 412, 415 (*see also* Toxic torts);
private parties, 198, 408–9, 474n1
(chap. 9) (*see also* The *Exxon
Valdez* Case); public resources, 9,
408–9 (*see also* 1991 civil settlement; Reopener for Unknown
Injury)
studies of: seawater, 196–98; sediment,
364–65; beach ecology, 208–14;
fish, 266–68; 337–38; marine birds,
246–48; marine mammals, 228
(*see also How to Lie with Statistics;*
Statistical tricks)
Exxon Valdez Case, The, 142, 474n1 (chap.
9)
Exxon Valdez oil spill.
See also Controversy; Ecological damage;
Ecological recovery; Illness claims;
Illness symptoms
legacy of, 387–415
social ramifications of, 417–444
spread of oil from, xx
volume dispute, 4–7
Exxon Valdez Oil Spill (EVOS) Trustee
Council, 9, 275, 294–95, 317–18,
334–35.
Exxon Valdez Oil Spill Legacy Educational
Fund (EVOS ELF), vi, 432–33, 444
Exxon Worker Safety Program.
See also Illness claims; Illness symptoms;
NIOSH Health Hazard Evaluation
air quality monitoring program, 45–46,
49, 56, 58, 103, 137, 389–90, 392,
412, 440, 450
clinical data, 125–27 (*see also* Upper
Respiratory Infections)
failings of, 39–53, 390, 438
federal involvement in, 28, 34, 135–137,
392, 438
risk of exposure to dangerous chemicals
despite, 59–67

F

Faro, Jim, 309, 312–13
Fatigue. *See* Illness symptoms

Fat (lipid)-soluble compounds, xvii, **143**,
149, 193, 346.
Fetal damage. *See* Illness symptoms
Fish.
See also Capelin; Forage fish; Pacific herring; Pink salmon; Sand lance
oil toxicity and, 343–58
shifts in population of, 238–39, 321–23,
340, 378, 381–82, 478n2 (chap. 20)
Fishermen's blockade, 274, 275
Flu-like symptoms. *See* Illness claims
Forage fish/species, 238, 247, 266, 276,
285, 305–6, 315, 318–34, 339–41,
378, 381–83
Freedom of Information Act, 360, 409,
435–36
Frost, Kathy, 224, 225–28, 233, 331–34,
338, 463
Fucus (rockweed, sea plant).
See also Beach; Ecological damage
ecology of, 202–3
status of: in 1993, 205; in 2003, 375, 378,
381–82
studies of spill effects on, 6, 26, 30–31,
201–5, 206–12, 359–60, 366, 392,
394–95, 475n1 (chap. 13)

G

Gag orders. *See* Confidential/confidential-
ity
Gay, Shelton, 276–80, 463
Genetic damage. *See* Illness symptoms
Golet, Greg, 304–8, 330–31, 463
Gulf Ecosystem Monitoring (GEM) pro-
gram, 287, 341, 403

H

Haptoglobin. *See* Biomarkers
Harbor seals.
See also Ecosystem studies, on apex
predators; Natural Resource
Damage Assessment
ecology of, 222, 226, 238–39, 317–18,
331–34, 336
estimated number of killed by spill, 226
studies of spill effects on, 224–28,
338–39, 347, 393, 399
status of, in 2003, 375, 379, 381
Harlequin ducks.
See also Ecosystem studies, NVP Project;

Marine birds; Sentinel species
 ecology of, 241, 295, 297, 304–5, 314,
   328
 status of: in 1993, 248; in 2003, 375, 379
 studies of spill effects on, 6, 239–45, 248,
   264, 297–300, 315, 331, 349, 364,
   393, 401, 403, 405, 432, 477n1
   (chap. 15)
Harris, Pat, 193–96
Hazardous waste cleanups.
 See also Cleanup
 oil spills as, 11, 27–29, 389, 404
 recommendations for, 415, 419–22,
   422–27, 430, 437–39
 training for, 26, 28–29, 32, 44, 48–49, 72,
   100, 118, 120–21, 412
Headaches. See Illness symptoms
Health Hazard Evaluation. See National
   Institute of Occupational Safety
   and Health (NIOSH)
Heintz, Ron, 344–47, 349, 463
Hemoglobin. See Biomarkers
Herring. See Pacific herring
Hickel, Walter, 244, 264, 275, 312
Holland, Russel, 142, 155–56, 410, 433,
   474nn1–2 (chap. 9)
Holland-Bartels, Leslie, 293–97, 301, 304,
   308, 315–16, 464
Hong, Steven, 91–94, 148–51
Hose, Jo Ellen, 262, 353
Houghton, Jon, 201–5, 207–8, 211–12,
   359–61, 366, 464
How to Lie with Statistics (Huff), 197, 228,
   314, 367, 407, 435.
 See also Statistical tricks
 errors of omission in, and Exxon studies,
   211–12, 228, 339, 248, 353–54,
   356, 357, 365, 368–69
 improper choice of controls in: and
   Exxon studies, 212–13; and public-
   trust science studies, 281
 incorrect inferences in, and Exxon studies, 197, 213, 248, 289, 313–14,
   354, 355, 356, 365, 367, 476n4
   (chap. 13)
 semi-attached figure in, and Exxon studies, 267, 314, 366, 368
 small sample bias in, and Exxon studies,
   209–10, 247–48, 266, 476n3 (chap.
   13), 477n3
 well-chosen average in, and Exxon studies, 210–11, 197, 266–68, 354, 407,
   477n4
Huff, Darrell, 197, 199, 209–10, 212–14,
   228, 247, 266–67, 314–15, 338–39,
   367–68, 407–8, 412

I

Illness claims.
 See also Alaska Workers' Compensation
   Board; Exxon Worker Safety
   Program; OSHA injury/illness
   codes; Toxic torts
 dismissal of, 131–38,142–43, 148,
   170–75, 410–11
 as exemptions to OSHA reporting
   requirements, 66, 391
 records of: with Alaska Workers'
   Compensation Board, 35, 125,
   127–35, 452–54; with Exxon,
   125–27 (see also Upper Respiratory
   Infections)
 underreporting of possible, 47–48, 58,
   66, 127, 130, 137, 391, 412, 451
Illness symptoms.
 See also specific chemicals and products;
   Chemical-induced illnesses;
   Environmental pollutants
 anemia, 11, 111, 162–64, 389–90
 arrhythmia, 79, 117
 asthma, xvi, 12–13, 47, 53, 55, 59–62, 77,
   116, 134, 166–167
 blood disorders, 11–13, 28, 53, 162–63,
   169, 390
 brain fog, 91, 110, 117, 133
 cancer (and carcinogen), xvi, 12, 17,
   27–28, 56, 59–61, 93, 121, 140,
   161–62, 362, 388–89, 396–97, 398,
   406, 473n2
 central nervous system (CNS), 11–13,
   27–28, 99, 121, 129–30, 143,
   162–64, 165–67, 224, 256, 304,
   347, 390, 393, 396, 405, 439
 chemical sensitivity, 39, 59, 74–83, 92,
   108, 133, 151
 depression, 72, 77, 79, 99, 121, 133,
   167–68
 dermatitis, 11–13, 53, 61, 115, 162, 390
 diabetes, 92–94
 dizziness, 11–12, 23–24, 27–8, 74, 87, 93,
   99, 115, 130, 175, 224, 347, 399,
   439
 endocrine disruption, 93, 117, 162–63,
   169, 173, 389, 396–97, 436–37
 fatigue (tiredness), 76, 78–80, 102, 111,
   117, 160, 168, 310 (bind heme)
 fetal damage, 11, 162, 169, 390
 flu-like symptoms, 58, 66–67, 91, 107,
   116, 120–21, 130, 147, 150
 genetic damage, 61, 262–63, 343, 362,
   352, 388
 headaches, 11–12, 24–26, 33, 50, 72, 77,

79, 88, 91–4, 99–100, 115, 121, 130, 133, 147–48, 150, 160, 175, 392, 425, 439
immune system: normal function of, 169, 309, 436; suppression of, 60, 91, 162–63, 264, 336, 351, 357–58, 379, 389, 437
kidney damage, 11–13, 27–28, 53, 81, 99, 117, 162–63, 169, 256, 390
leukemia, xvi, 11–12, 28, 162–63, 390, 411
liver damage, 11–13, 27–28, 53, 81, 93–94, 99, 117, 162–64, 169, 224, 256, 264, 304, 390
media reports of, 126
memory loss, 121, 146–47, 166, 168
mood swings, 72–73, 77–78, 80, 143, 160
nausea, 11, 27–28, 74, 91, 100, 115, 150,160, 175, 392, 425
from other oil spills, 160–61
rash (skin), 24–26, 51–52, 79, 82, 88, 93–94, 100, 107, 160, 163, 168
respiratory problems, 12–13, 27, 32–34, 42–43, 46–48, 57–58, 60, 92, 107, 115, 126, 128–130, 143, 146–47, 162–66, 217, 390–92, 398, 405
sore throat, 33, 88, 91, 94, 160, 392
Immune system suppression. *See* Illness symptoms
Industrial hygiene, 43–48, 59, 67, 130, 154
Inipol (EAP 22), 36, 204, 420.
See also Bioremediation; Dispersants; Solvents; 2-butoxyethanol
approval process for use of, 15, 97–111, 121
dispute over effectiveness of, 97–98, 100–1, 120, 473n1 (chap. 6)
health concerns about, 12–13: public, 404, 425–26; wildlife, 99, 243, 477n1 (chap. 15); worker, 51–52, 60–61, 64, 87–88, 99–100, 103–11, 114–15, 118–21, 132, 159, 162–64, 166, 170, 205, 390, 422–26, 438, 477n1 (chap. 15)
workers' stories about, 101–11, 113–21 (*see also* Clarke; Lange; Moeller; Nagel)
Interleukin-6. *See* Biomarkers
Intertidal zone, **202**. *See* Beach
Irons, Dave, 318–19, 328–31, 338, 464
Isleib, Pete, 221, 239–41, 245, 305, 328–29
Iverson, Sara, 331–33, 464

K
Karinen, John, 183, 191
Katalla oil seep, 185, 364–65, 368–69. *See also* Coal, vs. oil controversy
Keystone species, 211, 257, 324–25, 328, 402
Kidney disorders. *See* Illness symptoms
Kittiwakes. *See* Black-legged kittiwakes
Kocan, Dick, 262–63, 350, 358
Kuletz, Kathy, 304–5, 330–31

L
La Joie, Phyllis "Dolly," 45, 46, 50, 85–95, 132, 148, 166, 169, 464 *See also* Toxic Torts, *Roberts v. Exxon* (1999)
Lange, Evan (pseudonym), 102–3, 108–111, 464–65
Laws. *See also individual laws;* Recommendations
Clean Water Act, 182–83, 186, 429–30
Occupational Health and Safety Act, 11, 63, 93, 126, 144, 437–40
Oil Pollution Act of 1990 (OPA 90), 207, 418–20, 429–30
Superfund (Comprehensive Environmental Response, Compensation, and Liability Act), 206–7, 373, 429–30
Lawsuits. *See also* Confidentiality; 1991 civil settlement; Reopener for Unknown Injury; Toxic torts
Exxon Valdez Case, The, 142, 474n1 (chap. 9)
State of Alaska v. Exxon (1991), 9
U.S.A. v. Alaska (1991), 9
U.S.A. v. Exxon (1991), 9
Lees, Dennis, 201–4, 207, 360, 465
Leukemia. *See* Illness symptoms
Liability. *See also* Controversy; Dispersants; Reopener for Unknown Injury
associated with dangerous chemicals, 140, 172, 422–23, 426–28
for oil spill damages to wildlife, 198, 206, 225, 369, 408–9
for workers' health, 33, 143, 392, 412, 415, 422
use of, to deter spills, 418–21
Limonene, 46–47, 61, 115, 390

Liver disorders. *See* Illness symptoms
Lowe, Randy, 71–78
  See also Toxic torts, *Payne v. Exxon*
    (1997)
Lowry, Lloyd, 224, 225–28, 331–34, 338,
  463

M

Marine birds, 6, 231.
  See also Black-legged kittiwakes; Black
    oystercatchers; Harlequin ducks;
    Murres; Pigeon guillemots
  shifts in populations of, 237–39, 304,
    317–20, 322–23, 328–31, 337–38,
    381
Marine ecosystems.
  See also Alaska, Gulf of; Cook Inlet;
    Prince William Sound
  climate regime shifts and, 321, 326, 340,
    373, 381, 395, 399, 402–3
  environmental disturbances and, 247,
    341, 373, 402–3
  environmental pollutants and, 401–4
Marine mammals, 199, 215, 220, 225–26,
  232, 276, 318, 331.
  See also Harbor seals; Orcas; River otters;
    Sea otters
  shifts in populations of, 226, 238–39,
    322–23
Marty, Gary, 262–63, 350, 353, 465
Material Safety Data Sheet (MSDS), **11**
  excerpts from, 12–13, 53, 99, 104–5, 115,
    149, 424
  workers and, 11, 22, 49, 103, 108, 111,
    114, 119, 121, 139
Matkin, Craig, 221–25, 334–36, 465
Med-Tox, 56, 58, 64–65, 89, 103, 389
Meggert, Ed, 30, 32–33, 35, 98, 133–35,
  465
Memory loss. *See* Illness symptoms
Merrill, Richard, 139, 143
Mestas, Dennis, 56, 59, 465–66
Methylpentane, 75, 81, 92, 146–47
Miller, Pam, 39, 53, 55, 67, 113, 115, 125,
  130–31, 139, 159–62, 164, 167,
  170, 174–75, 405, 466
Moeller, Don, 101–8, 166, 466
Monnett, Chuck, 2, 215–21, 228, 334,
  476n1
Mood swings. *See* Illness symptoms
Murres.
  See also Ecosystem studies, on apex
    predators; Natural Resources

Damage Assessment
  ecology of, 232, 297, 306, 317–18,
    323–27, 330, 395
  estimated number killed by spill,
    236–37, 326–27
  status of: in 1992, 248; in 2003, 326–27,
    375, 377
  studies of spill effects on, 6, 231–39,
    246–48, 331, 337, 402
Mussels.
  See also Keystone species; Sentinel
    species
  ecology of, 202, 208, 210, 240, 242, 244,
    298, 302
  oil contamination of, 31, 219, 241–44,
    363, 377, 393 (*see also* Oiled food)
  status of: in 1993, 243–44; in 2003, 375,
    378
  studies of, 191, 193–95, 206, 353–54,
    363–64, 475n1 (chap. 13)
  use of, avoiding use of, to track oil pollu-
    tion, 185, 194, 197, 199, 243–44,
    260–63, 268, 353–54, 363–64, 378

N

Nagel, Richard, 118–21, 466
National Institute of Occupational Safety
  and Health (NIOSH), 28
  failure to obtain Exxon's records, 34, 67,
    126, 137–38, 176, 404, 438, 440
  Health Hazard Evaluation, 39, 46–48,
    135, 137, 474n1 (chap. 8)
  monitoring of cleanup workers by, 34,
    44–46, 48, 50–52, 60, 135–37
  recommended exposure levels (RELs),
    62–63, 65, 389–90, 414, 473n1
    (chap. 3)
National Marine Fisheries Service
  (NMFS). *See* U.S. National
  Oceanographic and Atmospheric
  Administration (NOAA)
National Research Council, 10, 22, 170,
  232, 395, 417–18, 423, 479n1, n2
  (chap. 24)
Natural Resource Damage Assessment
  (NRDA): of beach ecology,
  201–14; of fish, 252–59, 260–64; of
  marine birds, 233–37; of marine
  mammals, 216–220.
  See also Exxon studies
  confidentiality of, 9, 206, 267, 408
Nausea. *See* Illness symptoms
Nearshore Vertebrate Predator (NVP)

Project. *See* Ecosystem studies
1991 civil settlement (for damages to
    wildlife and habitat), 9, 198, 220,
    335, 343, 369, 380, 436–37, 412,
    431, 433, 447.
  See also Lawsuits; Reopener for
    Unknown Injury

O

Occupational asthma. *See* Illness symp-
    toms
Occupational medicine (and physicians),
    40–43, 47, 52–53, 59,, 92, 121, 130,
    143, 150, 397, 438–39
Occupational Safety and Health Act
    (OSHA), 63, 93, 128, 144.
  See also Illness claims; Permissible expo-
    sure limits
  cleanup monitors (inspectors), 45, 89,
    126
  exemption to reporting work-related
    injuries/illnesses, 66, 391
  injury/illness codes, 62–65, 93, 128–31,
    144, 389–91, 397–98, 414, 473n1
    (chap. 3) (*see also* Permissible
    Exposure Limits)
  recommendations for, 176, 436–40,
    459–60
  underreporting of possible work-related
    injuries/illnesses during cleanup,
    47–48, 58, 66, 127–28, 130, 137,
    391, 412, 451
Oil politics
  demand-side economics of, 440–44
  supply-side economics of, 419–440
Oil mist. *See* Crude oil mist
Oil pollution
  amount/sources of, in North America,
    417–19
  prevention of, 417–44
Oil Pollution Act of 1990, 207, 418–19,
    429–30
Oil Spill Liability Trust Fund, 420–21, 427
Oil spills
  federalizing cleanup of, 419–21
  financial liability for, 418–19
  recommendations to deter, 411–13, 415
Oil toxicology
  changing paradigms in, 349–350, 363,
    387–415
  fish studies and, 343–58
  human health and, 388–92, 396–98
  social ramifications of (*see*

Recommendations)
  wildlife health and, 392–96, 399–406
O'Neill, Annie, 164–67, 391–92
Orcas.
  See also Ecosystem studies, on apex
    predators; Natural Resource
  Damage Assessment
  ecology of, 221–22, 227, 395
  estimated number missing after spill,
    229
  status of: in 1993, 229; in 2003, 375,
    378–79, 381
  studies of spill effects on, 6, 223–25,
    334–36, 339–41
Otters. *See* River otters; Sea otters
Oystercatchers. *See* Black oystercatchers

P

Pacific Decadal Oscillation, 321, 326, 340,
    373, 381, 395, 399, 402–3
  effect on wildlife of: alone, 321, 323, 395
    (*see also* Fish; Marine birds;
    Marine mammals); in concert
    with other disturbances, 247, 341,
    373, 402–3
  studies on, 320–23
Pacific herring.
  See also Ecosystem studies, SEA
    Program; Forage fish; Keystone
    species
  ecology of, 202, 250–51, 277, 283,
    285–91, 305–6, 329–30, 332, 340,
    381
  economic impacts of disrupted fisheries,
    380
  status of: in 1993, 269; in 2003, 375, 379
  studies of spill effects on, 6, 15, 259–68,
    291, 350–55, 357–58, 395
PAHs.
  See also Crude oil; Cytochrome P450-1A;
    Ecological damage; Illness symp-
    toms; PBT pollutants;
    Petrochemical problem
  breakdown of, by vertebrates, 296–97,
    343–44
  oil toxicity and, 398–99
  parts per billion: as Alaska water quality
    standard, 184, 345, 475n1 (chap.
    11); biomarker sensitivity to, 137;
    as federal water quality guideline,
    184, 345, 431; in mussel tissue,
    354; in North America rivers, 418,
    479nn1–2 (chap. 24); as unsafe for

wildlife, 184, 363, 345, 347, 349, 352, 355, 393–95, 400, 414, 431
parts per million: in intertidal shellfish, 244; in mussels, 243–44; in sediments, 243–44, 399–400; in water soluble fraction, 183–84, 392, 394; in weathered crude oil, 136
parts per thousand: in sediments, 244
parts per trillion: as recommended standard, 431; in seawater, 344; as unsafe for wildlife, 263, 352, 431
as PBT pollutants, 388–89, 428–29
symptoms of overexposure to, 159–67
monitoring, during cleanup, 44–45 (see also Permissible exposure limits; Exxon Worker Safety Program)
recommendations for water quality standards for, 431
Paradigm shifts, xv–xvi, **388**, 406, 410, 414–15.
examples of: chemical sensitivity, 153, 167, 396–98; ecosystem studies, 286; oil toxicology, 349–50, 363, 392–96, 398, 403, 409; world is flat, 388
public interest and, 406–13, 415
Patrick, Vince, 280, 284, 286, 466
Patten, Sam, 239–46, 467
Payne v. Exxon (1997). *See* Toxic torts
PBT (persistent, bioaccumulative, and toxic) pollutants, 388–89, 395–96, 400, 428–29
PCBs, xvi–ii, 144, 296, 309, 336, 341, 375, 381, 388, 405, 478n1 (chap. 23)
Permissible exposure limits (PELs) (OSHA).
See also Public health; Recommended exposure limits
extended work hours and, 62–64, 397
maximum exposure levels and, 64–66, 389–90, 450
use of surrogate for compound of concern, 62, 397
vs. recommended exposure limits, 63–64, 389–391, 473n1 (chap. 3)
voluntary strengthening of, 63, 390–91
Personal injury lawsuits. *See* Toxic torts
Personal protective equipment (PPE), 24, 27, **45**, 50–52, 77, 99, 109, 427.
See also Exxon Worker Safety Program
Peterson, Charles "Pete," 199–14, 353, 360, 366, 467
Petrochemical problem. *See* Chemical-induced illnesses
Phytoplankton. *See* Plankton
Plankton, 267, 273–74, 276, 280–91

Piatt, John, 231–39, 318–27, 377–78, 495, 467
Pigeon guillemots.
See also Ecosystem Studies, NVP Project; Marine birds; Sentinel species
ecology of, 295, 305–6, 317, 324, 330–31, 363
status of, 375, 379, 381
studies of spill effects on, 234, 307–8, 315, 331, 337–38, 393, 401
Pink salmon.
See also Bioassays; Ecosystem studies, SEA Program; Forage fish; Natural Resource Damage Assessment
ecology of, 202, 222, 250–51, 274, 280–87, 290–91, 321, 332–33, 4774n1 (chap.16)
economic impacts of disrupted fisheries, 275, 376
status of: in 1993, 268; in 2003, 374–77
studies of spill effects on, 6, 15, 256–59, 266–67, 275, 290, 343–350, 352, 355–56, 393, 401, 478n1 (chap. 20)
wild and hatchery, differences between, 250–51, 290
Polycyclic aromatic hydrocarbons. *See* PAHs
Porphyrins. *See* Biomarkers
Pressurized hot water wash 14–15, 31–32, 44, 204.
See also Crude oil; Exxon Worker Safety Program; Illness claims; Illness symptoms; PAHs
approval of, 28, 31
creation of PAH aerosols by, 44, 160, 391
ecological damage from, 6, 31, 36, 203–7, 359–60, 381
human health concerns with, 32, 45, 49–50, 60, 65, 165, 389, 391, 420
Prince William Sound, xxviii (map).
See also Ecological damage; Ecology recovery
physical oceanography of, 276–80, 288, 328–29
Prince William Sound Aquaculture Corporation, 24, 118, 193, 251, 280–81, 289 (*see also* Pink salmon)
Prince William Sound Science Center, 274, 276, 278, 281–82
Public health
environmental pollutants and, xv–xviii, 388–392, 404–6, 414
protection of, 16, 38, 396–398, 400–1, 404–6, 414–15, 425–27, 436–40
risk assessment of, 396–98, 400–6, 396–400, 414–15

undermining protection of, 173–75, 411
Public interest
    controversy and, 206–7, 353, 408–10,
        412, 415
    protecting, 413, 427, 430, 436, 440
Public-trust science/scientists, **9**.
    See also Alaska Department of Fish and
        Game; Auke Bay Lab (NOAA);
        Ecosystem studies; Exxon Valdez
        Oil Spill Trustee Council; Natural
        Resource Damage Assessment;
        Prince William Sound Science
        Center; U.S. Fish and Wildlife
        Service; U.S. Geologic Services;
        U.S. National Oceanographic and
        Atmospheric Administration;
        University of Alaska, Fairbanks

R

Rash. *See* Illness symptoms
Rawl, Lawrence, 35
Rea, William, 74–83, 145–48, 152–55, 167,
    467, 474n1 (chap. 9)
Recommended exposure limits, 62–63, 65,
    389–90, 414, 473n1 (chap. 3).
    See also Permissible exposure limits;
        Public health
Recovery. *See* Ecological recovery
Recommendations, 417–44
    for corporate accountability, 417–19
    for personal accountability: be empow-
        ered, 443–44; be energy literate,
        442–43; be socially responsibility,
        441–42
    to strengthen environmental protection:
        federalize oil spill cleanups,
        419–21; reopen the civil settle-
        ment, 431–33; revise wildlife treat-
        ment guidelines, 428; separate pol-
        itics and science, 433–36; strength-
        en outdated oil pollution laws,
        428–31
    to strengthen protection for public
        health and worker safety, 436–40:
        provide redress for sick workers,
        440; require stockpiling of protec-
        tive gear, 427; revise dispersant use
        guidelines, 422–28; revise risk
        assessment, 436–37; separate poli-
        tics and science, 433–36
Reopener for Unknown Injury (in 1991
    civil settlement), **9**.
    See also Exxon *Valdez* Oil Spill Education

Legacy Fund (EVOS ELF)
    fulfillment of, 373–382
    recommendation to claim, 380, 431–33,
        447, 455–56
Residual oil, 214, 348, 375, 378, 395
    estimated amount of, 349, 361–64
    lingering harm, and 35, 208, 291, 302,
        308, 363, 369, 374–75, 387, 393,
        402, 432
Respirators, 45, 50, 52, 57, 87, 90, 99, 105,
    109–10, 165, 425.
    See also Exxon Worker Safety Program
    lack of, 25, 32, 42, 49–51, 87, 89, 100,
        103, 115, 120, 425
Rice, Stanley "Jeep," 181–84, 192, 194, 225,
    257–59, 343–45, 348–51, 356,
    361–64, 467
Risk assessment, 43–45, 184, 396–97, 414,
    431.
    See also Bioassays; Permissible exposure
        limits; Recommended exposure
        limits
River otters, 375, 377, 400–1.
    See also Ecosystem studies, NVP Project;
        Sentinel species
    ecology of, 202, 295, 308–9
    estimated number of killed by spill, 309
    status of, in 2003, 374–75, 377
    studies of spill effects on, 224, 308–13,
        315, 401
Robbins, Clyde, 13, 31
Roberts, Scott, 133, 142
    See also Toxic torts, *Roberts v. Exxon*
        (1999)
Roberts v. Exxon (1999). *See* Toxic torts
Robertson, Tim, 49, 100
Roby, Dan, 312–13, 319, 330, 468
Rotterman, Lisa, 215–21, 334, 476n1

S

Salmon. *See* Pink salmon
Sand lance, 238, 305–7, 311, 324–25,
    328–30, 332–33, 383, 478n1 (chap.
    19).
    See also Forage fish
Seabirds. *See* Marine birds
Seals, 225, 238.
    See also Harbor seals; Marine mammals
Sea otters.
    See also Ecosystem studies, NVP Project;
        Natural Resource Damage
        Assessment; Sentinel species
    ecology of, 202, 219, 295, 302

estimated number of killed by spill,
    218–19, 300–1
status of: in 1992, 229; in 2003, 374–75,
    378
studies of spill effects on, 216–21,
    228–29, 244, 300–4, 315, 331, 347,
    349, 361, 364, 393, 401–2, 432,
    476n1 (chap. 14), 476nn3–4 (chap.
    14)
Seawater
oil (PAH) levels in, 183–84, 344–45,
    392–95, 418, 431, 475n1 (chap. 11)
sampling, 191–94, 196–98
Sediments.
    See also Residual oil
oil (PAH) levels in intertidal, 195,
    243–44, 399–400, 475n3 (see also
    Coal)
source of PAHs in seafloor, 195, 197–98,
    303, 307, 311, 365, 418
Sentinel species, 295, 309, 402, 405
Sharr, Sam, 249–59, 262, 264, 266, 343,
    346–50, 355–56, 468, 475n2
Shoreline. See Beaches
Short, Jeff, 183, 185–86, 192–97, 260–61,
    344, 346–49, 352, 362, 364–65,
    368, 468
Simple Green, 13, 46, 61, 72, 88, 90, 143,
    160, 162, 166, 170, 390.
    See also 2-butoxyethanol
Smith, Ron, 46, 71–83, 92, 113, 166, 468
    See also Toxic torts, Roberts v. Exxon
    (1999)
Solvents.
    See also Benzene; Citra-Solv; CitroKleen;
        Corexit; De-Solv-It; Dispersants;
        Inipol; PAH aerosols; 2-
        butoxyethanol
health concerns with, 11, 13, 34, 46–47,
    61, 88, 92–94, 104, 113–115, 137,
    143, 146–49, 153, 160–62, 164,
    166–70, 172, 176, 389–92, 396–98,
    422, 424–27
monitoring of, during cleanup, 45–46,
    51, 77, 143, 312
OSHA injury/illness codes and, 128, 130,
    134
other (than listed) sources of, 144, 146,
    166, 389
use of, during cleanup, 14–15, 32, 72, 88,
    90, 98–99, 115, 134, 192
Sore throat. See Illness symptoms
SPECT brain scans, 146–47, 153
Standard Oil Company, 445–46
Statistical tricks, 214, 247, 266, 481.
    See also Advocacy science; Exxon studies;

How to Lie with Statistics
improper preservation of samples, 186,
    196–97, 407
invalid assumptions, 337, 477n4, 478n4
statistical power, 209, 213–14: uncon-
    trolled Type 2 error and, 214;
    weak, 247, 338, 478n5
Steiner, Rick, 222, 225, 335, 432, 468
Stenson, Daniel, 142, 150, 152–56
Stubblefield, Garry, 42, 49, 53, 55–56,
    59–61
    See also Toxic torts, Stubblefield v. Exxon
    (1994)
Stubblefield v. Exxon (1994). See Toxic
    torts
Subtidal communities
oil and, 195, 303, 307, 311
status of, in 2003, 382–83
wildlife uses of, 202, 219, 240, 306, 325
Superfund, 206–7, 373, 429–30

T

Tarbell, Ida, 445–46
Teitelbaum, Daniel, 469
on dangerous chemical on Exxon Valdez
    cleanup, 59–67, 160, 397
on failures of Exxon's worker safety pro-
    gram, 39–53, 160
Thorne, Lynn, 24–26, 469 (under
    Weidman)
Trimethybenzene, 75, 81, 146
Tobacco science, 407, 410, 415.
Toxic chemicals.
    See also Benzene; Corexit, Crude oil;
        Diesel exhaust; Inipol; PAHs;
        Simple Green; 2-butoxyethanol;
        Volatile organic carbonsin human
        bodies, xvi–xvii, 147, 436–37
present during cleanup, 59–67
Toxic torts.
    See also Illness claims
confidentiality of claims in, 56, 58, 67,
    126, 173–74, 440
Daubert v. Merrel Dow Pharmaceuticals
    (1993), 171–73, 411
Payne v. Exxon (1997), 75, 141–42
Roberts v. Exxon (1999), 71, 85, 133,
    139–56, 172, 411
Stubblefield v. Exxon (1994), 39–53,
    55–67, 174
Trans-Alaska Pipeline System (TAPS), 3,
    16, 38, 182–87, 475n2 (chap. 11)
2-butoxyethanol, 12–13, 65, 88, 98–100,

105, 162–63, 169–70, 390, 397–98,
474n1 (chap. 10).
See also Corexit; Inipol; Simple Green;
Solvents

U

U.S. Coast Guard, 3, 10, 11, 13, 14, 21–23,
26, 31–33, 47, 89, 99–100, 128,
135, 175, 204, 401, 415, 438, 479n1
U.S. Department of Health and Human
Services, 63, 43
See also Centers for Disease Control and
Prevention (CDC); National
Institute of Occupational Safety
and Health (NIOSH)
U.S. Fish & Wildlife Service (USFWS),
216–17, 219–20, 224, 231–34, 237,
293, 300, 304, 315, 318, 320, 328,
330, 337, 425, 476n4 (chap. 14)
U.S. Geologic Services (USGS), 220, 297,
318, 364, 461, 464, 467
U.S. National Oceanographic and
Atmospheric Administration,
31–32, 36, 181, 1992, 194, 204–5,
207, 220, 223, 239, 244, 274–75,
317–18, 323, 335, 355, 360–61,
366, 409, 422, 435
National Marine Fisheries Service
(NMFS), 181, 185, 191–92, 198,
233, 238–39, 252, 254, 259, 263,
275, 381
University of Alaska, Fairbanks (UAF),
221–22, 308, 312
Upper respiratory infections (URIs), 42,
58, 66–67, 125, 127–28, 136–37,
391, 412

V

Valdez Crud, 33, 66, 391.
See also Illness symptoms
Vaughan, Shari, 276–80
VECO, 25–37, 41, 50, 55–58, 66–67,
71–72, 75, 82, 85~–90, 99, 103–11,
114–~15, 126~–27, 136–48,
152–56, 176, 391, 412
VOCs (volatile organic carbons). *See*
Crude oil

W

Water soluble fraction (WSF), 182–84,
186, 346–47, 352, 392, 396, 399,
475n1 (chap. 11).
See also Bioassays; Crude oil
Water quality standards for PAHs
Alaska, 182–84, 186, 197, 345–46
federal (guidelines only), 184, 352
recommendations to strengthen, 352–54
Whales. *See* Marine mammals; Orcas
Wildlife. *See* Ecological damages;
Ecological recovery; Fish; Marine
birds; Marine mammals
Wildlife rescue operations, 217–21, 428
Willette, Mark, 284–87, 469, 477n1 (chap.
16)
Williams, Barbara, 131, 459–60
Wilson, Jim, 129–30
Workers' Comp Board. *See* Alaska
Workers' Compensation Board
Workers' compensation laws, 439–40
Worker safety program. *See* Exxon Worker
Safety Program
Wright, Bruce, 239, 274, 317–20, 323,
339–41, 469

Y

Yergin, Daniel, 445–47
Yost, Paul, 14–15

Z

Zooplankton. *See* Plankton

Also by Riki Ott

Alaska's Copper River Delta

About The Author

Since 1987 Dr. Ott has dedicated her academic training in marine biology and toxicology to help the general public understand the effects of oil, mining, and timber industry activities on water quality and marine and aquatic ecosystems. She has helped citizens use this new knowledge to redefine business practices and government accountability to improve the general quality of life through environmental protection, social justice, and economic stability. She has written numerous white papers and one book (*Alaska's Copper River Delta*, University of Washington Press, 1998), served on several state-appointed advisory or working groups, worked with many non-profit organizations in various capacities (founding director to volunteer), testified often before Congress and the Alaska State Legislature, and received state and national recognition for her work. Her second book, *Sound Truth and Corporate Myth$: The Legacy of the Exxon Valdez Oil Spill* (Dragonfly Sisters Press, 2005), frames the new understanding of oil toxicity to humans and wildlife from the stories of key witnesses to and participants in this drama.

Books Published by Lorenzo Press

A DOCTOR'S VISIT
by Siegfried Kra, M.D.

Three novellas and five short stories by the author of *The Three-Legged Stallion* ("Combines the best of art and the best of science . . . It is hard to put this book down" —Associated Press) exploring the emotional complexities of preparing for and practicing one of our society's most demanding professions.

"The tears and laughter which this book will lead you to will make you more aware of what we all experience in our lifetime."
—Bernie Siegel, MD, author of *Help Me to Heal*

ADRIANA'S EYES AND OTHER STORIES
by Anthony Maulucci

Twelve stories of love, family and the search for identity. Some poignant portrayals of the Italian American experience.

"Stories are well crafted and deserve wider attention . . . shouldn't be missed."
—*Primo Magazine*

"Natural storytelling skills . . . 'The Carpenter's Son' shines like a gem."
—*Voices in Italian Americana (VIA)*

"Reminds us of why reading was so pleasurable a couple of decades ago . . . The essence of good stories by Hemingway and Cheever is recaptured by Maulucci in much the same way that a younger musician pays homage to John Coltrane or Miles Davis."
—James Coleman, *The Red Fox Review*

THE ROSSELLI CANTATA
by A. S. Maulucci

A novel inspired by the true story of a search for revenge that lasted 35 years and turned into forgiveness.

"Written with the strength and simplicity of a folk tale . . .
Maulucci has a fine grasp of Italian American life."
—Eugene Mirabelli, author of *The World at Noon*

"Swiftly told, poetic prose . . . Maulucci has a gift for storytelling."
—*Voices in Italian Americana*

"A compelling new novel . . ."
—*The San Diego Union Tribune*

THE DISCOVERY OF LUMINOUS BEING,
a novel
by Anthony Maulucci

Set in Montreal during the Vietnam war, this is the story of one crucial week in the life of a young American man.

"We come away convinced that [the father and son] are flesh-and-blood people, and their issues are very much contemporary."
—Norm Goldman, *The Best Reviews*

"A highly lyrical, bittersweet, romantic story."
—Bill Brownstein, Montreal journalist

"Excellent work . . . The story has the quality of a Picasso line drawing."
—Herb Gerjuoy, *The Red Fox Review*

more information on these titles at
www.lorenzopress.com